STRATIGRAPHICAL ATLAS OF FOSSIL FORAMINIFERA

BRITISH MICROPALAEONTOLOGICAL SOCIETY SERIES

This series, published for the British Micropalaeontological Society, will gather together knowledge for a particular faunal group for specialist and non-specialist geologists alike. The scope of the series has been broadened to include the common elements of the fauna, whether index or long-ranging species, and to convey a broad impression of the fauna and allow the reader to identify common species as well as those of restricted stratigraphical range.
The synthesis of knowledge presented in the series will reveal its strengths and prove its usefulness to the practicing micropalaeontologist, and those teaching and learning the subject. By identifying some of the gaps in the knowledge, the series will, it is believed, promote and stimulate further active research and investigation.

A STRATIGRAPHICAL INDEX OF CALCAREOUS NANNOFOSSILS
Editors: G. B. HAMILTON, Consultant Micropalaeontologist, and A. R. LORD, Department of Geology, Unversity College London

A RESEARCH MANUAL OF FOSSIL AND RECENT OSTRACODS
Editors: R. HOLMES BATE, British Museum of Natural History, London, E. ROBINSON, Department of Geology, University College London, and L. SHEPPARD, British Museum of Natural History, London

STRATIGRAPHICAL ATLAS OF FOSSIL FORAMINIFERA
Editors: G. JENKINS, The Open University, and J. W. MURRAY, Professor of Geology, University of Exeter

MICROFOSSILS FROM RECENT AND FOSSIL SHELF SEAS
Editors; J. W. NEALE, Professor of Micropalaeontology, University of Hull, and M. D. BRASIER, Lecturer in Geology, University of Hull

ELLIS HORWOOD SERIES IN GEOLOGY

Editor: D.T.DONOVAN, Professor of Geology, University College London

This series aims to build up a library of books on geology which will include student texts and also more advanced works of interest to professional geologists and to industry. The series will include translations of important books recently published in Europe, and also books specially commissioned.

FAULT AND FOLD TECTONICS
W. JAROSZEWSKI, Department of Geology, University of Warsaw

A GUIDE TO CLASSIFICATION IN GEOLOGY
J. W. MURRAY, Professor of Geology, University of Exeter

THE CENOZOIC ERA
C. POMEROL, Professor, University of Paris VI.
Translated by D. W. HUMPHRIES, Department of Geology, University of Sheffield, and E. E. HUMPHRIES. Edited by Professors D. CURRY and D. T. DONOVAN, University College London

INTRODUCTION TO PALAEOBIOLOGY: GENERAL PALAEONTOLOGY
B. ZIEGLER, Professor of Geology and Palaeontology, University of Stuttgart, and Director of the State Museum for Natural Science, Stuttgart

First published in 1981 by
ELLIS HORWOOD LIMITED
Market Cross House, Cooper Street, Chichester, West Sussex, PO19 1EB, England

The publisher's colophon is reproduced from James Gillison's drawing of the ancient Market Cross, Chichester.

Distributors:
Australia, New Zealand, South-east Asia:
Jacaranda-Wiley Ltd., Jacranda Press,
JOHN WILEY & SONS INC.,
G.P.O Box 859, Brisbane, Queensland 40001, Australia
Canada:
JOHN WILEY & SONS CANADA LIMITED
22 Worcester Road, Rexdale, Ontario, Canada
Europe, Africa:
JOHN WILEY & SONS LIMITED
Baffins Lane, Chichester, West Sussex, England.
North and South America and the rest of the world:
Halsted Press: a division of
JOHN WILEY & SONS
605 Third Avenue, New York, N.Y. 10016, U.S.A.

British Library Cataloguing in Publication Data
Stratigraphical atlas of fossil foraminifera
(Ellis Horwood series in Geology)
1. Foraminifera, Fossil
I. Jenkins, D. G. II. Murray, J. W.
III. British Micropalaeontological Society
563'.12 QE772

Library of Congress Cataloguing in Publication Data
81-6227 AACR2

ISBN 0-85312-210-5 (Ellis Horwood Ltd.)
ISBN 0-470-27191-4 (Halsted Press)

Typeset in Press Roman by Ellis Horwood Ltd.
Printed in Great Britain by R. J. Acford, Chichester.

STRATIGRAPHICAL ATLAS OF FOSSIL FORAMINIFERA

Editors:

D. G. JENKINS, B.Sc., Ph.D., D.Sc., F.R.S.N.Z.
Department of Earth Sciences
The Open University, Milton Keynes

and

J. W. MURRAY, B.Sc., A.R.C.S., Ph.D., D.I.C.
Professor of Geology
University of Exeter

Published by
ELLIS HORWOOD LIMITED
Publishers · Chichester
for
THE BRITISH MICROPALAEONTOLOGICAL SOCIETY

Table of Contents

*These authors publish with permission of the Director, Institute of Geological Sciences.

Preface

The British Micropalaeontological Society may be said to have come of age when it published *A Stratigraphical Index of British Ostracoda* in 1978. This was the first of a series of special publications aimed at bringing together knowledge for a particular faunal group in order to make it more useful to specialists and non-specialists alike. The present volume has been written with the same objectives in mind.

Once the idea had been mooted, a group of prospective contributors met at the Institute of Geological Sciences in Leeds during the course of a Foraminifera Group weekend conference. General agreement was reached on the format and content which to a large extent follows that of the ostracod precursor. However, one major difference of approach is that we agreed to link our foraminiferal data to lithostratigraphical rather than biostratigraphical successions. The reasons for this are that it is easier to re-sample accurately with reference to a lithological succession at a specific locality; also biostratigraphical divisions (zones, subzones) may have their boundaries revised in the light of new evidence. Where possible we have correlated our lithostratigraphical stratigraphy with zones and stages.

Our original intention was to restrict the volume to include only species of short time range, i.e. true index species. However, this proved to be impractical for two reasons. First, most of our faunas are benthic, tied to facies, and therefore not truly index forms. Second, strict application of this approach led to some fossiliferous parts of the succession appearing to be barren because their faunas are made up of long-ranging species; It was therefore decided to broaden the scope to include the common elements of the fauna whether index or long-ranging species. In this way we hope to convey a general impression of the fauna, allowing the user to identify the common species as well as those of restricted stratigraphical range. For economy of space the original name is given for each species so that readers may have ready access to the type description of the Catalog of Foraminifera (Ellis and Messina, 1940 et seq.).

There has been a general policy not to include too much data from offshore because many users have no access to such material and also because problems of confidentiality limit the detail which can be given. Nevertheless, in the chapters on the Mesozoic, similarities and differences between onshore and offshore faunas are discussed and for the Cenozoic we have a chapter summarizing the principal features of the North Sea succession. The chapter on the Neogene depends heavily on offshore material as the successions on land are incomplete.

Inevitably a synthesis of knowledge such as we present here reveals not only its strengths but also its weaknesses. We hope it will prove to be useful to the practising micropalaeontologist and also to those who are being trained in the subject. We believe that by identifying some of the gaps in our knowledge we shall promote further active research.

All the contributors to this volume volunteered their services. The editors are indebted to them for their help and their patience during the two year period of gestation.

The editors wish to record their gratitude for the interest, help and encouragement given by Dr. C. G. Adams during the preparation of this work.

D. Graham Jenkins
October, 1980 John W. Murray

1 Some early students of Foraminifera in Britain

J. W. Murray

The scientific discoveries of one generation are built upon the foundations laid by earlier workers. The synthesis of data presented in this book may appear to rest largely on the labours of micropalaeontologists working in the twentieth century, but we should not overlook the contributions made by earlier workers, many of whom earned their living in business or in other fields of science or medicine. These early micropalaeontologists made contributions to science which were fundamental and their value is recognised throughout the world. Fortunately many of the collections on which these contributions were based are still available for examination (Adams *et al.*, 1980).

It is perhaps natural that many of these earlier studies should have been on Recent material from the shores of the British Isles, from the continental shelf and from the ocean basins. There were many naturalists who took part in scientific voyages of discovery during the nineteenth century and who were prepared to exchange material with colleagues. Early publications reveal a great deal about this collaboration.

As can be seen from the brief survey presented below, the contribution of British workers to the study of Recent foraminifera was large. However, the fossil forms were somewhat neglected especially in comparison with the labours of European workers in the same time period. By the end of the nineteenth century something was known of Carboniferous, Permian, Lower Cretaceous, Palaeocene and Plio-Pleistocene foraminifera of Britain but very little was known of the Mesozoic in general, or of the Eocene and Oligocene. Indeed, it is really only since the 1940's that much of the British stratigraphical record has received serious attention.

In each chapter there is a historical survey of past work and here it is intended only to emphasise the role of prominent researchers in the period up to the 1930's.

The study of foraminifera started as a hobby for eighteenth century gentlemen who took an interest in microscopy. The earliest of these was William Boys of Sandwich in Kent, a surgeon and ardent naturalist who first studied foraminifera from shore sand. Together with Walker he published descriptions of various small animals (Walker and Boys, 1784) to which binomial names were applied by Walker and Jacob (1794). Colonel George Montagu (1803, 1808) described foraminifera from the south coast. However, the really major contributions were made by W. C. Williamson, W. B. Carpenter, and H. B. Brady.

William Crawford Williamson, Professor of Natural History at Owen's College, Manchester, published several papers on foraminifera, two of which are of great significance. The first (1848) dealt with the taxonomy of Recent species of *Lagena*. The second, entitled *'On the Recent Foraminifera of Great Britain'* is the only attempt to monograph the fauna (1858). The plates are of outstanding quality and hand-coloured.

William Benjamin Carpenter, 1813-1885, was trained in medicine but practised for only a short period. He became Lecturer in Physiology at the medical school of the London Hospital, then Professor of Medical Jurisprudence at University College, and finally Registrar of the University of London. In the period up to this last appointment he devoted much time to writing to earn a living. Among the topics he discussed were microscopy and the physiology of the brain. However, as Registrar, a post he held for 23 years, he had time to develop his interests in foraminifera and in the period 1849-1885 he published numerous papers on *Orbitolites* and other large foraminifera (e.g. 1883) and on deep sea foraminifera (see Murray, 1971, for bibliography). His *'Introduction to the study of the Foraminifera'* (1862) still makes interesting reading.

Henry Bowman Brady, 1835-1891, is best known for his major contribution on the Recent foraminifera collected during the Challenger Expedition (Brady, 1884). This work was largely carried out during his retirement following a very active career as a pharmacist (see Adams, 1978, for details). The high quality of both text and plates in the Challenger Report makes this an important reference work even now. Carpenter and Brady joined forces to study *Parkeria* (which is now known not to be foraminiferal) and *Loftusia* (Carpenter and Brady, 1870).

William Kitchen Parker, , 1823-1890, at the age of fifteen was apprenticed to a pharmacist and three years later to a medical practitioner. During this time he collected and named plants, studied dissections and prepared skeletons of various animals. In 1844 he went to London where he trained to become a Licentiate of the Society of Apothecaries and then practised in Pimlico. He made significant contributions to the study of foraminifera. Together with Thomas Rupert Jones, he published during the period 1857 to 1871 numerous papers on the taxonomy of Recent species, especially those erected by Linnaeus, Gmelin, Walker, von Fichtel and von Moll, Lamarck, Denys de Montfort, de Blainville, Defrance, d'Orbigny, Ehrenberg and Batsch. He also assisted W. B. Carpenter in the preparation of *Introduction to the study of the Foraminifera*. Apart from this he published widely on the skeletal anatomy of birds, both modern and fossil, and various other vertebrates. He was elected Fellow of the Royal Society in 1865 and had numerous other honours bestowed upon him. In 1873 he became Hunterian Professor of Comparative Anatomy and Physiology at Kings College, London, (Jones, 1891).

Fortescue William Millett, 1833-1915, became interested in foraminifera in 1883 when he retired to south-west England. In later years he became a recluse (Sherborn, 1915). He is best known for his work on the Malay Archipelago (1899-1904) and for his studies of the St. Erth Clays. For the latter he was awarded the Bolitho Medal by the Royal Geological Society of Cornwall (Anon. 1915(a)). On his death his collection of 10,000 slides was bought by Heron-Allen. (Anon. 1915(b)).

Edward Heron-Allen and Arthur Earland published numerous joint works. Heron-Allen, 1861-1943, trained as a solicitor and succeeded his father as head of the family firm of solicitors. He retired at 50 and devoted himself to his hobby, microscopy. In 1907 he contacted Earland to seek help in identifying foraminifera from the shore sands of Bognor. This partnership continued for twenty-five years and led to many publications on Recent foraminifera from our own coasts (e.g. 1913, 1930) and from various other parts of the world including the Kerimba Archipelago, the Falkland Islands, and Egypt. Much of this work was carried out in the British Museum (Natural History) but, in Earland's words '. . . conditions were not favourable, there were too many other distractions at the Museum for rapid progress, and they resulted in friction between us.' (Earland, 1943). In 1932 the partnership broke up and Heron-Allen withdrew from scientific study.

Arthur Earland, 1866-1958, entered the Civil Service in 1885 as a 'boy clerk' and retired in 1926 (due to ill health) as an assistant controller. Like Heron-Allen his heart was not in his work and in

1887 he took up microscopy as a hobby. Hawkyard encouraged him to study foraminifera and later he received help from Millett, Wright, Lister and d'Arcy Thompson (Hedley, 1958). Macfadyen (1959) described Earland as 'the last representative of the Brady era in this country'.

Apart from their joint publications, Heron-Allen and Earland gathered together valuable reference collections and an extensive library, all of which now form part of the collections in the Protozoa Department of the British Museum (Natural History).

The first major account of British fossil foraminifera is that by Brady (1876) on Carboniferous and Permian forms. This monograph is beautifully illustrated showing both complete specimens and thin sections. The material studied came not only from Britain but also from Ireland, Belgium, Russia, United States, Canada and Germany.

The monograph of the Foraminifera of the Crag (Jones *et al.*, 1866-1897) was a major collaborative exercise. Part 1, published in 1866, was prepared by T. R. Jones, W. K. Parker and H. B. Brady. After its completion the authors were busily engaged in other matters (e.g. Brady prepared the 'Challenger' report) and so the preparation of subsequent parts was delayed. Then both Brady and Parker died. Jones retired from his post as Professor at the Military College, Sandhurst, in 1881 and moved to Chelsea (Robinson *in* Bate and Robinson, 1972). There he spent his retirement preparing the later parts of the monograph. Because of failing eyesight he took on Charles Davies Sherborn as a young assistant, (Elliott, 1978). An incomplete manuscript for Part II had been left by Brady and this was completed by Jones and Sherborn with assistance from H. W. Burrows, F. W. Millett, R. Holland and F. W. Chapman. All these authors contributed to Parts III and IV, these being published in 1895, 1896 and 1897 respectively.

The Foraminifera of the Crag is the largest single work on fossil foraminifera in Britain. Apart from the systematic description and illustration of the species, their geographical and stratigraphical distributions are discussed and brief comments are made on the use of the foraminifera in determining the environments of accumulation of the Crags.

Burrows and Holland also made a significant contribution to the study of the Thanet Beds at Pegwell Bay. They systematically sampled the succession and recorded the stratigraphical distribution of the foraminifera (Burrows and Holland, 1897).

Sherborn published three short papers with Chapman on foraminifera from the London Clay from drainage works in Piccadilly (1886, 1889) and from the cliffs at Sheppey (Chapman and Sherborn, 1889).

Frederick Chapman, during the period 1891-1896 produced a ten part illustrated account of the Foraminifera of the Gault. He also described faunas from the Rhaetic, the Bargate Beds and the Cambridge Greensand. Chapman left England in 1902 to take up a post at the Museum in Melbourne and from there he published on a wide variety of Recent and fossil material (Bate and Robinson, 1978).

This early research was carried out by men with scientific curiosity. Their objectives were to describe, record, and to define principles where appropriate. Thus, by the 1920's this pure science was ready to be applied to the practical problems of stratigraphical correlation and palaeoecology in the newly established oil exploration industry.

REFERENCES

Anon, 1915(a). Fortescue William Millett, *J. Quekett microsc. Club,* (2) **12,** 559-560.

Anon, 1915(b). *Nature,* **95,** 180.

Adams, C. G. 1978. Great names in Micropalaeontology. 3. Henry Bowman Brady, 1835-1891. *Foraminifera,* **3,** 275-280. Academic Press.

Adams, C. G., Harrison, C. A. and Hodgkinson, R. L. 1980. Some primary type specimens of foraminifera in the British Museum (Natural History). *Micropaleontology,* **26,** 1-16.

Bate, R. H. and Robinson, E. 1978. A stratigraphical index of British Ostracoda. *British Micropalaeontological Society, Special Publ.,* **1,** 538 pp., Seel House Press, Liverpool.

Brady, H. B. 1876. A monograph of Carboniferous and Permian Foraminifera (the genus *Fusulina* excluded). *Palaeontogr. soc. Monogr.,* London, 1-66, pls. 1-12.

Brady, H. B. 1884. Report on the Foraminifera dredged by H.M.S. Challenger during the years 1873-76. *Rep. Scient. Results Challenger Exped. Zoology,* **9,** 1-800, 115 pls.

Burrows, H. and Holland, R. 1897. The Foraminifera of the Thanet Beds of Pegwell Bay. *Proc. Geol. Assoc.,* **15,** 19-52.

Carpenter, W. B. 1862. Introduction to the study of the Foraminifera. *Ray Soc.,* 319 pp. London.

Carpenter, W. C. 1883. Report on the specimens of the genus *Orbitolites* collected by H.M.S. Challenger during the years 1873-1876. *Rep. Scient. Results Challenger Exped. Zoology,* 7, (21), art. 4, 1–47, pls. 1–8.

Carpenter, W. B. and Brady, H. B. 1870. Description of *Parkeria* and *Loftusia* two gigantic types of arenaceous Foraminifera. *Phil. Trans. r. Soc.,* **159,** 721–754.

Chapman, F. 1891–1898. The Foraminifera of the Gault of Folkestone. *Jl. R. microsc. Soc.* 10 parts.

Chapman, F. and Sherborn, C. D. 1889. The Foraminifera from the London Clay of Sheppey. *Geol. Mag.* Dec. 111, **6,** 497–499.

Elliott, G. F. 1978. Charles Davies Sherborn, 1861–1942: an appreciation. *Foraminifera,* **3,** 267–273. Academic Press.

Earland, A. 1943. Edward Heron-Allen, F.R.S. *Jl. R. microsc. Soc.* (3), **63,** 48–50.

Heron-Allen, E. and Earland, A. 1913. Clare Island Survey Foraminifera. *Proc. R. Ir. Acad.,* **31,** (64), 1–188.

Heron-Allen, E. and Earland, A. 1930. The Foraminifera of the Plymouth District. *Jl. R. microsc. Soc.,* **50,** 46–84, 161–191.

Hedley, R. H. 1958. Mr. Arthur Earland. *Nature,* **181,** 1440–1441.

Jones, T. R. 1891. William Kitchen Parker. *Proc. Roy. Soc.,* **48,** XV–XX.

Jones, T. R., Parker, W. K. and Brady, H. B. 1866–1897. A monograph of the Foraminifera of the Crag. *Palaeontogr. Soc. Monogr.,* 1–402.

Macfadyen, W. A. 1959. Arthur Earland. *Jl. R. microsc. Soc.,* 77, 146–150.

Millett, F. W. 1899–1904. Report on the Recent Foraminifera of the Malay Archipelago. *Jl. R. microsc. Soc.,* (17 parts).

Montagu, G. 1803. *Testacea Britannica, or natural history of British shells marine, land and freshwater, including the most minute.* pp. 1–606, pls. 1–16. Romsey.

Montagu, G. 1808. *Testacea Britannica: Supplement,* pp. 1–183, pls. 1–30, Exeter.

Murray, J. W. 1971. The W. B. Carpenter Collection. *Micropaleontology,* **17,** 105–106.

Sherborn, C. D. 1915. Fortescue William Millett. *Geol. Mag.* (6), **11,** 288.

Sherborn, C. D. and Chapman, F. 1886. On some microzoa from the London Clay exposed in the drainage works, Piccadilly, London, 1885. *Jl. R. microsc. Soc.,* **6,** 737–767.

Sherborn, C. D. and Chapman, F. 1889. Additional note on the Foraminifera of the London Clay exposed in the drainage works, Piccadilly, London, 1885. *Jl. R. microsc. Soc.,* **483–488.**

Walker, G. and Boys, W. 1784. *Testacea minuta rariora,* London.

Walker, G. and Jacob, E. 1979, *In:* Kanmacher, *Adam's essays on the microscope.* Ed. 2, London.

Williamson, W. C. 1848. On the Recent species of the genus *Lagena. Ann. Mag. nat. Hist.,* ser. 2, **1,** 1–20.

Williamson, W. C. 1858. On the Recent Foraminifera of Great Britain. *Ray. Soc.,* pp. 1–107, pls. 1–7.

2

Pre-Carboniferous faunas

J. W. Murray

The stratigraphic range of the Order Foraminiferida is ?Precambrian, Cambrian to Recent (Loeblich and Tappan, 1964). However, in the Cambrian only simple Lagnyacea and Astrorhizidae are known. In the Ordovician the Saccaminidae and Moravamminidae appear followed by the Ammodiscidae and Nodosinellidae in the Silurian. The Devonian saw the arrival of the Ptychocladiidae, Semitextulariidae and Endothyridae.

In Britain the Lower Palaeozoic rocks have been tectonically folded and cleaved in the Caledonian orogenic belt. Only in the marginal areas of the Welsh borderlands are the rocks less deformed. The Devonian succession of most of the British Isles is non-marine, and in south-west England where marine successions are known and rocks underwent deformation in the Hercynian orogeny. Thus, the likelihood of there being good stratigraphic index fossils is poor because of the simple nature of the foraminifera, and the chances of foraminiferal faunas being recovered are poor because of the tectonic deformation. Nevertheless, there are a few records of foraminifera from this time period.

There are no authenticated records of British Cambrian forms. Chapman (1904) reported foraminifera from a loose block of limestone thought to be of Cambrian age but Wood (1947) has shown that this record is erroneous. The limestone is most probably of Triassic (Rhaetian) or Lower Jurassic age.

Saccamminopsis cf. *fusuliniformis* (M'Coy) has been recorded (sometimes as *Saccammina carteri* Brady) from the Ordovician of Girvan (Nicholson and Etheridge, 1878; Lapworth, 1882; Cummings, 1952). This is the oldest occurrence of foraminifera in Britain. *Saccamminopsis* has also been recorded from the Silurian of Shropshire (Smith, 1881; Brady, 1888, as *Lagena*; Cummings, 1952). Ireland (1967) mounted >11,000 specimens from acid residues from Silurian limestones. Most were attached forms with brown, iron-stained tests, of the family Saccamminidae. Aldridge *et al.* (1979) reported *Ammodiscus, Lagenammina, Metamorphina* and *Bathysiphon* in abundance in washed shale and dissolved limestones of Silurian age from the Welsh borderland. Eisenack (1977) has also reported microfossils from these limestones. Records of *Dentalina communis* (Blake, 1876), *Rotalia* (?) and *Textularia* (Keeping, 1882) from the Llandovery of central Wales have been disputed by Wood (1949) who on further investigation concluded that the structures are inorganic mineral growths.

The only Devonian occurrence is of 'Fixo-sessile arenaceous foraminifera' attached to bivalve shells in Upper Devonian limestones of Chudleigh, Devon (Tucker and Van Straaten, 1970).

Apart from these published records simple foraminifera and possible foraminifera are sometimes encountered in the insoluble residues of calcareous rocks prepared for conodont extraction or palynology. Thus, the foraminifera so far recorded from the pre-Carboniferous rocks of Britain are of little value for stratigraphical correlation.

REFERENCES

Aldridge, R. J., Dorning, K. J., Hill, P. J., Richardson, J. B. and Siveter, D. J. 1979. Microfossil distribution in the Silurian of Britain and Ireland. In: Harris, A. L., Holland, C. H. and Leake, B. L. 1974. *The Caledonides of the British Isels – reviewed.*

Blake, J. F. 1876. Lower Silurian Foraminifera. *Geol. Mag.*, **3**, 134–135.

Brady, H. B. 1888. Notes on some Silurian Lagenae. *Geol. Mag.*, **5**, 481–484.

Chapman, F. 1904. Foraminifera from an Upper Cambrian horizon in the Malverns; together with a note on some of the earliest known Foraminifera. *Quart. J. geol. Soc. Lond.*, **54**, 257.

Cummings, R. H. 1952. *Saccamminopsis* from the Silurian. *Proc. Geol. Ass.*, **63**, 220–226.

Eisenack, A. 1977. Microfossilien in organischer substanz aus den Middle Nodular Beds (Wenlock) von Dudley, England. *N. Jb. Geol. Paläont.* Mh. 25–35.

Ireland, H. A. 1967. Microfossils from the Silurian of England. *Bull. Am. Assoc. Petrol. Geol.*, **51**, 471.

Keeping, W. 1882. On some remains of plants, Foraminifera and Annellida, in the Silurian rocks of central Wales. *Geol. Mag.*, **9**, 485–491.

Lapworth, C. 1882. The Girvan succession. *Quart. J. geol. Soc. Lond.*, **38**, 537–666.

Loeblich, A. R. Jr., and Tappan, H. 1964. Sarcodina, chiefly 'thecamoebians' and Foraminiferida. In: Moore, R. C., Ed., *Treatise on Invertebrate Paleontology. Geol. Soc. Amer.*, New York, pt. **C. 1-2**, 900 pp.

Nicholson, H. A. and Etheridge, R. 1879. *A monograph of Silurian fossils of the Girvan district in Ayrshire,* Fasc. 1, 135 pp. Blackwood & Sons, Edinburgh.

Smith, J. 1881. Notes on a collection of bivalved Entomostraca and other Microzoa from the Upper Silurian strata of the Shropshire District. *Geol. Mag.*, **8**, 70–75.

Tucker, M. E. and Van Straaten, P. 1970. Conodonts and facies on the Chudleigh schwelle. *Proc. Ussher Soc.*, **2**, 160–170.

Wood, A. 1947. The supposed Cambrian Foraminifera from the Malverns. *Quart. J. geol. Soc. Lond.*, **102**, 447–460.

Wood, A. 1949. The supposed Silurian Foraminifera from Cardiganshire. *Proc. Geol. Ass.*, **60**, 226–228.

3
Carboniferous

M. D. Fewtrell, W. H. C. Ramsbottom, A. R. E. Strank

3.1 INTRODUCTION

A British Carboniferous *Endothyra,* illustrated by John Phillips (1846) was the first fossil ever to be figured from a thin section. However, no extensive study of the group was made until H. B. Brady produced his *Monograph on Carboniferous and Permian foraminifera,* published by the Palaeontographical Society in 1876. This book, which forms a foundation for most subsequent work on late Palaeozoic foraminifera, is remarkable for its time in that there was generally a close stratigraphical control and fairly full details of collecting localities. Mostly this is because a large proportion of the material studied by Brady came from the Yoredale facies rocks of Northumberland and Scotland, where the stratigraphy was well known.

Brady was supplied with specimens by several of the numerous amateur collectors of the time – notably James Bennie in Scotland (later to join the Geological Survey there) and by Walter Howchin in Northumberland. Howchin's story is remarkable. A Primitive Methodist minister, he travelled around Northumberland on foot collecting the weathered debris from the numerous small working quarries then available, later sieving and washing the foraminifera. In 1881 he was advised to go to Australia for the sake of his health, which was very poor; actually, he lived to 92 and only died in 1937. He took his collection with him to Australia but much of it has recently been returned and is now in the British Museum (Natural History). Although he published only one paper on Carboniferous foraminifera (1888), he left some uncompleted manuscripts. The achievements of these early pioneers are all the more remarkable in that they usually worked with only a hand-lens and did not always investigate the internal structure of the specimens.

Virtually no more work was done in Britain until the 1940s, and this postdated the revolution in Carboniferous foraminiferal studies introduced by Russian authors who used only thin sections for their systematics. In Britain, A. G. Davies did some work, mostly on boreholes (1945, 1951) and R. H. Cummings published several papers (1955–61), mainly on Palaeotextulariidae, in which he began to tackle the problem of using randomly orientated thin sections. He devised and used his own zonal scheme, the details of which were never published but which is quoted (in the form F.Z.1, F.Z.2, etc.) by other authors (e.g. Sheridan, 1972). In Belgium, pioneer studies by R. Conil and M. Lys and their collaborators have used foraminifera in Dinantian stratigraphy with considerable success. Detailed studies are being made in the British Isles and enough has been done to show that foraminifera have very considerable potential here too. The ultimate aim of integrating the results as they continue to improve has implications for the refinement of Dinantian stratigraphy in the British Isles. Recent papers on the British faunas include those of Hallett (1970), Conil and George (1974), Marchant (1974), Fewtrell and Smith (1978), and Conil, Longerstaey and Ramsbottom (1980). New studies on British Carboniferous foraminifera are now in progress, and the present account is only an interim report.

The authors have contributed principally as follows: Courceyan to Arundian, M.D.F.; Holkerian to Brigantian, A.R.E.S.; Namurian, W.H.C.R.

3.2 LOCATION OF IMPORTANT COLLECTIONS

The majority of the specimens figured by Brady (1876) are in the British Museum (Natural History). The whereabouts of the specimens figured in Howchin's paper of 1888 is unknown, but the residue of Howchin's collections is also in the British Museum (Natural History), as are the specimens figured by Hallett (1970). Cummings' specimens are mostly in the Hunterian Museum, Glasgow. The largest collection of Carboniferous foraminiferal thin sections in Britain is in the Institute of Geological Sciences, Leeds, from which many of the specimens figured here came. Others are mainly from the Sedgwick Museum, Cambridge. Specimens figured by Conil and Longerstaey (1980) are mostly at the Geological Institute, Louvain la Neuve, Belgium – some are from the collections of the Institute of Geological Sciences, Edinburgh.

3.3 STRATIGRAPHIC DIVISIONS

British Carboniferous stratigraphy and correlations have been recently reviewed in two Special Reports of the Geological Society of London (George *et al.* 1976; Ramsbottom *et al.* 1978) and the limits and divisions used here are those proposed and used in these reports (see Fig. 3.1).

Although Foraminifera were used in choosing the local boundaries of some of the stages of the Dinantian proposed by George *et al.* (1976), no foraminiferal zones have yet been proposed in Britain. The foraminiferal zones erected by Conil *et al.* (1978) for Belgium (Fig. 3.1) are, in effect, assemblage zones. They are not directly adopted here because several discrepancies exist and our studies have not yet established whether this zonation is the most suitable that could be devised for Britain. At present, foraminiferal faunas are assigned to a chronostratigraphical stage framework, though usually it is possible to assign a fauna to the early, middle or late part of a stage.

3.4 OCCURRENCE

Calcareous foraminifera are more or less restricted to the shallow water shelf areas where limestone or highly calcareous mudstones were being deposited. Generally, foraminifera are found in the bioclastic limestones which contain a limited proportion of finer crinoid debris. They may also be common and well-preserved in oolitic and pellet limestones. They are usually rarer in micritic or arenaceous limestones, although some families seem to have been well adapted to the environment, e.g. small Archaediscidae. It has been observed that an abundance of bryozoa in a limestone means that there will be few foraminifera.

Nearly all work is now done on thin sections and, indeed, this is the only way that the wall structure, much used in classification, can be seen. Thus, although most work has been done with the foraminifera found in limestones, and complete foraminiferal tests may readily be washed from calcareous shales, the determination of whole specimens is difficult or impossible. It is known, however, that the foraminifera present in limestones are not necessarily identical with those found in adjacent shales, and that each fauna contains ecologically restricted elements.

The arenaceous forms found commonly in the Westphalian marine bands are closely indicative of a particular facies (Calver, 1969), but since they have not proved to be stratigraphically diagnostic they will not be mentioned further here.

3.4.1 Courceyan

Courceyan strata are widespread in the British Isles but are often in non-foraminiferal facies. This is particularly true in the lower part, and the base-Courceyan boundary stratotype in Co. Cork lacks foraminifera altogether. The main areas of study so far are Ireland (Midlands and Dublin Basin), and the Craven Basin of northern England.

The lowest known fauna recorded in the British Isles has been found in the Irish Midlands (Marchant, pers. comm. 1980). The fauna is dominated by *Tuberendothyra* sp. and *Eoforschia* sp. which are also present in later Courceyan rocks. The restricted fauna is not related to any obvious facies control. Later faunas found elsewhere in the British Isles also tend to be sparse and low in diversity. Forms present in the Swinden No. 1 borehole (the lowest horizon reached in the Craven Basin) include *Septabrunsiina* sp., *Septatournayella, Latiendothyra* cf. *parakosvensis, Endothyra* sp., *Tournayella discoidea, Spinoendothyra* sp., and *Palaeospiroplectammina tchernyshinensis.* Similar forms are present in the stratigraphically higher Haw Bank Limestones in the Craven Basin but with the addition of rare *Chernyshinella glomiformis,* and *Septaglomospiranella primaeva*; other more common forms include *Brunsia spirillinoides, B. pseudopulchra, Lugtonia concinna, Pseudolituotubella* sp., *Earlandia vulgaris, Archaesphaera, Palaeospiroplectammina mellina, Spinoendothyra recta, S. mitchelli, S. costifera, Spinobrunsiina ramsbottomi, Latiendothyra* cf. *latispiralis, Endothyra* including *E. bowmani, E. laxa, E. prokirgisana, E. prisca, E.* cf. *freyri, E.* cf. *nebulosa, E.* cf. *tumida,* and several previously undescribed forms are also present. The genera *Endothyra,* and *Septabrunsiina* become the dominant elements of the fauna in the Craven Basin.

Higher beds in the Haw Bank Limestones contain *Eblanaia michoti* which is a useful marker. Other species of *Eblanaia* were recorded by Conil (1976) at a similar level in the Dublin Basin with *Tournayella kisella* (in Marchant, 1974). Both *Tetrataxis* sp. and *Eotextularia* sp. have been recorded from the late Courceyan, the former from the S.W. Province and Dublin Basin, the latter from north of Dublin (George *et al.*, 1976). Conil *et al.* (1980) have illustrated faunas from the Black Rock Limestone of the Bristol area which contain *Eblanaia, Septabrunsiina,* and poorly preserved endothyrids which they referred to *Granuliferella.* The Black Rock Limestone is of late Courceyan age.

3.4.2 Chadian

Limestones of Chadian age are more widely developed in the British Isles and foraminifera are recorded from a somewhat greater number of localities. Published records however are from similar areas to those for the Courceyan, i.e. the S.W. Province (Mendips and Gower); the Dublin Basin and elsewhere in Ireland; the Craven Basin and Ravenstonedale in northern England.

Early Chadian faunas have much in common with those of the high Courceyan particularly in the Craven Basin where the transition is best developed in a thick sequence. The boundary, which has its stratotype in this area, is much less clearly marked in terms of foraminifera than it is in Belgium where there are probably substantial non-sequences at the Tournaisian-Viséan boundary. The Bankfield East beds which immediately overlie the base of the Chadian in the Chatburn road cutting near Clitheroe, Lancashire, contain the following forms; *Tournayella discoidea, T. kisella, Septabrunsiina* sp., *Brunsia,* including *B. pseudopulchra* and *B. spirillinoides, Glomospiranella* spp., *Spinoendothyra costifera, S. recta* and *S. mitchelli.* Many of the endothyrids mentioned in the Haw Bank Limestones are present, in particular *E. bowmani, E. freyri, E. laxa, E. danica,* and several undescribed forms. *L.* cf. *parakosvenis, Septabrunsiina* sp., *Spinobrunsiina ramsbottomi* and *Eblanaia* sp. are also characteristic forms in this type of section. *Palaeospiroplectammina* sp. and *Eotextularia diversa* are represented. In addition to these forms however, probable *Dainella,* a Chadian marker is present in the lower part of the section.

It has not been possible to confirm a report (Conil pers. comm.) of *Eoparastaffella* in these beds; a thorough study of the Clitheroe sequence is needed and has now been initiated. *Dainella* and *Eoparastaffella* enter together at the base of the Belgian Viséan, along with many of the elements of the above fauna. A 'basal' Chadian fauna with both *Dainella* and *Eoparastaffella* was reported by Conil and George (1973) from the Caninia (now the Gully) Oolite in Gower, S. Wales, but *Dainella* appears without *Eoparastaffella* in several sequences in the Craven Basin, including Ravenstonedale (Stone Gill Beds, with *Endothyra similis*; Ramsbotton in Holliday, Neves and Owens, 1979) as well as in the base-Chadian stratotype. Conil *et al.* (1980) have illustrated further material

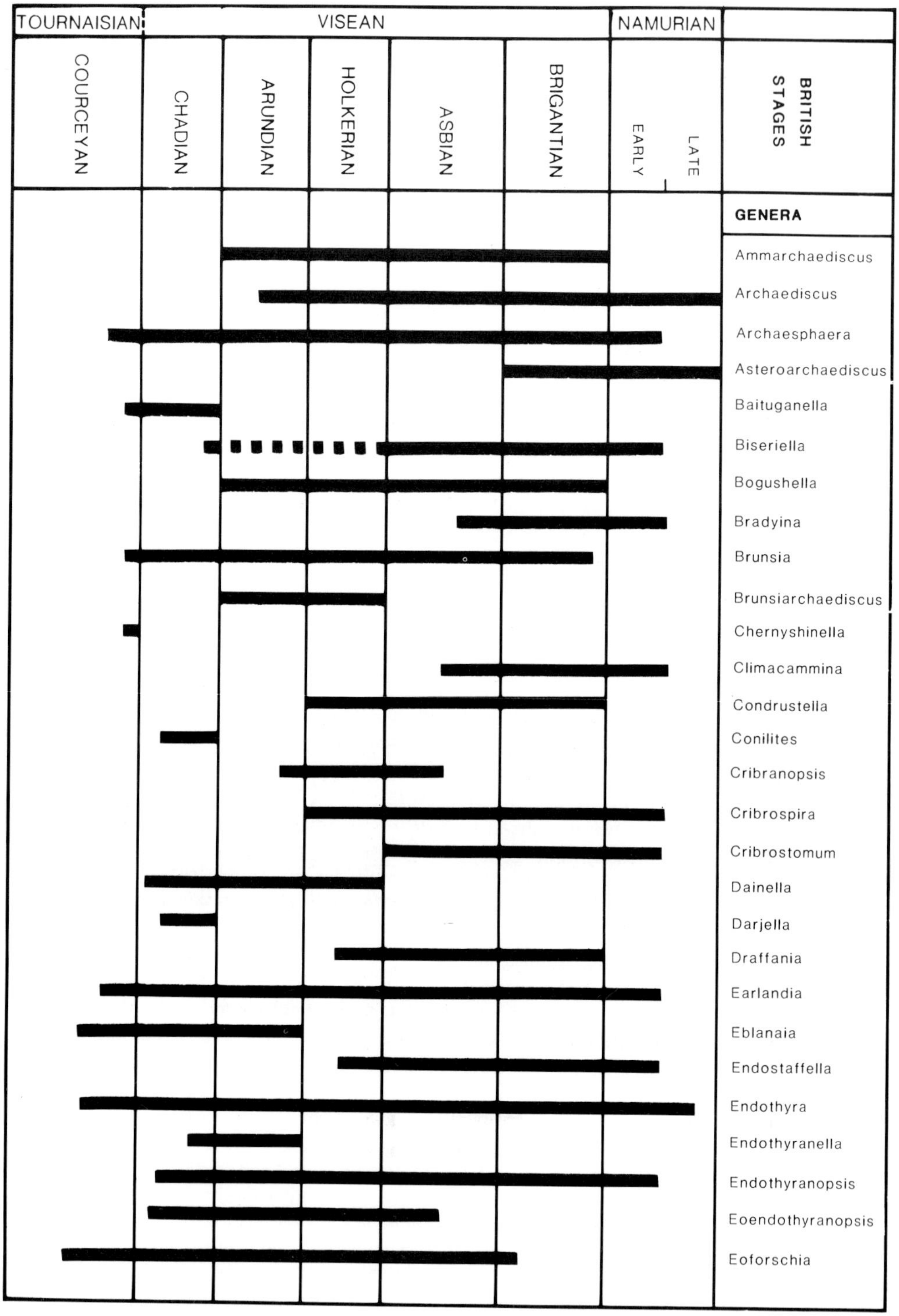

Fig. 3.1 – Range chart of Carboniferous genera.

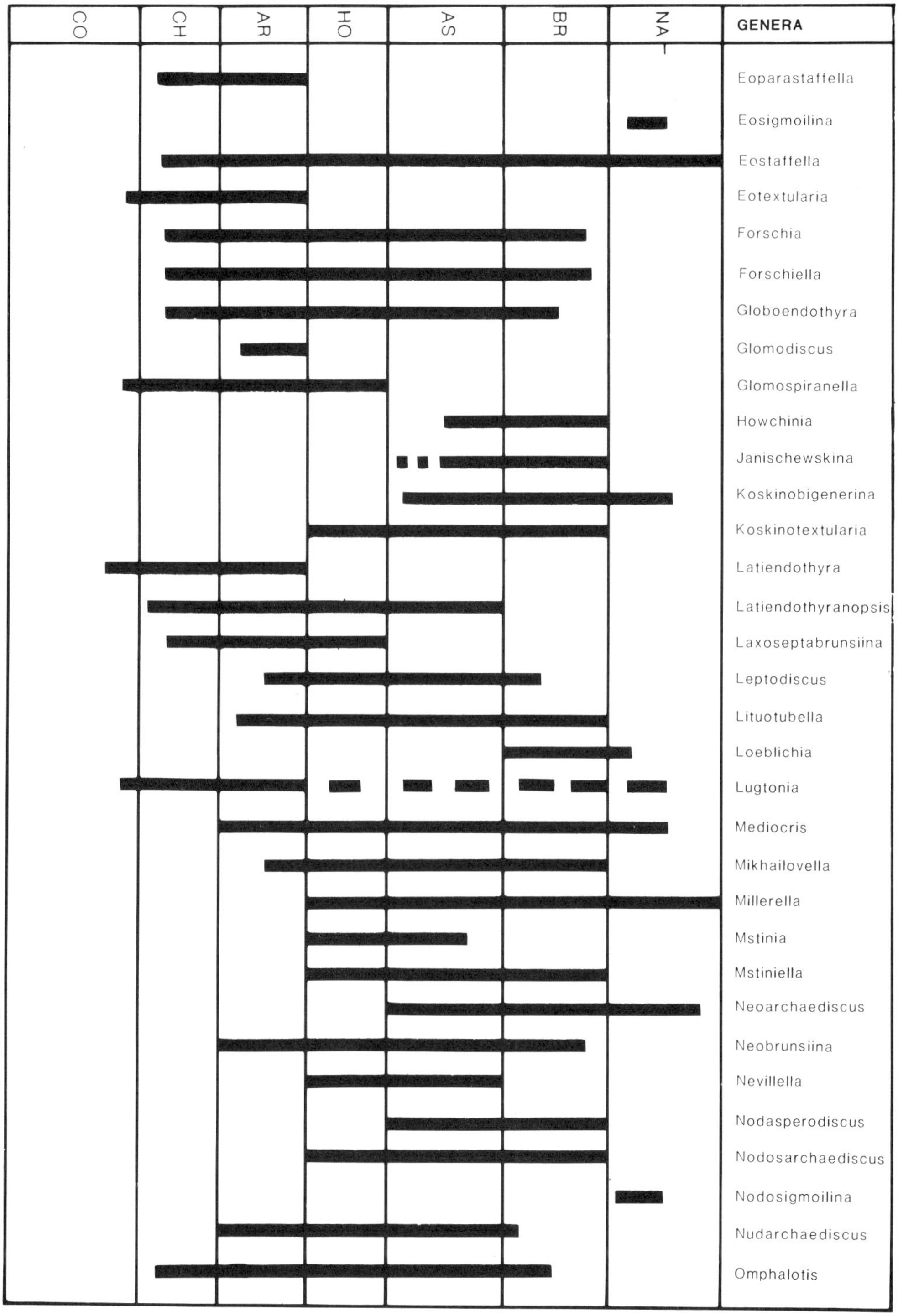

Fig. 3.1 – Range chart of Carboniferous genera.

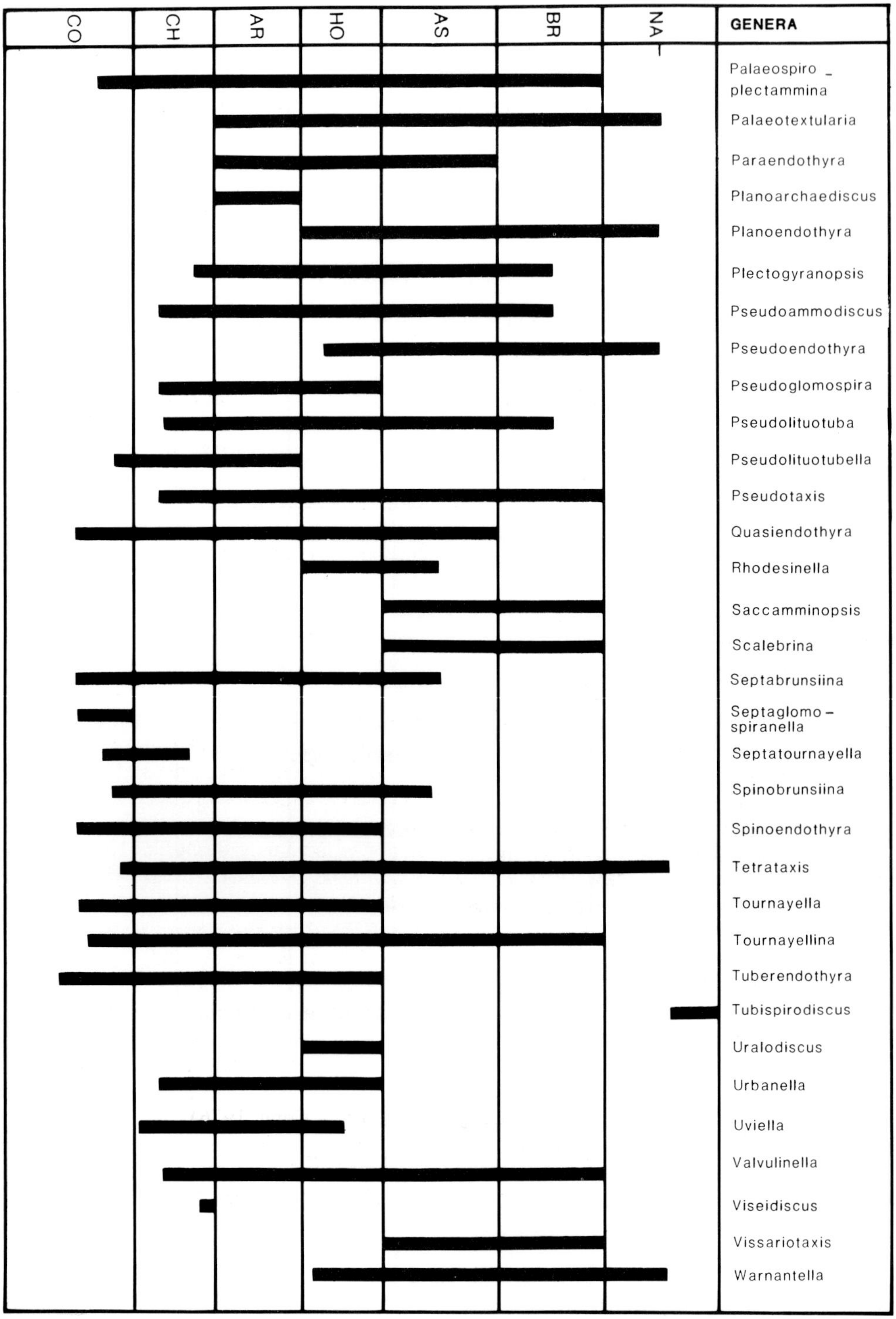

Fig. 3.1 – Range chart of Carboniferous genera.

from the Stone Gill Beds including *Spinoendothyra mitchelli, Brunsia spirillinoides,* and cf. *Eblanaia.* There is a good possibility that the sequence in northern England is more complete than in Belgium (see Fewtrell and Smith (1981) for further discussion of this problem). The implication by Conil *et al.* (1980, caption to Plate 3.2) that the topmost Courceyan at Chatburn is basal Viséan is not supported by the available evidence.

In the eastern Craven Basin, the Thornton Limestones (Clitheroe Formation) are particularly noteworthy, although the junction with Courceyan strata is not seen. The beds contain a rich, diverse and well preserved fauna, which characteristically includes the following: *Earlandia vulgaris, E. minor, Lugtonia, Darjella, Tetrataxis, Pseudotaxis, Brunsia pseudopulchra* and *B. spirillinoides, Forschia, Forschiella prisca, Septatournayella lebedevae, Septabrunsiina* sp., *Pseudolituotubella, Conilites dinanti, Eotextularia diversa, Palaeospiroplectammina mellina, Spinoendothyra recta, Latiendothyra latispiralis, Latiendothyra parakosvensis, Endothyra danica, E. lensi, E. costifera, Omphalotis* cf. *chariessa, Endothyranopsis compressa, E. crassa, Eoendothyranopsis, Tuberendothyra tuberculata magna, Dainella cognata, D. fleronensis* and *Urbanella fragilis. Eoparastaffella* is only doubtfully present. Similar faunas have been described from Ireland at this level, e.g. Conil and Lees (1974), Austin *et al.* (1973), Marchant (1974). *Eoparastaffella* appears in abundance, with *Eostaffella,* in several localities in the Craven area in which Chadian beds are immediately overlain by Arundian (e.g. Embsay Limestone, Haw Crag, Hetton Beck Limestones; Fewtrell and Smith, 1978). These faunas also include taxa appearing for the first time, such as *Plectogyranopsis exelikta, E. apposita, E. campinei, E. convexa, E. delepinei, Biseriella bristolensis,* and *Pseudolituotuba gravata. Biseriella bristolensis* is also found, along with *Brunsia* spp., *Dainella* spp., *Eoparastaffella* sp. and endothyrids in the Gully Oolite in Gower, S. Wales (Conil and George, 1973). Further material of Chadian age has been illustrated by Conil *et al.* (1980) from Furness, Derbyshire, and the S.W. Province.

3.4.3 Arundian

Arundian faunas have been studied in both shelf and basin facies in the S.W. Province, N. England and in Ireland. The base of the Arundian Stage is defined at the first lithological change below the entry of archaediscid foraminifera in the boundary stratotype section at Hobbyhorse Bay, Pembrokeshire (George *et al.*, 1976, p. 9). The entry of this particularly distinctive and biostratigraphically important group is therefore a useful marker in correlating this boundary in sections elsewhere in the British Isles. Above the base, the evolution of the Archaediscidae continues to provide biostratigraphical data upwards into the Namurian.

At Hobbyhorse Bay, the earliest archaediscids to appear are *Brunsiarchaediscus* and *Ammarchaediscus. Uralodiscus* and *Archaediscus* enter rather higher in the sequence. The fauna also includes *Eotextularia, Forschia, Omphalotis,* cf. *minima, Latiendothyranopsis menneri* and *Mediocris* including *M. breviscula* and *M.* cf. *carinata.* In the higher part of the Arundian in the same area, the fauna is characterised by distinctively small forms of *Archaediscus* at the *'involutus'* stage of evolution (Pirlet and Conil, 1978), *L. menneri* also being present.

The earliest Archaediscidae in the Craven Basin are found in the higher part of the Embsay Limestone (Clitheroe Formation), in the limestones in the core of the Lothersdale anticline and in the calcareous shales overlying the reef limestones of Haw Crag (cf. Fewtrell and Smith 1978, p. 267); these units are thereby dated as Arundian. *Brunsiarchaediscus* and *Ammarchaediscus* are among the first archaediscids to appear, as at the boundary stratotype, followed by large *Uralodiscus (Permodiscus* of Fewtrell and Smith, 1978, George *et al.*, 1976). *Glomodiscus* sp. and *Ammarchaediscus* are also present as are *Omphalotis, Urbanella fragilis, Endothyra* spp. including *E. prisca, Pseudolituotubella, Pseudotaxis, Tetrataxis, Pseudoammodiscus, Eostaffella. L. menneri* and *Mediocris,* common in the S.W. Province are very rare in the Craven Basin. *Valvulinella* has been found in the Arundian only in Ireland (Conil and Lees, 1974, Conil, 1976).

The Arundian of the Bristol area generally lacks the earliest archaediscids because of widespread non-sequence at this level. A fauna from the upper Burrington Oolite in the Mendips (Austin *et al.*, 1973) includes *Brunsia, Eotextularia, Endothyra bowmani, Dainella* spp., *Eostaffella* sp., *Archaediscus stilus* and *Nodosarchaediscus* (i.e. *Neoarchaediscus*). A 'nodose archaediscid' was also reported at a similar (high Arundian) stratigraphic horizon in Gower by Conil and George (1973); nodose forms are normally associated with Asbian and later faunas (see below). Marchant (1974), Conil and Lees (1974) and Conil (1976) have all recorded Arundian faunas from Ireland, with characteristics similar to those described already. Reference was made by George *et al.* (1976) to Arundian faunas from the south Askrigg Block and from localities in Ireland.

Additional taxa recorded from the Arundian by Conil *et al.* (1980) include the following. From the Bristol area: *Eoparastaffella, Uviella, Spinoendothyra recta, Endothyra* spp., and *Glomodiscus.* From northern England, *Rectoseptaglomospiranella, Globoendothyra, Pseudoammodiscus, Uralodiscus* (as *Rectodiscus*), *Plectogyranopsis* spp., *Glomodiscus, Uralodiscus* (as *Tubispirodiscus*) *settlensis, Archaediscus varsanofievae,* and *Quasiendothyra* sp.

3.4 Holkerian

The more primitive foraminifera of the two earlier Viséan stages are rapidly replaced by a new assemblage brought in with the Holkerian transgression. Not only is this a period of rapid development and colonisation by certain taxa but also a period when characteristically restricted foraminiferal assemblages existed. A comparable occurence of restricted faunas among the macrofossils has been taken to indicate abnormal salinity, probably hypersalinity.

Several new genera make their appearance in the Holkerian including *Koskinotextularia, Mikhailovella* and *Archaediscus* at the concavus stage. Many others use this stage as a period of development and diversification, including: *Omphalotis, Lituotubella, Nevillella, Bogushella, Globoendothyra, Endothyranopsis, Plectogyranopsis, Rhodesinella, Mstiniella,* and *Nodosarchaediscus.* By way of contrast, *Endostaffella, Millerella* and *Pseudoendothyra* make only an occasional appearance in the Holkerian but are all much more common and stratigraphically useful in the younger Viséan stages.

All the genera so far mentioned can be common in the Holkerian as a whole but they are frequently totally absent in the restricted assemblages. Essentially these consist of *Eostaffella parastruvei, E. mosquensis, Endothyra, Quasiendothyra nibelis, Dainella holkeriana, Rhodesinella avonensis, Septabrunsiina (Spinobrunsiina)* and the alga *Koninckopora.* Individually these forms may occur in older and/or younger rocks, but this particularly characteristic association is very typical of the Holkerian. Here, *Eostaffella* and *Koninckopora* are very abundant but *Q. nibelis* in particular is never better developed or present in greater numbers than in this stage.

3.4.5 Asbian

This stage can be divided into two distinct mesothems, the early and late Asbian. Of the two, it would appear that the early Asbian transgression was much less widespread as it is often missing in British stratigraphical successions.

The early Asbian can be distinguished from the underlying Holkerian by the appearance of several new forms. Amongst these the genus *Vissariotaxis* is the most striking but unfortunately it is not very common and its usefulness is therefore restricted. Further new genera appear in the form of primitive *Nodasperodiscus* and *Neoarchaediscus* but these are also relatively uncommon, often poorly defined, and not well preserved at this level. An additional introduction to the rapidly expanding Archaediscidae is the development of *Archaediscus* at the angulatus stage. This form is not only widely distributed but also fairly abundant and easy to identify. *Nodosarchaediscus* also becomes very widespread at this level. Several genera become less common in the Asbian compared with a relative abundance in the underlying Holkerian. Of these *Quasiendothyra nibelis* is the most notable example.

In contrast to the rest of north-west Europe the early Asbian in Britain contains double walled Palaeotextulariidae in the form of primitive *Cribrostomum,* a genus which was not introduced until the late Asbian transgression in Belgian stratigraphy.

A typical early Asbian assemblage in the British Isles might consist of *Eostaffella mosquensis, Omphalotis minima, Endothyranopsis crassa, E. compressa, Forschiella prisca, Forschia, Globoendothyra globula, Endostaffella fucoides, Pseudoendothyra sublimis, Valvulinella latissima, Palaeotextularia, Koskinotextularia, Lituotubella, Bogushella ziganensis, Mstinia, Palaeospiroplectammina, Archaediscus* at the angulatus stage, *Neobrunsiina, Nevillella, Mikhailovella, Scalebrina, Millerella, Endothyra spira* and other species of the above genera.

The late Asbian introduces a rich variety of fauna including several new genera. The most striking change is in the Palaeotextulariidae where plentiful large *Koskinobigenerina* and *Climacammina* appear for the first time; *Cribrostomum* now becomes more abundant and well preserved. The new forms *Howchinia* and *Bradyina* have unique wall or septal structures which can be readily identified even when only small fragments are available and are thus very useful stratigraphically.

A marked increase in the number of small stellate *Neoarchaediscus* and *Nodasperodiscus* is evident at this level together with *Archaediscus* at the angulatus stage. *Pseudoendothyra, Biseriella* and *Endostaffella* tend to be present in increasing quantities.

A typical late Asbian assemblage may include elements of the above mentioned genera together with species of Forschiinae, *Lituotubella, Cribranopsis, Bogushella, Koskinotextularia, Palaeotextularia, Nodosarchaediscus, Omphalotis, Vissariotaxis, Mikhailovella, Plectogyranopsis, Millerella, Palaeospiroplectammina, Cribrospira, Valvulinella,* and another new genus *Sacamminopsis.*

3.4.6 Brigantian

This stage yields a foraminiferal fauna which is in many ways similar to that of the Late Asbian, but the introduction of widely distributed new forms makes it easily distinguishable. During this stage there is a marked variation in foraminiferal faunas from province to province. In most areas of the British Isles the Brigantian may be divided into two mesothems (D6a and D6b of Ramsbottom, 1979) which yield characteristically different faunas. Brigantian foraminifera tend to be rich and varied and to occur in abundance, and are frequently excellently preserved.

Stratigraphically the most useful development is that of the small stellate *Nodosarchaediscus (Asteroarchaediscus),* which is widely distributed and often very abundant. These tend to be well preserved even in oolitic, dolomitic or siliceous limestones where all other genera are frequently destroyed or absent. Thus they are particularly useful in this stage where transitions to clastic beds are common. Another new form, slightly less useful due to its relative scarcity, is the very distinctive *Janischewskina.* Its characteristically complex septal areas render it easily identifiable when only small fragments of the test are preserved. The numerous subquadrate chambers of *Loeblichia,* another new genus introduced in the Brigantian, are very distinctive in thin section.

Associated Brigantian foraminifera may include giant tetrataxids, *Bradyina, Howchinia bradyana, Koskinobigenerina, Climacammina, Cribrostomum, Warnantella, Valvulinella,* large *Archaediscus* at the angulatus stage, *Loeblichia paraammonoides, Saccamminopsis, Biseriella, Endothyranopsis, Millerella, Lituotubella, Pseudoendothyra* and *Endostaffella.* Also frequently present as an aid to distinguishing these assemblages from those of the late Asbian is the alga *Calcifolium.*

In contrast to these flourishing genera a general and rapid decline of the Tournayellidae, Endothyridae, Calcisphaeridae, *Earlandia, Brunsia* and *Mediocris* began in the Brigantian.

3.4.7 Namurian

Foraminifera are abundant in the lower Namurian limestones of northern England and Scotland, but very little has been published about them. Foraminifera have also been found recently in the few thin limestones of upper Namurian age in northern England.

The lower Namurian (Pendleian and Arnsbergian stages – Early Namurian of the range chart) contain many of the genera and species which are also present in the Brigantian, and there is no really sharp change in the foraminiferal fauna at the base of the Namurian. Particularly common are forms such as *Endothyranopsis, Bradyina, Eostaffella, Asteroarchaediscus* and *Neoarchaediscus.* A totally new element in the fauna of the early part of the Namurian is *Eosigmoilina,* used by Conil *et al.* (1976) as the zonal index of their zone Cf7. In England, however, this fossil has not been found except in the E_{2a} Zone, and the same appears to be true in Scotland. The closely related *Nodosigmoilina* has been found in some of the E_{2b} limestones in the Alston Block area. Compared with faunas of the same age in the U.S.S.R. the early Namurian fauna is rather restricted in the number and variety of taxa present.

Late Namurian faunas are recorded here for the first time in Britain, but are restricted to beds of R_1, R_2 and G_1 zone age (Kinderscoutian to Yeadonian stages). No foraminifera have been found as yet in the Chokierian and Alportian stages, and the few limestones known of these ages are of an unsuitable facies. In the Kinderscoutian a few archaediscids and endothyrids together with *Millerella* s.s. are known in the N7 mesothem in the Throckley Borehole in Northumberland, and foraminifera are also known in the N6 and N8 mesothems, though they are poor and fragmentary. The highest foraminiferal fauna yet found is in the *Gastrioceras cumbriense* Marine band in Lincolnshire; small archaediscids occur here, almost to the exclusion of other foraminifera.

3.5 DESCRIPTION OF GENERA

Although species are illustrated in Plates 3.1 to 3.12, at the present state of knowledge of British Carboniferous foraminifera the most practicable taxa for biostratigraphical studies are genera. For most taxa the ranges are given on Fig. 3.1. Two genera recorded by Conil *et al.* (*Bessiella* and *Florenella*) are not yet validly published and are not included below. The classification followed is that of Loeblich and Tappan (1964).

3.5.1 Suborder Fusulinina Wedekind, 1937

Test calcareous, microgranular to granular, advanced forms with two or more layers in the wall. Within this suborder, Dinantian foraminifera are divided into the superfamilies Parathuramminacea, Endothyracea, and Fusulinacea. In the following generic descriptions the test is free unless otherwise stated.

Ammarchaediscus Conil and Pirlet, 1978. Type species *A. bozorgniae* Conil and Pirlet, 1978. Test lenticular, with planispiral coiling throughout. Wall double-layered, internal dark microgranular layer always present; external fibrous layer developed in central axial region only, or weakly throughout the test. Aperture simple, terminal.

Archaediscus Brady, 1873, emend. Conil and Pirlet, 1978. Type species *A. karreri* Brady, 1873. Test lenticular. Proloculus followed by nonseptate tubular chamber coiled in various planes. Evolute. Wall composed of clear finely fibrous outer layer and thin microgranular inner layer. Aperture a simple opening at end of tube.

Archaesphaera Suleimanov, 1945, emend. Conil and Lys, 1976. Type species *Diplosphaera inaequalis* Derville, 1931. Very simple microgranular test, varying from hemispherical or spherical to diplospherical (with two intersecting spheres of different size). Not certainly a foraminiferid, but conventionally regarded as such in much of the literature.

Asteroarchaediscus Miklukho-Maklai, 1956. Type species *Archaediscus baschkiricus* Krestovnikov and Theodorovich, 1936. Test lenticular or discoidal, often with an uneven serrated surface. Proloculus followed by irregularly coiled tubular chamber which is almost completely occluded by pseudofibrous nodosities throughout ('stellate' condition). Wall with inner dark micrograngular layer weakly developed and outer fibrous layer well developed throughout.

Baituganella Lipina, 1955. Type species *B. chernyshinensis* Lipina, 1955. Test highly irregular, consisting of a number of irregular chambers with angular constrictions and pseudosepta. Lacks a proloculus. Wall micrognanular to granular, secreted, including agglutinated material. Aperture simple, terminal.

Biseriella Mamet, 1974. Type species *Globivalvulina parva* Chernysheva 1948. Test subglobular. Proloculus followed by biserially arranged chambers with marked expansion of coiling. Initial biseriamminid coil followed by open helical coil. Wall microgranular with tectum. Aperture simple, lobate.

Bogushella Conil and Lys, 1978. Type species *Mstinia ziganensis* Grozdilova and Lebedeva, 1960. Test irregularly coiled with raised terminal whorl. Pseudochambers well developed. Wall generally well-differentiated into two layers; external layer irregularly granular and agglutinated; internal layer thin, dark, microgranular. Aperture terminal, cribrate.

Bradyina Moeller, 1878. Type species *Nonionina rotula* Eichwald, 1860. Test robust, planispiral and nautiloid, involute. Not more than 4 whorls, with 3-9 chambers in last whorl. Chamberlets or canals formed by converging septal lamellae or infolding of outer wall to form septa. Wall thick, microgranular, coarsely perforate with distinct radial lamellae. Aperture interiomarginal, with additional large areal pores forming a terminal cribrate aperture; sutural apertures well developed.

Brunsia Mikhailov, 1939. Type species *Spirillina irregularis* Moeller, 1879. Test discoidal, consisting of proloculus followed by coiled nonseptate tubular chamber. Early stages irregularly coiled, later portion planispiral, or with slight oscillations. Wall thin, microgranular. Aperture simple, terminal.

Brunsiarchaediscus Conil and Pirlet, 1978. Type species *Propermodiscus contiguus* Omara and Conil, 1965. Test *Brunsia*-like. Internal dark layer clearly developed, external fibrous layer forming umbilical lateral fillings. Coiling involute, becoming evolute in final whorls. Aperture simple, terminal.

Brunsiina Lipina, 1953. Type species *B. uralica* Lipina, 1953. Test small, discoidal, consisting of proloculus followed by poorly septate tubular chamber. Early portion irregularly coiled, later whorls planispiral. Slow expansion of the whorl with coiling. Wall thin, microgranular. Aperture simple, terminal. The attempt by Conil and Lys (1978) to synonymise *Brunsiina* with *Glomospiranella* Lipina 1953 confused it with *Brunsia*. Although *Brunsiina* is probably therefore still available, *Glomospiranella* is used herein.

Chernobaculites Conil and Lys, 1978. Type species *C. beschevensis* (Brazhnikova, 1965, al. *Ammobaculites sarbaicus* (Malakhova, 1956) var. *beschevensis* Brazhnikova 1965). Test initially coiled, later part straight, uniserial; uncoiled portion predominates. Wall thick, differentiated; outer layer thick, agglutinated; inner layer thin, microgranular. Chambers of coiled portion chernyshinellid in form; chambers of straight portion drop-shaped, with protruding septal necks. Aperture simple, terminal. Recorded from the low Chadian Bankfield East beds, Clitheroe. Conil *et al.*, 1980.

Chernyshinella Lipina, 1955. Type species *Endothyra glomiformis* Lipina, 1948. Small to medium test, with 1–4 simple to highly skewed whorls. Septa short, periphery lobate. Chambers typically 'droplet' shaped or 'chernyshinellid'. Later chambers inflated. Wall microgranular.

Climacammina Brady, 1873. Type species *Textularia antiqua* Brady, 1871. Test large, initial portion biserial, terminal portion uniserial and cylindrical. Chambers increase gradually in size through the biserial portion. Sutures depressed. Wall consists of inner fibrous radiate layer and outer granular layer with agglutinated particles. Aperture slit-like in biserial portion and cribrate in uniserial part.

Condrustella Conil and Longerstaey, 1980. Type species *Mstinia modavensis*, Conil and Lys, 1980. Coiling irregular. Chambers chernyshinellid, well defined and few in number. 2-3 whorls. 4-5 chambers per whorl. Wall thick, coarsely agglutinated with thin microgranular internal layer. Aperture low, simple.

Conilites Vdovenko, 1970. Type species *Ammobaculites dinantii* Conil and Lys, 1964. Initial part of test planispirally coiled for 3½–4 whorls, followed by an uncoiled straight uniserial portion. Pseudoseptate throughout, with 8-9 chambers in last spiral whorl. Wall thick, internal layer microgranular, external layer agglutinated. Aperture cribrate, terminal.

Cribranopsis Conil and Longerstaey, 1980. Type species *Cribrospira fossa* Conil and Naum, 1977. Test involute, planispiral or with slight initial oscillations. Wall granular and finely perforate, initially thin, becoming thicker. Septa short, truncated, thick and massive. Chambers subquadrate, numerous, 8–13 in final whorl, with tendency to uncoil in terminal chambers. Aperture cribrate.

Cribrospira Moeller, 1878. Type species *C. panderi* Moeller, 1878. Test nearly involute, coiling somewhat irregular. Chambers increase rapidly in size; whorls few in number. Septa short. Wall microgranular to granular with tectum. Aperture cribrate, consisting of large pores on apertural face. Intercameral opening large.

Cribrostomum Moeller, 1897. Type species *C. textulariforme* Moeller, 1879. Test large. Chambers broad and low, biserially arranged, increasing rapidly in size; sutures depressed. Wall calcareous, with an inner radiating fibrous layer, and an outer granular layer with agglutinated particles. Cribrate aperture in last chambers.

Dainella Brazhnikova, 1962. Type species *Endothyra chomatica* Dain, 1940. Test globular, irregularly coiled with rapid changes in coiling axis. Early whorls in particular have numerous closely packed chambers; outer whorls with larger chambers, often with high angled changes in coiling direction between whorls. Chomata well developed. Wall microgranular, single layered.

Darjella Malakhova, 1963. Type species *D. monilis* Malakhova, 1963. Test uniserial, numbers of chambers variable, usually 2–4. Chambers spheroidal, increasing in size. Each chamber envelops the preceding one at the connecting neck. Wall agglutinated, although the amount of agglutinated material in the cement is variable.

Draffania Cummings, 1957. Type species *D. biloba* Cummings, 1957. Test flask-shaped or pyriform. Central thin tube around which are arranged two hemispherical chambers. Tube extends into long neck. Periphery rounded; surface smooth and with minute pore openings. Wall calcareous, perforate, thick, layered. Aperture terminal, circular.

Earlandia Plummer, 1930. Type species *E. perparva* Plummer, 1930. Spherical proloculus, followed by cylindrical or flaring tube. Wall microgranular. Aperture simple, terminal.

Eblanaia Conil and Marchant, 1976. Type species *Plectogyra michoti* Conil and Lys, 1964. Test large, discoidal, with large and well marked umbilici. Evolute. Coiling planispiral to oscillating initially, outer whorls usually planispiral. Supplementary deposits in the form of basal nodes, becoming spines, projecting forwards, in the final chambers; corner fillings may also be present. Chambers arched, periphery lobate, septa endothyroid; early stages chernyshinellid with pseudochambers. Wall poorly to clearly differentiated, with a tectum, a medial thick granular layer, and an inner darker microgranular layer.

Endochernella Conil and Lys, 1978. Nomen nudum (no type species designated, although Conil and Lys apparently intended it to be *Endothyra quaestita* Ganelina, 1966). Test entirely enrolled, the inner part of *Chernyshinella*-like form, outer part *Endothyra*-like. Wall tends to show differentiation into 3 layers; no basal deposits; corner fillings accentuate the chernyshinellid form of the chambers. Recorded from the *Michelina grandis* beds (Arundian), Ravenstonedale and from the Chadian at the Eyam borehole, Conil *et al.,* 1980.

Endostaffella Rozovskaya, 1961. Type species *Endothyra parva* Moeller, 1879. Test lenticular or discoidal, concave or bilaterally symmetrical. Coiling involute, final whorls may be evolute. Initial whorls coiled at large angle to terminal portion. Chomata or pseudochomata in later whorls. Aperture simple, basal. Wall thin, microgranular.

Endothyra Phillips, 1846 emend. Brady, 1876. Type species *Endothyra bowmani* Phillips, 1846, emend. Brady, 1976. Test septate, involute, coiling irregular to planispiral, normally of 3–4 whorls, usually less than 10 chambers in last whorl, periphery lobate. A variety of secondary deposits may be present (e.g., spines, tubercules). Wall dark, microgranular, with thin tectum.

Endothyranella Galloway and Harlton, 1930. Type species *Ammobaculites powersi* Harlton, 1927. Early portion coiled and endothyroid; younger portion uncoiled, uniserial, and much larger. Wall microgranular. Aperture simple, basal and crescentiform in spiral part; terminal, circular to oval in rectilinear part. Septate throughout.

Endothyranopsis Cummings, 1955. Type species *Involutina crassa* Brady, 1869. Test globular, involute. Coiling planispiral. Septa thick and long; chambers large, subquadrate or slightly rounded. Periphery smooth; sutures poorly developed. Wall thick, coarsely granular, agglutinated and finely perforate. Aperture simple, lunate.

Eoendothyranopsis Reitlinger and Rostovceva, 1966. Type species *Parastaffella pressa* Grozdilova, 1954. Test discoidal, laterally compressed, markedly biumbilicate. Coiling involute, planispiral, may have feeble initial oscillations. Periphery rounded to subrounded. Chambers subrounded to subquadrate with well defined sutures. Wall granular, with inclusions, two-layered. Terminal forward-projecting spine well developed. Aperture simple, basal.

Eoforschia Mamet, 1970. Type species *Tournayella moelleri* Malakhova, 1953. Test large, thick-walled, predominantly planispirally coiled, with pseudosepta. Wall two-layered; inner layer may have agglutinated grains. Aperture a simple basal opening.

Eoparastaffella Vdovenko, 1954. Type species *Pseudoendothyra simplex* Vdovenko, 1954. Test involute, lenticular to discoidal, planispirally coiled, may have slight initial oscillations. Secondary deposits present in the form of simple paired chomata. Wall microgranular.

Eosigmoilina Ganelina, 1956. Type species *E. explicata* Ganelina, 1956. Test small, lenticular. Proloculus followed by coiled, non-septate chamber; coiling sigmoidal. Wall porcellanous, lacks fibrous layer.

Eostaffella Rauser-Chernousova, 1948. Type species *Staffella parastruvei* Rauser-Chernoussova, 1948. Test lenticular to nautiloid, laterally compressed, biumbilicate, involute, planispiral. Periphery often keeled in early whorls. Chambers subquadrate, numerous, 12–20 in last whorl. Septa nearly straight and at right angles to the wall. Wall dense, granular to microgranular, with pseudochomata or chomata throughout. Aperture simple, basal.

Eotextularia Mamet, 1970. Type species

Palaeotextularia diversa Chernysheva, 1948. Initial portion chernyshinellid in character with very few chambers. Terminal portion uncoiled and biserial; chambers increase rapidly in size. Wall thick with coarse agglutinated outer layer and thin, dark, microgranular inner layer. Aperture simple, basal.

Forschia Mikhailov, 1939. Type species *Spirillina subangulata* Moeller, 1879. Test planispiral. Tubular chamber without distinct septation; pseudosepta only. Wall differentiated; outer layer granular, may have agglutinated grains; inner layer dark, microgranular. Aperture terminal, cribrate where tube flares terminally.

Forschiella Mikhailov, 1935. Type species *F. prisca* Mikhailov, 1935. Coiling initially planispiral and *Forschia*-like. Later stage is uncoiled, uniserial and partially septate. Aperture cribrate.

Globoendothyra Reitlinger, 1959. Type species *Nonionina globulus* Eichwald, 1860. Test nautiloid. Coiling oscillating or aligned; involute or with last whorl evolute. Wall thick with outer tectum; thick granular layer. Septa inclined. Spine-like projection in last chamber. Corner and lateral fillings with low nodosities in some cases. 3–5 whorls, 8–10 chambers in last whorl. Aperture simple.

Glomodiscus Malakhova, 1973. Type species *G. biarmicus* Malakhova, 1973. Test discoidal to lenticular. Proloculus followed by an open tubular chamber, involutely coiled, with oscillating coiling axis as in *Archaediscus*. Wall consisting of two layers; inner dark microgranular layer well developed and thickened laterally; outer fibrous radiate layer also well developed and completely enveloping each whorl. Aperture simple, terminal.

Glomospiranella Lipina, 1953. Type series *G. asiatica* Lipina, 1953. Proloculus followed by irregularly coiled tubular chamber, initially undivided, then divided by pseudosepta. Wall dark, microgranular, single-layered. Aperture simple, terminal.

Granuliferella E. J. Zeller, 1957. Type species *Endothyra rjausakensis* Chernysheva, of which the original type species *Granuliferella granulosa* (Zeller, 1957) is generally regarded as a junior synonym. *E. rjausakensis* is alternatively regarded as belonging to *Latiendothyra* (e.g. Vachard, 1978)*. Test endothyroid. Wall tending to be differentiated into a thin internal dark microgranular layer, and an outer, more or less coarsely granular layer; an outer tectum may be present. Septate throughout. Recorded as *Granuliferella avonensis* sp. nov from the Black Rock Limestone of the Avon Gorge, and the Bankfield East beds Clitheroe (Chadian) by Conil *et al.*, 1980.

*This would leave *Granuliferella* as a *nomen nudum*.

Haplophragmella Rauser-Chernoussova, 1936. Type species *Endothyra panderi* Moeller, 1879. Early portion of test irregularly coiled, later portion uncoiled and uniserial. Coiled portion of about 2 whorls; uncoiled portion of about 7 chambers. Wall thick, coarse, agglutinated, with dark inner layer. Aperture simple and interiomarginal in coiled stage; terminal and cribrate in uncoiled stage.

Howchinia Cushman, 1927. Type species *Patellina bradyana* Howchin, 1888. Test trochoid, consisting of non-septate tube coiled in a high spire around a slightly depressed umbilicus filled with supplementary deposits of microcrystalline calcite. Vertical tubular structures developed in the umbilicus of some species. Spiral suture depressed, bridged by many extensions of shell matter, with small pits between. Wall with dark microgranular inner layer, and outer clear radiate layer. Aperture terminal.

Janischewskina Mikhailov, 1935. Type species *J. typica* Mikhailov, 1935. Test planispiral. Coiling involute with few chambers and whorls. Wall dark, microgranular, thin, and finely perforated. Septal chamberlets formed by infolding of outer wall. Cribrate aperture in apertural shield. Secondary sutural openings.

Koskinobigenerina Eickhoff, 1968. Type species *K. breviseptata* Eickhoff, 1968. Test biserial in early portion, uniserial in terminal portion. Wall single-layered, dark, granular with some agglutinated grains. Cribrate aperture in uniserial chambers.

Koskinotextularia Eickhoff, 1968. Type species *K. cribriformis* Eickhoff, 1968. Test biserial with cribrate aperture. Wall single-layered, granular, with some agglutinated grains. Septa with inflated ends.

Latiendothyra Lipina, 1963. Type species *Endothyra latispiralis* Lipina, 1955. Small to medium endothyrids with skew coiling in first few whorls and planispiral outer whorls, increasing rapidly in size. Wall microgranular, single layered (unlike other endothyrids).

Latiendothyranopsis Lipina, 1977. Type species *Endothyra latsipiralis grandis* Lipina, 1955. Large involute endothyranopsinae with oscillating coiling. Wall coarsely agglutinated with dark internal layer and traces of a tectum. Chambers subquadrate, usually more than 8 per whorl, septa thick. Sutures poorly developed. Aperture simple.

Laxoseptabrunsiina Vachard, 1978. Type species *L. valuzierensis* Vachard, 1978. Test irregularly coiled, with rapid expansion of whorl height. Tubular juvenarium leading to pseudochambers and true chambers in later portions of the test. Wall simple microgranular to granular. Aperture simple, basal.

Leptodiscus Conil and Pirlet, 1978. Type species *Permodiscus umbogmaensis* Omara and Conil, 1965. Test planispiral and lenticular. Internal dark microgranular layer clearly to weakly developed. External radiate layer forms a thick umbilical, lateral filling and thin peripheral layer around test. Chamber has convex floor.

Lituotubella Rauser-Chernoussova, 1948. Type species *L. glomospiroides* Rauser-Chernoussova, 1948. Test non-septate and irregularly coiled in early stages; uniserial, uncoiled, with regular constrictions of the wall forming pseudochambers in later stages. No true septa. Wall differentiated; outer layer granular with or without agglutinated particles; inner layer dark and microgranular.

Loeblichia Cummings, 1955. Type species *Endothyra ammonoides* Brady, 1873. Test discoidal, planispiral and evolute with numerous whorls. Wall finely granular. Chambers small, numerous (13-20 in final whorl) and quadrate. Sutures distinct, radial. Pseudochomata weakly developed. Septa straight, inclined towards the aperture. Aperture simple, basal, and crescent shaped.

Lugtonia Cummings, 1955. Type species *Nodosinella concinna* Brady, 1876. Uniserial test; proloculus followed by several chambers divided by convex septa and expanding in size. Wall microgranular; aperture simple.

Mediocris Rozovskaya, 1961. Type species *Eostaffella mediocris* Vissarionova, 1948. Test lenticular, discoidal or oval to subspherical. Periphery usually rounded. Coiling planispiral; may oscillate slightly in early whorls. Involute, rarely evolute in outer part. Wall microgranular, undifferentiated. Extensive lateral fillings. Aperture basal.

Mikhailovella Ganelina, 1956. Type species *Endothyrina gracilis* Rauser-Chernoussova, 1948. Initial part of test endothyroid, involute. Later part uncoiled, uniserial, subcylindrical, with few chambers. Wall microgranular to granular. Aperture simple in early stages, cribrate in last chambers of spiral portion and in uniserial portion.

Millerella Thompson, 1942. Type species *Millerella marblensis* Thompson, 1942. Test lenticular, nautiloid, laterally compressed, biumbilicate, planispiral. Evolute in ultimate and sometimes penultimate whorl. Periphery usually rounded. Septa arcuate, forward-pointing, regular and numerous. 16-20 chambers in last whorl. Chomata present. Wall microgranular. Aperture simple.

Mstinia Mikhailov, 1939. Type species *Mstinia bulloides* Mikhailov, 1939. Test irregularly coiled with small number of whorls and well developed chambers. Involute. Subglobular with lobulate periphery. Thick wall with coarse agglutinated outer layer and dark microgranular inner layer. Chambers chernyshinellid. Septation in early stages not always developed. Aperture terminal, cribrate.

Mstiniella Conil and Lys, 1978. Type species *Mstinia fursenkoi* Mikhailov *in* Dain, 1953. Test irregularly coiled. Initial tubular portion may be divided into pseudochambers, developing into later endothyroid chambered portion. Wall thick, well differentiated with external agglutinated layer and compact microgranular layer. Aperture basal, cribrate.

Neoarchaediscus Miklukho-Maklai, 1956. Type species *Archaediscus incertus* Grozdilova and Lebedeva, 1954. Coiling irregular. Wall double layered; inner dark microgranular layer very thin, outer radiating fibrous layer thick. Central whorls occluded, resulting in confused stellate flaring. Final 1-2 whorls well defined and without nodosities. Aperture simple, terminal.

Neobrunsiina Lipina, 1965. Type species *Glomospiranella finitima* Grozdilova and Lebedeva, 1954. Test large, globular to discoidal and umbilicate. Initial portion irregularly coiled with occasional terminal alignment of whorls. 4-7 whorls. Pseudochambers developed in terminal portion of test. Wall thick, outer layer granular with agglutinated grains; inner layer dark, microgranular. Aperture simple, basal.

Nevillella Conil, in press. (= *Nevillea* Conil and Longerstaey, 1980, = *Georgella* Conil and

Lys, 1978) Type species *Haplophragmella dytica* Conil and Lys, 1977. Initial coiling chernyshinellid, terminal portion uncoiled, uniserial. Wall thick, clearly differentiated into a thin, dark, microgranular layer and an external coarsely granular agglutinated layer. Aperture cribrate.

Nodasperodiscus Conil and Pirlet, 1978. Type species *Archaediscus saleei* var. *saleei* Conil and Lys, 1964. Test similar to that of *Nodosarchaediscus* but having a stellate central region, caused by occlusion of the central whorls and almost total disappearance of the dark internal layer. Final whorls with thick fibrous layers and thin, feeble microgranular layers. Aperture simple, terminal.

Nodosarchaediscus Conil and Pirlet, 1978. Type species *Archaediscus maximus* Gozrdilova and Lebedeva, 1954. Test lenticular to globular. Wall bilayered: internal dark microgranular layer very thin; outer radiating fibrous layer well developed. Nodosities present in part or all of tubular chamber, but no central stellate flaring developed. Aperture simple, terminal.

Nodosigmoilina nom. nud. Conil *in* Conil *et al.,* 1980. Type species apparently intended to be *Eosigmoilina rugosus* Brazhnikova, 1964, but not designated. Like *Eosigmoilina* but chambers more or less occluded by nodosities.

Nudarchaediscus Conil and Pirlet, 1978. Type species *Planoarchaediscus concinnus* Conil and Lys, 1964. Test irregularly coiled. Wall initially consisting of a thick, dark, microgranular layer and an outer fibrous layer. The latter is absent in the final whorls, where the microgranular layer is thick and well developed. Convex floor to tubular chamber. Aperture simple, terminal.

Omphalotis Schlykova, 1969. Type species *Endothyra omphalota* Rauser-Chernoussova and Reitlinger, 1936. Large endothyrid, involute; small number of whorls, large chambers. Distinctive wall structure, with external thin dark layer (tectorium), thick microcrystalline tectum, and continuous secondary deposits lining the chambers and projecting in places in the form of spines or tubercles.

Palaeospiroplectammina Lipina, 1965. Type species *Spiroplectammina tchernyshinensis* Lipina 1948. Test elongate. Initial coiled portion, small, followed by larger uncoiled biserial portion. Wall microgranular to granular, undifferentiated. Aperture simple, basal.

Palaeotextularia Schubert 1921. Type species *P. schellwieni* Galloway and Rynicker, 1930 (by subsequent designation). Test biserial. 5–12 pairs of biserially arranged chambers; septa straight or gently arched, with lunate ends. Double layered wall; inner layer fibrous, radiate, outer layer granular. Aperture simple, lunate, on interiomarginal arch.

Paraendothyra Chernysheva, 1940. Type species *P. nalivkini* Chernysheva, 1940. Test coiled throughout; predominantly planispiral, initially irregular. Wall structure distinctive, thick, granular, with inclusions. Chambers inflated, periphery lobate, septa long and curved. Prominent basal deposits increasing in height and becoming hook-shaped towards the aperture, which is raised.

Planoarchaediscus Miklukho-Maklai, 1956. Type species *Archaediscus spirillinoides* Rauser-Chernoussova, 1948. Test lenticular. Proloculus followed by coiled tubular chamber, irregular at first, then planispiral. Wall dark, microgranular, with a fibrous layer restricted to the early, irregularly coiled whorls. Aperture simple, terminal.

Planoendothyra Reitlinger, 1959. Type species *Endothyra aljutovicha* Reitlinger, 1950. Test discoidal, biconcave. Initial irregular whorls followed with sharp change in coiling axis by later, slightly oscillating whorls. Initial coils involute, later evolute. 8–11 chambers in last whorl. Wall microgranular. Pseudochomata, lateral fillings and floor linings present. Aperture a simple slit.

Plectogyranopsis Vachard, 1978. Type species *Plectogyra convexa* Rauser-Chernoussova, 1948. Test nautiloid, septate. Coiling irregular, may be aligned or oscillating. Septa thick and inflated. Involute. Wall microgranular to granular, thin in initial whorls, thicker in final whorl; some species finely perforated; wall may be coarse and agglutinated. No basal layer.

Pseudoammodiscus Conil and Lys, 1970. Type species *Ammodiscus priscus* Rauser-Chernoussova, 1948. Test planispiral, umbilicate. Proloculus followed by non-septate tubular coiled chamber with only very slight increase in whorl height. Wall thin, microgranular and compact. Aperture simple, terminal.

Pseudoendothyra Mikhailov, 1939. Type species *Fusulinella struvei* von Moeller, 1879. Test lenticular, laterally compressed, biumbilicate, involute, mainly planispiral; early whorls may deviate slightly. Wall with four layers (including diaphanotheca). Periphery often keeled. Chambers regular, subquadrate, very numerous in last whorl. Septa straight. Pseudochomata or low chomata present. Aperture simple, low crescentiform.

Pseudoglomospira Bykova, 1955. Type species *P. devonica* Bykova, 1955. Test consisting of globular proloculus and tubular undivided chamber which is streptospirally coiled. Wall dark, thin, microgranular. Aperture simple, terminal.

Pseudolituotuba Vdovenko, 1971. Type species *Haplophragmella gravata* Conil and Lys, 1965. Test fixed or encrusting, tubular, partly coiled. Septa absent or in the form of rudimentary internal protuberances. Wall thick, and clearly agglutinated, with a dark internal micrograngular layer.

Pseudolituotubella Vdovenko, 1967. Type species *P. multicamerata* Vdovenko, 1967. Test large. Early portion irregularly coiled with a few septa; later portion uniserial, uncoiled, and with septa throughout. Wall thick, outer layer granular or agglutinated; inner layer dark and microgranular. Aperture terminal, cribrate.

Pseudotaxis Mamet, 1974. Type species *Tetrataxis eominima* Rauser-Chernoussova, 1948. Test trochospirally coiled, with concave base. Low chambers arranged in helical spiral; 3-5 chambers per whorl visible around broad umbilicus in transverse section. Wall micrognanular to granular with tectum but no fibrous layer. Aperture terminal.

Quasiendothyra Rauser-Chernoussova, 1948. Type series *Endothyra kobeitusana* Rauser-Chernoussova, 1948. Test discoidal or nautiloid. Coiling involute and glomospiral in early stage, evolute and aligned or irregular in later stages. Septa commonly inflated, curved. Wall microgranular to granular, chomata well developed. Aperture simple.

Rectoseptaglomospiranella Reitlinger, 1961 (subgenus of *Septaglomospiranella*). Type species *S.* (*R.*) *asiatica* Reitlinger, 1961. The emendation of Conil and Lys (1978), by which *Rectoseptaglomospiranella* is removed from *Septaglomospiranella* which itself becomes a synonym of *Septabrunsiina,* is not accepted here. Test coiled initially, then straight uniserial. Initial portion irregularly coiled; septate throughout. Aperture simple, terminal. The form recorded as *Rectoseptaglomospiranella* sp., by Conil *et al.* (1980, pl. IX, Fig. 1) from the *Michelinia grandis* beds, Ravenstonedale (Arundian) is poorly septate and has a cribrate aperture and thus conforms neither to the original diagnosis nor to the emended one.

Rhodesinella Conil, in press. (= *Rhodesina* Conil and Longerstaey, 1980). Type species *Cribrospira pansa* Conil and Lys, 1965. Coiling planispiral, often with slight initial oscillations. Wall thick, granular to coarsely granular with agglutinated elements. Chamber size increases rapidly towards the final whorl. Terminal chamber raised into a short, almost uncoiled section. Aperture cribrate.

Saccamminopsis Sollas, 1921. Type species *Saccammina carteri* Brady, 1973, = *Nodosaria fusulinaformis* McCoy, 1849. Test uniserial, with globular to ovate chambers and strongly constricted sutures. Wall thin, microgranular. Aperture terminal, simple, rounded.

Scalebrina Conil and Longerstaey, 1980. Type species *S. compacta* Conil and Longerstaey, 1980. Test fixed for encrusting. Wall moderately thick, microgranular. Tubular chamber irregularly coiled. Septa absent or rudimentary. *Scalebrina* is distinguished from *Pseudolituotuba* Vdovenko, 1971 only by the wall structure and smaller size.

Septabrunsiina Lipina, 1955. Type species *Endothyra krainica* Lipina, 1948. Test coiled. Poorly septate or non-septate in early portion; later planispiral portion with well developed septa. Wall microgranular with basal supplementary deposits; terminal projections and nodosities. Aperture a simple opening at the end of tube.

Septaglomospiranella Lipina, 1955. Type species *Endothyra primaeva* Rauser-Chernoussova, 1948. Test small to medium, with 1–4 skew-coiled whorls. Periphery smooth to slightly lobate. Pseudosepta or septa present. Wall microgranular, single-layered. Aperture a simple opening. Conil and Lys (1978) suggested synonymising this genus with *Septabrunsiina;* this is not followed here although further consideration of this question is needed.

Septatournayella Lipina, 1955. Test species *Tournayella segmentata* Dain, 1953. Test planispiral, with pseudosepta or septa which increase in importance towards the outer whorls. Wall microgranular. Lacks secondary deposits. Aperture simple.

Spinobrunsiina Conil and Longerstaey, 1980 (subgenus of *Septabrunsiina*). Type species *S.*

ramsbottomi Conil and Longerstaey, 1980. Test coiled, non-planispiral. Wall microgranular, non-differentiated. Inner whorls have pseudochambers; outer whorls have true chambers. Supplementary deposits in the form of nodosites, arches, basal projections, basal and lateral thickenings. Aperture simple.

Spinochernella Conil and Lys, 1978. Type species *S. brencklei* Conil and Lys, 1978) (probable junior synonym of *Pseudochernyshinella* Brazhnikova, 1974, see Conil *et al.*, 1980), p. 58). Test coiling nearly planispiral. Wall more or less granular, tending to show differentiation into an inner, more compact layer. Septa and chambers chernyshinellid in inner whorls, becoming endothyrid in outer whorls. Supplementary deposits in the form of persistent thin basal projections. Distinguished from *Eblanaia* only by lack of a particularly prominent projection in the outermost chamber. (Recorded as *S.* cf. *brenklei* Conil and Lys) by Conil *et al.* (1980) from the Chadian of the Chatburn bypass section.

Spinoendothyra Lipina, 1963, emend. Mamet, 1976. Type species *Endothyra costifera* Lipina, 1955. Endothyrid with near planispiral or variable coiling. Wall microgranular. The distinctive feature is the presence of secondary deposits in the form of a spine at the base of each chamber. Aperture simple.

Tetrataxis Ehrenberg, 1854. Type species *T. conica* Ehrenberg, 1854. Test free or attached, trochospirally coiled, with a concave base, consisting of low to flattened chambers arranged in a helical spiral; four chamber per whorl. Umbilical cavity broad. Wall differentiated into two layers; outer layer dark, microgranular; inner layer fibrous, radiate. Simple umbilical aperture.

Tournayella Dain, 1953. Type species *T. discoidea* Dain, 1953. Test discoidal, planispirally coiled, lacking septa but sometimes having small pseudosepta. Evolute. Wall microgranular. Aperture simple.

Tournayellina Lipina, 1955. Type species *T. vulgaris* Lipina, 1955. Small number of whorls of which the terminal whorl is markedly expanded. Few chambers per whorl, usually 3-5, globular (chernyshinellid) in shape. Wall dark, granular. Aperture low, simple, terminal.

Tuberendothyra Skipp, 1969, emend. Mamet, 1976. Type species *Endothyra tuberculata* Lipina, 1948. Endothyrid with irregular coiling and a highly lobate periphery. Each chamber has a large rounded tubercle of secondary material at its base. Aperture simple.

Tubispirodiscus Browne and Pohl, 1973. Type species *T. simplissimus* Browne and Pohl, 1973. Test evolute. Coiling planispiral. Tubular chambers without nodosities. Dark micrugranular layer of wall very thin or absent. Radiating fibrous layer thick and well defined. Very gradual increase in size of tubular chamber towards simple aperture.

Uralodiscus Malakhova, 1973. Type species *U. librovichi* Malakhova, 1973. (= *Rectodiscus* Conil and Pirlet, 1978). Test discoidal to lenticular. Proloculus followed by a simple tubular chamber more or less planispirally coiled. Involute. Wall consisting of two layers; inner dark microgranular layer well developed, with lateral thickenings; outer fibrous radiate layer also well developed, and completely enveloping each whorl. Aperture simple.

Urbanella Malakhova, 1963. Type species *Quasiendothyra urbana* Malakhova, 1954. Test small, essentially planispiral. Numerous chambers, rather square in cross-section; periphery slightly lobate. Wall dark, microgranular. Differs from *Loeblichia* only in flatter axial section; *Urbanella* may be a junior synonym of *Loeblichia.*

Uviella Ganelina, 1966. Type species *U. aborigena* Ganelina, 1966. Test with 3½-6½ whorls, initial whorls irregularly coiled and skew to remainder, which are evolute and planispiral. Slight constriction in inner whorls, developing into rudimentary septa in outer whorls. Wall thick and commonly containing agglutinated grains. Aperture a narrow opening.

Valvulinella Schubert, 1907. Type species *Valvulina youngi* Brady, 1876. Test conical. Chambers trochospirally arranged with only 2-3 chambers per whorl. Chambers subdivided into additional chamberlets by horizontal and vertical partitions. Chamberlets vary from rudimentary, to distinct and regular. Wall microgranular, single-layered. Aperture simple, umbilical.

Vissariotaxis Cummings, 1966. Type species *Monotaxis exilis* Vissarionova, 1948. Test trochoid, consisting of non-septate tube coiled in an elevated spire around a depressed umbilical region. Wall microgranular. Aperture simple, terminal.

Viseidiscus Mamet 1975. Type species *Permodiscus* (?) *primaevus* Pronina, 1963. (= *Parapermodiscus transitus* Reitlinger, 1969). Proloculus followed by semi-cylindrical planispiral tube, deviating slightly only in inner two whorls. Evolute. 5-7 whorls in all. Outer pseudofibrous wall layer of archaediscid type developed only in umbilical region. Aperture simple, terminal. N.B. *Ammarchaediscus* Conil and Pirlet, not published until 1978, is a probable junior synonym in part at least.

Warnantella Conil and Lys, 1978. Type species *Glomospira tortuosa* Conil and Lys, 1964. Test irregularly coiled with zig-zag convolutions. Wall microgranular, compact and very dark. Tubular non-septate chamber enlarges gradually in diameter. Periphery irregular, angular, subquadrate. Aperture terminal.

PLATE 3.1

All at magnification × 75. Specimen numbers refer to collection of M. D. Fewtrell of which representative rock samples are held in the Sedgwick Museum, Cambridge, unless otherwise stated.

Earlandia vulgaris (Rauser Chernoussova and Reitlinger, 1937)
Plate 3.1, Fig. 1. HOW 43 Haw Bank Limestone, Chatburn Formation, Skipton, Yorkshire. Description: Large *Earlandia*, up to 2.5 mm long, with simple cylindrical chamber. Wall thick, about 70 μm, dark microgranular.
Range: Late Courceyan – early Namurian.

Archaesphaera inaequalis (Derville, 1931)
Plate 3.1, Fig. 2. HOW 124 Haw Bank Limestones, Chatburn Formation, Skipton, Yorkshire. Description: Diplosphaerid form of *Archaesphaera;* overall size up to 250 μm; smaller sphere up to 100 μm, located within the wall of the larger sphere; outer wall dark microgranular.
Range: Late Courceyan – Brigantian.

Archaesphaera reitlingerae (M. Maklai, 1958)
Plate 3.1, Fig. 3. FS 514 Haw Bank Limestones, Chatburn Formation, Skipton, Yorkshire. Description: Test consists of a flattened sphere attached to a basal disc which may be flat or convex. Overall size, up to 250 μm, wall dark microgranular.
Range: Late Courceyan – Brigantian. Remarks: Considered to be growth stage of *A. inaequalis* by Conil, Groessens and Lys, 1976.

Archaesphaera firmata (Conil and Lys, 1964)
Plate 3.1, Fig. 4. FS 514 Haw Bank Limestones, Chatburn Formation, Skipton, Yorkshire. Description: Test consists of single chamber semicircular to lenticular in cross-section. Overall size, up to 500 μm, wall dark microgranular. Remarks: Differs from *A. reitlingerae* in the overall shape, and relative thickness of the test.
Range: Late Courceyan – Asbian.

Tournayella discoidea (Dain, 1953)
Plate 3.1, Fig. 5. FS 186 Haw Bank Limestone, Chatburn Formation, Skipton, Yorkshire. Description: Planispiral tubular chamber with slight pseudoseptation. Microgranular wall of medium thickness.
Range: Late Courceyan – early Chadian.

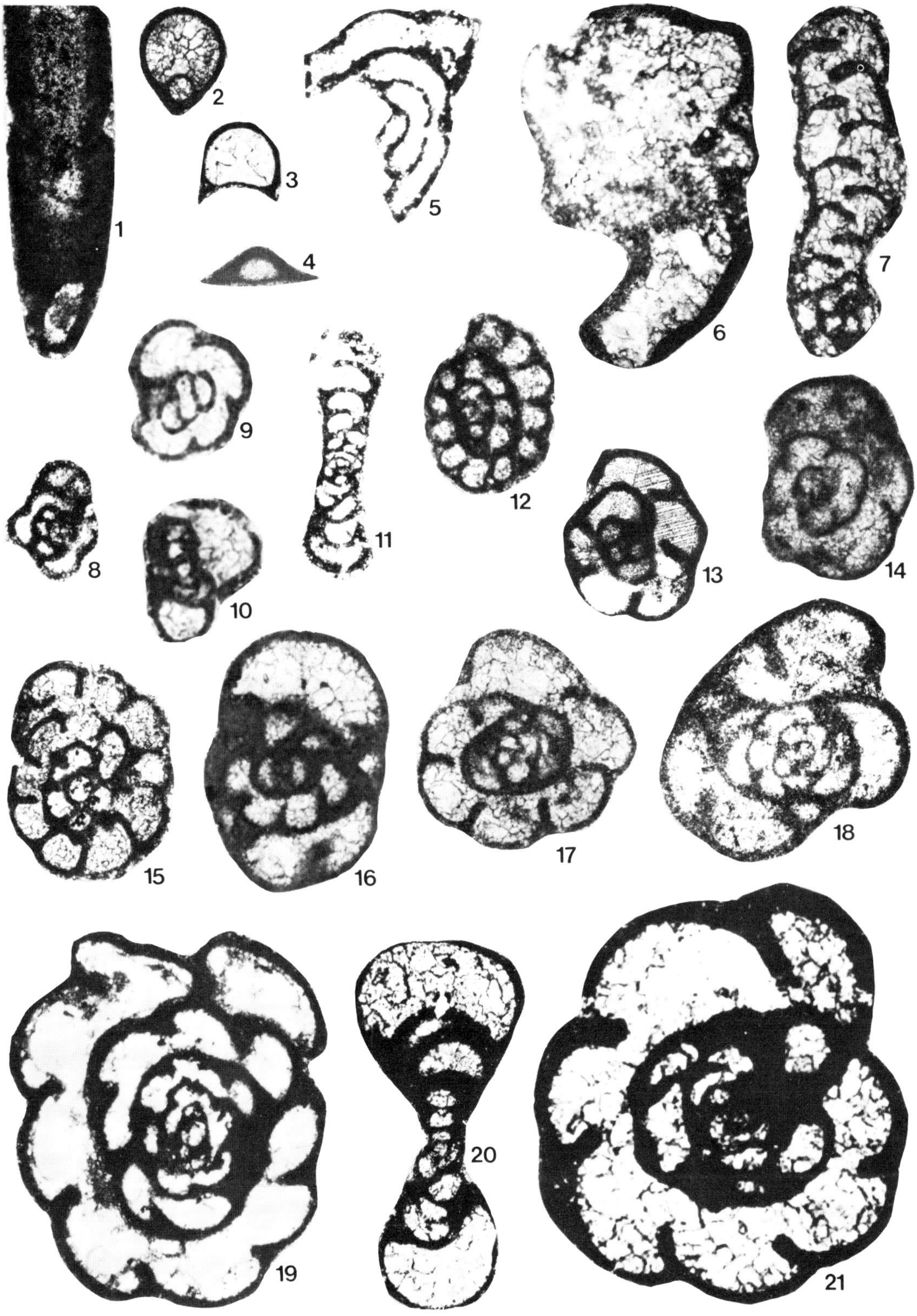

Plate 3.1

Baituganella sp.
Plate 3.1, Fig. 6. HOW 8, Haw Bank Limestone, Chatburn Formation, Skipton, Yorkshire. Description: Large, irregular sac-like form, thick-walled test of microgranular calcite; irregular constrictions or pseudosepta.
Range: Late Courceyan to late Chadian.

Palaeospiroplectammina tschernyshinensis (Lipina, 1948)
Plate 3.1, Fig. 7. I.G.S. Coll. Swinden Borehole (Chatburn Formation), Yorkshire. Description: Test elongate, small initial coiled portion followed by biserial portion of greater breadth and up to 9 arched chambers separated by curved and interlocking septa. Wall microgranular.
Range: Late Courceyan to late Chadian.

Septaglomospiranella primaeva (Rauser-Chernoussova, 1948)
Plate 3.1, Fig. 8. CA 72 Haw Bank Limestones, Chatburn Formation, Skipton, Yorkshire. Description: Small test of about 3 irregularly coiled whorls. Thick short septa poorly developed. Lobate periphery. Wall dark, granular.
Range: Late Courceyan.

Chernyshinella glomiformis (Lipina, 1948)
Plate 3.1, Fig. 9–10. FS 263 Haw Bank Limestones, Chatburn Formation, Skipton, Yorkshire. Description: Small; 2–2½ whorls, with rectangular changes in coiling axis; short, forward-pointing septa; highly lobate periphery. Wall calcareous, more or less granular.
Range: Late Courceyan.

Brunsia spirillinoides (Grozdilova and Glebovskaia, 1948)
Plate 3.1, Fig. 11. HOW 51, Haw Bank Limestones, Chatburn Formation, Skipton, Yorkshire. Description: Discoidal without central hump, 7–8 whorls, evolute irregularly coiled for first 4–5 whorls, then planispiral. Pseudosepta may be present in outer whorl. Wall dark, microgranular.
Range: Courceyan – Chadian.

Endothyra sp.
Plate 3.1, Fig. 12. HOW 4, Haw Bank Limestones, Chatburn Formation, Skipton, Yorkshire. Description: Small endothyrid of 3 whorls, nearly planispiral except initially. Wall rather thick, dark microgranular. Chambers numerous (about 12 in outer whorl), squarish. Septa at right angles to wall, sutures slightly depressed.
Range: Late Courceyan. Remarks: Wall thicker and sutures clearer than in *E. prisca* (Plate 3.3, Fig. 13).

Endothyra nebulosa (Malakhova, 1956)
Plate 3.1, Fig. 13. HOW 81B Haw Bank Limestones, Chatburn Formation, Skipton, Yorkshire. Description: Test of 2½–3 whorls irregularly coiled; 6½–7½ chambers in outer whorl. Periphery slightly lobate. No secondary deposits.
Range: Late Courceyan – early Chadian.

Endothyra tumida (Zeller, 1957)
Plate 3.1, Fig. 14. HOW 23, Haw Bank Limestones, Chatburn Formation, Skipton, Yorkshire. Description: Small to

medium endothyrid, tightly and irregularly coiled at first, then looser and approximately planispiral. Septa short. Periphery lobate. Septa may be secondarily thickened posteriorly.
Range: Late Courceyan – early Chadian. Remarks; resembles *E. laxa* (which is also present in this fauna) but differs in being essentially planispiral.

Spinoendothyra recta (Lipina, 1955)
Plate 3.1, Fig. 15. SK 30, Haw Bank Limestones, Chatburn Formation, Skipton, Yorkshire. Description: Test more or less planispirally coiled, 4–5 whorls, about 10 chambers in last whorl. Periphery nearly smooth. Septa straight. Basal deposits in the form of nodes, tending to become spines or hooks in ultimate chambers.
Range: Late Courceyan – early Chadian.

Endothyra cf. *freyri* (Conil and Lys, 1964)
Plate 3.1, Fig. 16. HOW 55, Haw Bank Limestones, Chatburn Formation, Skipton, Yorkshire. Description: Small endothyrid of 3–3½ whorls of which the last ½–1 whorl is at right angles to the earlier, planispiral part. Corner fillings may be present, otherwise secondary deposits are absent.
Range: Late Courceyan – early Chadian.

Endothyra laxa (Conil and Lys, 1964)
Plate 3.1, Fig. 17. HOW 14, Haw Bank Limestones, Chatburn Formation, Skipton, Yorkshire. Description: Endothyrid of 3½–4 whorls irregularly coiled initially, planispiral in outer 1½–2 whorls. 6 chambers in last whorl. Chambers large, separated by short, inclined septa. No secondary deposits. Wall thin, microgranular.
Range: Late Courceyan – Arundian.

Eblanaia michoti (Conil and Lys, 1964)
Plate 3.1, Fig. 19–20. HOW 143, Haw Bank Limestones, Chatburn Formation, Skipton, Yorkshire. Description: Large, evolute, planispiral with occasional deviations. Up to 5 whorls; 7–9 chambers in outer whorl. Pseudosepta becoming fully developed septa in outer whorl and pointing strongly forward. Lobate periphery.
Range: Late Courceyan – early Chadian – Remarks: Axial section (e.g. Fig. 20) shows deep umbilici and evolute, nearly planispiral coiling.

Septabrunsiina sp.
Plate 3.1, Fig. 18 and 21. FS 263, Haw Bank Limestone, Chatburn Formation, Skipton, Yorkshire. Description: Test large, initial whorls skew-coiled outer whorls nearly planispiral. Septa short and forward pointing; 5½–8 chambers in outer whorl.
Range: Late Courceyan – Mid Chadian. Remarks: basal deposits to chambers subdued as compared to spinose development in specimens 1 and 2, Plate 3.2. Large specimens of *Latiendothyra parakosvensis* are distinguished from forms such as this by the absence of spines, slightly thinner walls, and endothyrid septa, having said this the identification cannot always be made with certainty.

PLATE 3.2

All at magnification × 75. Specimen numbers refer to collection of M. D. Fewtrell, of which representative rock samples are held in the Sedgwick Museum, Cambridge unless otherwise stated.

Septabrunsiina krainica (Lipina, 1948)
Plate 3.2, Fig. 1–2. Fig. 1 I.G.S. Coll. SAD 812A, Chatburn bypass, Lancashire. Fig. 2 TLS 48, Lower Thornton Limestone, Clitheroe Fm, Broughton, Yorkshire. Description: Test large, up to 4 whorls; inner whorls skew coiled; outer planispiral. Septa short, very poorly developed in initial chambers, anteriorly directed. Spines well developed in outer chambers.
Range: Late Courceyan – early Chadian. Remarks: Conil *et al.* (1980 pl. V, Fig. 1–2) illustrate a rather similar form as *Endothyra* cf. *kosvensis* Lipina.

Endothyra danica (Michelsen, 1971)
Plate 3.2, Fig. 3. TLS 46, Lower Thornton Limestone, Clitheroe Formation, Broughton, Yorkshire. Description: Small endothyrid, up to 4 whorls, of which the outer two are planispiral. Septa pointed and directed forwards. Chambers increase moderately in size. Periphery slightly lobate.
Range: Late Courceyan to Mid Chadian.

Septatournayella cf. *lebedevae* (Poyarkov, 1961)
Plate 3.2, Fig. 4. SQB 80, Thornton Limestone, Clitheroe Formation, Broughton, Yorkshire. Description: Small *Septatournayella* with up to 3 planispiral whorls following a rather large proloculus. Simple forward-pointing septa throughout; chamber size increases only slowly.
Range: Late Courceyan – Mid Chadian. Remarks: The initial whorls of *Eblanaia* have a similar appearance.

Spinoendothyra cf. *praeclara* Conil and Longerstaey, 1980
Plate 3.2, Fig. 5. I.G.S. Coll. LL2033, Stone Gill Limestone, Ravenstonedale. Description: Irregularly coiled endothyrid of moderate size. Lobate periphery. Chambers numerous, up to 12 in outer whorl. Septa strong, straight. Secondary deposits in form of basal spines, one to each chamber in outer whorl. Thick dark microgranular wall.
Range: Early Chadian.

Spinoendothyra cf. *mitchelli* Conil and Longerstaey, 1980
Plate 3.2, Fig. 6. I.G.S. Coll. SAD806, Chatburn bypass, Clitheroe, Lancs. Description: Test of 4–5 whorls streptospirally coiled; 8½–9 chambers in last whorl. Periphery slightly lobate; septa forward pointing and often thickened terminally. Secondary deposits in the form of ridges of which those in the last chambers may be hook-shaped in cross-section. Wall rather thin compared to Fig. 5.
Range: Late Courceyan – early Chadian.

Endothyra bowmani (Phillips, 1846) emend. Brady, 1876
Plate 3.2, Fig. 7. TLS 89 Lower Thornton Limestone, Clitheroe Formation, Broughton, Yorkshire. Description: Irregularly coiled endothyrid becoming planispiral in outer whorls. Septa tend to point forward, separating subglobular chambers. Periphery lobate. Secondary deposits various, including nodes in last few chambers.
Range: Late Courceyan – Arundian.

Lugtonia concinna (Brady, 1876)
Plate 3.2, Fig. 8. SQB 1, Lower Thornton Limestone, Clitheroe Formation, Broughton, Yorkshire. Description: Slightly flaring cylindrical test with constrictions at the septa. Uniserial; chambers pear-shaped, increasing moderately in size. Septa pointing markedly towards aperture.
Range: Late Courceyan – Chadian – Arundian. Remarks: not previously recorded earlier than Namurian.

Tuberendothyra tuberculata (Lipina, 1948) subsp. *magna* Lipina and Safonova, 1967
Plate 3.2, Fig. 9. OQB 34, Lower Thornton Limestone, Clitheroe Formation, Broughton, Yorkshire. Description: Small endothyrid with about 4 whorls, coiling highly skewed throughout, periphery moderately lobate. Septa convex; may be secondarily thickened. Internal elements in form of pronounced rounded tubercles at base of each chamber.
Range: Mid Chadian.

Endothyra delepinei (Conil and Lys, 1964)
Plate 3.2, Fig. 10. HC 31, Haw Crag, Clitheroe Formation, Eshton anticline, Yorkshire. Description: Test of about 2½ whorls of moderate size, 7–8 chambers in outer whorl, periphery smooth. Secondary deposits in the form of floor deposits and nodes which become produced into large spines in last few chambers.
Range: Mid-late Chadian.

?*Dainella fleronensis* (Conil and Lys, 1964)
Plate 3.2, Fig. 11. I.G.S. Coll. LL 1906T$_2$, Chatburn bypass, Clitheroe, Lancs. Description: Test 3½–4½ whorls, irregularly coiled with rectangular change in coiling direction between last two whorls. 12 chambers in last whorl. Apparently lacking in chomata.
Range: Chadian.

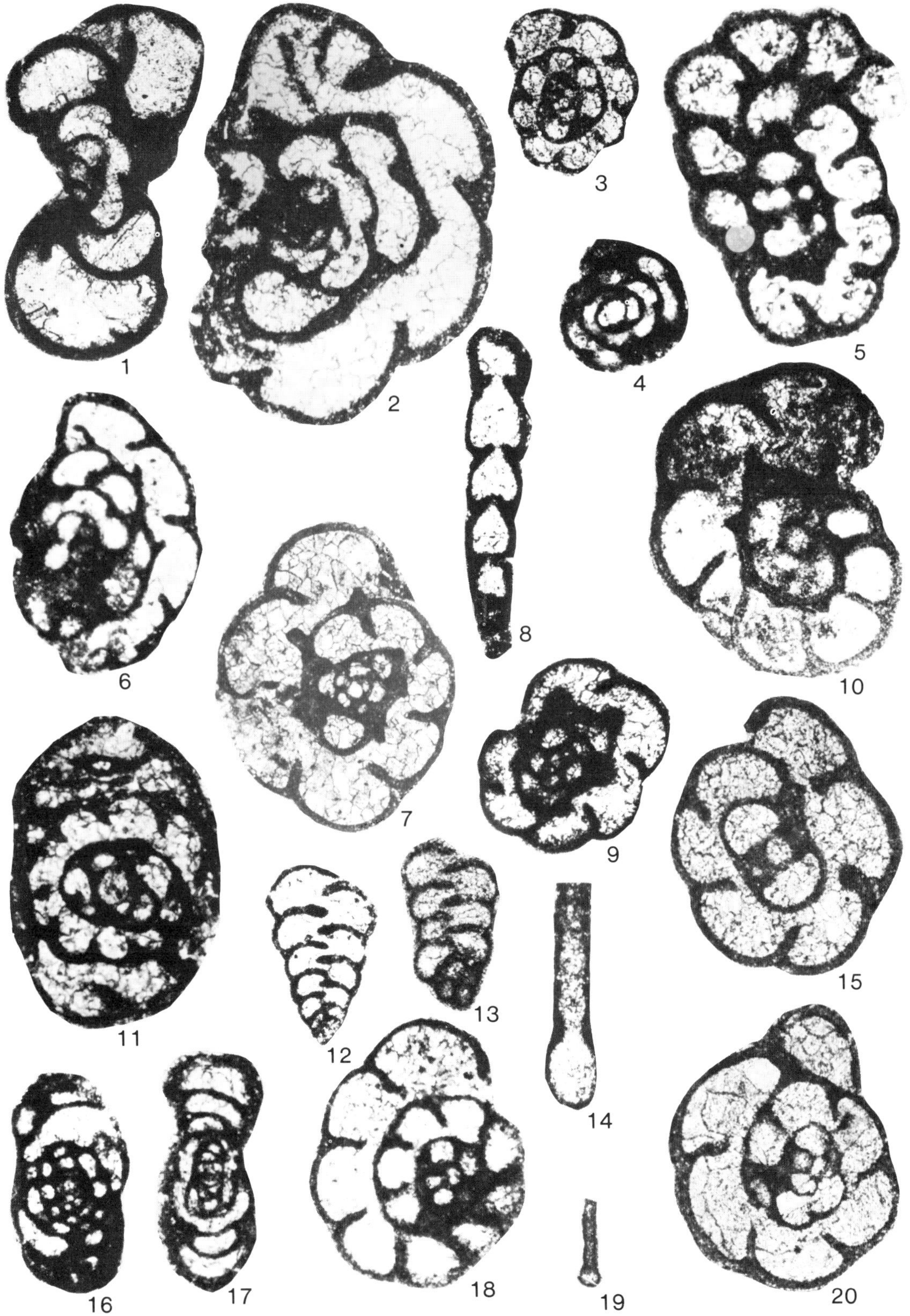

Plate 3.2

Palaeospiroplectammina mellina (Malakhova, 1965)
Plate 3.2, Fig. 12–13. 12, HED21B, Embsay Lst, Clitheroe Fm, Skipton. 13, TLS48, Lower Thornton Limestone, Clitheroe Formation, Skipton, Yorkshire. Description: Small biserial test with initial coiled portion of about 5 chambers; coiled portion small and usually indistinct.
Range: Late Courceyan – Arundian.

Earlandia minor (Rauser-Chernoussova, 1948)
Plate 3.2, Fig. 14. SQB 80. Lower Thornton Limestone, Clitheroe Formation, Skipton, Yorkshire. Description: Small *Earlandia*; wall of medium thickness (about 20 μm) but thickened around junction between proloculus and tubular chamber.
Range: Mid Chadian – Brigantian.

Endothyra sp.
Plate 3.2, Fig. 15. TLS 89, Lower Thornton Limestone, Clitheroe Formation, Broughton, Yorkshire. Description: Moderately large endothyrid, initially irregularly coiled but planispiral outer 1–1½ whorls. Axial section tends to have oblong outline. 8–10 chambers in last whorl. Wall rather thick, dark microgranular. Septa blunt and forward-pointing; may be secondarily thickened.
Range: Late Courceyan – early Chadian.

Dainella cf. *elegantula* (Brazhnikova, 1962)
Plate 3.2, Fig. 16. OQB 30, Lower Thornton Limestone, Clitheroe Formation, Broughton, Yorkshire. Description: Involute, skew-coiled subglobular test with many chambers. Secondary deposits in the form of chomata. Microgranular, single-layered wall.
Range: low to mid Chadian. Remarks: Sections through *Dainella* are very variable and specific identification is not always possible.

Glomospiranella cf. *barsae* (Conil and Lys, 1968)
Plate 3.2, Fig. 17. I.G.S. Coll. LL 1898, Chatburn bypass, Clitheroe, Lancs. Description: Test of many whorls, irregularly coiled initially, becoming essentially planispiral in outer whorls. Tubular chamber of low profile, pseudoseptate in outer whorls. Wall thin, dark, microgranular.
Range: Late Courceyan – early Chadian.

Endothyra sp.
Plate 3.2, Fig. 18. TLS 89, Thornton Limestone, Clitheroe Formation, Broughton, Yorkshire. Description: Endothyrid of about 2½ whorls, planispiral outer whorl following sharp change in coiling axis from inner whorls. 5–6 chambers in outer whorl, periphery slightly lobate. Septa blunt, pointing slightly forwards. Secondary deposits lacking except for possible septal thickenings.
Range: Late Courceyan – Chadian. Remarks: Simple endothyrids of this character are common in the lower part of the Craven Basin succession; they are not readily divided into species.

Earlandia elegans (Rauser-Chernoussova and Reitlinger, 1937)
Plate 3.2, Fig. 19. HE D21 Embsay Limestone, Clitheroe Formation, Skipton, Yorkshire. Description: Spherical proloculus followed by rectilinear tubular chamber, frequently with an internal constriction at the junction of the two. Maximum dimensions 330 μm × 120 μm.
Range: Mid Chadian – Holkerian.

Endothyra cf. *prokirgisana* Rauser
Plate 3.2, Fig. 20. FS263, Haw Bank Limestone, Chatburn Formation, Skipton, Yorkshire. Description: Endothyrid of moderate size. Coiling irregular to nearly planispiral. 6 to 8 chambers in last whorl. Septa slightly anteriorly directed. Chambers subquadrate; periphery slightly lobate.
Range: Late Courceyan – early Chadian. Remarks: Morphologically close to *Latiendothyra* of the group *latispiralis* (Lipina, 1954). Assignment to *Latiendothyra* depends on recognition of wall structure, which is not a reliable character in the available material which is subject to diagenetic alteration.

PLATE 3.3

All at magnification × 75. Specimen numbers refer to Fewtrell Collection of which representative rock samples are held in the Sedgwick Museum, Cambridge, unless otherwise stated.

Conilites dinantii (Conil and Lys, 1964)
Plate 3.3, Fig. 1. TLS 55, Lower Thornton Limestone, Clitheroe Formation, Broughton, Yorkshire. Description: Initial part of test nearly planispiral, evolute, of 3½ to 4 whorls, divided into chambers by pseudosepta becoming more clearly defined in last whorl. Later part of test uncoiled, straight, septate. Chambers increasing in size throughout. Periphery lobate. Wall thick, aperture cribrate.
Range: Mid to late Chadian.

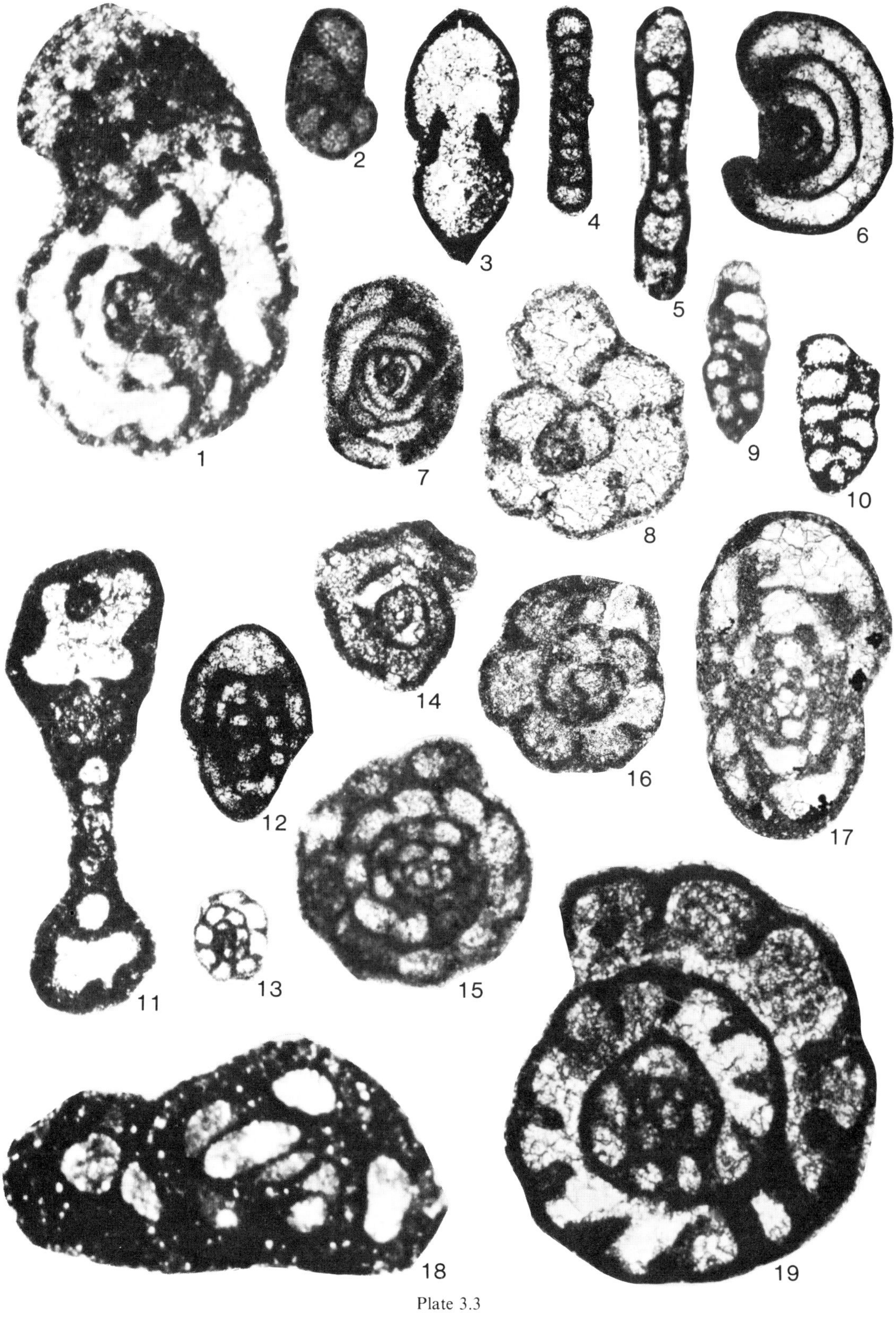

Plate 3.3

Biseriella bristolensis (Reichel, 1946)
Plate 3.3, Fig. 2. HEF 140, Embsay Limestone, Clitheroe Formation, Skipton, Yorkshire. (= *Globivalvulina bristolensis* Reichel, 1946. Eclog. Geol. Helv. 38 (2), 524–560, pl. 19). Description: Small *Biseriella,* rarely more than 300 μm high. Axial section resembles an orange segment; biserial character is readily apparent only in tangential sections. Periphery smooth, non-lobate. Chambers increase in height rapidly between tightly coiled inner and loosely coiled outer portions of test.
Range: Late Chadian.

Darjella monilis (Malakhova, 1963)
Plate 3.3, Fig. 3. TLS 26, Lower Thornton Limestone, Clitheroe Formation, Broughton, Yorkshire. Description: Test uniserial. Chambers large, flask-shaped, with a septal neck protruding into succeeding chamber. Wall thick agglutinated, cement dark, microgranular.
Range: mid-late Chadian. Remarks: Differs from *Lugtonia* in characteristically having few chambers, which are larger and more globular in outline.

Brunsia pseudopulchra (Lipina, 1955)
Plate 3.3, Fig. 4. RDC Lower Thornton Limestone, Clitheroe Formation, Broughton, Yorkshire. Description: Discoidal, evolute, 6–9 whorls, planispiral in outer 1½–5 whorls, with central hump where the coiling is irregular. Slight constrictions may be present in the outer whorl. Wall microgranular.
Range: late Courceyan – mid Chadian.

Viseidiscus sp.
Plate 3.3, Fig. 5. QDR 18 Dogber Rock Quarry, Clitheroe Formation, Yorkshire. Description: Primitive archaediscid with essentially planispiral coiling, deviating only in first 2 whorls. Evolute throughout. Microgranular wall layer well developed throughout; fibrous layer developed only in umbilical region.
Range: Late Chadian – early Arundian. Remarks: This genus, transitional to those with fully developed fibrous layer, is the only archaediscid believed to appear below the Arundian.

Tournayella kisella (Malakhova, 1965)
Plate 3.3, Fig. 6. HED 21, Embsay Limestone (late Chadian), Clitheroe Formation, Skipton, Yorkshire. Description: Planispiral, about 4 whorls, moderate to small in size. Wall microgranular, of moderate thickness. There may be a slight suggestion of wall constrictions in some specimens.
Range: Courceyan – Chadian.

Pseudoglomospira curiosa (Malakhova, 1956)
Plate 3.3, Fig. 7. SQB 1, Upper Thornton Limestone, Clitheroe Formation, Broughton, Yorkshire. Description: Test comprising a proloculus and an irregularly coiled non-septate tubular chamber. About 6 whorls, increasing very gently in height. Wall dark, microgranular. Overall size up to 500 μm.
Range: Chadian – Holkerian.

Plectogyranopsis exelikta (Conil and Lys, 1964)
Plate 3.3, Fig. 8. QDR 18, Dogber Rock Quarry, Clitheroe Formation, Coniston Cold, Yorkshire. = *Plectogyra exelikta* Conil and Lys, 1964, p. 185, Pl. XXVII, Fig. 555–563. Description: Test about 2½ whorls, planispiral except initially. Periphery lobate. 5–5½ chambers in outer whorl; chambers globular. Septa well developed, straight or slightly curved. Wall rather thick, microgranular to agglutinated: no supplementary deposits.
Range: Late Chadian. Remarks: the wall structure and the form of the septa and chambers require transfer of this species to *Plectogyranopsis* Vachard.

Palaeospiroplectammina mellina (Malakhova) subsp.
Plate 3.3, Fig. 9–10. Plate 3.4, Fig. 9. RCD 42, Lower Thornton Limestone, Clint Rocks Quarry, Clitheroe Formation, Broughton, Yorkshire. Plate 3.4, Fig. 10. OQB 34, Lower Thornton Limestone, Clitheroe Formation, Old Quarry, Broughton, Yorkshire. 1976 cf. *Palaeospiroplectammina mellina* (Malakhova) – Conil, pl. II, Fig. 15. Description: Subspecies of *P. mellina* in which the biserial portion is twisted half way along its length. Initial coiled portion succeeded by wider and gently flaring biserial portion of up to about 10 chambers. About half way along the biserial portion, there is a slight but distinct twist in the biserial plane. The chambers in the biserial portion are roundly crescentic; the periphery is slightly lobate. Max. length about 300–500 μm, breadth 150–250 μm.
Range: mid Chadian – Arundian. Remarks: Conil's specimen from the Lane Limestone, Co. Dublin appears very similar to subspecies. T. R. Marchant (pers. comm.) has also reported a similar form from the Dublin Basin at a similar stratigraphic level.

Forschiella prisca (Mikhailov, 1935)
Plate 3.3, Fig. 11. SQB 1, Upper Thornton Limestone, Clitheroe Formation, Broughton, Yorkshire. Description: Test discoidal, planispiral, evolute, of 4–5 whorls of rapidly increasing cross-sectional diameter. Last whorl partly uncoiled. Aperture cribrate.
Range: mid Chadian – Asbian.

Eoparastaffella simplex (Vdovenko, 1954)
Plate 3.3, Fig. 12. HED 14B, Embsay Limestone, Clitheroe Formation, Skipton, Yorkshire. Description: Lenticular to ovoid test, involute planispiral coiling. Numerous chambers in each whorl. Secondary deposits forming simple paired chomata.
Range: Chadian – Arundian.

Endothyra prisca (Rauser-Chernoussova and Reitlinger, 1936)
Plate 3.3, Fig. 13. HEH 47, Embsay Limestone, Clitheroe Formation, Skipton, Yorkshire. Description: Small, mainly planispiral endothyrid of rather variable form. 2–4 whorls, 6–13 chambers in last whorl. Outer 2–3 whorls planispiral, initial part less regularly coiled. Septa straight or forward-pointing. Chambers low, periphery more or less smooth. Wall thin, dark, microgranular, secondary deposits absent.
Range: Courceyan – Brigantian. Remarks: The variability of this species has led to the proposal of a new genus, *Priscella* by Mamet (1974), but this is not used here.

Forschia cf. *parvula* (Rauser-Chernoussova, 1948)
Plate 3.3, Fig. 14. OQB 34, Thornton Limestone, Clitheroe Formation, Broughton, Yorkshire. Description: Test discoidal, planispirally coiled, about 3 whorls. In the later part of the outer whorl there are several constrictions. Aperture slightly flared and apparently cribrate. Wall thick, probably agglutinated. Size of test about 500 μm.
Range: mid-late Chadian. Remarks: Not certainly identified with *F. parvula*, which is an Upper Viséan form.

Spinobrunsiina cf. *ramsbottomi* (Conil and Longerstaey, 1980)
Plate 3.3, Fig. 15. LL 1930, Clitheroe Formation, Chatburn bypass, Lancs. Description: Small initial 'pellet' followed by 2–3 aligned evolute whorls; 5½–7 whorls together, 8½–10 chambers in outer whorl, pseudochambers in initial part. Secondary nodes developed, particularly in last half whorl. Diameter 400–700 μm. Undifferentiated microgranular wall.
Range: late Courceyan to Arundian. Remarks: It should be noted that this form also closely resembles *Eoendothyranopsis spiroides* (Zeller, 1957).

Endothyra convexa (Rauser-Chernoussova, 1948)
Plate 3.3, Fig. 16. HEG 20, Embsay Limestone, Clitheroe Formation, Skipton, Yorkshire. Description: Endothyrid with two skew-coiled whorls followed by a more or less planispiral outer whorl. Chambers increase rapidly in size, being globular in outer whorl: lobate periphery. Septa may be thickened; secondary deposits otherwise absent.
Range: late Chadian. Remarks: This form bears some resemblance to the inner whorls of *Plectogyranopsis settlensis* Conil and Longerstaey, 1980.

Eostaffella sp. cf. *E. parastruvei* (Rauser-Chernoussova, 1948)
Plate 3.3, Fig. 17. HED 21A, Embsay Limestone, Clitheroe Formation, Skipton, Yorkshire. Description: Test ovoid in cross-section, keels well-rounded, flanks slightly umbilicate. 4–5 whorls, involute throughout. Chomata small but distinct, paired. Remarks: Differs from *E. parastruvei* in its more rounded keel and less sharply defined umbilicus.
Range: mid Chadian – Arundian.

Pseudolituotuba gravata (Conil and Lys, 1965)
Plate 3.3, Fig. 18. QDR 18, Dogber Rock Quarry, Coniston Cold, Yorkshire. Description: Large highly irregular, probably fixed, test. Size up to 2.5 mm. Test comprises an irregularly coiled non-septate tube with numerous whorls. Wall very thick (often 200 μm) and agglutinated, with large included grains.
Range: Chadian – Holkerian.

Endothyranopsis crassa (Brady, 1876)
Plate 3.3, Fig. 19. IGS coll. HR 2553, Peach Quarry Limestone, Clitheroe Formation, Clitheroe, Lancs. Description: Test large. 2½ to 3½ whorls, 10–12 chambers in last whorl, subspherical. Involute, Whorl height increases steadily throughout the test. Coiling planispiral: may be slightly irregular initially. Septa massive, often pointing towards the aperture. Sutures depressed only in last half whorl. Wall thick and with agglutinated grains.
Range: Chadian – Brigantian.

PLATE 3.4

All at magnification × 75. Specimen numbers refer to Fewtrell Collection of which representative samples are held in the Sedgwick Museum, Cambridge, unless otherwise stated.

Pseudoammodiscus sp.
Plate 3.4, Fig. 1-2. Embsay Limestone, Clitheroe Formation, Skipton, Yorkshire. Fig. 1, HEH 17, equatorial section, Fig. 2, axial section, ?megalospheric form, HEF 69. Description: Test discoidal, comprising a proloculus followed by a planispirally coiled tubular chamber, increasing very little in height, 3–4 whorls. Overall diameter small, between 100 μm and 200 μm.
Range: Mid Chadian – Arundian.

Uralodiscus rotundus (Chernysheva, 1948)
Plate 3.4, Fig. 3. LOT LEQ, Clitheroe Formation, Lothersdale, Yorkshire. Description: Test involute, planispirally coiled, roundly lenticular. Up to 6 whorls. Wall two-layered: inner dark microgranular layer moderately well-developed, outer fibrous layer enveloping each whorl. Inner layer thickened on either side of the base of the aseptate chamber, which has a convex floor. Maximum diameter 300–400 μm.
Range: Arundian.

Glomodiscus sp.
Plate 3.4, Fig. 4, 10. Fig. 4, LOT 100, Clitheroe Formation, Skipton, Yorkshire. Fig.10, LOT MEQ, Clitheroe Formation, Lothersdale, Yorkshire. Description: skew-coiled initially, more nearly planispiral in outer whorls, roundly lenticular in overall shape. Wall two-layered, inner dark microgranular layer well developed and thickened laterally; outer fibrous layer envelopes each whorl. Chamber floor is convex. Differs from *Uralodiscus* spp. in its non-planispiral mode of coiling and in the better development of the inner dark layer. Size 250–350 μm.
Range: Arundian. Remarks: Most closely resembles *Propermodiscus oblongus* Conil and Lys, 1964, pl. XX, Fig. 406, not here transferred to *Glomodiscus*, as *Propermodiscus* appears to remain a valid taxon.

Ammarchaediscus spirillinoides (Rauser-Chernoussova, 1948)
Plate 3.4, Fig. 5, 11. Fig. 5, HEH 39 Embsay Limestone, Clitheroe Formation, Skipton, Yorkshire, Fig. 11, HE H17, Embsay Limestone, Clitheroe Formation, Yorkshire. Description: Test discoidal, small. Planispirally coiled, deviating slightly only in initial 1-2 whorls: 5-6 whorls altogether. Fibrous layer developed only in central portion of test, not covering outer 1-2 whorls, and forming parallel to concave flanks to the test. Microgranular layer well developed, sometimes developing corner fillings.
Range: Arundian. Remarks: Pirlet and Conil (1978) did not formally transfer this species to *Ammarchaediscus* although they indicated that they regarded it as belonging here. The name is therefore proposed here as *Ammarchaediscus spirillinoides* (Rauser-Chernoussova, 1948) Fewtrell comb. nov.; basionym *Archaediscus spirillinoides* Rauser-Chernoussova, 1948, Pl. 2, Figs. 7-8.

Brunsiarchaediscus sp.
Plate 3.4, Fig. 6. HE H17, Clitheroe Formation, Embsay Limestone, Skipton, Yorkshire. Description: Test discoidal with a slight axial swelling. Proloculus followed by coiled aseptate tubular chamber, irregularly coiled for first few whorls, then planispirally coiled for about 3 whorls. Wall microgranular, with a fibrous layer partially enveloping the inner, irregularly coiled whorls only. Test moderately large, about 400 μm.
Range: Arundian.

Planoarchaediscus cf. *concinnus* (Conil and Lys, 1964)
Plate 3.4, Figs. 7-8. LOT MEQ, Clitheroe Formation, Lothersdale, Yorkshire. Description: Test discoidal with a marked axial swelling. Coiling initially irregular or oscillating, then planispiral in outer 3–4 whorls, becoming progressively more evolute. Fibrous layer developed mainly in axial region. Microgranular layer well developed throughout.
Remarks: Differs from *P. concinnus* in having an axial swelling, and in its earlier range.

Latiendothyranopsis menneri (Bogush and Juferev, 1962)
Plate 3.4, Fig. 9. ING 15, Horton Limestone, Thornton Force, Yorkshire. Description: 2–3 whorls, increasing progressively in height; 7-10½ chambers in last whorl. Chambers rather arched, sutures weak to clear. Septa massive, may be swollen at tips, usually at right angles to wall. Corner and septal thickenings frequent. Coiling more or less planispiral.
Range: Arundian.

Archaediscus krestovnikovi (Rauser-Chernoussova, 1948), ('*involutus* stage' of Pirlet and Conil, 1978)
Plate 3.4, Figs. 12-14. IGS coll, Fig. 12, HR 2869; Fig. 13, KR 3343; Fig. 14, HR 2885 Pen-y Holt Formation, Hobbyhorse Bay, Penbrokeshire. Description: *Archaediscus* with dark microgranular inner layer moderately well developed and chamber floors flat to convex. Coiling oscillates more or less markedly.
Range: Arundian – Brigantian (*involutus* stage ranges from Arundian to Asbian).

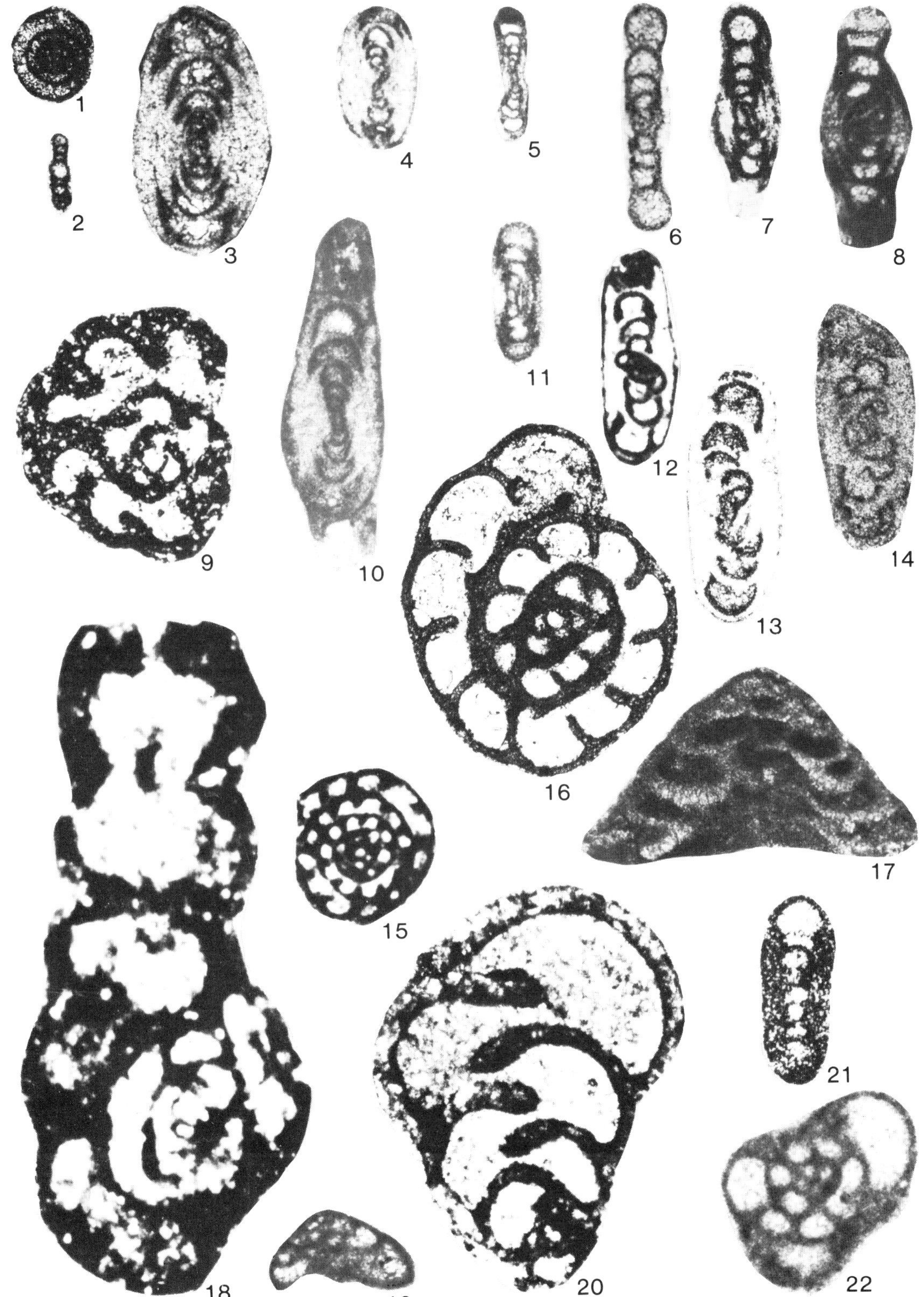

Plate 3.4

Urbanella fragilis (Lipina, 1951)
Plate 3.4, Fig. 15. HE H18, Embsay Limestone, Clitheroe Formation, Skipton, Yorkshire. Description: Test small. Planispiral with very low rate of whorl expansion. Numerous squarish chambers, blunt septa. Wall thin, dark, microgranular.
Range: Mid Chadian – Arundian.

Omphalotis minima (Rauser and Reitlinger, 1936)
Plate 3.4, Fig. 16. IGS coll. KR 3351, Pen-y-holt Formation, Hobbyhorse Bay, Pembrokeshire. Description: Endothyrid with skew-coiled initial whorls, becoming planispiral in outer 2–3 whorls. Periphery lobate; septa curved. Some secondary internal thickenings.
Range: mid-Chadian – Arundian.

Tetrataxis conica s.l. Ehrenberg, 1854
Plater 3.4, Fig. 17. LOT MEQ, Lower Thornton Limestone, Clitheroe Formation, Broughton, Yorkshire. Description: Test conical with apical angle between 80° and 100°. Flanks generally smooth, wall with well developed fibrous layer. Umbilical area pronounced. Chambers increase gradually in size.
Range: Late Courceyan – Arundian – Brigantian.

Pseudolituotubella sp.
Plate 3.4, Fig. 18. HE H17 Embsay Limestone, Clitheroe Formation, Skipton, Yorkshire. Description: Large. Coiled portion of 2–3 whorls, straight uniserial portion with several chambers. Septate throughout. Cribrate aperture.
Range: Courceyan – Arundian.

Pseudotaxis sp.
Plate 3.4, Fig. 19. LOT 16, Clitheroe Formation, Lothersdale, Yorkshire. Description: Test conical, trochospiral, with at least 5 whorls. Apical angle obtuse, flanks smooth; small, wall microgranular, single-layered.
Range: Mid Chadian – Arundian. Remarks: The form shown here most closely resembles *Tetrataxis pusillus* Conil and Lys, 1964, the name of which is preoccupied by *T. pusillus* Golubsov, 1954.

Eotextularia diversa (Chernysheva, 1948)
Plate 3.4, Fig. 20. IGS coll. KR 3354/1, Pen-y-Holt Formation, Hobbyhorse Bay, Pembrokeshire. Description: Large test: biserial with a coiled initial part of a single whorl and about 5 chambers; biserial part of 5–10 chambers, increasing rapidly in breadth. Periphery rounded, slightly lobate. Apical angle rather variable. Septa thick, curved, and terminally thickened.
Range: ?Courceyan – Arundian.

Mediocris breviscula (Ganelina, 1956)
Plate 3.4, Fig. 21. ING 21 Horton Limestone, Thornton Force, Yorkshire. Description: Test small, discoidal with flattened flanks and rounded keel, 3–4 whorls. Secondary deposits well developed in the form of lateral fillings.
Range: Arundian – Asbian.

Endothyra sp.
Plate 3.4, Fig. 22. LOT 33, Clitheroe Formation, Lothersdale, Yorkshire. Description: Test small, irregularly coiled, slightly lobate periphery. Septa gently curved to the anterior, of moderate length. Secondary deposits not developed.
Range: High Courceyan – Arundian. Remarks: small endothyrids of this type are numerous at this level in the Craven Basin.

PLATE 3.5

All the figured specimens are in the IGS collection at Leeds and are × 75, except where stated otherwise.

Dainella holkeriana (Conil and Longerstaey, 1980)
Plate 3.5, Fig. 1. AL 1364, Holkerian, Stackpole Limestone, South Bay, Tenby, S. Wales, 74 m below top of section. Description: Test very irregularly coiled, involute. Whorl height initially low; two final whorls increase rapidly in size and their division into chambers much less numerous than in earlier whorls. Septa well inclined towards aperture.
Range: Holkerian.

Millerella excavata (Conil and Lys, 1974)
Plate 3.5, Fig. 2. ARE 923. Holkerian, Garsdale Limestone, R. Clough, near Sedbergh, Yorkshire. Description: Test deeply and widely umbilical: periphery well rounded. Coiling often slightly distorted at origin, becoming planispiral. 4 whorls. Final whorl very large, evolute. Chomata or pseudochomata very well developed.
Range: Holkerian – Asbian.

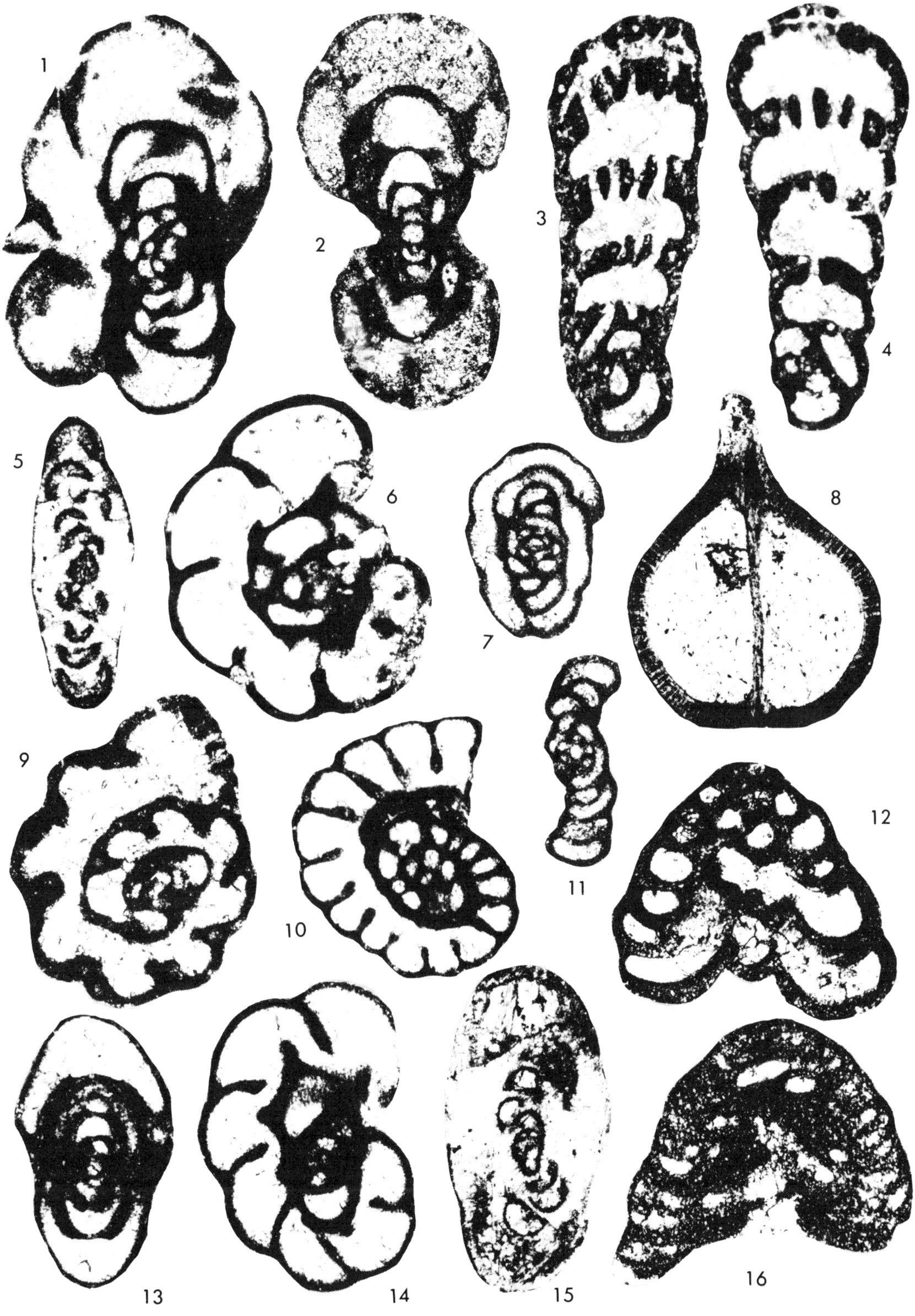

Plate 3.5

Nevillella tetraloculi (Rauser-Chernoussova)
Plate 3.5, Figs. 3, 4. Fig. 3, AL 1363 × 38; Fig. 4, AL 1353 × 38. Holkerian, Stackpole Limestone, South Bay, Tenby, S. Wales. = *Haplophragmella tetraloculi* Rauser-Chernoussova, 1948. Description: Test initially irregularly coiled, laterally compressed. Later portion straight cylindrical or expanding slightly. Coiling irregular, 2–3 whorls, 4 chambers in final whorl. Chambers slightly convex in spiral portion, sub-rectangular in straight portion. 3–4 chambers in uniserial section. Chamber width is twice its height. Diam. spiral portion 0.62–0.85 mm: width of straight portion 0.67–0.80 mm, length of straight portion up to 1.6 mm. Wall up to 75–80 μm thick in uniserial section. Aperture low, slit-like initially, cribrate in final 2 chambers of spiral portion and all of uncoiled portion. Convex or almost flat apertural face. Range: Arundian – Asbian.

Nodosarchaediscus sp.
Plate 3.5, Fig. 5. AL 1367. X140. Holkerian, South Bay, Tenby. Description: Test laterally compressed, lenticular, irregularly coiled. Well developed nodosities. Well rounded hemispherical roof to chamber. 6–7 whorls. Dark micro-granular wall well developed.

Endothyra ex gr. phrissa (D. Zeller)
Plate 3.5, Fig. 6. AL 1368. Holkerian, Stackpole Limestone, South Bay, Tenby, S. Wales. = *Plectogyra phrissa* Zeller, 1953. Description: Test discoidal, umbilicate. Proloculus large. Chambers large and strongly inflated. Secondary deposits well developed, a hook being present in the final chamber and nodes in the preceding chambers with connecting basal coverings. Total rotational distortion very high. Septa moderately long and strongly arcuate, with some evidence of light secondary deposits on posterior surfaces.

Glomospiranella aff. *barsae* (Conil and Lys)
Plate 3.5, Fig. 7. = *Brunsiina barsae* Conil and Lys, 1968. ARE 571. Holkerian, Tunstead Quarry, Derbyshire. Description: Coiling initially irregular, final whorls oscillate slightly from one stabilised plane. Final 2–2½ whorls almost planispiral. 7–8 whorls in total. Tubular chamber initially thin, slowly widening in course of coiling. Pseudochambers formed by inflexions of the wall well developed in final two whorls. Initial whorls free of all divisions. 12–15 pseudo-chomata in total. Wall microgranular – may enclose some larger grains.

Draffania biloba Cummings, 1957.
Plate 3.5, Fig. 8. ARE 554. Early Asbian, Timpony Limestone, near Stump Cross Cavern, Grassington, Yorkshire.
Range: Holkerian – Brigantian.

Rhodesinella avonensis (Conil and Longerstaey)
Plate 3.5, Fig. 9 (see also Plate 3.7, Fig. 5). = *Rhodesina avonensis* Conil and Longerstaey, 1980. ARE 923. Holkerian Tunstead Quarry, Derbyshire. Description: Evolute, 2½–3 whorls; chamber height increases gradually towards aperture, chambers swollen, sutures well defined: septa short, thick, wedge-shaped, inclined towards aperture. 7½ chambers in

final whorl.
Range: Holkerian – Early Asbian.

Quasiendothyra nibelis (Durkina, 1959)
Plate 3.5, Fig. 10 (see also Plate 3.6, Fig. 1; Plate 3.8., Fig. 10). ARE Holkerian, Garsdale Limestone, R. Clough, near Sedbergh, Yorkshire.
Range: Holkerian – Asbian.

Brunsia sp.
Plate 3.5, Fig. 11. ARE 571, Holkerian, Tunstead Quarry, Derbyshire.

Tetrataxis sp.
Plate 3.5, Fig. 12. ARE 552. Early Asbian, Timpony Limestone, near Stump Cross Cavern, Grassington, Yorkshire.

Eostaffella parastruvei (Rauser-Chernoussova)
Plate 3.5, Fig. 13. AL 1368, Holkerian, Stackpole Limestone, South Bay, Tenby, S. Wales. = *Staffella parastruvei* Rauser-Chernoussova, 1948. Description: Test slightly compressed laterally with moderate umbilici. Periphery carinate or slightly rounded, never pointed. Whorls increase gradually in size towards the final significantly larger revolution. 4–5 whorls. Pseudochomata in the form of bands. Chomata very rare. Diameter 450–700 μm. Width 250–350 μm. Width/diameter ratio 0.43–0.58.
Range: Holkerian – Brigantian.

Endothyra maxima (D. Zeller)
Plate 3.5, Fig. 1.14. ARE 549. Early Asbian, Timpony Limestone, near Stump Cross, Grassington, Yorkshire. = *Plectogyra maxima* Zeller, 1953. Description: Test large, elongate. Wall thin in relation to size of shell. Chambers large, swollen. Proloculus small. Supplementary deposits well developed in form of basal coverings and large forward-curving hooks directly behind each septum.
Remarks: Distinguished by its extreme size, the development of large hooks and its high angular distortion.
Range: Asbian – Namurian.

Archaediscus sp.
Plate 3.5, Fig. 15. AL 1354 X140. Holkerian, Stackpole Limestone, South Bay, Tenby, S. Wales. Description: Test laterally compressed. Coiling irregular. Tubular chamber increases slowly in size until the final whorl, where a massive expansion occurs, producing a chamber approximately twice the size of the penultimate whorl. 5–6 whorls.

Valvulinella cf. *conciliata* (Ganelina)
Plate 3.5, Fig. 16 (see also Plate 3.10, Fig. 15). = *Tetrataxis conciliatus* Ganelina, 1956. ARE 551. Early Asbian, Timpony Limestone, near Stump Cross, Grassington, Yorkshire.
Remarks: *V. conciliata* is distinguished from other species by its thin wall and 90° apical angle.

PLATE 3.6

All the figured specimens are in the IGS collections at Leeds, and are × 75, unless stated otherwise.

Quasiendothyra nibelis (Durkina, 1959)
Plate 3.6, Fig. 1. (See Plate 3.5, Fig. 10, and Plate 3.8, Fig. 10). ARE 341, Early Asbian, Tandinas Quarry, Anglesey.
Range: Holkerian – Asbian.

Cribrospira mira (Rauser-Chernoussova, 1948)
Plate 3.6., Fig. 2. BLE 4697 ×40, Late Asbian, Danny Bridge Limestone, Raydale Borehole, Askrigg, depth 13 m. Description: Test involute, increasing rapidly in height and less rapidly in width, giving an elongate, ovoid section. 2 whorls, regularly coiled, expanding markedly in final chamber. Total number of chambers 11–17 with 7–8 in final whorl. Septa long, hook-like, situated at almost equal distances from each other. Sutures distinct. Aperture cribrate in final chamber, high and slit-like in other chambers.
Remarks: differs from *C. panderi* Moeller in having a compressed test along shell axis, in its lower whorl and lower final chambers, in having the septa evenly spaced and fewer apertural openings.
Range: Holkerian – Brigantian.

Endothyranopsis sphaerica (Rauser-Chernoussova and Reitlinger)
Plate 3.6, Fig. 3 (= *Endothyra sphaerica* Rauser-Chernoussova and Retilinger, 1937). (See also Plate 3.9, Figs. 2, 3). ARE 1422, Early Asbian, Strandhal, Isle of Man.
Range: Asbian – Brigantian.

Mediocris mediocris (Vissarionova)
Plate 3.6, Fig. 4. (See also Plate 3.8, Fig. 5 and Plate 3.7, Fig. 4). ARE 1439, Early Asbian, Strandhall, Isle of Man. = *Eostaffella mediocris* Vissarionova, 1948.
Range: Tournaisian – Namurian.

Nevillella dytica (Conil and Lys)
Plate 3.6, Fig. 5. AL 1350, Holkerian, Stackpole Limestone, South Bay, Tenby, S. Wales; 8 m below top of section. = *Haplophragmella dytica* Conil and Lys, 1977. Description: test massive, uncoiled part subcylindrical with a diameter close to that of coiled part. 2–3½ whorls, 4–4½ chambers per whorl. Chambers initially chernyshinellid then swollen in uncoiled section. Wall relatively thin at the origin, becoming rapidly thicker. Sutures well defined. Aperture cribrate in last 1–2 chambers of uniserial section.
Range: Holkerian – Brigantian.

Palaeotextularia aff. *longiseptata* (Lipina, 1948)
Plate 3.6, Fig. 6. ARE 719 ×40. Asbian, Urswick Limestone, 1.5 m below Woodbine Shale, Stainton Quarry, Dalton-in-Furness, Cumbria. Description: Chambers slightly convex. Septa long, straight and slightly thickened at ends. Wall bilaminate, thickness in final chambers 36–50 μm. 6–9 chambers each side of test. Apertural face flat or slightly curved.
Remarks: distinguished by its long, straight septa.

Mstiniella sp.
Plate 3.6, Fig. 7. AL 1319, Holkerian, Stackpole Limestone, South Bay, Tenby, S. Wales; 64 m below top of section. Description: Wall initially fairly thin, increasing rapidly in thickness towards the final whorls. Sutures well defined. Septa fairly short, thick, blunt and forward pointing. Chambers swollen and well rounded. 8 chambers in final whorl. 4–5 whorls. Aperture low, cribrate in final chamber.

Endothyra excellens (D. Zeller)
Plate 3.6, Fig. 8. ARE 1156, Early Asbian, Potts Beck Limestone, Little Asby Scar, Cumbria. = *Plectogyra excellens* Zeller, 1953. Description: Test large, discoidal with rounded periphery. Wall thin, finely granular with tectum. Proloculus small. Chambers large. Secondary deposits massive, continuous, very well developed. Well defined hook present in last chamber. Septa short, directed slightly anteriorly. Aperture high.
Remarks: differs from *E. pandorae* (Zeller) in its strong supplementary deposits, more swollen chambers and thinner walls.
Range: Asbian – Lower Namurian.

Lituotubella magna (Rauser-Chernoussova)
Plate 3.6, Fig. 9. (see also Plate 3.9, Fig. 5). ARE 1244, Asbian, Oxwich Head Limestone, Mumbles, near Swansea. = *L. glomospiroides magna* Rauser-Chernoussova, 1948. Description: width of straight portion increases slightly. Glomospiral portion laterally compressed. Septa and septal sutures initially almost imperceptible. 3½–5 whorls in glomospiral portion. Up to 8 chambers in uniserial section, septa more distinct.
Range: Asbian – Brigantian.

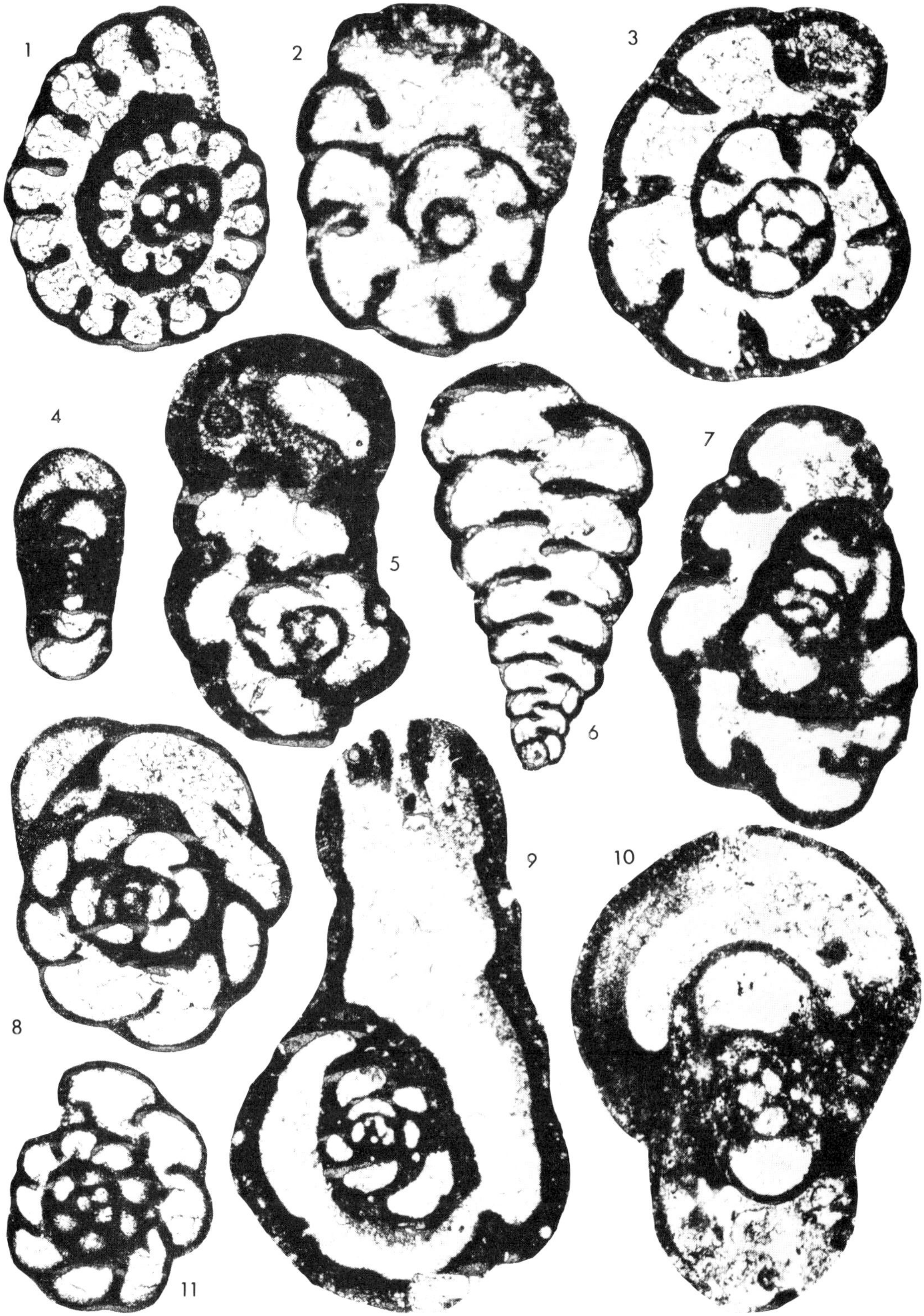

Plate 3.6

Endothyranopsis pechorica (Rauser-Chernoussova)
Plate 3.6, Fig. 10. ARE 1432, Early Asbian, Strandhall, Isle of Man. = *Endothyra crassa pechorica* Rauser-Chernoussova, 1936. Description: Deeply umbilicate. Chambers flat, sutures faint. 2½–4 whorls, 7–10 chambers in final whorl. W/d = 0.7–0.8. Heavy secondary infilling in axial region.
Range: Asbian – Brigantian.

Endothyra ex. gr. *spira* (Conil and Lys)
Plate 3.6, Fig. 11. ARE 643. Asbian, Trowbarrow quarry, Silverdale, Cumbria. Description: coiling planispiral; slight initial oscillations. Whorls increase fairly rapidly in height; 2½–3 whorls. Chambers swollen, becoming globular in final whorl. Sutures distinct, especially between terminal chambers; 8 in final whorl. Supplementary deposits well developed. Nodosities increase in importance until the final chamber where they are replaced by a spinal projection inclined towards aperture.
Range: Asbian – Brigantian.

PLATE 3.7

Unless otherwise stated, the figured specimens are in the I.G.S. collections at Leeds, and are ×75.

Endothyranopsis crassa (Brady) emend. Cummings, 1955.
Plate 3.7, Fig. 1. ARE 973. Late Asbian, Danny Bridge Limestone, R. Clough, Sedbergh, Cumbria. = *Endothyra crassa* Brady, 1876. Description: Test free, large, nautiloid, subglobular and slightly asymmetrical. 3 whorls present increasing moderately in height with complete embracement throughout. Approx. 10 chambers in final whorl. Sutures moderately well defined. Periphery broadly rounded with lobulation.
Range: Holkerian – Namurian.

Endothyra sp. ex gr. *phrissa* (D. Zeller, 1953)
Plate 3.7, Fig. 2, (See also Plate 3.5, Fig. 6). ARE 1434. Early Asbian, Strandhall shore, Isle of Man.

Nodosarchaediscus sp.
Plate 3.7, Fig. 3. ARE 539 ×140. Early Asbian, Stump Cross Limestone, near Stump Cross, Grassington, Yorkshire. Description: Test compressed laterally but slightly swollen. Periphery wall rounded and fairly even. Coiling initially oscillating, becoming more aligned in final whorls. 6½ whorls. Gradual increase in size of tubular chamber towards last two whorls, where dimensions are similar.

Archaediscus reditus (Conil and Lys)
Plate 3.7, Fig. 4. ARE 1194 ×140. Early Asbian, Groups Hollows, near Little Asby Scar, Cumbria, = *A. krestovnikovi reditus* Conil and Lys, 1964. Description: Test lenticular, slightly flattened. Coiling initially compact and oscillating, tending to become aligned and more spacious in final whorls. 5–7 whorls. L/d 0.42–0.55. Proloculus small, 25–40 μm diameter.
Remarks: Differs from *A. koktubensis* in mode of coiling and less compressed test.
Range: Holkerian – Late Asbian.

Rhodesinella avonensis (Conil and Longerstaey)
Plate 3.7, Fig. 5. (See also Plate 3.5, Fig. 9). AL 1344. Holkerian, South Bay, Tenby, 58 m below top section. = *Rhodesina avonensis* Conil and Longerstaey, 1980.
Range: Holkerian – Early Asbian.

Omphalotis cf. *volynica* (Brazhnikova, 1956.)
Plate 3.7, Fig. 6. ARE 18. Late Asbian, Trowbarrow Quarry, Cumbria. Description: Test irregularly coiled. Whorl height and chamber size increase rapidly from the third whorl towards the aperture. Second whorl perpendicular to the third. 3½–4 whorls. Chambers well rounded, convex, and divided by curved, almost hooked septa. 7 chambers in final whorl. Sutures deep and well defined. Supplementary deposits in the form of large nodosities grading into well developed spines in the last two chambers.

Planoendothyra sp.
Plate 3.7, Fig. 7. ARE 1010. Late Asbian, Oxwich Limestone, Pwll Ddu Head, Gower, S. Wales. Description: Test discoidal, compressed laterally. Proloculus followed by an initially irregularly coiled involute spire. Final whorls more regularly coiled, oscillating and evolute. Chambers rounded with forward pointing septa of same thickness as the wall. Septa generally short. Probably 10–13 chambers in final whorl.

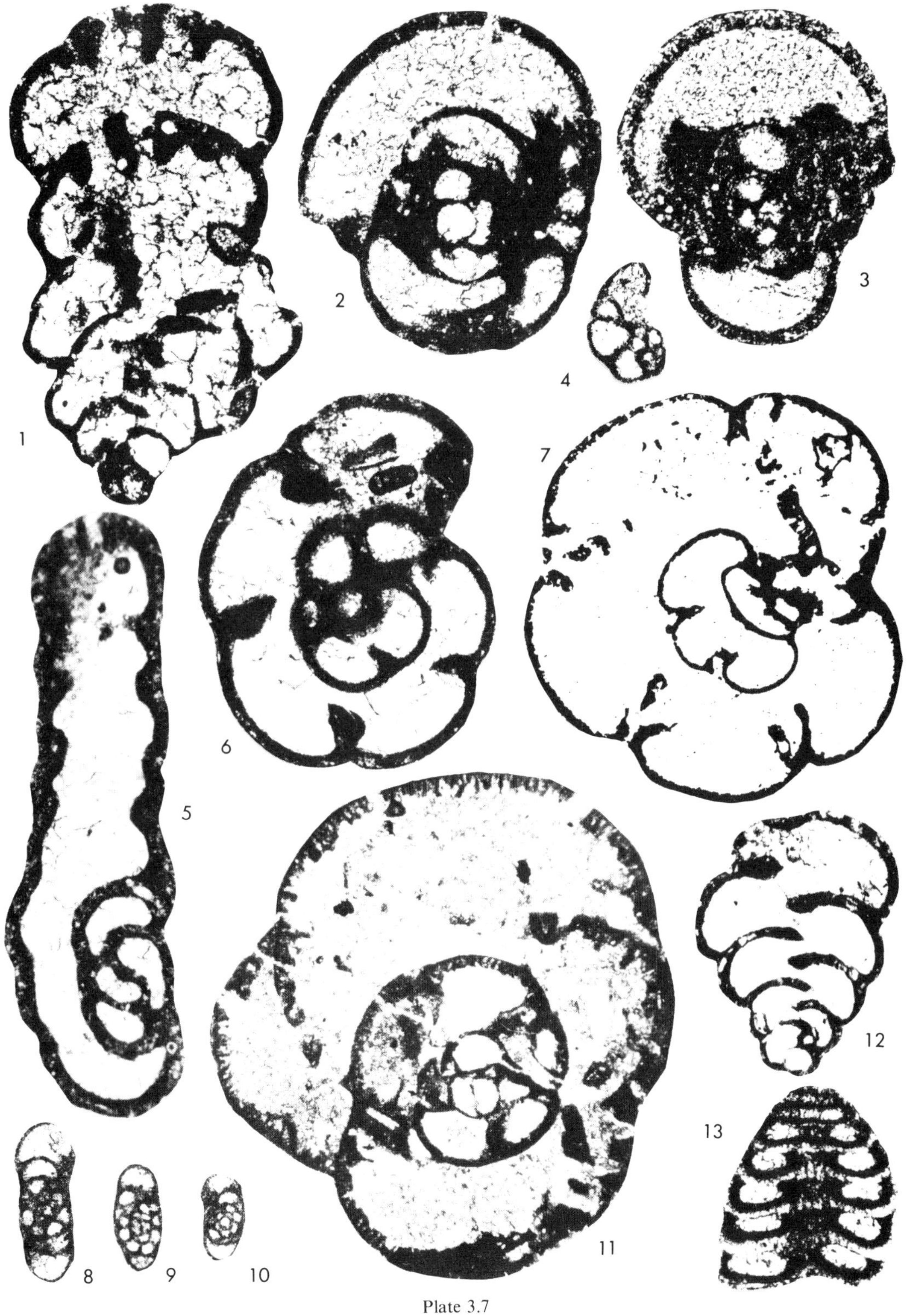

Plate 3.7

Vissariotaxis exilis (Vissarionova)
Plate 3.7, Fig. 8. ARE 719. Early Asbian, Urswick Limestone, 1-5 m below Woodbine Shale, Stainton Quarry, Dalton in Furness, Cumbria. = *Monotaxis exilis* Vissarionova, 1948. Description: Test small, conical with wide umbilicus. 7-8 whorls, rarely 9. Chambers distinct. Aperture wide, opening into umbilical cavity. Apical angle 73°-81°.
Remarks: Differs from *V. compressa* by smaller number of whorls, larger apical angle and less compressed form of test.
Range: Asbian.

Cribrospira pansa (Conil and Lys, 1965)
Plate 3.7, Fig. 9. ARE 1329. Late Asbian, Fifth Limestone, Yeathouse Quarry, Frizington, W. Cumbria. Description: Test initially coiled and compact, followed by a final whorl which grows very rapidly in height. 2-2½ whorls. 5-5½ chambers in final whorl. Chambers moderately swollen and separated by straight septa which tend to be slightly inclined towards the aperture. Sutures well defined. Supplementary deposits absent. Aperture cribrate over large area of final chambers.
Remarks: Differs from *C. rara* by tighter, more compact chambers, more regular coiling and a cribrate aperture found only in final whorl (present in last 3 in *C. rara*).
Range: Late Arundian – Asbian.

Bogushella ziganensis (Grozdilova and Lebedeva)
Plate 3.7, Fig. 10. ARE 10 ×40. Late Asbian, Trowbarrow Quarry, Cumbria. = *Mstinia ziganensis* Grozdilova and Lebedeva, 1960. Description: Involute, initial whorls undivided, later part with short pseudosepta. 6 whorls, 6 chambers in final whorl. Cribrate aperture.
Range: Holkerian – Asbian.

Mediocris mediocris (Vissarionova)
Plate 3.7, Fig. 11. (See also Plate 3.6, Fig. 4 and Plate 3.8, Fig. 5). ARE 1190. Early Asbian, Groups Hollows, opposite Little Asby Scar, Cumbria. = *Eostaffella mediocris* Vissarionova, 1948.
Range: Tournaisian – Namurian.

Koskinotextularia sp.
Plate 3.7, Fig. 12. LL 2142 ×38. Early Asbian, Little Asby Scar, Cumbria. Description: Chambers (9 in total) arranged biserially throughout. Initial spherical prololucus. Chambers swollen and well rounded. Sutures well defined. Septa with swollen and rounded extremities. Septa initially overstep central axial plane of shell but become more widely separated later and do not overlap the central position. Aperture cribrate in the last two chambers, where the apertural faces significantly overstep the central axial plane.

Loeblichiidae
Plate 3.7, Fig. 13. (In I.G.S. Edinburgh): PS 2197 (EA 3183). Late Asbian, Archerbeck Beds, Archerbeck Borehole, 2,028 ft.

Palaeospiroplectammina syzranica (Rauser-Chernoussova)
Plate 3.7, Fig. 14. AL 1319. Holkerian, South Bay, Tenby, 10 m below top of section. = *Spiroplectammina? syzranica* Rauser-Chernoussova, 1948. Description: Shell very small and delicate, elongate with subparallel walls. 8-16 chambers in uncoiled part of shell. Chambers well-rounded with long septa overlapping central axial plane. Septa same thickness as wall and uniform throughout. Length of shell up to 360 μm.
Range: Holkerian – Brigantian.

Vissariotaxis compressa (Brazhnikova)
Plate 3., Fig. 15. ARE 5. Late Asbian, Trowbarrow Quarry, Cumbria. = *Monotaxis exilis compressa* Brazhnikova, 1956. Description: Test tall, narrow and conical. 8-10 whorls. Umbilicus almost cylindrical, straight and very deep. Flanks very regular and sutures feeble.
Range: Asbian.

PLATE 3.8

Unless otherwise stated, the figured specimens are in the I.G.S. collections at Leeds and are ×75.

cf. *Omphalotis* sp.
Plate 3.8, Fig. 1. ARE 695. Late Asbian, Knipe Scar Limestone, Little Asby Scar, Cumbria. Description: Wall of constant thickness throughout. Coiling planispiral. 3-3½ whorls. Chambers smoothly rounded with septa curved towards the aperture. Final chamber very well rounded and curved inwardly. 9½ chambers in final whorl. Basal supplementary deposits and poor nodosities in last few chambers. Spine-like projection in final chamber. Aperture low, simple.

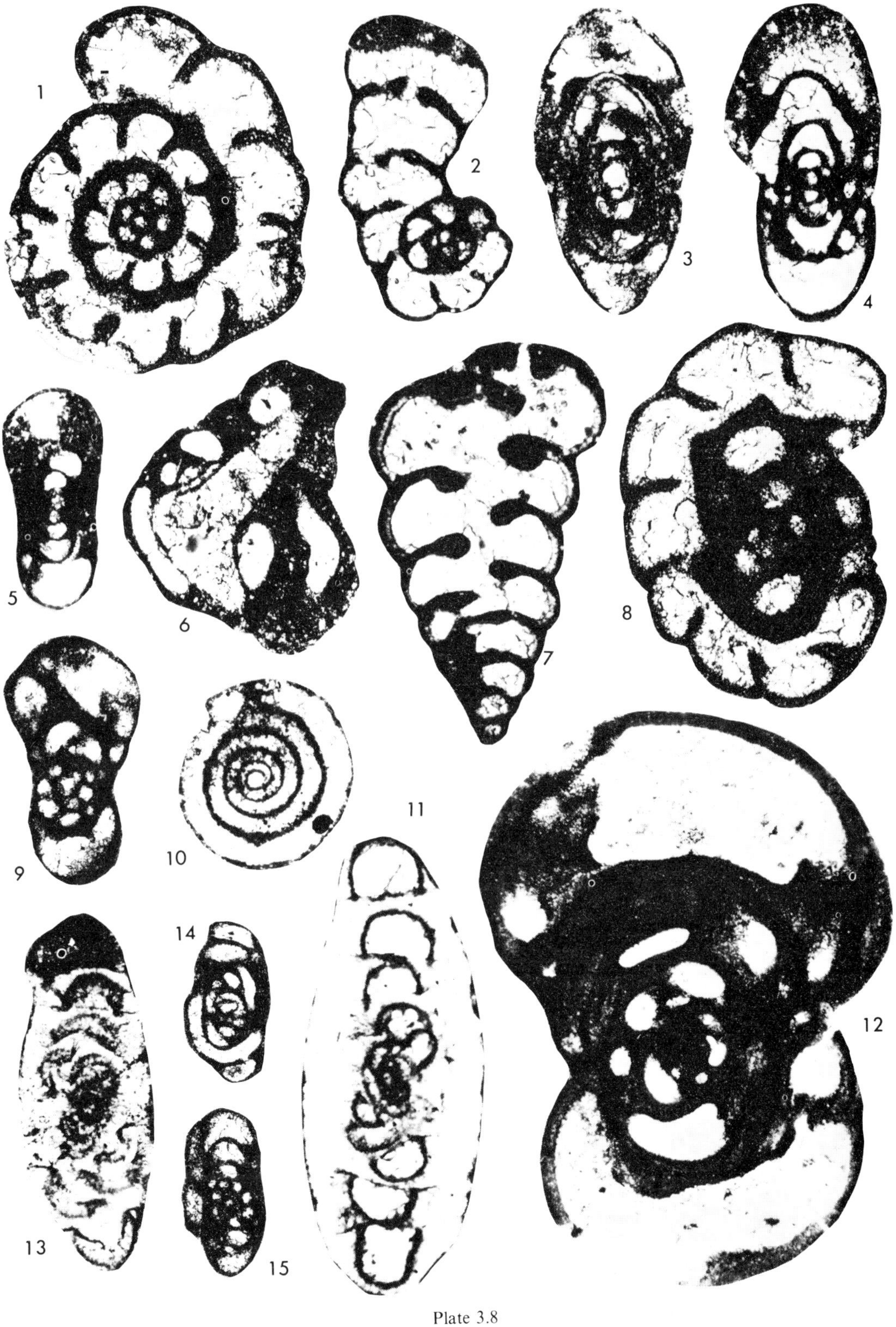

Plate 3.8

Mikhailovella gracilis caledoniae (Conil and Longerstaey, 1980)
Plate 3.8, Fig. 2. (in I.G.S. Edinburgh): PS 2094. Late Asbian, Cornet Limestone, Archerbeck Borehole, depth 1,652 ft. Description: Test initially coiled – aligned or oscillating, later uncoiled. Whorl initially low, growing regularly towards the final chamber. 3 chambers in uniserial section, 6 in last whorl or coiled section. Corner fillings may sometimes be present. Aperture cribrate in last few chambers.
Range: Late Asbian.

Pseudoendothyra sublimis (Schlykova, 1951)
Plate 3.8, Fig. 3. ARE 702. Late Asbian, Knipe Scar Limestone, Little Asby Scar, Cumbria. Description: Test laterally compressed: diaphanotheca and chomata very well developed. Whorl increases gradually in size, becoming slightly more keeled than rounded in the final revolutions. 4–6 whorls. w/d = 0.47–0.55.
Range: Asbian – Brigantian.

Eostaffella mosquensis (Vissarionova, 1948).
Plate 3.8, Fig. 4. ARE 598. Late Asbian, Knipe Scar Limestone, Little Asby Scar, Cumbria. Description: Test small, lenticular, laterally compressed with small shallow umbilici. Inner whorls with rounded periphery; outer whorls with rounded acute periphery. w/d = 0.5–0.56. Spire expands fairly rapidly throughout the 4–5 whorls. Pseudochomata well developed.
Remarks: Differs from *E. ikensis* Vissarionova by its notable lateral compression, the small umbilici and discontinuity of pseudochomata.
Range: Holkerian – Asbian.

Mediocris mediocris (Vissarionova)
Plate 3.8, Fig. 5. (See also Plate 3.6, Fig. 4; Plate 3.7, Fig. 11). ARE 1409. Early Asbian. Strandhall shore, Isle of Man. = *Eostaffella mediocris* Vissarionova, 1948.
Range: Tournaisian – Namurian.

Pseudolituotuba wilsoni (Conil and Longerstaey, 1980)
Plate 3.8, Fig. 6. (In I.G.S. Edinburgh): PS Late Asbian, Archerbeck Beds, Archerbeck Borehole, depth 1863.5 ft. Description: Test irregularly coiled on fine support around a central fixed point. Internal divisions of tubular chamber not observed. Max. diameter of internal tube 80 μm. Wall thickness up to 65 μm in known specimens.
Remarks: Differs from *P. gravata* Conil and Lys by smaller dimensions and thinner, less coarsely agglutinated walls, and from *P. berwicki* Conil and Longerstaey by larger dimensions and a thicker, coarser wall, giving it a more robust appearance.
Range: Late Asbian – Brigantian.

Cribrostomum lecomptei (Conil and Lys, 1964)
Plate 3.8, fig 7. (in I.G.S. Edinburgh): PS 2159 ×50. Late Asbian, Archerbeck Beds, Archerbeck Borehole, depth 1912 ft. Description: 15 biserial chambers. Septa overlap axial line of shell initially and spread apart gradually later. Septa swollen and club-shaped. Last two chambers very large and significantly overlap axis of test. Sutures often deep and well defined. Apical angle 30°.
Range: Late Asbian – Brigantian.

Endothyra sp.
Plate 3.8, Fig. 8. ARE 2. Late Asbian, Trowbarrow Quarry, Cumbria. Description: Test irregularly coiled, approximately 4 whorls. Whorl height increases gradually towards the final chamber. Septa long, projected slightly towards the aperture and of similar thickness to the wall. Chambers rounded but not swollen. 10 chambers in final whorl. Supplementary deposits very well developed in the form of thick basal layers, nodosities in final chambers and a terminal spine which is projected towards the aperture. Aperture low and simple.

Quasiendothyra nibelis (Durkina, 1959)
Plate 3.8, Fig. 9. (See also Plate 3.5, Fig. 10; Plate 3.6, Fig. 1). ARE 596. Late Asbian, Knipe Scar Limestone, Knipe Scar, Cumbria. Description: Coiling glomospiral initially followed by regular aligned terminal whorls. Whorl height increases gradually towards aperture. Chambers numerous, 15–17 in final whorl. Septa long, straight, perpendicular to the walls and slightly swollen at the ends. Sutures well defined between the bulbous chambers. Basal supplementary deposits thick, increasing towards the aperture. Strong chomata. 3 whorls. Small proloculus. Aperture simple, terminal.
Range: Holkerian – Asbian.

Pseudoammodiscus aff. *volgensis* (Rauser-Chernoussova)
Plate 3.8, Fig. 10. (In I.G.S. Edinburgh): PS 2170. Late Asbian, Archerbeck Beds, Archerbeck Borehole, depth 1,922 ft. = *Ammodiscus volgensis* Rauser-Chernoussova, 1948. Description: 4½ whorls from central proloculus. Tubular chamber grows gradually in height towards the final chamber, where it is approximately four times its original size.

Archaediscus stilus (Grozdilova and Lebedeva, 1954)
Plate 3.8, Fig. 11. (In I.G.S. Edinburgh): PS 2027 ×140. Archerbeck Borehole, depth 2,928 ft. Description: Test free, lenticular flattened or slightly swollen. Wall bi-layered. Coiling aligned but not planispiral. 4½–6 whorls. W/d 0.44. Diameter usually in the region of 170–380 μm.
Remarks: Differs from *A. stilus eurus* by its smaller size and smaller number of whorls.
Range: Asbian – Namurian.

Omphalotis samarica (Rauser-Chernoussova)
Plate 3.8, Fig. 12. (In I.G.S. Edinburgh): PS 2140. Late Asbian, Archerbeck Beds, Archerbeck Borehole, 1,863½ ft. = *Endothyra samarica* Rauser-Chernoussova, 1948. Description: Test coiled, umbilical. Last two planispiral whorls grow rapidly in height so that shell seems flattened. Initial whorls irregularly coiled. Septa almost perpendicular to wall, slightly inclined towards aperture. Wall thickness up to 40 μm. Aperture slit-shaped.
Range: Holkerian – Asbian.

Nodosarchaediscus cornua (Conil and Lys)
Plate 3.8, Fig. 13. ARE 1211 ×140. Early Brigantian, Ravensholme Limestone, Ravensholme Quarry, Downham, Lancashire. = *Archaediscus cornua* Conil and Lys, 1964. Description: Test lenticular, compressed. Coiling oscillating. 6–7 whorls. Large well-rounded nodosities. Tubular chamber very small at origin, grows regularly in size towards aperture. Small proloculus.
Range: Holkerian – Brigantian.

Brunsia jactata (Conil and Lys)
Plate 3.8, Fig. 14. ARE 602. Late Asbian, Knipe Scar Limestone, Knipe Scar, Cumbria. = *Glomospira jactata* Conil and Lys, 1964. Description: Test discoidal with very irregular coiling. 6–7 whorls. Wall of moderate thickness. Diameter 275–360 μm, width 125–200 μm.
Range: Holkerian – Asbian.

Endostaffella sp.
Plate 3.8, Fig. 15. ARE 702. Late Asbian, Knipe Scar Limestone, Little Asby Scar, Cumbria.

PLATE 3.9

Unless otherwise stated, all the figured specimens are in the I.G.S. collections at Leeds and are × 75.

Koskinobigenerina sp.
Plate 3.9, Fig. 1. HR 3454. Brigantian, Eelwell Limestone, Spittal shore, Northumberland. Description: Chambers biserially arranged from proloculus. Initial 8 chambers separated by septa overstepping the central axial plane of test. Septa well rounded and increasing gradually in thickness throughout the shell. Chambers rounded and swollen with well defined sutures. Last two chambers distinguished by cribrate apertural faces reaching over whole width of shell.

Endothyranopsis sphaerica (Rauser-Chernoussova and Reitlinger)
Plate 3.9, Figs. 2, 3. (See also Plate 3.6, Fig. 3). Fig. 2, PS 1879. Late Brigantian, Buccleuch Limestone, Archerbeck Borehole, depth 952 ft. Fig. 3, ARE 486. Early Brigantian, Lower Bath House Wood Limestone, River North Tyne. = *Endothyra sphaerica* Rauser-Chernoussova and Reitlinger, 1937. Description: w/d ratio 0.9–1.0 mm. Umbilicus poorly defined. Chambers separated by poorly defined thin septal furrows. 10–14 chambers in outer whorl. 3–3½ whorls slowly increasing in height. Max. diam. 1.33 mm. Aperture low, wide, semilunar at base of apertural face. Septa inflated at ends.
Range: Asbian – Brigantian.

Biseriella parva (Chernysheva)
Plate 3.9, Fig. 4. R. Conil Coll. (in I.G.S. Edinburgh). PS 1882 ×38. Late Brigantian, Buccleuch Limestone, Archerbeck Borehole, depth 955 ft. = *Globivalvulina parva* Chernysheva, 1948. Description: Test small, almost hemispherical with slightly concave apertural face and faintly lobulate peripheral margin. Initial portion of test coiled in trochoid spire; final whorl completely embraces earlier portion. Sutures well defined. External surface exhibits biserial arrangement of chambers. Aperture simple, opening into a curved apertural depression in each chamber.
Remarks: Distinctive features include small test and arrangement, convexity, and number of chambers.
Range: Late Asbian – lower Namurian.

Lituotubella aff. *magna* (Rauser-Chernoussova, 1948)
Plate 3.9, Fig. 5. (See also Plate 3.6, Fig. 9). PS 1882 in I.G.S. Edinburgh. Late Brigantian, Buccleuch Limestone, Archerbeck Borehole.

Plectogyranopsis ampla (Conil and Lys)
Plate 3.9, Fig. 6. Bk 3381. Early Brigantian, Rownham Hill Coral Bed, Ashton Park Borehole, Bristol, depth 1,638 ft. = *Plectogyra exelikta ampla* Conil and Lys, 1964. Description: Test large. Coiling irregular, becoming planispiral. 2–2½ whorls. Terminal whorl raised and divided into 5–5½ large swollen chambers. Sutures well defined. Septa thick, feebly inclined towards aperture. Wall granular, increasing in thickness with growth to 40–50 μm in final chambers. Supplementary deposits absent.
Range: Arundian – Brigantian.

Janischewskina operculata (Rauser-Chernoussova and Reitlinger)
Plate 3.9, Fig. 7. RC 0000 ×50. Petershill Quarry, Bathgate. = *Samarina operculata* Rauser-Chernoussova and Reitlinger, 1937. Description: Test subspherical, slightly compressed laterally. Wall perforate, thin, becoming progressively thicker (up to 25–55 μm) in later whorls. 2–3 whorls, 5–6 chambers per whorl. Chambers slightly inflated, with double septal sutures. Septa initially short and curved but later fused in pairs. Umbilicus closed. Apertural face covered with 3–4 rows of convex apertural plates. Supplementary septal apertures well developed. Diameter often up to 2.15 mm.
Remarks: Differs from *J. miniscularia* Ganelina 1956 by the shape of the test, larger size and greater number of whorls. Differs from *J. calceus* Ganelina, 1965, by the latter's asymmetrical position in the first whorl.
Range: Brigantian.

Endostaffella fucoides (Rozovskaya, 1963)
Plate 3.9, Figs. 8–10. Fig. 8, ARE 519. Late Brigantian, Cockleshell Limestone, Whitfield Gill, Askrigg. Fig. 9, 10, HR 3442. Early Brigantian, Oxford Limestone, Spittal Shore, Northumberland. Description: Test free, last whorls evolute, periphery rounded, w:d = 0.35–0.45. Axial region tends to be swollen. Last 1½ whorls aligned.
Range: Asbian – Brigantian.

Bradyina rotula (Eichwald)
Plate 3.9, Fig. 11. PS 1882 ×28. Late Brigantian, Buccleuch Limestone, Archerbeck Borehole, depth 955 ft. = *Nonionina rotula* Eichwald, 1878. Description: Test large; whorl low at the origin, increasing rapidly in height towards aperture; chambers large, slightly swollen and separated by thick septa inclined towards aperture. Wall reaches 120 μ in thickness; 2½ whorls; 6 chambers; diameter up to 1,200 μ.
Remarks: Differs from *B. cribrostomata* by larger number of chambers in final whorl.
Range: Late Asbian – Brigantian.

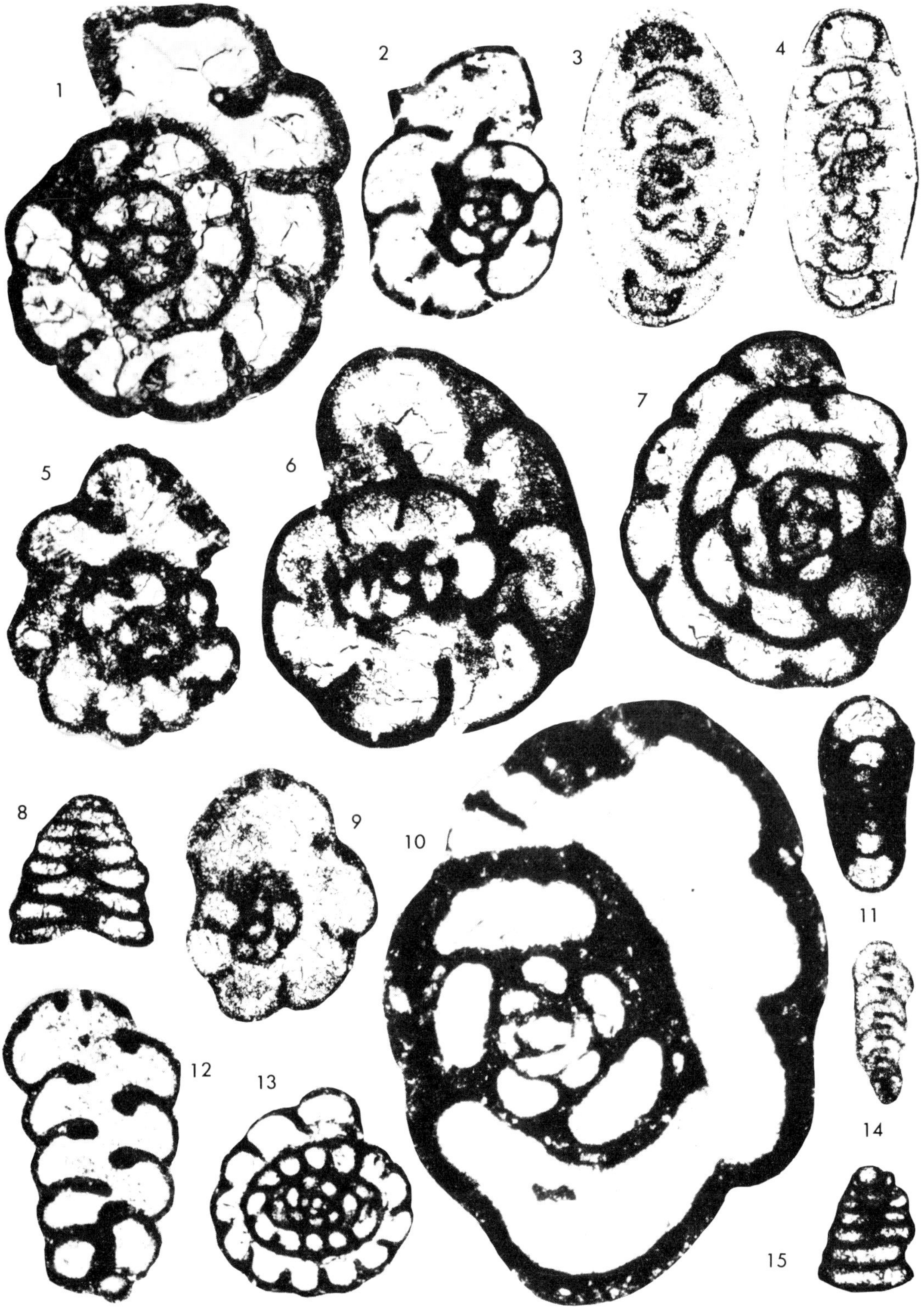

Plate 3.9

Palaeotextularia aff. *lipinae* (Conil and Lys, 1964)
Plate 3.9, Fig. 12. ARE 1358. Brigantian, Middle Limestone, Garth Gill, Askrigg Block. Description: Test conical. Septa well rounded and slightly overstepping the axial plane of shell leaving only small spaces between each other. Septa tend to be swollen at their extremities. Sutures well defined due to swollen nature of chambers. 8–11 chambers. Apical angle 30–45°.

Howchinia bradyana (Howchin)
Plate 3.9, Fig. 13. (See also Plate 3.10, Figs. 9, 10). ARE 430 ×80. Late Brigantian, Acre Limestone, Beadnell, Northumberland. = *Patellina bradyana* Howchin, 1888. Description: Test free, conical, trochoid. Sutures well defined. Shell consists of spiral tube wound round central umbilical core of microcrystalline calcite. 5–12 whorls. Aperture slit-like from periphery to concave umbilicus.
Range: Late Asbian – Brigantian.

PLATE 3.10

All the figured specimens (except Figs. 9–12) are in the IGS collection at Leeds and are ×75 unless stated otherwise.

Nodasperodiscus stellatus (Bozorgnia)
Plate 3.10, Fig. 1. HR 3695 ×105. Late Brigantian, Scar Limestone, Trout Beck, Cumbria. = *Rugosoarchaediscus stellatus* Bozorgnia, 1973. Description: Test lenticular, swollen, rounded periphery. Coiling irregular. Chamber size increases regularly towards very large final whorl. Inner whorls fully occluded forming a stellate flare. Outer 1–2 whorls without occluding nodosities.
Range: Asbian – Namurian.

Archaediscus karreri Brady, 1973 (at angulatus stage)
Plate 3.10, Figs. 2, 3. ARE 465 ×105. Late Brigantian, Beadnell, Northumberland. Description: Test large, lenticular swollen. Irregularlycoiled. 5–6½ whorls.
Range: Brigantian.

Forschia sp.
Plate 3.10, Fig. 4. ARE 1453. Late Asbian, Poyll Ritchie, near Strandhall, Isle of Man. Description: whorl increases in size gradually towards aperture. 3½–4 whorls.

Archaediscus stilus eurus (Conil and Longerstaey, 1980)
Plate 3.10, Fig. 5. ARE 458 ×140. Late Brigantian, Beadnell, Northumberland. Description: Test lenticular, strongly elongate along diameter, laterally compressed. Sides almost subparallel but slightly swollen. Rounded periphery. Small spherical proloculus. Initially involute, coiling oscillates by up to 40° per whorl. Final 2–3 whorls evolute and more or less aligned in same plane. 6–7 whorls. Diameter 450–550 μm, width 170–220 μm, w/d = 0.34–0.41. Dark microgranular wall developed throughout. Lighter, fibrous, radiating layer very thin in first whorls, becoming much thicker in later revolutions. Tubular chamber relatively small in first 3 whorls, becoming higher and wider in final 3 whorls.
Range: Brigantian.

Nodasperodiscus gregorii (Dain)
Plate 3.10, Fig. 6. ARE 495 ×140. Early Brigantian, Simonstone Limestone, Whitfield Gill, Askrigg. = *Archaediscus gregorii gregorii* Dain, 1953. Description: discoidal with rounded periphery. 5–6 whorls. Coiling oscillating. Final 3–4 whorls evolute.
Range: Brigantian – Upper Namurian.

Asteroarchaediscus occlusus (Hallett)
Plate 3.10, Fig. 7. HR 3697 ×140. Late Brigantian, Scar Limestone, Trout Beck, Cumbria. = *Rugosarchaediscus occlusus* Hallett, 1970.
Range: Brigantian.

Pseudoammodiscus aff. *buskensis* (Brazhnikova)
Plate 3.10, Figs. 8, 17. Fig. 8, ARE 1143. Early Asbian, Potts Beck Limestone, Little Asby Scar, Cumbria. Fig. 17, HR 3698. Late Brigantian, Scar Limestone, Trout Beck, Cumbria. = *Ammodiscus buskensis* Brazhnikova, 1956. Description: 4¾ whorls from small proloculus. Tubular chamber grows rapidly in height from 10–15 μm initially to c. 70 μm in final whorl. Wall dark, 10–15 μm. Diameter 340 μm.

Howchinia bradyana (Howchin)
Plate 3.10, Figs. 9, 10 (see also Plate 3.9, Fig. 13). (In I.G.S. Edinburgh): Fig. 9, PS 1813. Late Brigantian, Buccleuch Limestone, Archerbeck Borehole, depth 946 ft. Fig. 10, PS 1873, as Fig. 9 but depth 943 ft. = *Patellina bradyana* Howchin, 1888
Range: Late Asbian – Brigantian.

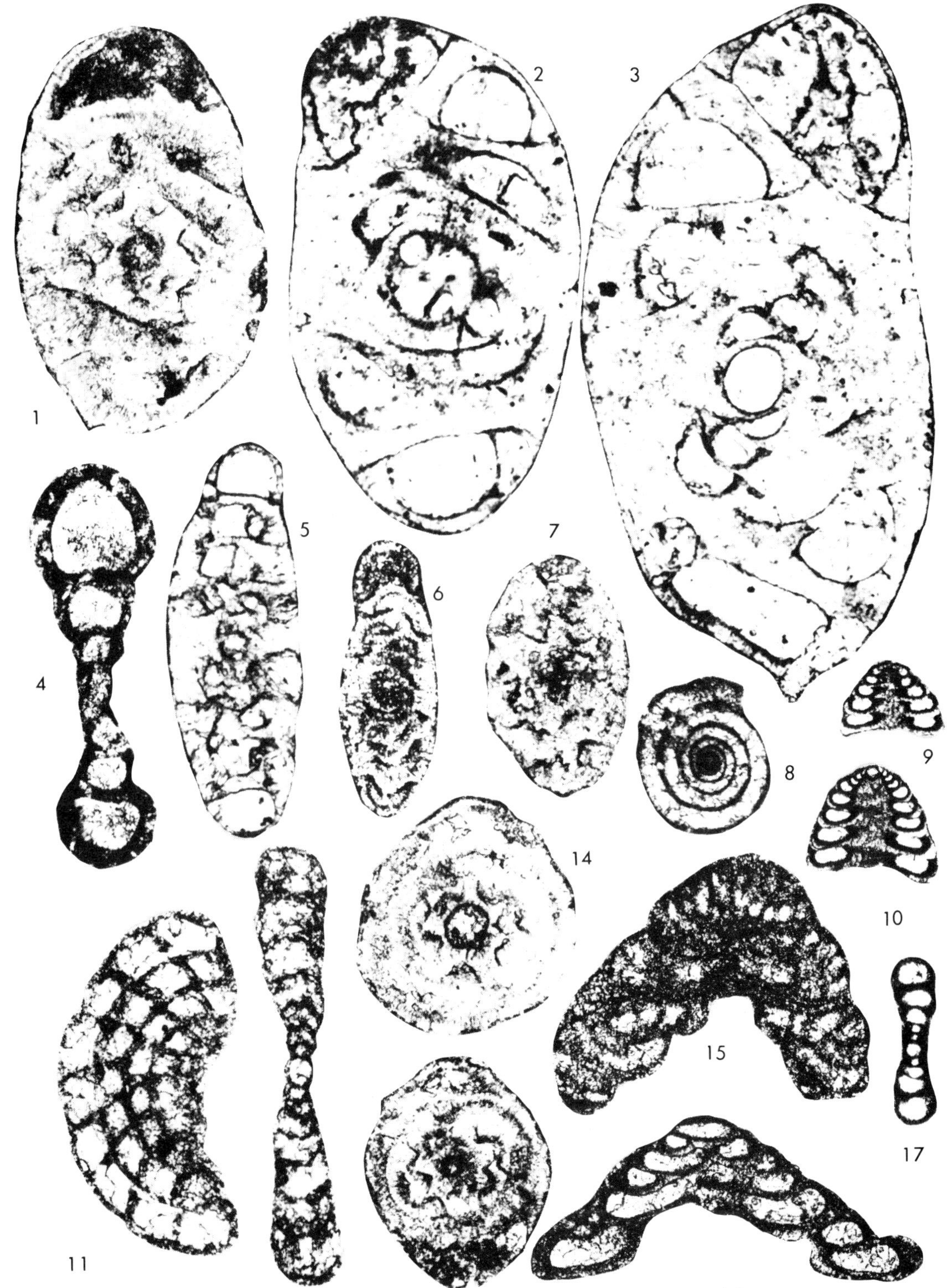

Plate 3.10

Loeblichia paraammonoides (Brazhnikova)
Plate 3.10, Figs. 11, 12. R. Conil Coll. ×140. = *Nanicella paraammonoides* Brazhnikova, 1956. Description: Shell regularly coiled, flat and discoidal: strongly compressed along axis of rotation. Wide, shallow umbilici. Coiling planispiral or almost so. Numerous quadrate chambers, long, straight septa, approx. 20 in final whorl. Initially, whorls very narrow, becoming wider and higher in the later whorls. Spiral and septal sutures very indistinct initially but deepening in last whorls. 4–7 whorls. w/d = 0.13–0.28. Supplementary deposits in the form of chomata, often weak. Aperture narrow, slit-like, basal.
Range: Brigantian.

Nodasperodiscus sp.
Plate 3.10, Fig. 13. HR 3697 ×105. Late Brigantian, Scar Limestone, Trout Beck, Cumbria. Description: Test rounded, slightly laterally compressed. Central small proloculus. Final whorls with poorly formed nodosities. Marked increase in size of tubular chamber with growth. 5–6 whorls.

Asteroarchaediscus? sp.
Plate 3.10, Fig. 14. HR 3697 ×140. Late Brigantian, Scar Limestone, Trout Beck, Cumbria. Description: Test subspherical, well rounded. Central large proloculus.

Valvulinella conciliata (Ganelina)
Plate 3.10, Fig. 15 (see also Plate 3.5, Fig. 16). ARE 527. Late Brigantian, Cockleshell Limestone, Whitfield Gill, Askrigg. = *Tetrataxis conciliatus* Ganelina, 1956. Description: Test conical, apex rounded, flanks convex. Umbilicus straight, occupying a quarter of basal diameter. Apical angle approx. 89°. 6 whorls.
Range: Brigantian.

Textrataxis pressula pressula (Malakhova, 1956)
Plate 3.10, Fig. 16. ARE 479. Late Brigantian, Upper Bath House Wood Limestone, River North Tyne, Barrasford, Northumberland. Description: Test in form of broad, low cone, flattened from the apex. Umbilicus large and quite deep. Wall dark, microgranular to granular with a thin fibrous layer. Coiling regular with whorls gradually increasing in size. 5–7 whorls. Diameter 800–960 μm. Height 315–500 μm. h/d 0.37–0.46. Apical angle 110°–115°.
Range: Brigantian.

PLATE 3.11

Unless otherwise stated the figured specimens are in the I.G.S. collections at Leeds and are ×75.

Koskinotextularia cribriformis (Eickhoff, 1968)
Plate 3.11, Fig. 1. ARE 378. Late Asbian, Trefor Rocks, Llangollen. Description: Proloculus spherical or oval, often slightly displaced laterally. Chambers 9–13 in number, biserially arranged. Septa in the first 6–8 chambers are less hooked and overlap the central plane to a greater extent than the septa in more adult chambers. Septa in latter chambers tend to be more rounded and slightly globular at the ends. Aperture slit-like, cribrate. Length 0.81–1.04 mm; width 0.58–0.63 mm.
Remarks: differs from *K. obliqua* (Conil and Lys) by its regular expansion in width, the more feeble arrangement of sutures and thinner wall.
Range: Asbian – Brigantian.

Septabrunsiina mackeei (Skipp, 1966)
Plate 3.11, Fig. 2. ARE 837. Brigantian, Ravensholme Limestone, Red Syke, Pendle, Lancashire. Description: Test small/medium, discoidal, largely evolute with broad shallow umbilicus. Proloculus small 10–15 μm. Coiling initially irregular, becomes planispiral in final 2–3 whorls. 3½–5 whorls altogether. Chambers low, elongate and slightly lobate. 7½–8 chambers in final whorl. Septa present throughout, less distinct initially, strong anterior orientation.
Range: Asbian – Brigantian.

Eostaffella sp.
Plate 3.11, Fig. 3. Early Asbian, Potts Beck Limestone, Little Asby Scar, Cumbria. Description: Test very large, elongate. Periphery well rounded. Well defined umbilici. Wall microgranular with tectum. Coiling planispiral. Spire expands gradually in height and rapidly in width throughout coiling. 5–6 whorls from central circular proloculus. Supplementary deposits in form of poorly developed pseudochomata on inner whorls.

Lituotubella glomospiroides (Rauser Chernoussova, 1948)
Plate 3.11, Fig. 4. BLG 9996 ×38. Late Asbian, Beckermonds Scar Borehole, depth 82 m. Description: Test large tubular, evolute, glomospirally coiled in early stages, uncoiled in later stages. Septal sutures not well developed. Glomospiral portion laterally compressed, 4–5 chambers/whorl. 5–6 pseudochambers in uniserial section. Aperture broad, coarsely cribrate in last 2 chambers. Flat apertural shield. Length 1.9 mms.
Range: Asbian.

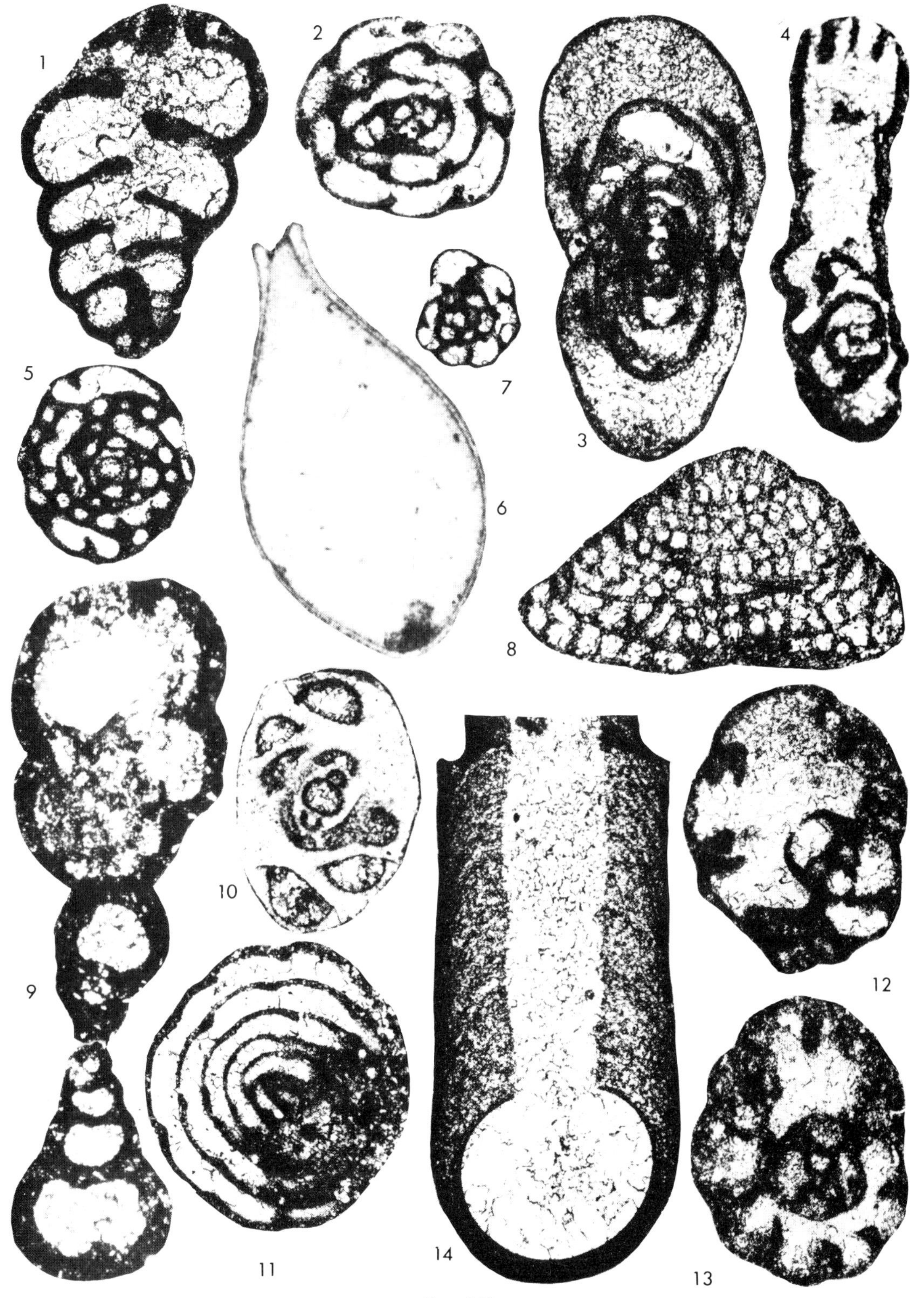

Plate 3.11

Urbanella miranda matura (Vdovenko, 1972)
Plate 3.11, Fig. 5. AL 1367. Holkerian, South Bay, Tenby. 82 m below top of section.
Range: Holkerian.

Saccamminopsis fusulinaformis (McCoy)
Plate 3.11, Fig. 6. HR 3737 ×37. Brigantian, Jew Limestone, Moulds Meaburn Quarry, near Appleby, Cumbria = *Nodosaria fusulinaformis* McCoy, 1849; = *Saccammina cateri* Brady.
Range: Brigantian.

Endothyra obsoleta (Rauser Chernoussova, 1948)
Plate 3.11, Fig. 7. ARE 973. Late Asbian, Danny Bridge Limestone, River Clough, near Sedbergh, Yorkshire. Description: Test involute, with a small number of convex chambers. 2–3 whorls, 6–7 chambers in final whorl. Septa long, broadly curved in the direction of coiling. Wall thin, dark, finely granular 10–15 μm thick. Supplementary deposits developed in the form of prominent nodosities on basal protruberances. Diameter 0.25–0.37 mm.
Remarks: Differs from *E. similis* Rauser Chernoussova and Reitlinger by having fewer, more convex chambers.
Range: Asbian – Brigantian.

Valvulinella latissima (Conil and Lys, 1964)
Plate 3.11, Fig. 8. ARE 1442. Early Asbian, Strandhall Shore, Isle of Man. Description: Test irregularly conical with base broader than height. Apical angle approx. 100°.
Remarks: differs from *V. lata* Grozdilova and Lebedeva by the larger apical angle of the cone.
Range: Asbian – Brigantian.

Forschiella prisca (Mikhailov, 1939)
Plate 3.11, Fig. 9. ARE 730. Asbian, Urswich Limestone, below Woodbine Shale, Stainton Quarry, Dalton in Furness, Cumbria. Description: Gradual increase in height and wifth of whorls resulting in two large, deep umbilical depressions. Proloculus small. Axial portion of test often very thin. 4–5 whorls. Final whorl large and robust. Wall thickness increasing regularly up to 75–1500 μm in the lat revolution. Aperture cribrate and complex. Diameter 900–1500 μm. Width 300–400 μm approx.
Range: Arundian – Lower Namurian.

Archaediscus itinerarius (Schlykova, 1951)
Plate 3.11, Fig. 10. BLG 9852 × 140. Early Brigantian, Beckermonds Scar Borehole, depth 4 m. Description: Test involute, subspherical and laterally compressed. Periphery well rounded. 4–5 whorls. $w/d = 0.61$ to 0.78. Proloculus fairly large. Coiling sigmoidal. Tubular chamber with concave base.
Range: Brigantian.

Glomospiranella sp.
Plate 3.11, Fig. 11. ARE 933, Holkerian, Garsdale Limestone, River Clough, near Sedbergh, Yorkshire. Description: Coiling irregular initially. Later whorls becoming generally orientated in the same plane. 6–7 whorls. Wall calcareous, microgranular to granular with some larger grains. Pseudosepta well defined in final whorls but quite indistinct or absent initially. Probably 10–13 pseudochambers in last revolution.

?*Janischewskina minuscularia* (Ganelina, 1956)
Plate 3.11, Fig. 12. ARE 335, Early Asbian, Tandinas Quarry, Anglesey. Description: Test coiled, laterally compressed, involute. Umbilical depressions large, periphery rounded. 1½–2 whorls. Chambers increase progressively in height from a spherical proloculus.
Remarks: Differs from *J. calceus* Ganelina in the nature of the wall, smaller dimensions, less numerous and symmetrical nature of the whorls.

Cribranopsis sp.
Plate 3.11, Fig. 13. Early Asbian, Tandinas Quarry, Anglesey.

Earlandia sp.
Plate 3.11, Fig. 14. ARE 1194. Early Asbian, Potts Beck Limestone, Groups Hollow, near Little Asby Scar, Cumbria. Description: Test very large. Proloculus large spherical, slightly compressed. Tubular chamber cylindrical, with parallel sides. Width of proloculus and tubular chamber about the same. Wall very thick with dark microgranular outer layer grading into a less dense granular layer. Orientation of grains in the lighter layer tends to follow the rounded shape of the proloculus throughout the shell. Inner wall of proloculus dark and microgranular. Maximum length observed on incomplete specimens 1300 μm. Maximum external diameter of proloculus 575 μm. Maximum internal diameter of proloculus 450 μm.
Remarks: This large *Earlandia* is restricted to the Early Asbian of the British Isles.

PLATE 3.12

Unless otherwise stated the figured specimens are × 75. Register numbers preceded by the letters HM are of specimens in the Hunterian Museum, Glasgow, the remainder are in the I.G.S. collections at Leeds.

Endothyra excellens (Zeller)
Plate 3.12, Fig. 1, HM P433/1. Arnsbergian, Plean Limestone, Craigburn, Uddington, Lanarkshire. = *Plectogyra excellens* Zeller, 1953. Description: secondary deposits continuous on floor of chambers, well developed hook in last chamber. Proloculus small.
Range: Late Asbian – early Namurian.

Bradyina cribrostomata (Rauser-Chernoussova and Reitlinger, 1937)
Plate 3.12, Fig. 2. Arnsbergian, Pike Hill Limestone, Throckley Borehole, Northumberland, depth 285 m. Description: slightly compressed laterally, chambers inflated, sutures depressed. 2–3 whorls, 6–7 chambers in last whorl; diameter 1.7–3.0 mm.
Remarks: differs from *B. potanini* Beninkov in greater number of chambers and more compressed test.
Range: Namurian.

Biseriella parva (Chernysheva)
Plate 3.12, Fig. 3 HM P421/1 Pendleian, Index Limestone, Kennox Water, Ayrshire. (See also Plate 3.9, Fig.4).
Range: Late Asbian – early Namurian.

Endothyranopsis crassa (Brady)
Plate 3.12, Fig. 4. HM P 47531. Arnsbergian, Castlecary Limestone, Westerwood, Lanarkshire. (See also Plate 3.7, Fig. 1).
Range: Holkerian – early Namurian.

Endostaffella sp.
Plate 3.12, Fig. 5. Wo 1844, Arnsbergian, Corbridge Limestone, Ouston Borehole, Northumberland.

Climacammina postprisca (Brazhnikova and Vinnichenko)
Plate 3.12, Fig. 6. Arnsbergian, Pike Hill Limestone, Throckley Borehole, Northumberland, depth 287 m. Description: biserial portion small compared with elongate uniserial, subparallel part. Septa short.
Range: Namurian.

Asteroarchaediscus sp.
Plate 3.12, Fig. 7. HR 3559, × 140. Pendleian, Great Limestone, Brunton Bank Quarry, Northumberland.

Earlandia pulchra (Cummings, 1955)
Plate 3.12, Fig. 8. HM P538/1. Arnsbergian, Orchard Limestone, River Nethan, Auchlochen House, Ayrshire. Description: tubular chamber with faint depressed sutures, no trace of septation internally.
Range: Early Namurian.

Cribrospira sp.
Plate 3.12, Fig. 9. HR 3577, Pendleian, Belsay Dene Limestone, Aydon near Corbridge, Northumberland.

Endothyra cf. *pandorae* (Zeller)
Plate 3.12, Fig. 10. HR 3558, same horizon and locality as Fig. 7. = *Plectogyra pandorae* Zeller, 1953. Description: test with broadly rounded periphery, umbilicate on one side only. Secondary deposits on floor of chambers with low nodes in last one or two chambers.
Remarks: differs from *E. excellens* in less extensive secondary deposits, thicker walls and less swollen chambers.
Range: Brigantian – early Namurian.

Warnantella subquadrata (Potievskaya and Vakarchuk)
Plate 3.12, Figs. 11–12. Fig. 11, HM P 529/2; Fig. 12, HM P 529/4. Arnsbergian, Castlecary Limestone, Westerwood, Lanarkshire. = *Glomospira subquadrata* Potievskaya and Vakarchik.
Remarks: characterised by its irregular tortuous coiling. Differs from *W. tenuiramosa* in its larger size.
Range: Brigantian – early Namurian.

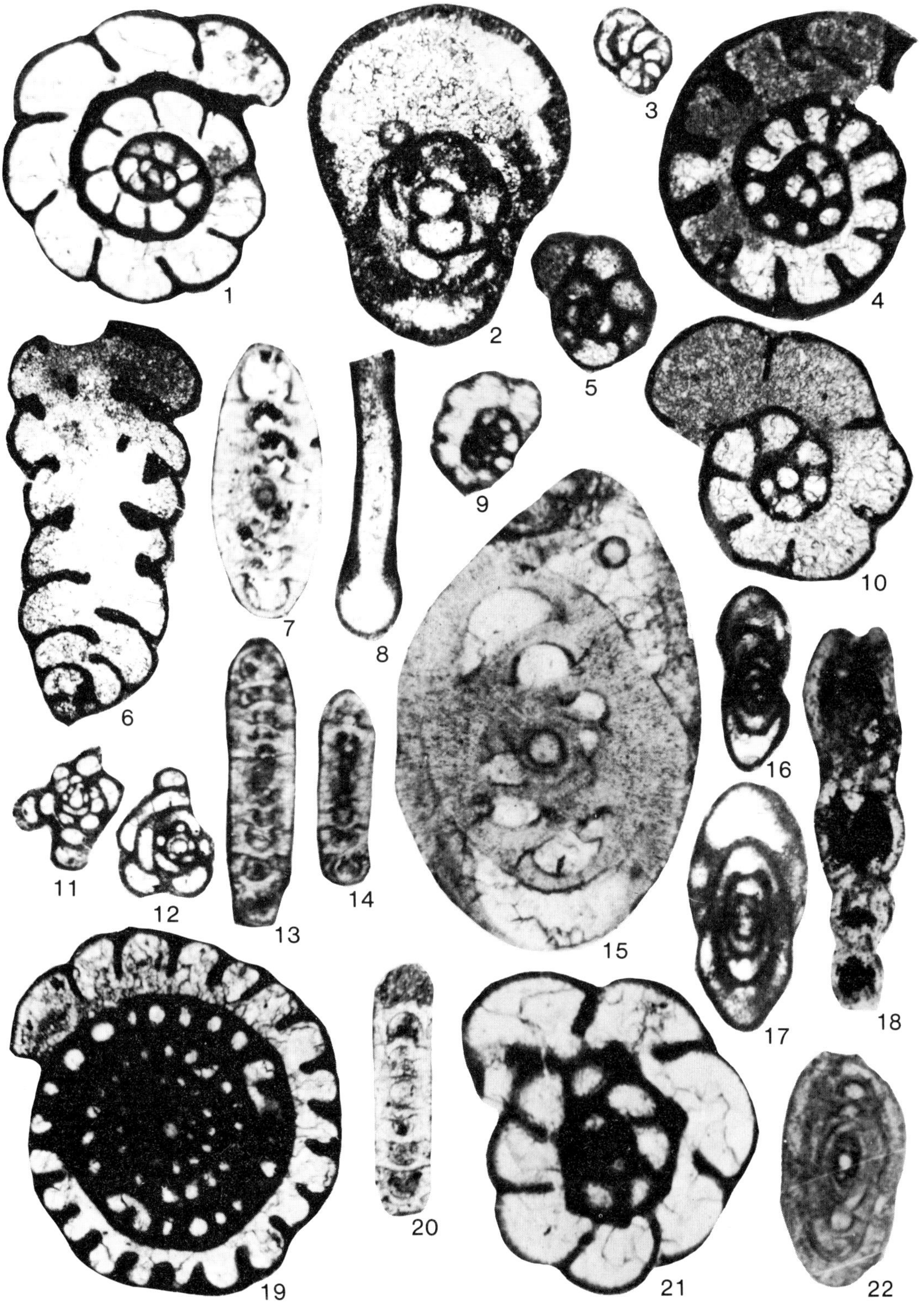

Plate 3.12

Neoarchaediscus incertus (Grozdilova and Lebedeva)
Plate 3.12, Figs. 13–20. GN 1184, × 140, Yeadonian, G. cancellatum Band, Morton No. 1 Borehole, Lincolnshire. = *Archaediscus incertus* Grozdilova and Lebedeva, 1954. Description: 4–6 whorls, diameter of test 0.2–0.31 mm, subparallel sides.
Range: Late Asbian – late Namurian.

Asteroarchaediscus gregorii (Dain)
Plate 3.12, Fig. 14. GN 1184, × 140. Yeadonian, G. cancellatum Band, Morton No. 1 Borehole, Lincolnshire. (See also Plate 3.10, Fig. 6).
Range: Brigantian – late Namurian.

Archaediscus moelleri (Rauser-Chernoussova, 1948)
Plate 3.12, Fig. 15, HR 3605, Arnsbergian, Styford Limestone, Styford, Northumberland. Description: coiling irregular, tubular chamber initially high, grows slowly with coiling. W/d ratio 0.7.
Remarks: differs from *A. convexus* Grozdilova and Lebedeva, 1954 in having high initial chamber from the origin, and in having fewer whorls and more swollen test.
Range: Brigantian – early Namurian.

Eostaffella sp.
Plate 3.12, Figs. 16, 17, 19. Arnsbergian. Fig. 16, HR 3600, Styford Limestone, Styford, Northumberland. Fig. 17, HR 3565, Corbridge Limestone, Corbridge, Northumberland. Fig. 19, BLG 6816, Newton Limestone, Hexham bypass Borehole No. 138a, Northumberland.

Lugtonia elongata (Cummings, 1955)
Plate 3.12, Fig. 18. HM P486, Arnsbergian, shale over Orchard Limestone, Poniel Water, Ayrshire. Description: about 5 pyriform linear chambers gradually increasing in size, separated by deep sutures.
Remarks: more elongate and cylindrical then *L. concinna* and with more chambers.
Range: Early Namurian.

Endothyra sp.
Plate 3.12, Fig. 21. ARE 633, Pendleian, Great Limestone, Greenleighton Quarry, Northumberland.

Eosigmoilina robertsoni (Brady)
Plate 3.12, Fig. 22, HM P515, Arnsbergian, Castlecary Limestone, Westerwood, Lanarkshire. = *Trochammina robertsoni* Brady, 1879.
Range: Early Namurian (E_{2a} Zone).

3.6 REFERENCES

Armstrong, A. K. and Mamet, R. L. 1977. Carboniferous microfacies, microfossils and corals, Lisburne Group, Arctic Alaska. *U.S. Geol. Surv. Prof. Paper,* No. 849.

Brady, H. B. 1873(a). On *Archaediscus karreri,* a new type of Carboniferous Foraminifera. *Ann. Mag. Nat. Hist.* (4) **12**, 286-90.

Brady, H. B. 1873(b). *In* Notes on certain genera and species. *Mem. Geol. Surv. Scotland,* sheet 23, 93-107.

Brady H. B. 1876. A monograph of Carboniferous and Permian Foraminifera (the genus *Fusulina* excepted). *Monogr. Palaeontogr. Soc.* 1-166, 12 Plates.

Brazhnikova, N. E. 1962. *Quasiendothyra* and related forms in the Lower Carboniferous of the Donetz Basin and other areas of the Ukraine. *Akad. Sci. Uk., Trud. Inst. Geol., Strat. Paleont.* **44**, 1-48.

Browne, R. G. and Pohl, E. R. 1973. Stratigraphy and genera of calcareous Foraminifera of the Frailey's facies (Mississippian) of central Kentucky. *Bull. Amer. Paleont.* **64**, 173-239.

Bykova, E. V. and Polenova, E. I. 1955. Foraminifera and Radiolaria from the Volga-Ural region of the central Devonian basin, and their stratigraphical importance. *Trud. VNIGRI* n.s. **87**, 1-141.

Calver, M. A. 1969. Westphalian of Britain. *C.R. 6me Cong. Int. Strat. Géol. Carb.* (Sheffield 1967), **1**, 233-54.

Chernysheva, N. E. 1940. On the stratigraphy of the Lower Carboniferous Foraminifera in the Makarovski district of the south Urals. *Bull. Soc. Nat. Mosc.* n.s. **48**, 113-35.

Conil, R. 1976. Contribution a l'étude des forminifères du Dinantien en Irlande. *Ann. Soc. Géol. Belg.* **99**, 129-41.

Conil, R. and George, T. N. 1973. The age of the Caninia Oolite in Gower. *C.R. 7me Cong. Int. Strat. Géol. Carb.* (Krefeld 1971), **2**, 323-32.

Conil, R., Groessens, E. and Pirlet, H. 1976. Nouvelle charte stratigraphique du Dinentien type de la Belgique. *Ann. Soc. géol. Nord,* **96**, 363-71.

Conil, R. and Lees, A. 1974. Les transgressions viséennes dans l'Ouest de l'Irlande. *Ann. Soc. géol. Belg.* **97**, 463-84.

Conil, R., Longerstaey, P. J. and Ramsbottom, W. H. C. 1980. Matériaux pour l'étude micropaléontologique du Dinantien de Grande-Bretagne. *Mém. Inst. géol. Univ. Louvain,* **30**, 1-186, 30 Plates (dated 1979).

Conil, R. and Lys, M. 1964. Matériaux pour l'étude micropaléontologique du Dinantien de la Belgique et de la France (Avenois), Algues et Foraminiféres. *Mém. Inst. géol. Univ. Louvain,* **23**, 1-279, 42 Plates.

Conil, R. and Lys, M. 1970. Données nouvelles sur les Foraminifères du Tournaisien inférieur et des couches de passage du Fammenien au Tournaisien dans l'Avenois. *Cong. Coll. Univ. Liège,* **55**, 241-65.

Conil, R. and Lys, M. 1977. Les foraminifères du Viséan moyen V2a aux environs de Dinant. *Ann. Soc. géol. Belg.* **99**, 109-42.

Conil, R. and Lys, M. 1978. Les transgressions dinantiennes et leur influence sur la dispersion et l'évolution des foraminifères. *Mém. Inst. géol. Univ. Louvain,* t. **29**, 9-55 (dated 1977).

Conil, R. and Pirlet, H. 1978. L'évolution des Archaediscidae viséens. *Bull. Soc. Belge Géol.,* **82**, 241-300 (dated 1974).

Conil, R. 1981. Note sur quelques foraminifères du Strunien et du Dinantien d'Europe occidentale. *Ann. Soc. géol. Belg.,* **103**, 43-53.

Cummings, R. H. 1955a. *Nodosinella* Brady 1976 and associated Upper Paleozoic genera. *Micropaleontology,* **1**, 221-38.

Cummings, R. H. 1955b. New genera of Foraminifera from the British Lower Carboniferous. *Washington Acad. Sci. J.,* **45**, 1-8.

Cummings, R. H. 1956. Revision of the Upper Paleozoic textulariid foraminifera. *Micropaleontology,* **2**, 201-42.

Cummings, R. H. 1957. A problematic new microfossil from the Scottish Lower Carboniferous. *Micropaleontology,* **3**, 407-9.

Cummings, R. H. 1961. The foraminiferal zones of the Archerbeck Borehole. *Bull. geol. Survey Gt. Br.* **18**, 107-128.

Cushman, J. 1927. An outline of the reclassification of the Foraminifera. *Contr. Cushman Lab. Foram. Res.,* **3**.

Dain, L. G. 1953. Tournayellidae *in* Fossil Foraminifera of the U.S.S.R. *Trud. VNIGRI,* n.s. **74**, 7-54.

Davies, A. G. 1945. Micro-organisms in the Carboniferous of the Alport Boring. *Proc. Yorks. geol. Soc.* **25**, 312-18.

Davies, A. G. 1951. *Howchinia bradyana* (Howchin) and its distribution in the Lower Carboniferous of England. *Proc. Geol. Assoc.,* **62**, 248-53.

Ehrenberg, C. 1854. *Microgeologie: Das Wirken des unsichtbaren kleinen Lebens auf der Erde.* Leipzig.

Eikhoff, G. 1968. Neue Textularien (Foraminifera) aus dem Waldecker Unterkarbon. *Pal. Zeitsch.,* **42**, 162-78.

Fewtrell, M. D. and Smith, D. G. 1978. Stratigraphic significance of calcareous microfossils from the Lower Carboniferous rocks of the Skipton area, Yorkshire. *Geol. Mag.,* **115**, 255-71.

Fewtrell, M. D. and Smith, D. G. 1981. The recognition and division of the Tournaisian Series in Britain. Discussion. *J. geol Soc. Lond.,* **138**, (1).

Galloway, J. J. and Harlton, B. H. 1930. *Endothyranella,* a genus of Carboniferous Foraminifera. *J. Paleont.* **4**, 2-10.

Ganelina, R. A. 1956. Foraminifera of the Viséan sediments of the north-west region of the Moscow Syncline. *Trud. VNIGRI* n.s. **98**, 61-159.

Ganelina, R. A. 1966. Tournaisian and Lower Viséan Foraminifera of the Kama-Kinel basin. *Trud. VNIGRI* **250**, 64-151.

George, T. N., Johnson, G. A. L., Mitchell, M., Prentice, J. E., Ramsbottom, W. H. C., Sevastopulo, G. and Wilson, R. B. 1976. A correlation of Dinantian rocks in the British Isles. *Geol. Soc. Lond. Spec. Rep.* 7, 1-87.

Hallett, D. 1970; Foraminifera and algae from the Yoredale "Series" (Viséan-Namurian) of Northern England. *C.R. 6me Cong. int. Strat. Géol. Carb.* (Sheffield 1967) **3**, 873-900.

Holliday, D. W., Neves, R. and Owens, B. 1979. Stratigraphy and palynology of early Dinantian (Carboniferous) strata in shallow boreholes near Ravenstonedale, Cumbria. *Proc. Yorks. geol. Soc.*, **42**, 343-56.

Howchin, W. 1888. Additions to the knowledge of the Carboniferous Foraminifera. *J. Roy. Micr. Soc.* pt. 2, 533-45.

Lipina, O. A. 1955. Foraminifera of the Tournaisian stage and uppermost Devonian of the Volga-Ural region and western slope of the Central Urals, *Akad. Sci. U.S.S.R., Trav. Inst. Sci. geol.*, **163**, 1-96.

Lipina, O. A. 1963(a). On the evolution of the Tournaisian Foraminifera. *Vop. Mikropaleont.*, **7**, 13-21.

Lipina, O. A. 1963(b). *In* Results of the second colloquium on the systematics of the endothyroid Foraminifera organised by the Commission on Micropaleontological Coordination. *Vop. Mikropaleont.* **7**, 223-7.

Lipina, O. A. 1965. Taxonomy of the Tournayellidae. *Akad. Sci. U.S.S.R. Trud. Geol. Inst.*, **130**, 1-115.

Lys, M. 1976. Valorisation par microfaunas du Bashkirian inférieur (Namurian B) sous-zone R2) dans le Bassin Houiller du Nord de la France (groupes de Douai et Valenciennes). *Ann. Soc. géol. Nord,* **96**, 379-85.

Malakhova, N. P. 1963. A new foraminiferal genus from the Lower Viséan of the Urals. *Paleont. Zh.* **3**, 111-2.

Malakhova, N. P. 1973. Tournaisian and Lower Viséan formations of the eastern slopes of the southern Urals. *Trav. Inst. Geol. Geochem. Akad. Sci. Central Urals SSR,* **82**, 5-14.

Mamet, B. 1970(a). Precisions sur l'age Viséan du calcaire d'Ardengost (Pyr. Centrales). *C.R. Som. Soc. géol. Fr.*, **4**, 127-8.

Mamet, B. 1970(b). Carbonate microfacies of the Windsor Group (Carboniferous), Nova Scotia and New Brunswick. *Geol. Surv. Canada, Paper,* **70-21**, 1-64.

Mamet B. 1970(c). Sur une microfaune tournaisienne du Massif Centrale. *C.R. Som. Soc. géol. Fr.*, **4**, 110.

Mamet, B. 1974(a). Taxonomic note on Carboniferous Endothyraceae. *J. Foram. Res.*, **4**, 200-4.

Mamet, B. 1974(b). Une zonation par foraminifères du Carbonifère inférieur de la Téthys occidentale. *C.R. 7me Cong. int. Strat. Géol. Carb.* (Krefeld 1971), **3**, 391-408.

Mamet, B. 1975. *Viseidiscus,* un noveau genre de Planoarchaediscinae (Archaediscinae, Foraminifères). *Bull. Soc. géol. France, Suppl.*, **17**, 48-9.

Mamet, B. 1976. An atlas of microfacies in Carboniferous carbonates of the Canadian Cordillera. *Bull. geol. Surv. Can.*, **255**, 1-131, 95 Plates.

Mamet, B., Choubert, G. and Hottinger, G. 1966. Notes sur la Carbonifère du Jebel Ouarkziz. Etude du passage du Viséan au Namurien d'après les foraminifères. *Notes Serv. géol. Maroc,* **27**, 7-21.

Marchant, T. R. 1974. Preliminary note on the micropalaeontology of the Dinantian, Dublin Basin, Ireland. *Ann. Soc. geol. Belg.* **97**, 447-61.

Mikhailov, A. V. 1935. Foraminifera from the Oka series of the Borovich District, Leningrad Region. *Geol. Razv. Leningr. Izvestia,* **2-3**, 33-6.

Mikhailov, A. V. 1939(a). On the characteristics of the genera of Lower Carboniferous Foraminifera. *Leningr. geol. Admin. Sbornik,* **3**, 47-62.

Mikhailov, A. V. 1939(b). On Palaeozoic Ammodiscidae. *Leningr. geol. Admin. Sbornik,* **3**, 63-9.

Miklucho-Maklay, A. D. 1956. On the systematics of Palaeozoic Foraminifera. *Vest. Univ. Leningr. (Geol. Geog.),* **1**, 57-66.

Möller, V., Von. 1878. Die Spiral-Gewundenen Foraminifera des Russischen Kohlenkalks. *Mém. Acad. Imp. Sci. St. Petersbourg,* **(7) 25**, 1-147.

Möller, V., Von. 1879. Die Foraminiferen des Russischen Kohlenkalks. *Mém. Acad. Imp. Sci. St. Petersbourg,* (7) **27**, 1-131.

Phillips, J. 1846. On the remains of microscopic animals in the rocks of Yorkshire. *Proc. Yorks geol. poly. Soc.*, **2**, 277-9.

Pirlet, H. and Conil, R. 1978. See Conil and Pirlet 1978.

Plummer, H. J. 1930. Calcareous Foraminifera in the Brownswood Shale near Bridgeport, Texas. *Bull. Univ. Texas,* **3019**, 5-21.

Ramsbottom, W. H. C., Calver, M. A., Eagar, R. M. C., Hodson, F., Holliday, D. W., Stubblefield, C. J. and Wilson, R. B. 1978. A correlation of Silesian rocks in the British Isles. *Geol. Soc. London Spec. Rep.* **11**, 1-81.

Rauser-Chernoussova, D. M., Beljaev, G. M. and Reitlinger, E. A. 1936. Upper Palaeozoic Foraminifera of the Pechora district. *Trud. geol. Inst.*, **28**, 1-127.

Rauser-Chernoussova, D. M. *et al.* 1948. [A symposium of papers on stratigraphy of the Lower Carboniferous based on Foraminifera]. *Trud. Inst. geol.* **62**, *ser. geol.* **19**, 1-243.

Reitlinger, E. A. 1958. The problem of the systemization and phylogeny of the superfamily Endothyridae. *Vop. Mikropaleont.* **2**, 55-73.

Reitlinger, E. A. 1959. Foraminifera of the border stage of the Devonian and Carboniferous in the western part of central Kazakhstan. *Dokl. Acad. Sci. U.S.S.R.* **127**, 659-62.

Rozovskaya, C. E. 1961. On the systematics of the Endothyridae and Ozawainellidae. *Pal. Zh.* **3**, 19-21.

Schlykova, T. I. 1969. A new genus of Foraminifera from the Lower Carboniferous. *Vop. Mikropaleont.*, **12**, 47-50.

Schubert, R. J. 1907. Beitrage zur einer naturlicheren Systematik der Foraminiferen. *Neues Jahrb. Min. Geol. Pal.* **B-Band, 25**, 233-60.

Schubert, R. J. 1921. Palaeontologische Daten zur Stammengeschichte der Protozoen. *Pal. Zeitschr.* **3**, 129-88.

Sheridan, D. J. R. 1972. Upper Old Red Sandstone and Lower Carboniferous of the Slieve Beagh Syncline and its setting in the north-west Carboniferous basin, Ireland. *Geol. Surv. Ireland Spec. Rep.*, **2**, 1-129.

Skipp, B. 1969. Foraminifera *in* History of the Redwall

Limestone of northern Arizona. *Mem. geol. Soc. Amer.*, **114**, 175–255.

Sollas, W. S. 1921. On *Saccamina carteri* Brady and the minute structure of the foraminiferal shell. *Quart. J. geol. Soc. London,* 77, 193–211.

Suleimanov, I. S. 1945. Some new species of small foraminfers from the Tournaisian of the Ishimbayevo oil-bearing region. *Dokl. Acad. Sci. U.S.S.R.,* **48**, 124–7.

Thompson, M. L. 1942. New genera of Pennsylvanian fusulinids. *Amer. J. Sci.,* **240**, 403–420.

Vachard, D. 1978. Etude stratigraphique et micropaléontologique (algues et foraminifères) du Viséen de la Montagne Noire (Hérault, France). *Mem. Inst. géol. Louvain,* **29**, 111–95 (dated 1977).

Vdovenko, M. V. 1954. Some new species of Foraminifera in the Lower Viséan deposits of the Donetz Basin. *Notes Sci. Univ. Schevchenko Kieve.,* **13**, 64–76.

Vdovenko, M. V. 1967. Some representative Endothyridae, Tournayellidae and Lituolidae of the Lower Viséan of the Grand Donbass. *Akad. Sci. Ukr. S.S.R. Inst. geol. Sci.,* 18–27.

Vdovenko, M. V. 1970. New information on the systematics of the family Forschiidae. *Geol. J.,* **30**, 66–78.

Vdovenko, M. V. 1971. New genera of Foraminifera from Viséan strata. *Dokl. Akad. Sci. Ukr. S.S.R.,* **B**, **33**, 877–9.

Wedekind, P. R. 1937. Einführung in die Grundlagen der historischen Geologie. Band II, *Mikrobiostratigraphie die Korallen und Foraminiferenzeit.* 136 pp. Stuttgart.

4 Permian

J. Pattison

4.1 INTRODUCTION

The only marine Permian rocks in the British Isles are deposits of either the late Permian Zechstein Sea, which covered most of north-central Eurpoe including northeast England, or the contemporaneous Bakevellia Sea of northwest England and Northern Ireland. The fauna in the latter was less varied than the Zechstein Sea's but similar enough for the term 'British Permian foraminifera' to be almost synonymous with 'British Zechstein foraminifera'. Any historical account of their study necessarily includes references to German and Polish work on Zechstein faunas.

The first references to Zechstein foraminifera were included in the separate, general Permian palaeontological works of H. B. Geinitz, R. Howse and W. King, all published in 1848, but all three assigned the foraminifera they described to the annelid genus *Serpula*. However, King's monograph on British Permian fossils (1850) contained descriptions by T. R. Jones of several foraminiferal species, including the first nodosariids to be recognised in Zechstein rocks. Reuss (1854), Richter (1855, 1861) and Schmid (1867) erected several more Zechstein nodosariid species, beginning a proliferation of names which was later criticised by Brady (1876). Schmid *op. cit.* also recorded more tubular forms, as did Jones, Parker and Kirkby (1869), and the confusion between the porcellanous and siliceous tests which started then has continued to the present-day. Brady's monograph (1876) included the first comprehensive review of Zechstein foraminifera; and the only one relating to the British faunas. In it he proposed the species *Nodosinella digitata*, the wall-structure of which, and its taxonomic significance, is another problem yet to be resolved.

The only published works on Zechstein foraminifera in the next eighty years came from Germany. Spandel (1898) produced a brief general summary which included several Zechstein 'firsts': records of *Ammobaculites, Lingulina* and *Frondicularia*; recognition of a distinction between the porcellanous and finely siliceous, tubular forms; and the discovery that the Zechstein *'Textularia'* of earlier authors were uniserial nodosariids (now referred to *Geinitzina*). Further studies of German Zechstein foraminiferal faunas were made by

Paalzow (1935) and Brand (1937).

A revival of interest in Zechstein faunas in general and the microfossils in particular followed the drilling of deep boreholes in the north European plain from the 1950s onwards. Scherp (1962) described 119 foraminiferal species and subspecies, 66 of which were new, from the Zechstein rocks of one borehole in the German Rhineland. Among several Polish works on the subject, the most notable are those of Peryt and Peryt (1977), in which the emphasis was on palaeoenvironments, and Woszczynska (1968) who recorded the vertical distribution of the Zechstein foraminifera in some boreholes in northern Poland. Stratigraphical distribution was also the prime concern of the work by Mikluhu-Maclay and Uharskaja *in* Suveizdis (1975) on the Zechstein faunas in the Baltic area of the U.S.S.R. Both the latter and the recent Polish studies have been based mostly on random thin-sections in contrast to the earlier British and German work on solid specimens.

The only publications on British Zechstein foraminifera in the last few decades have been limited studies of particular genera or small faunas. They include the work of Cummings (1955, 1956) on Upper Palaeozoic forms and wall-structures, which touched on some Zechstein species, Anderson (1964) on *Aschemonella,* and Pattison (1969) on the foraminifera of the Manchester Marl.

4.2 LOCATION OF COLLECTIONS OF BRITISH ZECHSTEIN FORAMINIFERA

1. The British Museum (Natural History) has
   (a) the W. K. Parker Collection including some of the specimens figured by Jones, Parker and Kirkby (1869)
   (b) the H. B. Brady Collection including type material of *Nodosinella digitata* and *Textularia jonesi*,
   (c) other material which may include specimens figured by Jones (1850) in King's monograph.

2. The Northern England office of the Institute of Geological Sciences at Leeds has Zechstein foraminifera from many localities in the British Isles within the general collections plus
   (a) some material presented by both T. R. Jones and W. King, and
   (b) specimens figured by Anderson (1964) and Pattison (1969, 1970).

3. Sunderland Museum has good collections of foraminifera from local Zechstein rocks including the Concretionary Limestone (Z_2) and classic Middle Magnesian Limestone reef localities, as well as Marl Slate foraminifera figured by Bell and others (1979).

4.3 STRATIGRAPHIC DIVISIONS

No formally-defined chronostratigraphic subdivisions are in use for the Zechstein but for several decades it has been recognised (see e.g. Richter-Bernberg 1955) that Zechstein rocks are the deposits of four or more sedimentary cycles identified as Z_1, Z_2 etc. The most complete Zechstein succession in the British Isles is in Yorkshire and the lithostratigraphic groups proposed for it by Smith and others (1974) (from base to top: Don, Aislaby, Teesside, Staintondale and Eskdale) are the respective products of cycles Z_1 to Z_5. The successions in the areas of eastern England from which most of the known British Zechstein foraminifera have been collected are shown in Fig. 4.1 together with the equivalent standard German sequence. All of the Permian formations in northwest England and Northern Ireland to have yielded marine fossils are correlated with Z_1, with the exception of the Belah Dolomite in the Eden Valley which has been tentatively dated as Z_3.

Correlation of Zechstein successions with Permian chronostratigraphic units established elsewhere in the world is hampered by the restricted nature of Zechstein faunas in general and the absence, in particular, of ammonoids and fusulinids. Recent work however, most notably on the conodonts, suggests that the cycles Z_1 to Z_3 are equivalent to the Abadehian Stage of the Upper Permian (Kozur 1978).

4.4 FACIES AND ITS CONTROL ON FAUNAS

Zechstein invertebrate fossils, including the foraminifera, have a very uneven distribution and are especially sparse above Z_1. This is probably mainly due to palaeoenvironmental characteristics: hypersalinity must have been a major limiting factor; but diagenetic processes may also have been responsible for much of the patchiness. It is

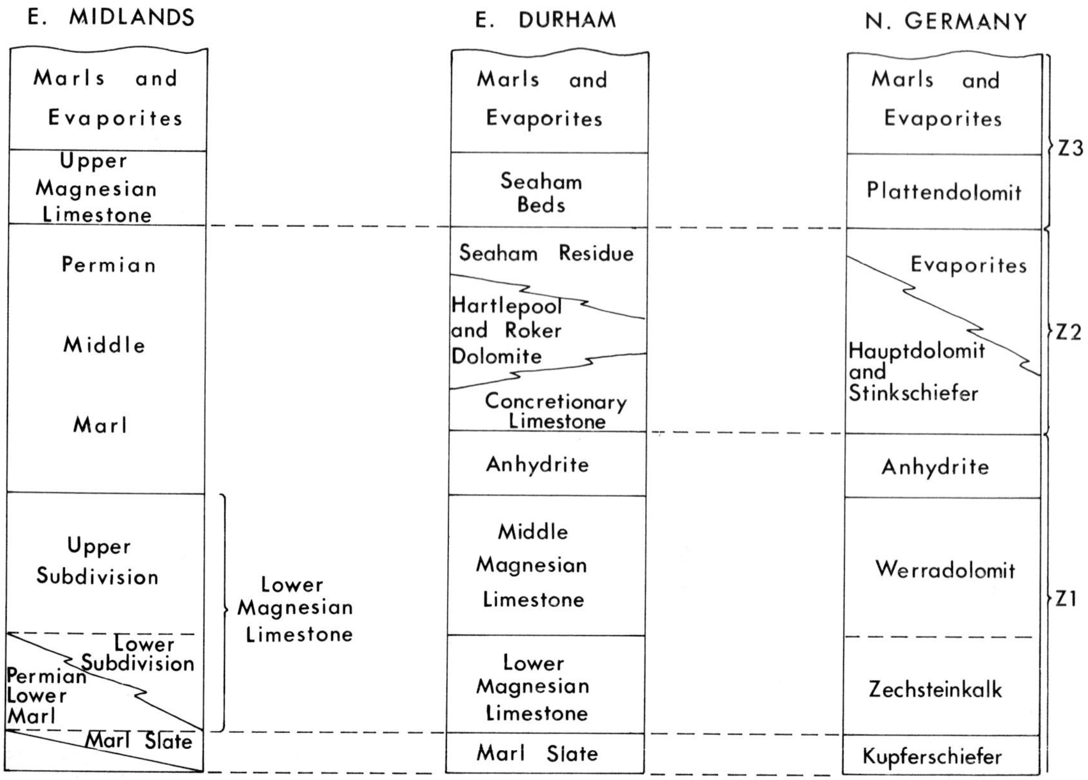

Fig. 4.1 – Permian successions in eastern England correlated with that in North Germany.

quite likely that the dissolution and recrystallisation which accompanied dolomitisation destroyed all the calcitic organic remains in some beds and in others they have left only moulds of the original structures (see Plate 4.1,. Figs. 4, 5, 7).

Random sampling of Zechstein rocks for foraminifera is, therefore, largely unrewarding and both palaeoecological and biostratigraphical studies suffer as a result. For the former, it is difficult to obtain an appreciation of foraminiferal distribution patterns at particular horizons, and virtually impossible in rocks of post-Z_1 age. As to Zechstein biostratigraphy, despite fairly extensive sampling by the writer and others, with existing knowledge it is impossible to produce meaningful foraminiferal species range-charts and that is why they are missing from this section.

The effects of hypersalinity are probably most manifest west of the Pennines where the known foraminiferal faunas, except in the Manchester Marl, are restricted to *Agathammina* and other porcellanous tubular genera. The Zechstein 1 rocks east of the Pennines have yielded a number of distinct foraminiferal faunas. Foraminifera were absent from the typically euxinic conditions of the Marl Slate sea-bottom but conditions ameliorated enough, periodically and locally, during the deposition of that formation for the introduction of a shelly benthos including foraminifera (Bell and others 1979). Strata of what Smith (1970) called 'standard marine facies' within the Z_1 carbonate rocks contain sparse but relatively varied assemblages including nodosariids and porcellanous tubular forms (especially *Agathammina*) associated with brachiopods and bryozoa. Reef dolomites, both in the Lower Magnesian Limestone patch reefs of Yorkshire and the Middle Magnesian Limestone reef barrier of east Durham, have abundant *Orthovertella? gordiformis* and *Calcitornella* spp.

The largest and most varied foraminiferal faunas extracted from the British Zechstein have come from the grey marls and argillaceous limestones of the Permian Lower Marl and Lower

Magnesian Limestone in north Nottinghamshire. They include abundant *Agathammina, Ammobaculites, Ammodiscus, Hyperammina* and a large number of nodosariid forms. In that area there appears to be some faunal variation in response to changes in substrate and depth, as indicated in the systematic section on p. 000. It is noteworthy that the north Nottinghamshire fauna is one of several rich and varied Z_1 foraminiferal faunas to have been recorded from more or less argillaceous beds deposited in what were apparently comparable marginal positions in broad bights on the south side of the Zechstein (and Bakevellia) Seas: The others are from the Manchester Marl (Pattison 1969), the Rhineland (Scherp 1962) and the Fore-Sudetic monocline in southwest Poland (Peryt and Peryt 1977). It prompts the conjecture that in these areas the restrictive effects of hypersalinity, general in the Zechstein and Bakevellia Seas, were offset by the influx of estuarine water.

4.5 DISTRIBUTION

Because range charts would be misleading, the distribution of each species is indicated, under the respective descriptions, by the following symbols referring to recordings of the species or a synonym from particular formations. An asterisk indicates that the species is known by the writer to be common and/or widespread.

A	Mutterflöz of Poland (Peryt 1976)	both sub-Kupferschiefer
B	"Upper Rotliegend' of north Germany (Plumhoff 1966)	

- C Manchester Marl
- D Other Z_1 age deposits of the Bakevellia Sea
- E Lower Magnesian Limestone/Permian Lower Marl of N. Nottinghamshire in the E. Midlands
- F Lower Magnesian Limestone of Durham
- G Other Lower Magnesian Limestone of eastern England
- H Middle Magnesian Limestone of Durham

I	Z_1 rocks of Germany	above Kupferschiefer
J	Z_1 rocks of Poland	

- K Z_1 rocks in Baltic area of U.S.S.R. (in Suveizdis 1975)
- L Concretionary Limestone
- M Z_2 rocks of Germany
- N Z_2 rocks of Poland (Woszczynska 1968)
- O Z_2 rocks in Baltic area of U.S.S.R. (in Suveizdis 1975)
- P Z_3 rocks of Poland (Woszczynska 1968)

The classification employed here is that of Loeblich and Tappan (1964) except in the use of *Geinitzina* as noted below.

PLATE 4.1

4.5.1 Suborder Textulariina

Ammobaculites eiseli (Paalzow)
Plate 4.1, Fig. 6 (× 40), Fig. 18 (thin-sect. × 50). = *Ammobaculites eiseli* (Spandel) Paalzow, 1935. *Haplophragmium eiseli* Spandel, 1898 was a *nomen nudum.* (syn. *A. procera* Scherp, 1962 but not *A. eiseli* (Spandel) Scherp). Description: Early coiled part is 1.5 to 2 times the width of the first of the succeeding chambers and is eccentrically positioned in relation to rectilinear part; chambers in latter are broader than long; sutures distinct; wall, coarse-grained, bound by siliceous cement.
Remarks: Rectilinear part has narrower chambers then in *A. eiseli* (Spandel) Scherp 1962. Distribution: Z_1 (B, ?C, E*, H, I*, K), Z_2 (?O). Deep-water and relatively stenohaline.

Ammodiscus roessleri (Schmid)
Plate 4.1, Fig. 2 (× 40), Fig. 16 (thin sect. × 130). = *Serpula roessleri* Schmid, 1867. (syns. ? *Trochammina incerta* (d'Orbigny) [part] Brady, 1876; *T. bradyna* s.l. Spandel, 1898). Description: discoidal, round to elliptical; proloculus spherical; tubular second chamber wound in evolute plane-spiral with up to 6 whorls, increasing slowly in thickness; aperture is open end of second chamber; finely siliceous.
Remarks: It is a thinner disc and finer-grained than *A. robustus* Vangerow, 1962 and lacks that species' involute final whorl. Distribution: Z_1 (A, B, C, E*, G, I*, J, K), Z_2 (?L, M). Associated with low-energy environments.

Hyperammina acuta (Scherp, 1962)
Plate 4.1, Fig. 1 (× 40). (syns. *H. acuta conica* Scherp and *H. crescens* Scherp). Description: An elongate, straight or gently arcuate, flattened cone with a fine-grained, siliceous wall; width increases uniformly from the roughly pointed proximal end; aperture formed by the entire internal diameter of distal end.
Remarks: Differs from *H. recta* Scherp, 1962 in its narrow proximal end. Distribution: Z_1 (?C, E*, G, I*, ?J, ?K), Z_2 (?M). Most common in argillaceous rocks.

4.5.2 Suborder Miliolina

Agathammina milioloides (Paalzow) *non* (Jones, Parker and Kirkby)
Plate 4.1, Fig. 5 (int. mould × 25), Fig. 9 (× 40), Fig. 19 (thin-sect. × 50). = *Glomospira milioloides* (Jones, Parker and Kirkby) [part] Paalzow, 1935. Description: As *A. pusilla* (see below), but early milioline whorls are followed by planispiral coiling. Distribution: Z_1 (D, E*, F, G, H. I, K). Facies as *A. pusilla* but probably more stenohaline.

Agathammina pusilla (Geinitz)
Plate 4.1, Fig. 3 (× 25), Fig. 4 (int. mould × 25) = *Serpula pusilla* Geinitz, 1848 [part]. (syn. *Trochammina milioloides* Jones, Parker and Kirkby, 1869). Description: Elongate, fusiform test consisting of tubular chamber with a hemispherical cross-section wound entirely in a milioline manner; with or without a globular proloculus; calcareous, porcellanous test. Distribution: Z_1 (A, D, E*, F*, G*, H*, I*, J*, K*). Shallow-water and euryhaline.

Cyclogyra kinkelini (Spandel)
Plate 4.1, Fig. 15 (× 40), Fig. 17 (thin-sect. × 40). = *Cornuspira kinkelini* Spandel, 1898. (syn. *Trochammina incerta* d'Orbigny. Jones, Parker and Kirkby, 1869). Description: Calcareous, porcellanous test consisting of a spherical proloculus and a tubular second chamber with a hemispherical cross-section, which increases slowly and uniformly in diameter, wound in an evolute plane spiral with two to seven whorls. Distribution: Z_1 (B, ?D, E*, F, G, H, I, ?J, K). Characteristic of low-energy environments.

Orthovertella? gordiformis (Spandel)
Plate 4.1, Figs. 7 (int. mould) and 8 (× 40). = *Ammodiscus gordiformis* Spandel, 1898. (syns. *Trochammina gordialis* Jones, Parker and Kirkby, 1869; *Hemigordius?* sp. Pattison, 1969). Description: Compact and globular, comprising a calcareous, porcellanous tube with a hemispherical cross-section wound tightly in a streptospiral; tube-width increases uniformly towards aperture but there may be shallow constrictions; suture not incised.
Remarks: More globular and tightly-coiled than *O.? mutablis* (Scherp, 1962). Distribution: Z_1 (B, C, D, E*, F, G, H*, I, ?J, K). Common in high-energy environments.

4.5.3 Suborder Rotaliina

Dentalina permiana (Jones, 1850)
Plate 4.1, Figs. 12, 13 (× 40), Fig. 22 (thin-sect. × 50). (syns *D. communis* d'Orbigny [part] Brady, 1876; *D. fallax* Franke and *D. farcimen* Soldani. Paalzow, 1935; *D. sp.* Pattison, 1969): Description: 7–9 barrel to bead-shaped chambers in a slightly curved arc; sutures distinct and incised; non-radiate aperture in line with test axis; wall is calcareous, hyaline and simple.
Remarks: Smaller and has more distinct sutures than *D. lineamargaritarum* Scherp, 1962. Distribution: Z_1 (?B, C, E*, F, G, H, I*, J, K), Z_2 (L). Widespread but especially common in low-energy and ? deeper-water environments.

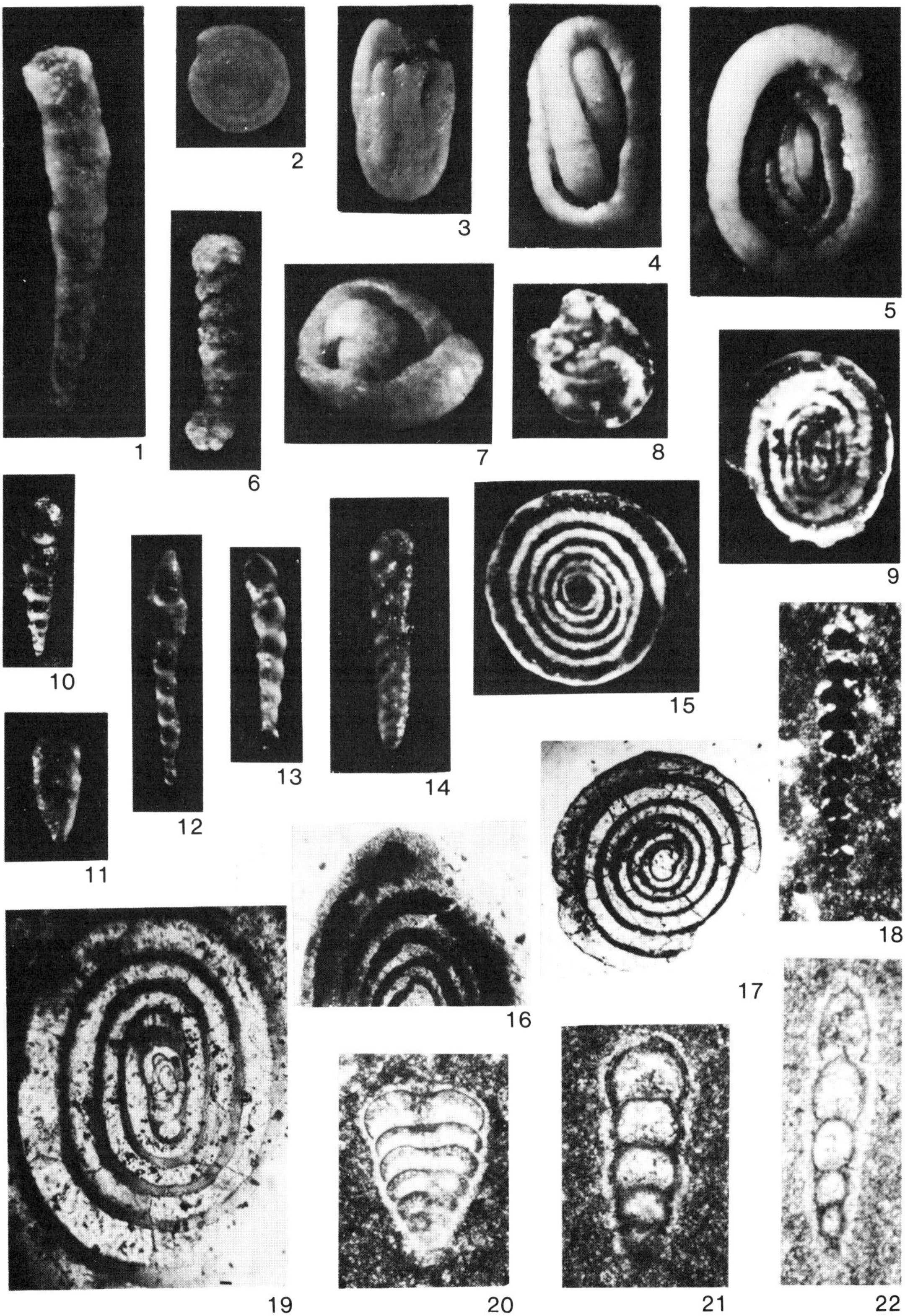

Plate 4.1

Geinitzina acuta (Spandel)
Plate 4.1, Fig. 11 (× 40), Fig. 20 (thin-sect. × 50). = *Geinitzella acuta* Spandel, 1898. (syns. *Textularia cuneiformis* Jones and *T. triticum* Jones. Richter, 1855; *T. jonesi* [part] Brady, 1876; *G. postcarbonica* Spandel. Scherp, 1962 and Woszczynska, 1968). Description: Uniserial, rectilinear, flattened and centrally-constricted with 8–10 chambers; spherical proloculus; most of other chambers are hour-glass shaped in transverse section but earliest and latest may lack the central constriction; from front, at least the early part of the test is wedge-shaped but latest part may have parallel sides; aperture is centrally-placed slit on top of last chamber; calcareous, hyaline, double-layered wall.
Remarks: *G. kirkbyi* (Richter, 1861) is narrower and has more incised sutures and central constriction. The status of the taxon *Geinitzina* is in doubt but the name is used here in order to conform with current usage among Zechstein foraminifera workers. Distribution: Z_1 (?A, ?B, ?C, E, F, H, I, J, K), Z_2 (?L), Z_3 (P). Usually found with other nodosariids but more tolerant of high-energy environments.

Nodosaria geinitzi (Reuss, 1854)
Plate 4.1, Fig. 10 (× 40), Fig. 21 (thin-sect. × 50). (syns. *N. radicula* (Linné) [? part] Brady, 1876; *N. cushmani* Paalzow, 1935; *?N. scherpi* Mikluho-Maclay, 1975). Description: Globular proloculus; width and length of intermediate chambers about equal; final chamber elongate and may have pouting aperture; sutures distinct and moderately incised; calcareous, hyaline wall which is simple or, more rarely, double-layered.
Remarks: Longer chambers and more incised sutures than *N. permiana* (Spandel, 1898). Distribution: Z_1 (?A, B, E*, F, H, I*, J, ?K), Z_2 (L). Associated with low-energy and ? deeper-water environments.

Frondicularia cavernula (Paalzow)
Plate 4.1, Fig. 14 (× 40). = *Spandelina cavernula* Paalzow, 1935. (syn. *Monogenerina n.* sp. a Scherp, 1962). Description: Uniserial, rectilinear, flattened back and front; spherical proloculus; each of other chambers is sharply arched and clasps the preceding one; distinct sutures; aperture round to oval; calcareous, hyaline wall. Distribution: Z_1 (?A, B, E, I, J). One of the many rare members of the nodosariid assemblages in low-energy environments.

4.6 REFERENCES

Alexandrowisz, S. F. and Barwicz, W. 1970. Pozycja stratygraficzna; paleogeograficzna mikrofauny cechsztynu monokliny przedsudeckiej. *Acta Geol. Pol.*, **20**, 287–324.

Anderson, F. W. 1964. *Aschemonella longicaudata sp. nov.* from the Permian of Derbyshire, England. *Geol. Mag.*, **101**, 44–47.

Bell, J., Holden, J., Pettigrew, T. H. and Sedman, K. W. 1979. The Marl Slate and Basal Permian Breccia at Middridge, Co. Durham. *Proc. Yorks geol. Soc.*, **42**, 439–460.

Brady, H. B. 1876. Carboniferous and Permian foraminifera. *Palaeontogr. Soc. (Monogr.).*

Brand, E. 1937. Über Foraminiferen im Zechstein der Wetterau. *Senckenbergiana*, **19**, 375–380.

Cummings, R. H. 1955. *Nodosinella* Brady, 1876, and associated Upper Palaeozoic genera. *Micropaleontology*, **1**, 221–238.

Cummings, R. H. 1956. Revision of the Upper Palaeozoic textulariid foraminifera. *Micropaleontology*, **2**, 201–242.

Geinitz, H. B. 1848. *Die Versteinerungen des deutschen Zechsteingebirges.* Arnoldische. Dresden and Leipzig.

Howse, R. 1848. A catalogue of the fossils of the Permian system of the counties of Northumberland and Durham. *Trans. Tyneside Nat. Field Club*, **1**, 219–264.

Jones, T. R., Parker, W. K. and Kirkby, J. W. 1869. On the nomenclature of the foraminifera. XIII. The Permian *Trochammina pusilla* and its allies. *Ann. Mag. nat. Hist.* ser. **4**, **4**, 386–392.

King, W. 1848. *A catalogue of the organic remains of the Permian rocks of Northumberland and Durham.* Newcastle upon Tyne.

King, W. 1850. A monograph of the Permian fossils of England. *Palaeontogr. Soc. (Monogr).*

Kozur, H. 1978. Beiträge zur Stratigraphie des Perms. Teil II: Die Conodontenchronologie des Perms. *Freiberger Forsch. H.*, **C334**, 85–161.

Paalzow, R. 1935. Die Foraminiferen im Zechstein des östlichen Thüringen. *Jb. preuss. geol. Landesanst.* **56**, 26–45.

Pattison, J. 1969. Some Permian foraminifera from northwestern England. *Geol. Mag.*, **106**, 197–205.

Pattison, J. 1970. A review of the marine fossils from the Upper Permian rocks of Northern Ireland and northwest England. *Bull. geol. Surv. Gt. Br.*, **32**, 123–165.

Peryt, T. M. 1976. Ingresja morza Turynskiego (Gorny Perm) na obszarze monokliny Przedsudeckiej. *Rocznik Pol. Towarzystwa geol.*, **46**, 455–465.

Peryt, T. M. and Peryt, D. 1977. Otwornice cechsztynskie monokliny Przedsudeckiej i ich paleokologia. *Rocznik Pol. Towarzystwa geol.*, **47**, 301–326.

Plumhoff, F. 1966. Marines Ober-Rotliegendes (Perm) im Zentrum des nordwest-deutschen Rotliegend-Beckens. Neue Beweise und Folgerungen. *Erdöl. Kohle Erdgas Petrochem.*, **10**, 713–720.

Reuss, A. E. 1854. Ueber Entomostracen und Foraminiferen im Zechstein der Wetterau. *Jber. Wetterauer Ges.*, **59–77**.

Richter, R. 1855. Aus dem thüringischen Zechstein. *Z. dt. geol. Ges.*, **7**, 523–533.

Richter, R. 1861. in Geinitz, H. B *Dyas*. Wilhelm Engelmann. Leipzig.

Richter-Bernberg, G. 1955. Stratigraphische Gliederung des deutschen Zechsteins. *Z. dt. geol. Ges.*, **105**, 843–854.

Scherp, H. 1962. Foraminiferen aus dem Unteren und Mittleren Zechstein Nordwestdeutschlands. *Fortschr. Geol. Rheinld. West.* **6**, 265–330.

Schmid, E. E. 1867. Über die kleineren organischen Formen des Zechsteinkalks von Selters in der Wetterau. *Neues Jb. Geol. Petrefacten-Kunde.* 576–588.

Smith, D. B. 1970. The palaeogeography of the British Zechstein. 20–23 in Third Symposium on Salt, *Northern Ohio geol. Soc.* Cleveland.

Smith, D. B., Brunstrom, R. G. W., Manning, P. I., Simpson, S. and Shotton, F. W. 1974. A correlation of Permian rocks in the British Isles. *Spec. Rep. geol. Soc., London,* **5**.

Spandel, E. 1898. *Die Foraminiferen des deutschen Zechsteines.* Nürnberg.

Suveizdis, P. (Ed.) 1975. Permian deposits of Baltic area (Stratigraphy and fauna) *Liet. geol. Moksly. Tyrimo Inst. Tr.*, **29**, 1–305. [in Russian with English summary].

Vangerow, E. F. 1962. Über *Ammodiscus* aus dem Zechstein. *Paläont. Z.*, **36**, 125–133.

Woszcynska, S. 1968. Wstepne wyniki badan mikrofauny osadow cechsztynu. *Kwart. geol.* **12**, 92–103.

5
Triassic

P. Copestake

Records of foraminifera from the British Triassic are few and are confined to the uppermost, Rhaetian stage which in Britain is developed in a marine facies (the Penarth Group). The Penarth Group (Kent 1970, Ivimey-Cook 1974, Warrington *et al.* 1980) (equivalent to the Lower and Upper Rhaetic of Richardson 1905 and the Westbury, Cotham and Langport Beds of Richardson 1911) represents the initial inundation of the Triassic continent by a shallow epeiric sea which continued to be present throughout the following Early Jurassic.

On the whole British Rhaetian assemblages are rather sparse and have been largely neglected by workers. The earliest recorded are by Chapman (1895) from the early Rhaetian (lowermost Westbury Formation) of Wedmore (Somerset); this microfauna is dominated by agglutinating forms several of which Chapman refers to Carboniferous species of *Haplophragmoides* and *Ammodiscus*, for example *H. emaciatum* (Brady), *A. anceps* (Brady), *A. centrifugus* (Brady), *A. pusillus* (Geinitz). Also present are specimens allocated to the Carboniferous endothyracean genera *Nodosinella* and *Stacheia* including several 'new species'. Unfortunately, the descriptions and figures, although good for their time, do not permit an adequate assessment of these 'Palaeozoic' taxa. Also reported in the Wedmore Stone at the base of Chapman's studied section are the nodosariids *Nodosaria* and *Marginulina,* genera which become common in the Lower Jurassic. The Palaeozoic taxa in the Wedmore fauna do seem at present to be anomalous, in the light of other known Rhaetian assemblages and may prove to be reworked occurrences. Until the Wedmore fauna is restudied in detail, these reported occurrences must be viewed with reservation (recent sampling at Wedmore of the presumed Wedmore Stone equivalent has yielded no foraminifera; H. C. Ivimey-Cook, personal communication). The whereabouts of Chapman's collection is not known at present.

The only other known record from the Westbury Formation is of *Eoguttulina liassica* (Strickland), 1.7 m above the base at Lavernock (H. C. Ivimey-Cook, personal communication). Subsequent British records are from the late

Rhaetian Lilstock Formation, at the base of which (in the Cotham Member) *Eoguttulina liassica* (Strickland) occurs in numbers at Lavernock, Glamorgan (Ivimey-Cook, 1974, personal communication, 1980). At this locality, this species is accompanied above, in the lower part of the Langport Member, by *Lingulina tenera tenera* (Bornemann), *Nodosaria columnaris* Franke and *Dentalina* cf. *ventricosa* Franke (Ivimey-Cook, personal communication). In addition, from the upper part of the Langport Member and from the Pre-planorbis beds at Lavernock, Banner *et al.* (1971) recorded *L. tenera tenera* and *L. tenera tenuistriata* (Nørvang) (≡ *Geinitzinita* spp. sensu Banner). These Glamorgan late Rhaetian assemblages are of undoubted Lower Jurassic affinity overall, although Banner *et al.* did figure a weakly striated *Eoguttulina* (probably a new species, F. T. Banner, personal communication) from the upper Langport Member, which is not at present known from the Lower Jurassic.

Additional published observations of late Rhaetian foraminifera include records of *Nodosaria nitidana* Brand and *Lingulina tenera pupa* (Terquem) from the Langport Member and Pre-planorbis beds, respectively, of the Dorset coast (Barnard 1950), of *Eoguttulina liassica, Lenticulina* aff. *muensteri* (Roemer) and *Nodosaria* cf. *nitidana* from the Cotham Member of the Kineton area, Warwickshire and of *Dentalina pseudocommunis* Franke, *E. liassica* and *Lingulina lanceolata* (Haeusler) from the Langport Member of Lighthorne near Banbury (Ivimey-Cook *in* Edmonds *et al.* 1965). Each of these species is most common in the Lower Jurassic.

These latter accounts all report the presence only of long-ranging Jurassic species in the Rhaetian, and none can be considered as an index for this stage. The shortest ranging taxon known at present from the Rhaetian is *Lingulina tenera collenoti* (Terquem) which has been found in the uppermost Rhaetian Pre-planorbis beds of the Platt Lane Borehole (Cheshire) and Tolcis quarry, Axminster (Devon) (unpublished data). This subspecies is the nominate taxon for the latest Rhaetian-early Jurassic foraminiferal zone (Copestake and Johnson in press).

To summarise, the lower part of the British Rhaetian (Westbury Formation) is poorly known but contains some agglutinating taxa (some of late Palaeozoic affinity, but not yet proven to be *in situ*) plus rare forms generally more typical of the Lower Jurassic, including *Eoguttulina liassica*. Several additional taxa appear in the late Rhaetian (Lilstock Formation), including the *Lingulina tenera* plexus; *E. liassica* also becomes common at this level. The latest Rhaetian (pre-planorbis beds) contains *Lingulina tenera collenoti* which continues into the Early Jurassic (Hettangian, mid-*Angulata* Zone). During the Rhaetian, therefore, new taxa successively appear, most of which range into and attain their maximum development in the Lower Jurassic.

Triassic foraminifera are better known from the continent of Europe, and have been described by, amongst others, Oberhauser (1960) (Ladinian and Carnian of Austria and Iran), Kristan-Tollmann (1964) (Rhaetian of Austria) and Zaninetti (1976) (a summary of non-nodosariid assemblages from Europe and Tethyan realm).

ACKNOWLEDGEMENTS

The author is grateful to the Institute of Geological Sciences for access to unpublished borehole information; to Dr. H. C. Ivimey-Cook of the Institute for advice on points of stratigraphy, for reading the original manuscript and for permission to utilise unpublished data; to Dr. B. Johnson (British National Oil Corporation) for kindly supplying material from Tolcis quarry, Axminster; and to the directors of Robertson Research International Limited for permission to publish.

REFERENCES

Banner, F. T., Brooks, M. and Williams, E. 1971. The geology of the Approaches to Barry, Glamorgan. *Proc. Geol. Ass.* **82**, (2), 231-247.

Barnard, T. 1950. Foraminifera from the Lower Lias of the Dorset coast. *Q. Jl. geol. Soc. London,* **105**, (3), 347-391.

Chapman, F. 1895. Rhaetic foraminifera from Wedmore in Somerset. *Ann. Mag. Nat. Hist.,* **16**, 307-329.

Copestake, P. and Johnson, B. (in press). Liassic (Hettangian-Toarcian) foraminifera from the Mochras Borehole, North Wales.

Edmonds, E. A., Poole, E. G. and Wilson, V. 1965. Geology of the country around Banbury and Edge Hill (sheet 201). *Mem. Geol. Surv.*

Ivimey-Cook, H. C. 1974. The Permian and Triassic deposits of Wales. *In:* Owen, T. R. (Ed.). *The Upper Palaeozoic and post-Palaeozoic rocks of Wales.* University of Wales Press, Cardiff.

Kent, P. E. 1970. Problems of the Rhaetic in the East Midlands. *Mercian Geologist,* **3**, 361-372.

Kristan-Tollmann, E. 1964. Die Foraminiferen aus den Rhatischen Zlambachmergeln der Fischerweise bei Aussee in Jalzkammergut. *Jb. Geol. B.A.,* **10**, 1-189.

Oberhauser, R. 1960. Foraminiferen und Mikrofossilien 'incertae sedis' der ladinischen und Karnischen Stufe der Trias aus den Ostalpen und aus Persien. *Jb. Geol. B.A.,* **5**, 5-46.

Richardson, L. 1905. The Rhaetic and contiguous deposits of Glamorganshire. *Q. Jl. geol. Soc. London,* **61**, 385-424.

Richardson, L. 1911. The Rhaetic and contiguous deposits of West, Mid and part of East Somerset. *Q. Jl. geol. Soc. London,* **67**, 1-72.

Warrington, G. *et al.* 1980. A correlation of Triassic rocks in the British Isles. *Geol. Soc. Lond.,* Special Report No. 13, 78 pp.

Zaninetti, L. 1976. Les foraminifères du Trias. *Riv. Ital. Paleont.,* **82** (1), 1-258.

6
Jurassic

6.1 THE HETTANGIAN TO TOARCIAN

P. Copestake, B. Johnson

6.1.1 Introduction

Foraminifera were first described from the British Lower Jurassic during the second half of the nineteenth century, in several small, but important, papers. The earliest, that of Strickland (1846), named two of the most common Early Jurassic species, namely *Spirillina infima* and *Eoguttulina liassica*, from the Hettangian (*planorbis* Zone) of Gloucestershire. In a subsequent paper Jones and Parker (1860) figured and discussed a small assemblage from a blue clay at Chellaston, near Derby. They believed the clay to be Triassic in age, though the microfauna is undoubtedly Early Jurassic, probably Toarcian, in character. Most important of these early works was that by Blake (*in* Tate and Blake 1876) on the Yorkshire Lias, since a considerable number of species are figured and indications are given of their stratigraphical distribution. Most of the forms described are from the Hettangian-Pliensbachian, probably because the Yorkshire Toarcian is generally impoverished in foraminifera. Much research on Early Jurassic foraminifera was being conducted around this time by Brady, though he published only two brief papers (*idem.* 1864, 1866). The second of these was based upon Late Pliensbachian and Toarcian material in Charles Moore's collection from south-west England. Brady indicated that he had prepared a monograph on British Early Jurassic foraminifera which had been accepted by the Palaenotolographical Society (Macfadyen 1941), but unfortunately it was never published. This nineteenth century interest was completed in the 1890's by two brief studies of Crick and Sherborn (1891, 1892) upon material from the Pliensbachian and Toarcian of Northamptonshire. Interest subsequently waned for fifty years or so, and little was published other than lists of species from Hock Cliff, Gloucestershire (Richardson 1908, Henderson 1934) and from localities in Lincolnshire (Trueman 1918, Bartenstein and Brand, 1937).

The value of many of these early works is considerably diminished by poor figures, brief descriptions and the use of Tertiary or Recent species names; recourse to the authors' original collections is necessary before their figures can be accurately identified. It is unfortunate that Blake's collection and the major part of Brady's have not yet been found.

The most comprehensive British works are those of Macfadyen (1941) and Barnard (1950(a)), on the Dorset coast Hettangian to Lower Pliensbachian, and of Barnard (1950(b)) on the Toarcian of Byfield (Northamptonshire). Macfadyen solved several taxonomic problems, whilst Barnard first applied the plexus concept to Jurassic foraminifera, an approach which he continued to pursue in several subsequent works (*idem.* 1956, 1957, 1960, 1963). These authors were among the first to grasp the problem of intraspecific variation in Jurassic foraminifera, and this aspect was further explored by Adams (1957) in the case of Toarcian microfaunas from Lincolnshire.

Documentation of British microfaunas has been continued more recently by Banner *et al.* (1971) and Warrington and Owens (1977), based on offshore investigations, and by Horton and Coleman (1978) on Toarcian borehole material from Leicestershire (formerly Rutland). Several unpublished theses also contain important results.

6.1.2 Location of collections of importance

British Museum (Natural History), London.
Collections of Jones and Parker (1860), Brady (undescribed material from Chellaston, Hock Cliff and Stockton), Macfadyen (1941), Wood and Barnard (1946), Barnard (1950(a), 1950(b), 1952) and Adams (1962), (Johnson, 1975, Upper Domerian and Toarcian foraminifera from the Llanbedr (Mochras Farm) Borehole, North Wales, University of Wales; Copestake, 1978, Foraminifera from the Lower and Middle Lias of the Mochras Borehole, University of Wales).

Department of Geology, University College of Wales, Aberystwyth.
Assemblages slides and unpublished material of Jenkins, Johnson and Copestake.

Department of Palaeontology, Institute of Geological Sciences, London.
Specimens described by Horton and Coleman (1978) and undescribed material from the Stowell Park, Wilkesley, Platt Lane, Burton Row, Hill Lane, Truch and Cocklepits Boreholes.

Institute of Geological Sciences, Leeds.
Figured specimens of Copestake and Johnson (in press).

Geology Museum, Queen Square, Bath.
Specimens described by Brady (1866), plus additional undescribed material, in the Charles Moore collection.

6.1.3 Stratigraphic divisions

The Lower Jurassic comprises the Hettangian, Sinemurian, Pliensbachian and Toarcian stages, and is represented lithostratigraphically by rocks known as the Lower, Middle and Upper Lias. Until recently, the Lower Jurassic was regarded as exactly equivalent to the Lias, but it is now proposed that the base of the Jurassic System should be taken a little higher than previously (Warrington *et al.,* 1980) to exclude the Pre-planorbis Beds, which are the basal strata of the Lias. The Pre-planorbis Beds, were previously included in the Hettangian on lithostratigraphical grounds. These beds lack ammonites and have a fauna transitional between the underlying Rhaetian (Upper Triassic, Penarth Group) and the Lower Jurassic. As the base of the Hettangian is now redefined by the first occurrence of the ammonite *Psiloceras,* the basal Lias Pre-planorbis Beds are regarded as being Rhaetian in age. Nevertheless because of their strong microfaunal affinity with the Lower Jurassic, the Pre-planorbis Beds are here included in the range charts. The top of the Lower Jurassic is taken, as usual, at the Toarcian/Aalenian stage boundary.

The most sophisticated ammonite zonation of the Jurassic has been established for the Lias (Dean, Donovan and Howarth, 1961), against which the foraminiferal ranges can be calibrated. A zonation based upon foraminifera has also been erected for the north-west European Lias, utilising species total known ranges, although it is not yet in print (Copestake and Johnson, in press). The one scheme which has to date been published, for the Empingham Upper Lias (Horton and Coleman 1978), seems to have little regional applicability, although it is undoubtedly workable on a local scale; this is also most likely the case for the ostracod zonation of the same area (Bate and Coleman 1975). No broad scheme is yet available for the Lower Jurassic on the basis of ostracods (Lord 1978), and the coccolith zonation proposed for Dorset and northern France (Barnard and Hay 1974) has yet to be applied regionally.

6.1.4 Faunal association and facies

During the late Triassic in Britain the prevailing

dominantly continental environment came to an end. A marine transgression led to the establishment, by the beginning of the early Jurassic, of fully marine conditions. In this shallow-shelf Liassic sea, which extended across north-west Europe, sedimentation took place within a régime of 'basins' and 'swells' (Hallam 1958) which controlled lithological and thickness variations. Thus thick, continuous, basinal sequences (as in the Midlands-Severn and Cardigan Bay Basins) are separated by thinner, condensed, often interrupted swell sequences (as in the Dorset and Radstock areas). Liassic facies variations were also influenced by several eustatic sea level changes, some of which initiated transgressions and regressions, as well as by local tectonic movements causing parochail lithological differences. In general the British Lower Jurassic is mainly represented by argillaceous deposits (mudstones, clays, shales and siltstones), interrupted by a limestone/shale alternation (the Blue Lias) in the Hettangian and Lower Sinemurian; oolitic ironstones, ferruginous limestones and sandstones (e.g. the Marlstone Rock-bed, Cleveland Ironstone Formation, Scalpay Sandstone) in the Upper Pliensbachian; and condensed limestones (e.g. the Lower and Upper Cephalopod Limestones), sandstones (e.g. the Bridport Sands) and ironstone (the Raasay Ironstone) in the Toarcian. Further modifications are seen, especially in the Lower Lias, adjacent to land areas such as the Mendips-Glamorgan and Scottish Islands, where shallow water bioclastic limestones and calcarenites (e.g. the Downside Stone, the Southerndown Beds and the Broadford Beds) were deposited.

The foraminiferal assemblages of the Lower Jurassic were also affected by the same environmental factors which caused the lithological diversification. Thus the succession of foraminiferal assemblages in the British Lias is a function of these environmental conditions, though certain variations are, of course, due to the progressive influences of evolution. Whilst evolution is not discussed here in detail, it should be remembered that extinctions and replacements are intimately bound up with environmental changes, and this is particularly well shown in the Lower Jurassic.

The deposits of the Rhaetian (Chapter 5) and basal Jurassic transgression contain many new species and subspecies, no doubt as a result of the opening up of many new vacant niches and their subsequent occupation, an example of adaptive radiation. This is demonstrated by the abundant and diverse microfaunas of the Hettangian and Lower Sinemurian. However, this phase was preceded by one of widespread stagnation of the sea bed, causing the restricted foraminiferal faunas of the *planorbis* Zone and the often-encountered flood of *Reinholdella* (especially *R? planiconvexa* (Fuchs)) in the early *liasicus* Zone. In some places the improved oxygenation which followed is marked by an influx of *Ophthalmidium* (mainly *O. liasicum* (Kübler and Zwingli)), showing that shallow depths persisted from area to area. This was followed by a gradual deepening and, for the rest of the Hettangian and Sinemurian, shelf environments dominated by diverse nodosariid assemblages were prevalent, though localised tectonic activity resulted in variations of limited areal extent.

The late Sinemurian in Britain was a time of widespread shallowing. This was a gradual and progressive process which began in the late *raricostatum* Zone and culminated at the base of the *jamesoni* Zone (Lower Pliensbachian), and foraminiferal assemblages at these horizons are in some areas dominated by nubeculariids (especially *Ophthalmidium* spp. together with species of *Nubecularia*). This regression was followed by the basal Plienbachian transgression which brought about renewed deepening. The late Sinemurian shallowing seems to have caused the extinctions of several species and subspecies, and these were replaced by new taxa following the subsequent transgression.

This pronounced faunal turnover across the Sinemurian-Pliensbachian boundary occurred throughout north-west Europe, and on mainland Europe it was usually succeeded by diverse microfaunas for the duration of the Early Pliensbachian. In Britain, however, Early Pliensbachian foraminiferal faunas are often comparatively sparse, particularly in the *jamesoni* and early *ibex* Zones. By *davoei* Zone times, British microfaunas had regained their former diversity and this situation was maintained through the succeeding *margaritatus* Zone. Little is known of *spinatum* Zone microfaunas in Britain, since this zone is usually represented by sandstones and ironstones of the late Domerian regressive phase, which foreran the early Toarcian transgression. However, the Zone occurs in an offshore argillaceous facies in the

Mochras Borehole where influxes of *Ophthalmidium (O. northamptonensis* Wood and Barnard. *O. macfaydeni* Wood and Barnard, *O. liasicum* (Kübler and Zwingli)) and *Spirillina* (*S. infima* (Strickland), *S. tenuissima* Gümbel) at the top of the Zone provide evidence of the regional shallowing. The proliferation of shallow-water foraminifera continued into the basal *tenuicostatum* Zone of the Toarcian, and is recognised as a marker horizon in several parts of Britain. Later in the Zone bottom conditions became stagnant and were maintained as such until the early *falciferum* Zone (*exaratum* Subzone); this interval is dominated by *Reinholdella dreheri* (Bartenstein), *R. pachyderma pachyderma* Hofker and, especially in the east Midlands and Yorkshire, *R. macfadyeni* (Ten Dam). Influxes of agglutinating foraminifera and miliolids later in the Subzone indicate an environmental change. It can be seen that the stagnations caused many typical Hettangian–Pliensbachian taxa to become extinct, including *Lingulina tenera tenera* Bornemann, *Berthelinella involuta* (Terquem), *Involutina liassica* (Jones), *Dentalina tenuistriata* Terquem, *Ophthalmidium liascum, O. macfadyeni* and the *Marginulina prima* d'Orbigny *Frondcularia terquemi* d'Orbigny plexuses. This event signalled the most marked microfaunal turnover of the Lower Jurassic, and the later Toarcian sediments contain many new appearances with the overall balance of species becoming progressively more Middle Jurassic in affinity, especially in the *thouarsense* and *levesquei* Zones.

Conditions from the *falciferum* Zone (*falciferum* Subzone) onwards are generally typified by deeper water assemblages than at earlier levels in the Jurassic, even though there are temporary (possibly local) shallowings before the Late Toarcian. Clearly the pronounced deepening was associated with the Toarcian transgression. As with that at the base of the Jurassic, the pattern of this transgression was of preceding phases of shallowing and stagnation followed by gradual deepening, with the latter bringing in replenished microfaunas.

Whilst we are slowly acquiring some understanding of the interrelationships between environments and foraminiferal assemblages in the Lower Jurassic, little is known of the microfaunas of certain facies types. These are mostly those of shallow-water origin such as calcareous sandstones, limestones and ironstones, rock types which do not yield whole specimens readily.

Although this publication lays emphasis upon short-ranging index species, these are always accompanied and usually outnumbered by a characteristic 'background' fauna. This comprises many forms which are helpful in the determination of broader ages, in those cases where index species may be absent, and they thus deserve some mention here. As stated above, a pronounced microfaunal change occurred in the Early Toarcian, and this effectively distinguishes Toarcian (Upper Lias) assemblages from those of the Hettangian, Sinemurian and Pliensbachian (Lower and Middle Lias). Assemblages from the latter three stages are usually dominated by members of the *Lingulina tenera* and *Lenticulina varians* plexuses, with large numbers of *Lingulina lanceolata* Haeusler, the *Marginulina prima* plexus, the *Frondicularia terquemi* plexus, *Spirillina infima, S. tenuissima, Cyclogyra liasina* (Terquem), and at certain horizons, *Ophthalmidium liasicum, O. northamptonensis, O. macfadyeni* and *Eoguttulina liassica.* In the Toarcian, the *Lenticulina muensteri* plexus became numerically dominant, although the species had been present since the late Triassic, and then continued as a common species into the Cretaceous. The *Lenticulina varians, Lingulina tenera, M. prima* and *F. terquemi* plexuses all continued into the Early Toarcian but in considerably reduced numbers; the latter four plexuses were virtually extinct by the Late Toarcian. *Spirillina, Cyclogyra, Eoguttulina* and *O. northamptonensis,* however, remained fairly common, although the other two species of *Opththalmidium* became extinct in the early part of the Stage. Also numerous in the Toarcian were *Reinholdella pachyderma pachyderma, R. dreheri, R. macfadyeni* and a broader range of arenaceous taxa. *Frondicularia brizaeformis* Bornemann, *Vaginulina listi* Bronemann and *Nodosaria dispar* Franke ranged virtually throughout Early Jurassic time, but these species were less numerically prominent in the Toarcian than at earlier levels.

6.1.5 Index species

The species included here are mainly those which are known at present to have restricted stratigraphical ranges in more than one area. There are many other species with short ranges which are at the moment recorded from single sections only, but these have for the most part been omitted until more information regarding their broader geographical distribution is at hand.

Two species, *Citharina* sp. A and *Saracenella* sp. A, which are only yet known from Toarcian of the Mochras Borehole, are included, however, because this is the only complete British Toarcian section which has to date been studied in detail, the Toarcian as a whole being poorly documented in Britain. Several of the index species have only single records from within Britain, but are also known from other countries (e.g. *Thurammina jurensis*).

For the British Lower Jurassic there is unfortunately little published information on whole successions. The only accounts are those of the Dorset Lower Lias (Barnard 1950(a); though he described only about one-third of the total foraminifera) and the Mochras Borehole (Copestake and Johnson *ibid.*; who described the whole foraminiferal fauna from the complete sequence). For the present work, several additional I.G.S. borehole sequences have been studied, namely those from Stowell Park (Gloucestershire), Burton Row and Hill Lane (Somerset), Wilkesley and Platt Lane (Cheshire), Cocklepits (Humberside) and Trunch (Norfolk). The authors are grateful to the Institute for permission to utilise data from these boreholes. Additional information from the authors' own researches has also been incorporated.

Attempts to establish the palaeoecological significance of individual Liassic species are comparatively rare (see Brouwer 1969, Johnson 1977). In view of the lack of information the authors do not feel that any definitive palaeoecological inferences should be made at this time.

Lithological logs and correlations are not included on Fig. 6.1.1, since the sections studied are largely developed in homgeneous argillaceous or silty sediments, which are fairly uniform throughout Britain. For the Toarcian (Fig. 6.1.2), the lithological log of the major British section, that of the Mochras Borehole, is given, since the sediments of this stage are considerably more variable within Britain.

The classification followed is that of Loeblich and Tappan (1964). References have been numbered to reduce the size of species entries. Where no number is quoted the source is the authors' research.

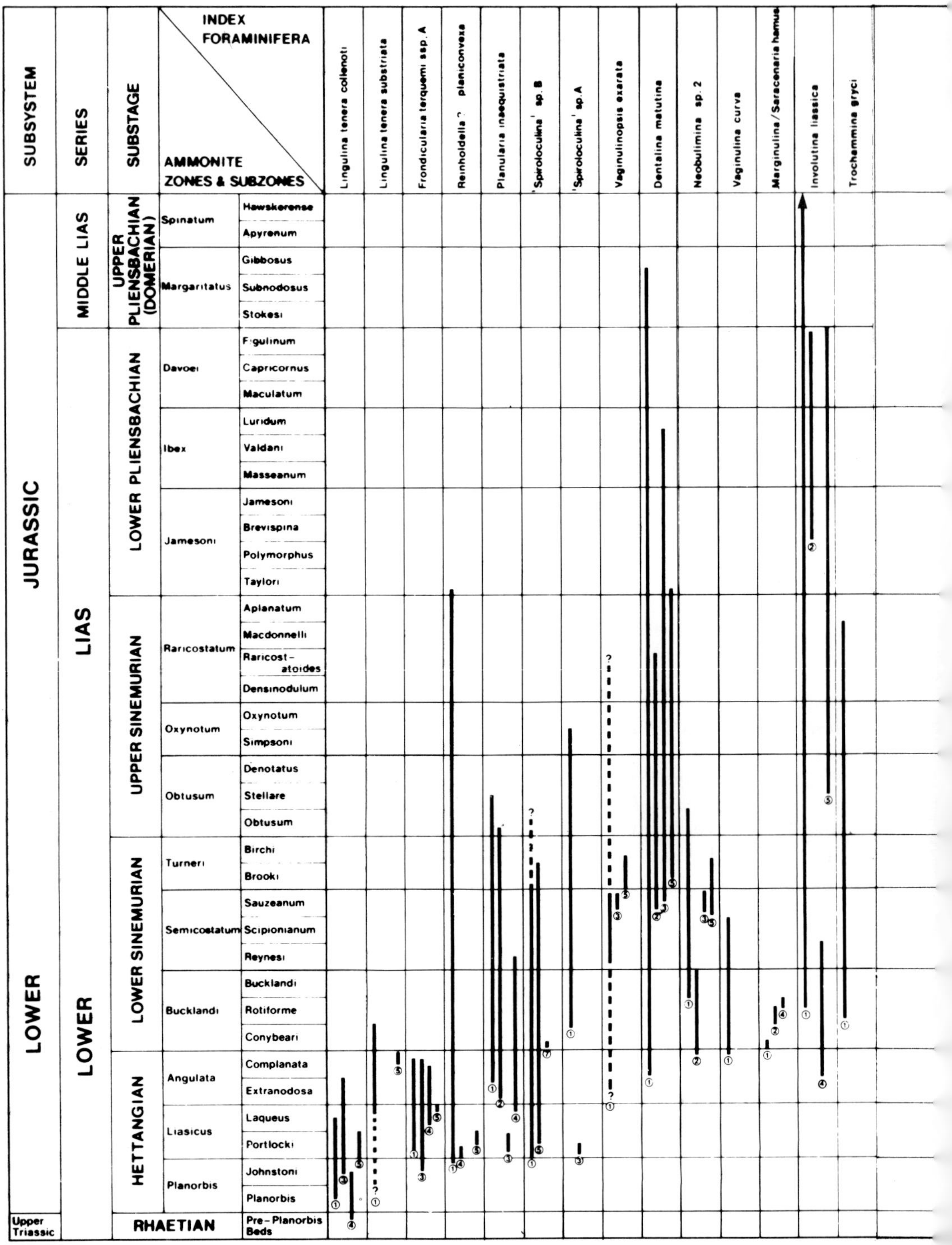

Fig. 6.1.1 – Range chart of Hettangian to Toarcian Foraminifera.

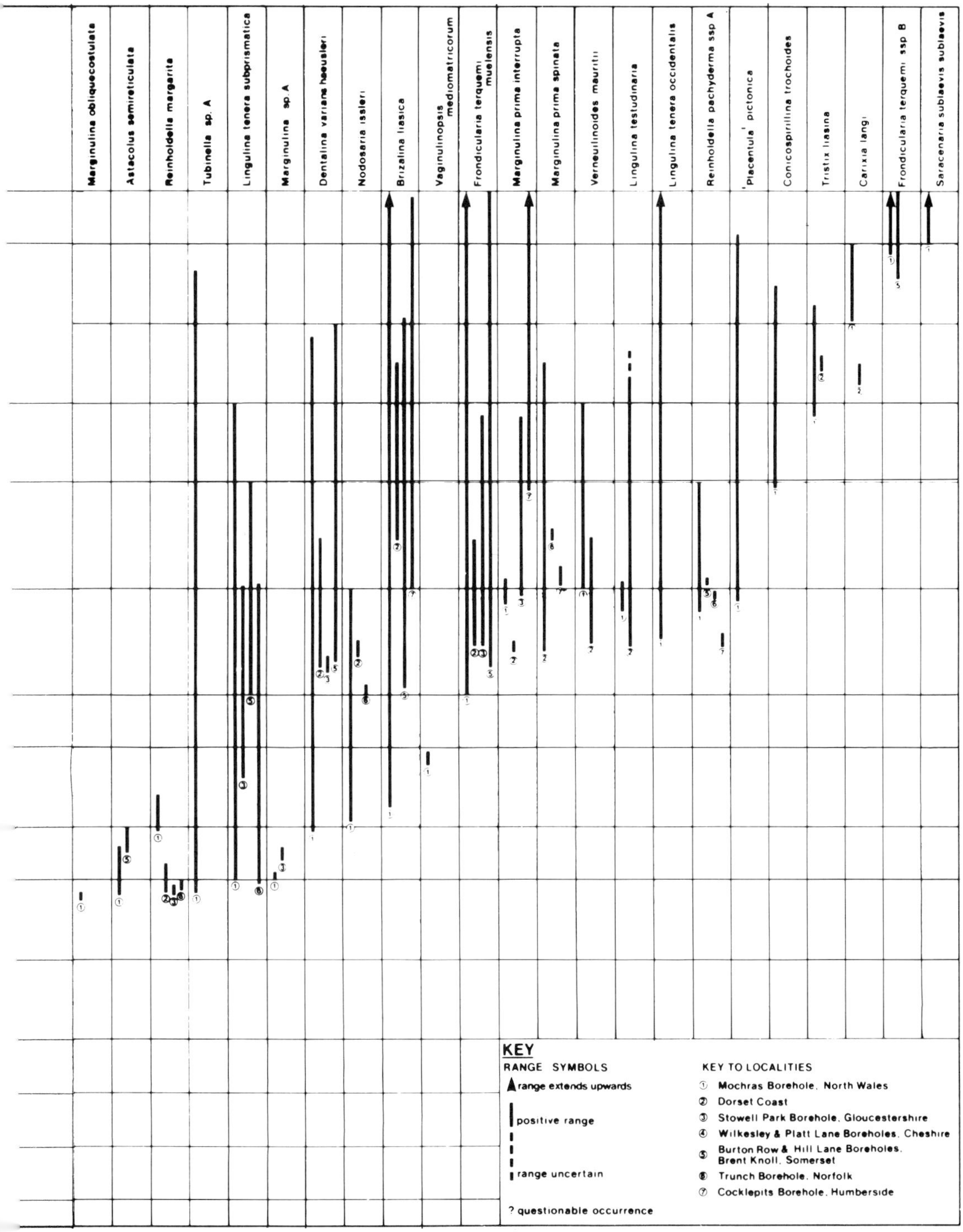

Fig. 6.1.1 – Range chart of Hettangian to Toarcian Foraminifera.

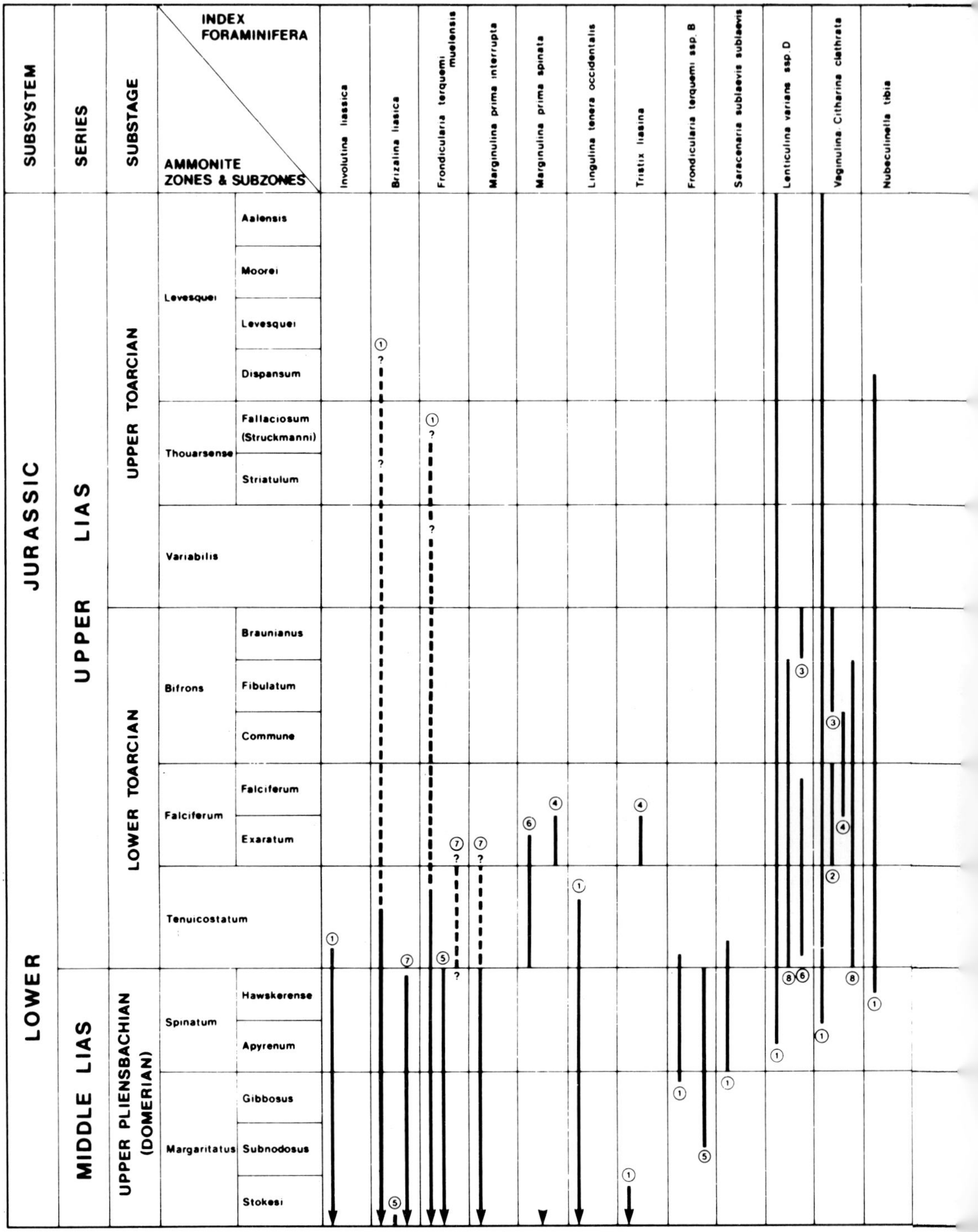

Fig. 6.1.2 – Range chart of Hettangian to Toarcian Foraminifera.

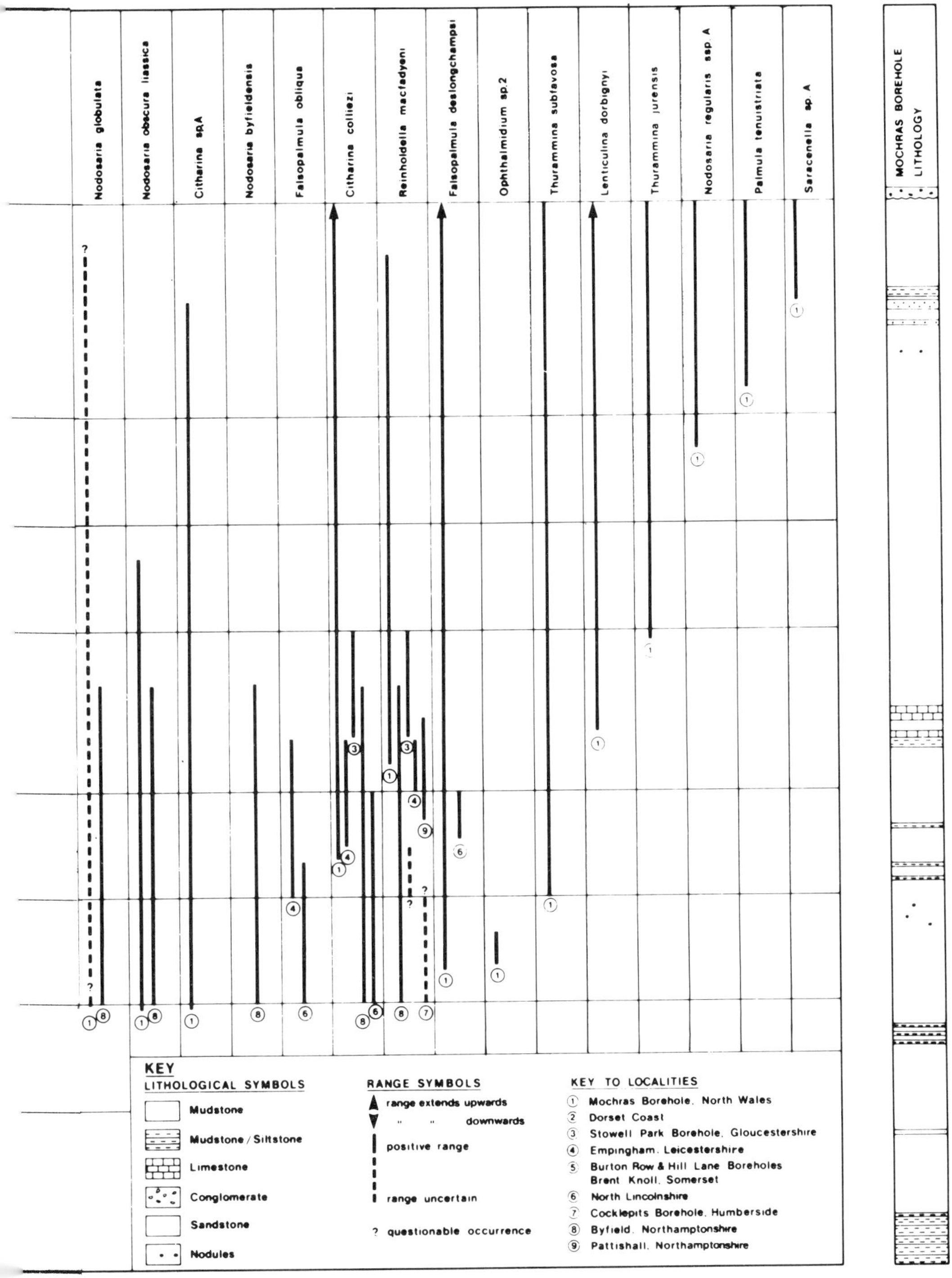

Fig. 6.1.2 – Range chart of Hettangian to Toarcian Foraminifera.

6.1.5.1 SUBORDER TEXTULARIINA

Thurammina jurensis (Franke)

Plate 6.1.1, Fig. 1, × 46, U Toarcian, Mochras Borehole, N. Wales. Fig. 6.1.2. = *Thyrammina jurensis* Franke, 1936. Description: Small, spherical, with large blunt, radiating spikes. Distribution: Toarcian/Aalenian from N.W. Germany/E. Holland, France, S. Germany/Austria/Swiss Jura [32]. Only British record from Mochras Borehole Upper Toarcian [35]. Restricted to latest Toarcian (*levesquei* Zone) in Portugal [42].

Thurammina subfavosa Franke, 1936

Plate 6.1.1, Fig. 3, × 85, U. Toarcian, Mochras Borehole, N. Wales. Fig. 6.1.1. Description: small, with honeycomb surface ornamentation, produced by short spikes connected by narrow ridges. Distribution: recorded from Toarcian of Mochras Borehole [35] and W. Germany [43].

Trochammina gryci Tappan, 1955

Plate 6.1.1, Fig. 4, dorsal view, × 86, L. Sinemurian, Mochras Borehole, N. Wales. Fig. 6.1.1. (syn. *Trochammina nana* form a Brand, 1937). Description: test agglutinated; chambers arranged in a high trochospire of 3 whorls; sutures radial, depressed between slightly inflated chambers.

Remarks: more chambers than other Liassic species, chambers less inflated than in *T. vanningensis* Tappan. Distribution: restricted to the Sinemurian in Britain (Mochras Borehole; [35]), West Germany [20], Sweden [64] and Denmark [5, 7].

Verneuilinoides mauritii (Terquem)

Plate 6.1.1, Fig. 5, ventral view, × 196, L. Pliensbachian, Dorset coast; Fig. 10, side view, same specimen, × 110; Fig. 6, dorsal view, × 140, L. Pliensbachian, Mochras Borehole, N. Wales. Fig. 6.1.1. = *Verneulina mauritii* Terquem, 1886a (syn. *Verneuilina georgiae* Terquem, 1866b). Description: test smooth, finely arenaceous, usually high trochospiral, circular in cross-section and sometimes nail-shaped in side view; sutures flush; chambers indistinct.

Remarks: not synonymous with *V. mauritii sensu* Franke (1936), Bartenstein and Brand (1937) and Usbeck (1952) from the LIas alpha (Hettangian) of West Germany. Test more conical and ventrally broader than *V. favus* (Bartenstein *in* Bartenstein and Brand 1937). Distribution: restricted to the top U. Sinemurian (*raricostatum* Zone) to L. Pliensbachian in Britain (Mochras Borehole [35] and Dorset coast (Jenkins, personal communication) and Paris Basin [25], and to Pliensbachian in Portugal [42].

6.1.5.2 SUBORDER MILIOLINA (Delage and Hérouard, 1896, emend. Brönnimann and Zaninetti, 1971)

Carixia langi Macfadyen, 1941

Plate 6.1.1, Fig. 2, × 200, U. Pliensbachian, Mochras Borehole, N. Wales. Fig.6.1.1. Description: test attached, comprising a reticulate network of straight, narrow, anastomosing tubes (white or brown) between which clacereous cement is sometimes secreted. Distribution: in Britain yet recorded only from the *davoei* Zone of the Dorset coast [62] and the *margaritatus* Zone of the Mochras Borehole [35]; also reported, though not figured from the Hettangian of West Germany [39].

'Spiroloculina' sp. A

Plate 6.1.1. Fig. 9, × 170, Figs 12 and 13, × 128, L. Sinemurian, Mochras Borehole, N. Wales. Fig. 6.1.1. Description: test smooth, symmetrical, compressed; periphery carinate; chambers coiled in two planes.

Remarks: species represents a new genus, soon to be fully described (Copestake and Johnson, in press). Distribution: in Britain the species is known in the Mochras [35] and Stowell Park Boreholes, at Hock Cliff, Gloucestershire (= *Spirophthalmidium acutimargo*, Henderson 1934) and at Stockton, Warwickshire (Brady Coll. and Jones and Parker Coll., B.M. (N.H)). The latter two occurrences are in the *bucklandi* Zone, at which level the species is most common in West Germany (= *Spirophthalmidium concentricum*, Bartenstein and Brand 1937, Plate 2B, Fig. 37 only).

'Spiroloculina' sp. B

Plate 6.1.1, Fig. 8, × 250, Fig. 11, × 223, L. Sinemurian, Mochras Borehole, N. Wales; Fig. 16, × 128; Fig. 17, × 205. Hettangian, same locality. Fig. 6.1.1. Description: shape irregular, surface uneven; sutures depressed; periphery rounded; chamber length variable; coiled in two planes.

Remarks: species belongs to a new genus, soon to be described (Copestake and Johnson, in press). Differs from *'S'* sp. A in its more elongate shape, rounded periphery, smaller size and uneveness of its external surface. Distribution: in Britain, the species occurs in the Mochras [35], Cocklepits and Burton Row Boreholes and is most common in the *liasicus* and *angulata* Zones, ranging between basal *liasicus* and mid-*turneri* Zones. Not yet reported outside Britain.

Nubeculinella tibia (Jones and Parker)

Plate 6.1.1, Fig. 14, × 70, L. Toarcian, Mochras Borehole, N. Wales. Fig. 6.2.2. = *Nubecularia lucifuga* Defrance var. *tibia* Jones and Parker, 1860 *partim*. Description: irregularly coiled initial portion with approximately 6 chambers; uncoiled portion has subtubular chambers which taper distally.

Remarks: coiled portion rarely found. Distribution: questionable Lower Lias records exist, but the British (Mochras Borehole [35] and Lincolnshire [2] and W. German acmes are in the Toarcian.

Ophthalmidium sp. 2 (Ruget and Sigal)

Plate 6.1.1, Fig. 7, × 90, L. Toarcian, Mochras Borehole, N. Wales. Fig. 6.1.2. = *Spirophthalmidium* sp. 2 Ruget and Sigal, 1970 (? syn. *Spiroloculina aspera* Terquem and Berthelin, 1875). Description: test with agglutinated external

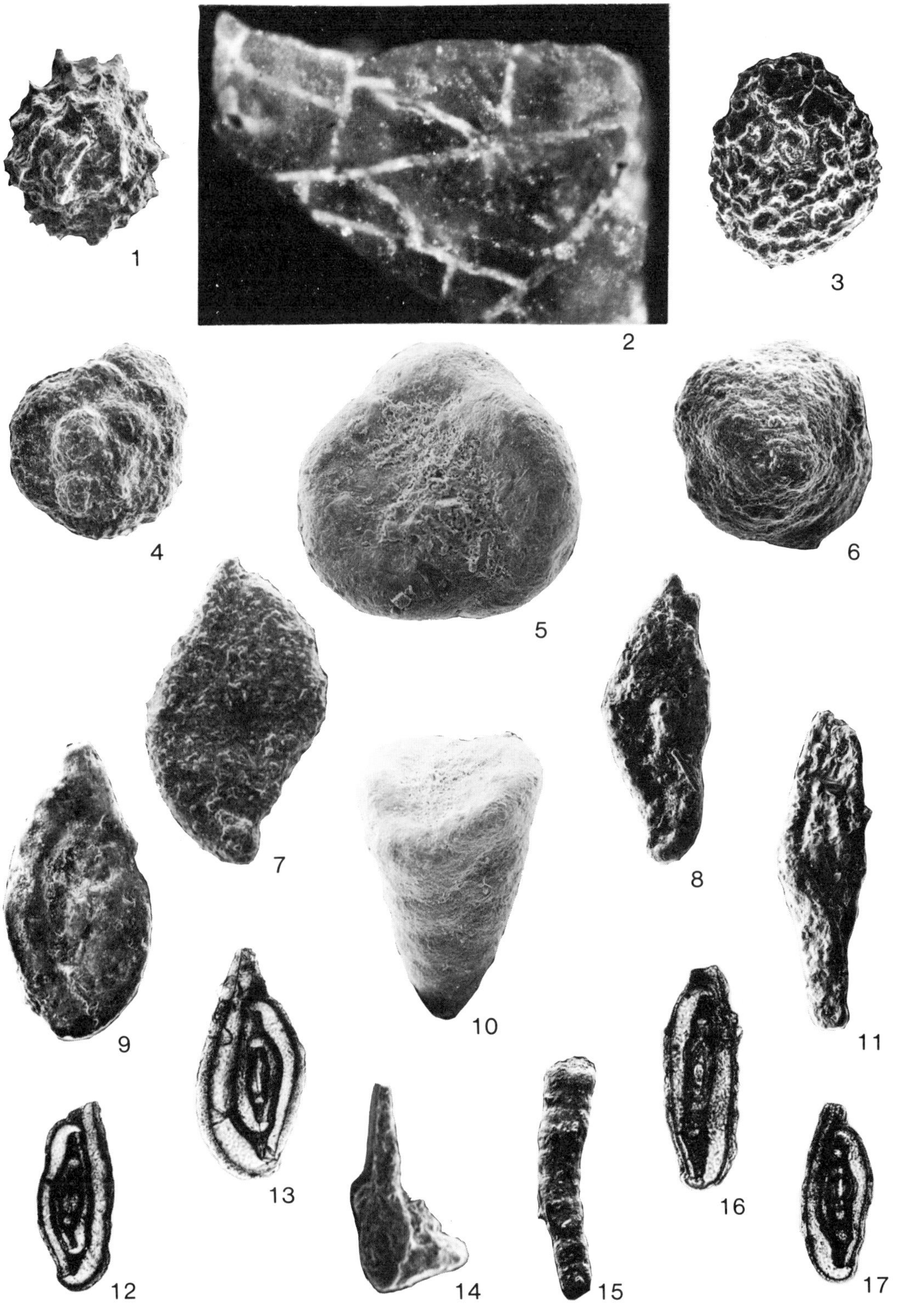

Plate 6.1.1

coating; up to 9 chambers (megalospheric generation), arranged symmetrically; sutures depressed; periphery rounded.
Remarks: distinguished from other species of *Ophthalmidium* by its thick test with an agglutinated outer layer. *Spiroloculina aspera* is said to have a rugose surface ornament (Terquem and Berthelin 1875), but is otherwise similar; it may have been incorrectly described. Distribution: recorded from the Pliensbachian to basal Toarcian (*tenuicostatum* Zone) of Portugal [70,42] (= *Ophthalmidium* sp. 1, Exton 1979), the basal Toarcian (*tenuicostatum* Zone) of the Mochras Borehole [35], and, if *S. aspera* is synonymous, also from the Pliensbachian of offshore England [87], West Germany [20] and France (*margaritatus* Zone [83]).

Tubinella sp. A
Plate 6.1.1, Fig. 15, × 60, U. Sinemurian, Mochras Borehole, N. Wales. Fig. 6.1.1. (syn *Tubinella inornata,* Bartenstein and Brand 1937, Frentzen, 1941, Klingler, 1962). Description: test smooth or transversely constricted, with a small bulbous initial portion followed by a large, flaring tubular portion.
Remarks: the only known Jurassic *Tubinella*. Distribution: only yet recorded in Britain from the Mochras Borehole [35] here being commonest in the late *raricostatum* Zone (*macdonnelli* and *aplanatum* Subzones), early *jamesoni* Zone (*taylori* Subzone) and mid *ibex* Zone. Known from the L. Pliensbachian to Middle Jurassic in West Germany [20, 47] and from the L. Pliensbachian of Poland [56].

6.1.5.3 SUBORDER ROTALINA

Astacolus semireticulata (Fuchs)
Plate 6.1.2, Fig. 1, × 93, L. Sinemurian, Mochras Borehole, N. Wales. = *Lenticulina* (*Lenticulina*) semireticulata Fuchs, 1970. (syn. *Lenticulina* sp. 26 Bang, 1968; *Astacolus semireticulata* Norling, 1972). Description: test compressed; early chambers with reticulate ornament, later chambers smooth with depressed sutures; aperture radiate, with apertural chamberlet.
Remarks: *Lenticulina dorbignyi* (Franke) differs in its more regular reticulate ornament which completely covers the test surface. Distribution: in Britain the species is known from the Inner Hebrides, Radstock (Avon) and in the Mochras [35] and Burton Row Boreholes, restricted to the late *semicostatum* Zone – later *turneri* Zone interval. In Sweden [64], Denmark [62] and Austria [45] it occurs between the late Hettangian and Lower Sinemurian.

Citharina colliezi (Terquem)
Plate 6.1.2, Fig. 2, × 35, U. Toarcian, Mochras Borehole, North Wales. Fig. 6.1.2. = *Marginulina colliezi* Terquem, 1866a (syn. *Marginulina flabelloides* Terquem, 1868). Description: large, compressed, multichambered (10 to 16); many longitudinal striae which are interrupted over the sutures; thick dorsal margin. Distribution: British records from Bristol Channel [59], probable age Toarcian to Lower Bajocian; possibly from undifferentiated 'Middle and Upper Lias' of Somerset (?= *Planularia reticulata,* Brady 1866) and from the Toarcian of the Mochras Borehole [35], Byfield (Northants) [12], Empingham (Leicestershire) [50], north Lincolnshire (Adams, personal communication) and Midlands/Yorkshire [32]. European records: Toarcian to Oxfordian of Sweden [64], W. Germany, France, Holland [32] and Portugal [42].

Citharina sp. A
Plate 6.1.2, Fig. 7, × 70, U. Toarcian, Mochras Borehole, N. Wales. Fig. 6.1.2. Description: small, with convex dorsal margin; 5 to 7 chambers; 13 to 14 continuous, longitudinal ribs; flush sutures; elongate neck.
Remarks: similar to *C. colliezi* (Terquem) but with continuous ribs. Smaller than *Vaginulina*/*C. clausa* (Terquem), with more ribs. Distribution: Mochras Borehole [35]; may be of local stratigraphical importance only. Included here because of the lack of recorded species restricted to the Upper Lias of Britain and its association with other ribbed *Citharina* species.

Dentalina matutina (d'Orbigny)
Plate 6.1.2, Fig. 9, × 20.4, U. Sinemurian, Mochras Borehole, N. Wales. Fig. 6.1.1. (syn. *D. primaeva* d'Orbigny, 1849). Description: robust, elongate, ornamented with coarse, oblique costae; aperture produced, radiate. Distribution: ranging as high as the *margaritatus* Zone in Britain and up to the top *spinatum* Zone in mainland Europe, though most common in U. Sinemurian. Widespread throughout north-west Europe.

Dentalina varians haeusleri (Schick)
Plate 6.1.2, Fig. 5, × 50, L. Pliensbachian, Dorset coast. Fig. 6.1.1. = *Nodosaria* (*Dentalina*) *hausleri* Schick, 1903 (syn. *D. varians* Terquem, 1866a *partim*). Description: ornamented with longitudinal ribs paralleling the test margin; sutures deeply constricted; aperture marginal, produced, radiate.
Remarks: distinguished from end chambers of *D. matutina* in having fewer ribs which are parallel, rather than oblique, to the test margin. Distribution: although long-ranging (L. Sinemurian–L. Toarcian), the subspecies is most often recorded in the U. Sinemurian in north-west Europe [32]. In Britain it is commonest in the *raricostatum* Zone (at which level it first appears at several localities) but has been recorded throughout the U. Sinemurian and L. Pliensbachian.

Falsopalmula deslongchampsi (Terquem)
Plate 6.1.2, Fig. 3, × 75, U. Toarcian, Dorset coast; Fig. 4, × 40, U. Toarcian, Mochras Borehole, N. Wales. Fig. 6.1.2. = *Flabellina deslongchampsi* Terquem, 1864 (syn. *Flabellina jurensis* Franke, 1936). Description: small lenticuline

Plate 6.1.2

portion followed by rapidly flaring, embracing flabelline portion; distinct, depressed sutures. Distribution: widespread in Europe, ranging from L. Pliensbachian to Kimmeridgian, but with an acme in the Toarcian [32, 64]. British records from Mochras Borehole [35], Bristol Channel [59], possibly from the undifferentiated 'Middle and Upper Lias' of Somerset (?= *Flabellina rugosa*, Brady 1866, Plate 3, Fig. 46), Midlands/Yorkshire [32], Lincolnshire (Adams, personal communication) and Dorset coast.

Falsopalmula obliqua (Terquem) is figured and described in the Jurassic, Part II (see p. 118).

Frondicularia terquemi subsp. A Barnard
Plate 6.1.2, Fig. 10, × 50, Fig. 11, × 120, Hettangian, Mochras Borehole, N. Wales. Fig. 6.1.1. = *F. sulcata* form A Barnard, 1957. Description: small subspecies with 4 to 5 weak longitudinal ribs; chambers 6 to 8 in number.
Remarks: distinguished from most members of the *F. terquemi* plexus by its small size and rib number. Close to small Hettangian forms of *F. terquemi bicostata* d'Orbigny with which it intergrades, but has more ribs. Distribution: British range from *planorbis* Zone (*johnstoni* Subzone) to *angulata* Zone (*complanata* Subzone); known from the Bristol Channel [8], Glamorgan (= *Ichthyolaria sulcata quadricosta* of Banner *et al.* 1971, Plate 8, Fig. A) and the Stowell Park [15], Mochras [35], Wilkesley and Burton Row Boreholes. Also recorded from *planorbis* Zone of West Germany (= *F. sulcata*, Bartenstein and Brand 1937, Plate 1A, Fig. 12) and Austria (?) (= *F. bicostata,* Brouwer 1969, Plate V, Fig. 10) and top *angulata* Zone of South Germany.

Frondicularia terquemi subsp. B
Plate 6.1.2, Fig. 6, × 40, Fig. 8, × 38, L. Toarcian, Mochras Borehole, N. Wales. Figs. 6.1.1, 6.1.2. Description: robust subspecies with 4 to 6 longitudinal costae confined to central region of test and bordered by lateral smooth areas; margin keeled.
Remarks: differs from other members of the *F. terquemi* plexus in its robustness and degeneration of ribs laterally away from the central region. Distribution: known in Britain from Tuffley (Gloucester), and in the Mochras [35] and Hill Lane Boreholes, with total range from *margaritatus* Zone (*subnodosus* Subzone) to base of *tenuicostatum* Zone. On the continent, also recovered from W. Germany (= *F. baueri,* Franke 1936, Plate 7, Fig. 10, Wicher 1938, Plate 25, Fig. 6), Austria (?) (= *F. bicostata,* Brouwer 1969, Plate 5, Figs. 13–15) and France (= *F. bicostata,* Espitalié and Sigal 1960, Plate 3, Fig. 9; *F.* sp. 3 Seronié-Vivien *et al.* 1961) in the *margaritatus* and *spinatum* Zones.

Frondicularia terquemi muelensis Ruget and Sigal
Plate 6.1.2, Fig. 12, × 98, Fig. 13, × 68, U. Sinemurian, Mochras Borehole, N. Wales. Fig. 6.1.1, 6.1.2. = *Frondicularia muelensis* Ruget and Sigal, 1970. (syn. *F. sulcata* form K Barnard, 1957). Description: test compressed, non-sulcate, ornamented with numerous fine, parallel striae. Distribution: widespread in Europe during the Pliensbachian, occurring in Britain, West Germany (= *F. bicostata sulcata,* Dreyer 1967, Plate 9, Fig. 3), France (= *F. bicostata,* Brouwer 1969, Plate 5, Figs. 5, 6), Portugal [70, 42] and South Germany (authors' obs.). First appears at several British localities in the *raricostatum* Zone; in the Cocklepits and Mochras Boreholes [35] upper limit of its range is top *tenuicostatum* Zone, (though questionably occurring at the latter locality as high as *thouarsense* Zone (*struckmanni* Subzone)).

Lenticulina dorbignyi (Roemer)
Plate 6.1.2, Fig. 14, × 70, U. Toarcian, Mochras Borehole, N. Wales. Fig. 6.1.2. = *Peneroplis d'Orbignii* Roemer, 1839. Description: loosely coiled, 6 to 8 chambers in final whorl; reticulate ornament.
Remarks: both astacoline (Franke 1936) and flabelline forms (Brouwer 1969, Copestake and Johnson, in press) reported. Ornament regular to irregular. Distribution: British records from Mochras Borehole [35], Bristol Channel [59], probable age Toarcian to Bajocian (Early) and possibly from the undifferentiated 'Middle and Upper Lias' of Somerset (?= *Cristellaria costata*, Brady 1866) [29]. Apart from an isolated Upper Pliensbachian record from N.W. Germany/E. Holland [32], European occurrences from W. Germany [43], Holland, France [32] and Switzerland [90] fall within a Toarcian to Lower Bajocian range.

Lenticulina varians subsp. D Barnard
Plate 6.1.2, Fig. 15, × 77, Fig. 16, × 105, U. Toarcian, Mochras Borehole, N. Wales. Fig. 6.1.2 = *Lenticulina varians* Bornemann form D, Barnard, 1950b. Description: test palmate, comprising small, loosely coiled lenticuline portion followed by flabelline portion of 1 to 5 chambers; sutures raised.
Remarks: differs from *Eoflabellina chicheryi* Payard in having raised sutures and no keel. Distribution: Toarcian of Mochras Borehole [35], Byfield (Northants.) [12] and possibly France (?= *Falsopalmula chicheryi,* Magné *et al.* 1961, Champeau 1961; *F.* cf. *centro-gyrata,* Magné *et al. ibid.*). May range into Lower Aalenian [32].

Lingulina tenera collenoti (Terquem)
Plate 6.1.3, Fig. 7, × 45, Hettangian, Galboly, N. Ireland. Fig. 6.1.1. = *Marginulina collenoti* Terquem, 1866a. (syn. *Lingulina striata* Blake, 1876; *L. tenera* form A, Barnard, 1957). Description' test elongated, ornamented with fine, often discontinuous striae; periphery usually rounded.
Remarks: the largest and most elongated form in the *L. tenera* plexus. Distribution: index for the basal Lias throughout Britain, e.g. South Wales [8], Yorkshire [26], and Stowell Park [15], Wilkesley (late Rhaetian–Hettangian), Burton Row and Mochras Boreholes [35], Watchet (Somerset), Mull (Scotland) and Northern Ireland. Total British range from uppermost Triassic (?Penarth Group and pre-*planorbis* Beds) to *angulata* Zone (*extranodosa* Subzone). Also occurs in the *planorbis* Zone of Denmark [65] and France [80, 23].

Plate 6.1.3

Lingulina tenera occidentalis (Berthelin)
Plate 6.1.3, Fig. 4, × 105, L. Pliensbachian, Mochras Borehole, N. Wales. Figs 6.1.1, 6.1.2. = *Frondicularia occidentalis* Berthelin, 1879. Description: test with transverse, raised sutural ribs between 2 longitudinal ribs; 7 to 9 chambers. Distribution: in Britain only yet recorded from Mochras Borehole [35] and possibly Radstock (Avon) (*raricostatoides-macdonnelli* Subzone). Also present between the *jamesoni* and *margaritatus* Zones in the Paris Basin [25].

Lingulina tenera subprismatica (Franke)
Plate 6.1.3, Fig. 5, × 81, L. Sinemurian, Mochras Borehole, N. Wales. Fig. 6, apertural view, same specimen × 269. Fig. 6.1.1. = *Nodosaria subprismatica* Franke, 1936. (syn. *Frondicularia tenera prismatica* Brand, 1937). Description: test narrow, elongate, parallel-sided, with hexagonal cross-section, basal spine and oval aperture. Remarks: test form (tending to become *Nodosaria*-like), aperture shape and basal spine separate the subspecies from other members of *L. tenera* plexus. Distribution: first appears and is commonest in the U. Sinemurian, ranging into the Early Pliensbachian. Widespread in Europe, occurring in Britain [35, 15], Denmark [65], Sweden [64], W. Germany [43], E. Germany [67], Portugal [42, 70], Poland [56], Italy and France [32].

Lingulina tenera substriata (Nørvang)
Plate 6.1.3, Fig. 8, × 94, Hettangian, Mochras Borehole, N. Wales. Fig. 6.1.1. = *Geinitzinita tenera substriata* Nørvang, 1957. Description: test keeled, indistinctly sulcate; ornamented with irregular, often discontinuous, non-parallel ribs, of which two are usually prominent.
Remarks: distinguished by its keel, sulcus and less elongated form from *L. tenera collenoti* and by its irregular ribbing from *L. tenera tenuistriata* (Nørvang). Distribution: observed in Britain in the Mochras [35] and Burton Row Boreholes, its known range in this country being from *planorbis* Zone (*planorbis* Subzone) to *bucklandi* Zone (topmost *conybeari* Subzone). In Denmark [65, 6] and Sweden [64] it ranges throughout the Hettangian (*planorbis* to top *angulata* Zones).

Lingulina testudinaria Franke, 1936
Plate 6.1.3, Fig. 1, × 107, Fig. 2, × 163, Fig. 3, × 128, U. Sinemurian, Dorset coast. Fig. 6.1.1. Description: test sharply keeled; ornamented with two longitudinal, central ribs (bordering a median sulcus) and transverse sutural ribs; cross-section elliptical/subrhomboidal; aperture slit-shaped and sometimes centrally constricted.
Remarks: ribbing pattern, combined with test shape and keeled margin, is unique. Distribution: in Britain the species is known to occur at Bracebridge (Lincoln), in the Mochras Borehole [35] and in Dorset (W. A. Macfadyen Coll.) with a total range of *raricostatum* Zone (topmost *raricostatoides* Subzone) to *tenuicostatum* Zone. The species occurs throughout the L. Pliensbachian in W. Germany [20] but in Austria, it is reported from about the Lower/Upper Sinemurian junction [45].

Marginulina/Saracenaria hamus (Terquem)
Plate 6.1.3, Fig. 14, × 72.5, L. Sinemurian, Mochras Borehole, N. Wales. Fig. 6.1.1. = *Marginulina hamus* Terquem, 1866b. Description: initial portion of test triangular in cross-section, with flush sutures and oblique ribs; later portion circular in section, smooth, and with depressed sutures. Distribution: non-triangular forms (mostly megalospheric tests) are longer-ranging, but triangular (saracenarian) forms, to which the description refers, are restricted to the *bucklandi* Zone (*conybeari* and *rotiforme* Subzones) in Britain (Dorset coast [11] and Mochras [35] and Wilkesley Boreholes). Observed in the Hettangian and L. Sinemurian [32] and in the *angulata* Zone (Lias alpha 2) [39] of W. Germany. Also occurs in the *angulata* Zone of Lorraine, France (= *Saracenaria* sp. 1 Bizon, 1961).

Marginulina obliquecostulata Hagenmeyer, 1959
Plate 6.1.3, Fig. 15, × 96, L. Sinemurian, Mochras Borehole, N. Wales. Fig. 6.1.1. Description: test curved, ornamented with numerous fine, oblique, occasionally bifurcating striations which disappear on final chamber.
Remarks: possibly synonymous with *Dentalina langi* Barnard, 1950a (known from top *angulata* Zone of Redcar and Dorset coast). Fine ribbing is distinctive. Distribution: a single record only from Britain, in the *semicostatum* Zone (*sauzeanum* Subzone) of the Mochras Borehole [35]. In W. Germany the species ranges between the *angulata* and *bucklandi* Zones [4].

Marginulina prima interrupta (Terquem)
Plate 6.1.3, Fig. 11, × 87, L. Pliensbachian, Mochras Borehole, N. Wales; Fig. 12, × 48, U. Sinemurian, same locality. Figs. 6.1.1, 6.1.2. = *Marginulina interrupta* Terquem, 1866a. Description: 10 to 12 ribs, interrupted across depressed sutures; chambers arranged in a straight series.
Remarks: interrupted ribs and nodosarian chamber arrangement are distinctive. Distribution: known British range is from *raricostatoides* Zone (late *raricostatoides* Subzone) to early *tenuicostatum* Zone. A notable member of early *tenuicostatum* Zone assemblages in eastern England (e.g. Cocklepits Borehole, Scunthorpe and Lincoln). Widespread in northern hemisphere, occurring in W. Germany [20], Sweden [64], Denmark [65], Portugal [70], France [23], Poland [56] and Alaska [73], with an identical stratigraphic range.

Marginulina prima spinata (Terquem)
Plate 6.1.3, Figs. 9, 10, × 100, U. Sinemurian, Dorset coast. Figs. 6.1.1, 6.1.2. = *M. spinata* Terquem, 1858. Description: 6 or 7 ribs, notched and typically projected as spines over sutures, but not interrupted.
Remarks: notched and spinose ribs are diagnostic. Also usually has fewer ribs and chambers than *M prima interrupta.* Distribution: main British range from *raricostatum* Zone (late *raricostatoides* Subzone) to *davoei* Zone (*capricornus* Subzone) as in Dorset [62], (Jenkins, personal communication), but also recorded from the Early Toarcian in north Lincolnshire [1] and Empingham (Leicestershire) [50] (*falciferum* Zone, top *exaratum* Subzone). Continental range mainly late U. Sinemurian to top L. Pliensbachian, occurring in Sweden [64], Denmark [5, 65], W. Germany [20], E. Germany [67] and France [76]. In South Germany [85] and Poland [46] the subspecies is said to occur in the Hettangian/L. Sinemurian.

Marginulina sp. A
Plate 6.1.3, Fig. 13, × 72, L. Sinemurian, Mochras Borehole, N. Wales. Fig. 6.1.1. Description: test smooth, comprising 5 to 7 chambers widening markedly with growth, arranged rectilinearly; sutures flush; aperture at dorsal margin, produced, radiate.
Remarks: *M. reversa* Blake differs in its dentaline test curvature. *Saracenella* sp. A has a triangular cross-section. Distribution: rare, restricted to the early to mid-*turneri* Zone interval in the Mochras [35] and Stowell Park Boreholes and at Robin Hood's Bay (Yorkshire). Not yet recorded outside Britain.

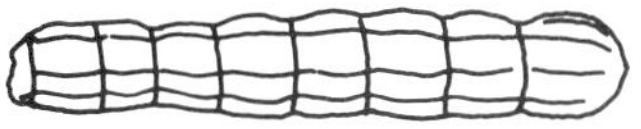

Fig. 6.1.3 – *Nodosaria byfieldensis* Barnard (drawing of holotype based on Barnard 1950, pl. 3, fig. 2) × 22.

Nodosaria byfieldensis Barnard, 1950b
Fig.s 6.1.2, 6.1.3. Description: large, elongate, parallel-sided, with 10 to 12 drum-shaped chambers; flush sutures; 8 coarse, continuous, longitudinal ribs. Distribution: recorded between Upper Sinemurian and mid-Toarcian in S. Germany/ Austria/Swiss Jura and France [32]. In Britain recorded only from the Lower and mid-Toarcian of Byfield (Northants.) [13] and Yorkshire/Midlands [32], and thus may be of local stratigraphical value.

Nodosaria globulata Barnard, 1950b
Plate 6.1.3, Fig. 16, × 60, U. Toarcian, Mochras Borehole, N. Wales. Fig. 6.1.2. (syn. *Nodosaria pulchra,* Wernli and Septfontaine, 1971). Description: test robust; 3 to 6 slightly flaring chambers; discontinuous, offset, longitudinal ribs which do not cross the sutures or the raised apertural face.
Remarks: *N. pulchra* Franke lacks both offset ribs and raised aperture. Distribution: British records from Byfield (Northants.) [12] and Mochras Borehole [35]. European records from the Toarcian/Aalenian of France and Switzerland [90], questionably from W. Germany (= *N. prima*, Paalzow 1917), and from the Toarcian of Portugal [42].

Nodosaria issleri Franke, 1936
Plate 6.1.4, Fig. 1, × 70, Fig. 2, × 86, U. Sinemurian, Mochras Borehole, N. Wales. Fig. 6.1.1. (syn. *N. aequalis*, Issler

1908). Description: 6 to 8 sharp ribs which pass uninterrupted over flush sutures, disappearing at middle of final chamber; test occasionally slightly curved; aperture produced on short, stout neck.
Remarks: combination of flush sutures, semi-smooth final chamber and produced aperture distinguish the species from the related *N. radiata* (Terquem), *N. mitis* (Terquem and Berthelin) and *Pseudonodosaria multicostata* (Bornemann). Distribution: ranges throughout U. Sinemurian in Mochras Borehole [35], and also present in this substage in the Trunch Borehole, and Dorset coast. Restricted to this substage also in W. Germany [20], E. Germany [67] and Denmark [65].

Nodosaria regularis subsp. A
Plate 6.1.4, Fig. 3, × 17.5, U. Toarcian, Mochras Borehole, N. Wales. Fig. 6.1.2. (syn. *N. regularis,* Franke 1936, Magné *et al.* 1961, Wernli and Septfontaine 1971, Exton 1979). Description: large subspecies with regularly enlarging, subspherical chambers connected by short, constricted necks.
Remarks: stratigraphically distinct and 3 to 4 times larger than *N. regularis regularis* Terquem. Distribution: British records from Mochras Borehole [35] and possibly from the undifferentiated 'Middle and Upper Lias' of Somerset (?= *N. radicula,* Brady 1866). European Range: Toarcian to Bathonian, of France [63], S. France/Switzerland [90] and W. Germany [30], but restricted to *levesquei* Zone in Portugal [42]. Its acme appears to be Upper Toarcian-Aalenian.

Palmula tenuistriata (Franke)
Plate 6.1.4, Fig. 4, × 70, U. Toarcian, Mochras Borehole, N. Wales. Fig. 6.1.2. = *Flabellina tenuistriata* Franke, 1936. Description: initial small coiled portion with 2 to 3 comma-shaped chambers followed by asymmetrical, flabelline portion with 4 to 5 chambers; raised sutures; subreticulate ornament. Distribution: excellent marker for Upper Toarcian and Aalenian. Recorded in Britain (Mochras Borehole) [35], W. Germany [20], France [89], Switzerland [90] and Portugal [42].

Planularia inaequistriata (Terquem)
Plate 6.1.4, Fig. 5, × 68, L. Sinemurian, Mochras Borehole, N. Wales. Fig. 6.1.1. = *Marginulina inaequistriata* Terquem, 1863 (syn. *P. choiseulensis* Ruget and Sigal, 1967). Description: test robust, ornamented with irregular, often bifurcating ribs; periphery keeled; sutures flush. Distribution: British range from *liasicus* Zone (*portlocki* Subzone) to *obtusum* Zone (*stellare* Subzone). Widespread in this country and Europe (Denmark, Sweden, W. Germany, Holland, Austria, France [32]), and also occurs in Alaska (= *Vaginulina curva*, Tappan 1955, Plate 22, Figs. 12, 16–19).

Saracenaria sublaevis sublaevis (Franke)
Plate 6.1.4, Fig. 6, × 35, U. Pliensbachian, Mochras Borehole, N. Wales. Fig.s. 6.1.1, 6.1.2 = *Cristellaria* (*Saracenaria*) *sublaevis* Franke, 1936. Description: test comprises a coil of 3 to 5 chambers followed by an uncoiled portion of 2 to 6 chambers; margins subangular; ventral surface slightly convex. Distribution: despite its reported Hettangian to Lower Toarcian range, its acme during the Upper Pliensbachian has been used zonally, e.g. in W. Germany [20] and Sweden [64]. Known in Britain from Mochras Borehole [35], Bracebridge (Lincoln) and Tilton (Leics.) between *spinatum* and topmost *tenuicostatum* Zones. Also recorded from Denmark [65], Holland, France [32], East Germany [66], Poland [56], Italy [9], Spain [32] and Portugal [42].

Saracenella sp. A
Plate 6.1.4, Fig. 7, × 55, U. Toarcian, Mochras Borehole, N. Wales. Fig. 6.1.2. Description: (microspheric) test smooth, elongate, sutures flush; 7 to 10 'pagoda-like' chambers; aperture produced; (megalospheric) smaller, more robust with fewer chambers and a shorter, more curved initial portion. Distribution: only yet recorded from U. Toarcian of Mochras Borehole [35] and may have local stratigraphical importance. Included here because of the lack of useful data on the uppermost British Toarcian.

Tristix liasina (Berthelin)
Plate 6.1.4, Fig. 8, × 106, L. Pliensbachian, Mochras Borehole, N. Wales. Figs. 6.1.1, 6.1.2. = *Rhabdogonium liasinum* Berthelin, 1879. Description: test triangular in cross-section with rounded or angular margins; sutures depressed or flush; test sometimes coarsely perforated; aperture produced, radiate.
Remarks: this is the only known Liassic *Tristix* species. It is common in the Swedish Pliensbachian (Norling 1972) but the Swedish material differs from that so far recorded from Britain in having mainly angular or keeled, but not rounded margins, flush rather than depressed sutures, and fewer chambers. These differences suggest there may be geographic differentiation within *T. liasina* in Europe. Distribution: British range is from *ibex* Zone (*luridum* Subzone) to *falciferum* Zone (*exaratum* Subzone), with occurrences in the Mochras Borehole [35], Dorset [62] and Empingham (Leicestershire) [50]. Swedish range is from *jamesoni* Zone to Toarcian [64]. Restricted to the Lower Pliensbachian in Poland [56], the *margaritatus* Zone in the Paris Basin [22] and the *davoei* to early *margaritatus* Zone interval in Portugal [42].

Vaginulina curva Franke, 1936
Plate 6.1.4, Fig. 9, × 92, Fig. 10, × 264, Hettangian, Mochras Borehole, N. Wales. Fig. 6.1.1 (syn. *Marginulina radiata* Terquem, 1866, Plate 21, Fig. 17 only *non* Terquem 1863, Plate 9, Figs. 10a, b). Description: test ornamented with longitudinal or oblique striae, compressed or rounded in section with either keeled or rounded margin respectively; sutures mostly flush; microspheric tests with loose initial coil.

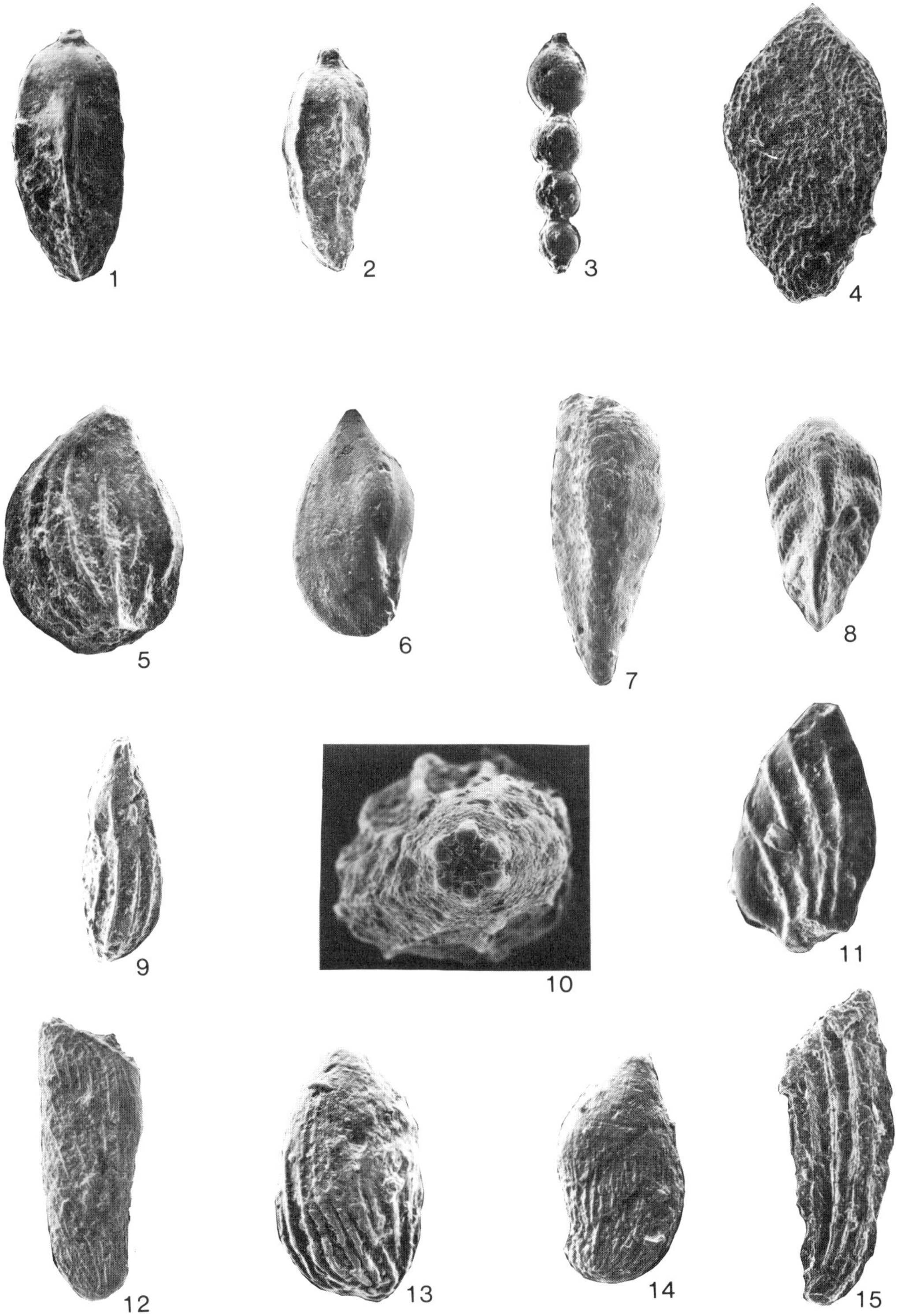

Plate 6.1.4

Remarks: *M./S. hamus* differs in being triangular and non-compressed, with a final smooth chamber. *Astacolus speciosa* (Terquem) has a larger coil. Distribution: total British range from *planorbis* Zone to *semicostatum* Zone (*scipionianum* Subzone), with recorded occurrences in Mochras Borehole [35], Yorkshire [26] and Bristol Channel (Evans, personal communication). Continental range comparable; reported from the West German *planorbis* Zone [43], Lias alpha (*bucklandi/semicostatum* Zones) of Denmark (= *Marginulina radiata,* Nørvang, 1957) and *semicostatum* Zone of France [81].

Vaginulina/Citharina clathrata (Terquem)
Plate 6.1.4, Fig. 11, × 70, L. Toarcian, Mochras Borehole, N. Wales; Fig. 15, × 70, U. Toarcian, same locality. Fig. 6.1.2. = *Marginulina longumari* var. *clathrata* Terquem, 1864 (syn. *Marginulina proxima* Terquem, 1868; *Vaginulina infraopalina* Brand, 1949). Description: test robust, keeled, triangular to curved-triangular; 3 to 6 continuous costae paralleling dorsal margin. Distribution: a characteristic Toarcian/Aalenian species (= *V. proxima,* Barnard 1948, *C. infraopalina,* Brand and Fahrion 1962). Norling (1972) (= *C. clathrata* and *C. infraopalina*) established a Zone with *C. clathrata* in the Swedish Toarcian/Aalenian. Although ranging from uppermost Pliensbachian to Upper Jurassic, its acme throughout Europe is in the Toarcian/Aalenian. British records from Mochras Borehole [35], Bristol Channel (Evans, personal communication), possibly from the undifferentiated 'Middle and Upper Lias' of Somerset [29] (?= *V. striata,* Brady 1866), Byfield (Northants.) [12], Empingham (Leicestershire) [50] and Midlands/Yorkshire [32].

Vaginulinopsis exarata (Terquem)
Plate 6.1.4, Fig. 12, × 57, Fig. 13, × 92, Fig. 14, × 79, L. Sinemurian, Mochras Borehole, N. Wales. Fig. 6.1.1. = *Marginulina exarata* Terquem, 1866b (syn. *V. subporrecta* Bizon, 1960). Description: test elongate, ornamented with numerous fine, oblique striations; long, rectilinear, parallel-sided uncoiled portion with flush early sutures becoming depressed between adult chambers.
Remarks: finer ornamentation and more elongate test than *Astacolus speciosa* (Terquem). Differs from *Dentalina langi* Barnard in having an initial coil. Distribution: common in the British *semicostatum* Zone (Mochras [35] and Stowell Park Boreholes), and thought to occur between the Hettangian and U. Pliensbachian in Europe. A marker for the L. Sinemurian in Sweden [64], North Germany [55], and the Paris Basin [25], but occurs in the U. Sinemurian of Denmark [65] and the Pliensbachian of South Germany [55].

Vaginulinopsis mediomatricorum (Ruget and Sigal)
Plate 6.1.5, Fig. 1, × 52, U. Sinemurian, Mochras Borehole, N. Wales. Fig. 6.1.1. = *Marginulinopsis mediomatricorum* Ruget and Sigal, 1967. Description: sutures strongly limbate, giving periphery a notched outline; test has an elongated uncoiled portion. Distribution: yet recorded only from the *obtusum* Zone in Britain (Mochras Borehole) [35] and France [69].

'Placentula' pictonica (Berthelin)
Plate 6.1.5, Fig. 2, dorsal view, × 185, L. Pliensbachian, Mochras Borehole, N. Wales; Fig. 3, dorsal view × 150, U. Pliensbachian, same locality; Fig. 6, edge view, × 215. U. Sinemurian, same locality. Fig. 6.1.1. = *Placentula pictonica* Berthelin, 1879. Description: test conical, patelline, with 4 to 6 chambers per whorl; small, apical depression occurs on the dorsal side; numerous curved growth lines radiate from the umbilicus on the ventral side.
Remarks: species represents a new genus soon to be fully described (Copestake and Johnson, in press). Distribution: in Britain, yet observed only in the Mochras Borehole (late *raricostatum* Zone, *aplantum* Subzone to basal *spinatum* Zone, *apyrenum* Subzone) [35] where it is common in the *margaritatus* Zone. Previous records from the *margaritatus* Zone of La Vendée, Paris Basin [22] and Portugal (J. Exton, personal communication).

Conicospirillina trochoides (Berthelin)
Plate 6.1.5., Figs. 4, 5, edge view, × 200, L. Pliensbachian, Mochras Borehole, N. Wales. Fig. 6.1.1. = *Spirillina trochoides* Berthelin, 1879. Description: test dome-shaped, with a small dorsal apical depression; whorls 5 to 8 in number, separated by a depressed spiral suture, with a single row or pores along the outside edge of each whorl. Distribution: the only known Liassic species of the genus. Only two known positive records, viz. from the Mochras Borehole (*jamesoni* Subzone to *margaritatus* Zone, *gibbosus/subnodosus* Subzone) [35] and from *margaritatus* Zone of La Vendée, Paris Basin [22]. Possibly conspecific with *C. pictonica sensu* Brouwer (1969) from the Toarcian–Aalenian of France, but this form is poorly figured and not described.

Involutina liassica (Jones)
Plate 6.1.5, Fig. 9, × 47, L. Sinemurian, Mochras Borehole, N. Wales. Figs. 6.1.1, 6.1.2. = *Nummulites liassicus* Jones *in* Brodie, 1853. (syn. *I. turgida* Kristan-Tollmann, 1957). Description: numerous irregularly-shaped pillars fill umbilicus on both sides of test, obscuring all but final whorl; test planispiral; surface rugose or smooth. Distribution: the only known Liassic *Involutina* species in N.W. Europe. Widespread in Britain and Europe, ranging from Rhaetian to Toarcian but most common between Hettangian and L. Pliensbachian [32]. In British areas north of the Mendips the species is most abundant in the *angulata-bucklandi* Zones, but in Dorset it is commonest in the *davoei* Zone [11, 62]. Total British range from *angulata* Zone to *tenuicostatum* Zone.

Reinholdella macfadyeni (Ten Dam) emend. Hofker
Plate 6.1.5, Fig. 7, × 57, Fig. 8, × 65, U. Toarcian, Mochras Borehole, N. Wales. = *Asterigerina macfadyeni* Ten Dam, 1947. Description: convex dorsal surface with up to 3½ whorls; raised, merging spiral and sutural ridges; ventral surface

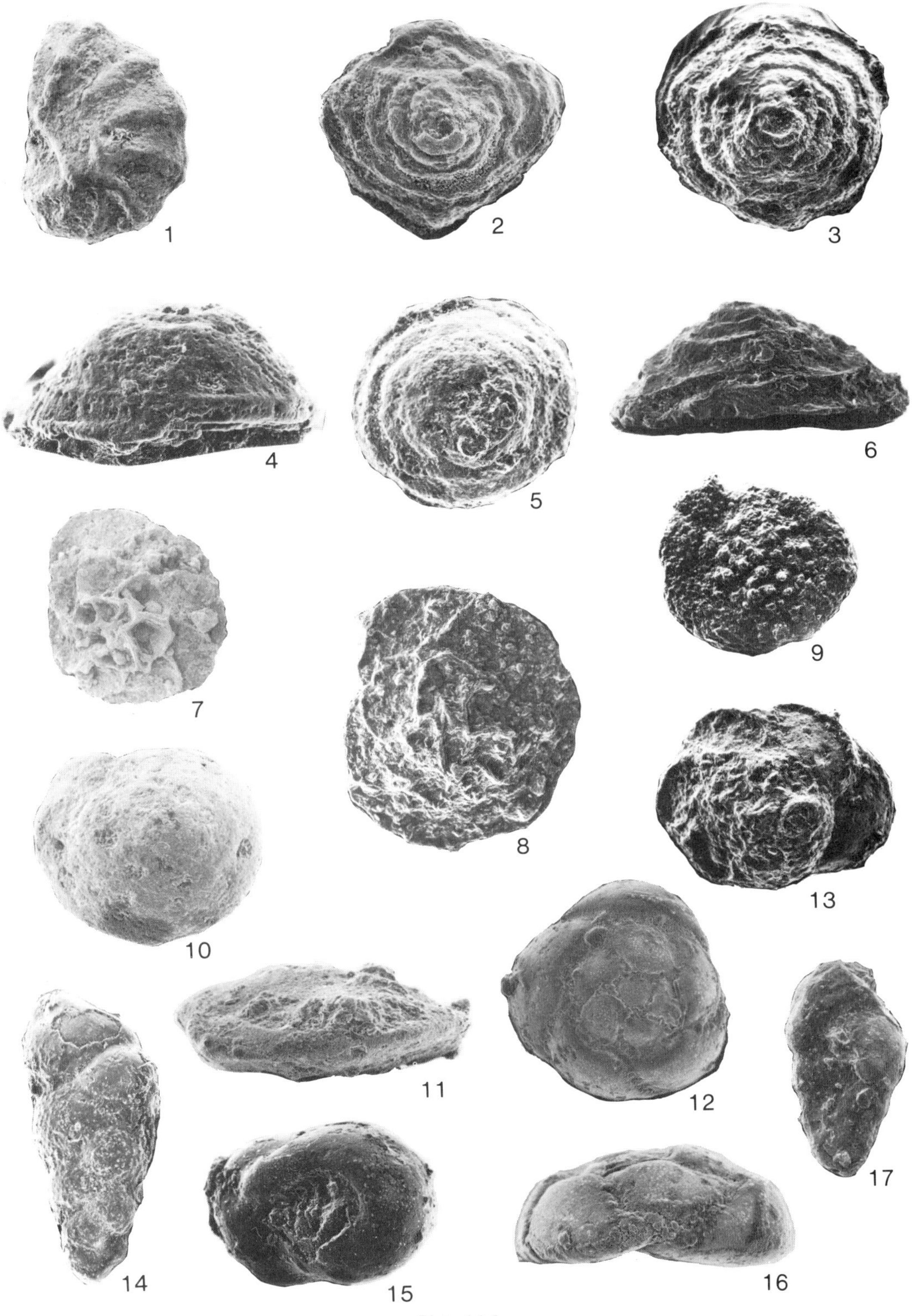

Plate 6.1.5

smooth, planar/slightly convex; limbate marginal keel. Distribution: British records from U. Pliensbachian of Yorkshire and Lincoln (Bracebridge) and Toarcian of Yorkshire, Lincoln, Tilton (Leicestershire), Mochras Borehole [35], Empingham (Leicestershire), Byfield (Northants.) [12], north Lincolnshire (Adams, personal communication) and from *falciferum-bifrons* Zones in a boring at Pattishall (Northants.) [75]. European range between Upper Pliensbachian and Lower Aalenian from S. Germany/Austria/Switzerland and France [32].

Reinholdella margarita (Terquem)
Plate 6.1.5, Fig. 13, dorsal view × 100, U. Sinemurian, Mochras Borehole, N. Wales. Fig. 6.1.1. = *Rotalina margarita* Terquem, 1866b (syn. *Epistomina liassica* Barnard, 1950a). Description: test high-spired, with elevated spiral and septal sutures and chamber margins; open, crater-like umbilicus on ventral side; 7 or 8 chambers in last whorl.
Remarks: *R. macfadyeni* (Ten Dam) differs in its thinner, less elevated sutures, less convex trochospire and closed umbilicus. Distribution: British range from late *semicostatum* Zone (*sauzeanum* Subzone) to early *obtusum* Zone (early *stellare* Subzone), with observed occurrences in the Mochras [35], Stowell Park and Truch Boreholes and on the Dorset coast (= *Epistomina liassica* Barnard, 1950a). Also recorded from France (range late *semicostatum* Zone to *obtusum* Zone) [24, 69] and Sweden (basal L. Sinemurian) [64].

Reinholdella pachyderma subsp. A
Plate 6.1.5, Fig. 10, dorsal view, × 70, Sinemurian/L. Pliensbachian, Mochras Borehole, N. Wales; Fig. 11, edge view, × 125, L. Pliensbachian, same locality. Fig. 6.1.1. Description: test low trochospiral, variable in shape, planar on ventral side with a partially open umbilicus, 6 or 7 chambers in final whorl.
Remarks: most similar to *R. pachyderma pachyderma* Hofker which differs in having a higher spire, a more convex ventral side with a closed umbilicus and 9 chambers in the final whorl. Distribution: only yet recorded from Britain. Observed in the Mochras [35], Burton Row, Trunch and Cocklepits Boreholes and Brant Broughton (Lincolnshire) (J. Exton, personal communication), with a total range of mid-*raricostatum* Zone (*raricostatoides* Subzone) to late *jamesoni* Zone (*jamesoni* Subzone). Most abundant between uppermost *raricostatum* Zone and basal *jamesoni* Zone (*taylori* Subzone).

Reinholdella? planiconvexa (Fuchs)
Plate 6.1.5, Fig. 12, dorsal view, × 150, Fig. 16, edge view, × 21. L. Sinemurian, Mochras Borehole, N. Wales. = *Oberhauserella planiconvexa* Fuchs, 1970 (syn. *?Conorboides* sp. Brouwer, 1969). Description: test smooth, shape discoidal to ovate; plano-convex in vertical section; marked umbilical hollow and usually clearly visible supplementary aperture on ventral side; 2 to 2½ whorls with 5 to 6 chambers in final whorl.
Remarks: differs from *R. dreheri* (Bartenstein) in having an open umbilicus and greater variability in shape. Distribution: in Britain the species is abundant in the early *liasicus* Zone, as in the Mochras [35], Wilkesley and Burton Row Boreholes, though ranging between the topmost *planorbis* Zone (*johnstoni* Subzone) and basal *jamesoni* Zone (*taylori* Subzone). Other British occurrences, the *planorbis* Zone of St. Audries Slip (Somerset) and N. Ireland, the *liasicus* Zone of Yorkshire, the *angulata* Zone of Watchet (Somerset) [32], Lavernock (Glamorgan) and probably the *planorbis* to *bucklandi* Zones of Yorkshire (= *Pulvinulina elegans*, Blake 1876). Restricted to the Hettangian in West Germany (= *Discorbis advena*, Bartenstein and Brand 1937, Klingler 1962), Austria [45] and France (= *?Conorboides* sp. Brouwer 1969). Abundant generally in the Hettangian and L. Sinemurian throughout north-west Europe [32].

Brizalina liasica (Terquem)
Plate 6.1.5, Fig. 17, × 143, U. Sinemurian, Mochras Borehole, N. Wales. Figs. 6.1.1, 6.1.2. = *Textilaria liasica* Terquem, 1958 (syn. *Bolivna rhumbleri* Franke, 1936; *Bolivina rhumbleri amalthea* Brand, 1937). Description: test smooth, compressed, sometimes with thickened median area; early sutures flush, later ones depressed; aperture an elongate slit; internal toothplate.
Remarks: the only known Liassic *Brizalina*, though possibly aragonitic in composition. Distribution: British range from *obtusum* Zone (*obtusum* Subzone) to *tenuicostatum* Zone and possibly *levesquei* Zone (*dispansum* Subzone), becoming widespread especially in Pliensbachian. Known occurrences in Dorset [62, 32], Yorkshire/Midlands [32] the Mochras [35], Burton Row, Hill Lane and Cocklepits Boreholes, Port Mulgrave (Yorkshire), Scunthorpe (South Humberside) and Bracebridge (Lincolnshire). Widespread and common in Pliensbachiam of Western Europe, but reported to range into the Aalenian in Sweden [64] and Switzerland [89].

Neobulimina sp. 2 Bang, 1968
Plate 6.1.5, Fig. 14, × 125, Fig. 15, × 215. L. Sinemurian, Mochras Borehole, N. Wales. Fig. 6.1.1. (syn. *Gaudryina gradata* form a Brand, 1937). Description: initial triserial portion comprising less than one third test length; surface hispid or smooth; test sides divergent throughout; periphery rounded; aperture wide, 'u' shaped; internal toothplate.
Remarks: the only known Liassic species of the genus, but probably aragonitic in composition. Distinguished from *Brizalina liasica* by its triserial initial portion, hispid ornament and thicker test with rounded margins. Distribution: main British range from early *bucklandi* Zone to *obtusum* Zone (*obtusum* Subzone), though occurring in the late *angulata* Zone in Dorset [64] and Larne (N. Ireland). Additionally observed in the *turneri* Zone of Morvern (Scotland), the *semicostatum* Zone at Hornblotton Mill (Somerset), and in the Mochras [35], Stowell Park, and Burton Row Boreholes. The species has an acme in the *semicostatum* Zone. Ranging through but restricted to L. Sinemurian on continent, i.e. Denmark [5, 6, 7], Sweden [64] and West Germany [20].

ACKNOWLEDGEMENTS

The authors are grateful to the Institute of Geological Sciences for making available for study material from the Stowell Park, Wilkesley, Platt Lane, Burton Row, Hill Lane, Trunch and Cocklepits Boreholes, and to the directors of Robertson Research International Limited and British National Oil Corporation for permission to publish.

6.1.6 REFERENCES

[1] Adams, C. G. 1957. A study of the morphology and variation of some Upper Lias foraminifera. *Micropaleontology*, **3**, 205-206.

[2] Adams, C. G. 1962. Calcareous Adherent foraminifera from the British Jurassic and Cretaceous and the French Eocene. *Palaeontology*, **5**, 149-170.

[3] Arbeitskreis Deutscher Mikropaläontologen, 1962. *Leitfossilien der Mikropalaontologie*. Gebrüder Borntraeger, Berlin-Nikolassee.

[4] Bach, H., Hagenmeyer, P. and Neuweiler, F. 1959. Neubeschreibung und Revision einiger Foraminiferenarten und – unterarten aus dem Schwabischen Lias. *Geol. Jb.* **76**, 427-452.

[5] Bang, I. 1968. Biostratigraphical investigation of the pre-Quarternary in the Oresund Boreholes mainly on the basis of foraminifera. *In:* Larsen, G. *et al.*, 86-88.

[6] Bang, I. 1971. Jura aflejringerne i Ronde Nr. 1 (2103-2164M). Biostratigrafi pa grundlag af foraminiferer. *Dan. Geol. Unders.* 3 Rapp., No. 39, 74-80.

[7] Bang, I. 1972. Jura-biostratigrafi i Novling Nr. 1 pa grundlag af foraminiferer. *Dan Geol. Unders.* 3 Rapp., No. 40, 119-123.

[8] Banner, F. T., Brooks, M. and Williams, E. 1971. The Geology of the Approaches to Barry, Glamorgan. *Proc. Geol. Assoc.*, **82**, 231-247.

[9] Barbieri, F. 1964. Micropaleontologia del Lias e Dogger del pozzo Ragusa (Sicilia). *Rev. Ital. Pal.*, **70**, 709-830.

[10] Barnard, T. 1948. The uses of foraminifera in Lower Jurassic stratigraphy. *Int. Geol. Congress* 'Report of the 18th Session, Great Britain 1948', **15**, 34-41. London.

[11] Barnard, T. 1950(a). Foraminifera from the Lower Lias of the Dorset Coast. *Quart. Jl. geol. Soc. London,* **105**, 347-391.

[12] Barnard, T. 1950(b). Foraminifera from the Upper Lias of Byfield, Northamptonshire. *Quart. Jl. geol. Soc. London,* **106**, 1-36.

[13] Barnard, T. 1952. Notes on *Spirillina infima* (Strickland Foraminifera. *Ann. Mag. Nat. Hist. ser.*, **12**, 5, 905-909.

[14] Barnard, T. 1956. Some Lingulinae from the Lias of England. *Micropaleontology,* **2**, 271-282.

[15] Barnard, T. 1957. *Frondicularia* from the Lower Lias of England. *Micropaleontology,* **3**, 171-181.

[16] Barnard, T. 1959. Some arenaceous foraminifera from the Lias of England. *Contrib. Cushman Found. Foramin. Res.*, **10**, 132-136.

[17] Barnard, T. 1960. Some species of *Lenticulina* and associated genera from the Lias of England. *Micropaleontology,* **6**, 41-55.

[18] Barnard, T. 1963. Evolution in certain biocharacters of selected Jurassic Lagenidae. *In:* Koeningswald *et al.* (Eds.) *Evolutionary trends in foraminifera,* 79-92, Elsevier, Amsterdam.

[19] Barnard, T. and Hay, W. W. 1974. On Jurassic Coccoliths: a tentative zonation of the Jurassic of Southern England and North France. *Eclogae Geol. Helvetiae,* **67**, 563-585.

[20] Bartenstein, H. and Brand, E. 1937. Mikropalaontologische Untersuchungen zur Stratigraphie des nordwestdeutschen Lias und Doggers. *Senckenb. Naturf. Ges., Abh.*, **439**, 1-224.

[21] Bate, R. H. and Coleman, B. E. 1975. Upper Lias Ostracoda from Rutland and Huntingdonshire. *Bull. Geol. Surv. G.B.*, **55**, 1-42.

[22] Berthelin, G. 1879. Foraminifères du Lias moyen de la Vendée. *Rev. Mag. Zool.*, **3**, 24-41.

[23] Bizon, G. 1960. Revision de quelques espèces-types de foraminifères du Lias du Bassin Parisien de la collection Terquem. *Rev. Micropaléont.*, **3**, 3-18.

[24] Bizon, G. 1961. Contributions a l'étude micropaléontologique du Lias du Bassin de Paris. Deuxième Partie. Lorraine, Region de Nancy et Thionville. *In: Colloque sur le Lias Francais,* 433-436.

[25] Bizon, G. and Oertli, H. 1961. Contribution a l'étude micropaléontologique du Lias du Bassin de Paris. Septième Partie – Conclusions. *In: Colloque sur le Lias Francais,* 107-119.

[26] Blake, F. J. 1876. Class Rhizpoda. *In:* Tate, R. and Blake, J. F., 449-473.

[27] Bornemann, J. G. 1854. *Uber die Liasformation in der Umgegend von Göttingen und ihre organichen Einschlusse.* A. W. Schade, Berlin.

[28] Brady, H. B. 1864. On *Involutina liassica (Nummulites liassicus,* Rupert Jones). *Geol. Mag.*, **1**, 193-196.

[29] Brady, H. 1866. Foraminifera. *In:* Moore, C. On the Middle and Upper Lias of the South West of England. *Proc. Somerset archaeol. nat. Hist. Soc.*, **13**, 119-230.

[30] Brand, E. and Fahrion, H. 1962. Dogger, N. W. – Deutschlands. *In; Leitfossilien der Mikropalaontologie,* 123-158.

[31] Brodie, P. B. 1853. Remarks on the Lias at Fretherne near Newnham and Purton near Sharpness; with an account of some new Foraminifera discovered there. *Proc. Cotteswold Nat. Fld. Cl.*, **1**, 241-246.

[32] Brouwer, J. 1969. Foraminiferal assemblages from the Lias of N.W. Europe. *Ver. K. Ned. Akad, Wet., Afd. Natuurk; 1 Reeks, Deel.*, **25**, 1-48.

[33] Champeau, H. 1961. Contributions a l'étude micropaléontologique du Lias du Bassin de Paris. Troisième Partie. Etude de la microfauna des niveaux marneux du Lias dans le sud-est du Bassin de Paris.

In: Colloque sur le Lias Francais, 437–443.

[34] Colloque sur le Lias Francais 1961. *Mém. Bur. Rech. Géol. Min.*, **5**.

[35] Copestake, P. and Johnson, B. (in press). *Liassic (Hettangian-Toarcian) foraminifera from the Mochras Borehole, North Wales.*

[36] Crick, W. D. and Sherborn, C. D. 1891. On some Liassic foraminifera from Northamptonshire. *Jour. Northampt. Nat. Hist. Soc.*, **6**, 1-15.

[37] Crick, W. D. and Sherborn, C. D. 1892. The Leda-ovum beds of the Upper Lias, Northamptonshire. *J. Northampt. Nat. Hist. Soc.*, **7**, 67-72.

[38] Dean, W. T., Donovan, D. T. and Howarth, M. K. 1961. The Liassic ammonite zones and subzones of the north-west European Province. *Bull. Br. Mus. nat. Hist.* (Geol.), **4**, 435-505.

[39] Drexler, E. 1958. Foraminiferen und Ostracoden aus dem Lias alpha von Siebeldingen/Pfalz. *Geol. Jb.*, **75**, 475-554.

[40] Dreyer, E. 1967. Mikrofossilien der Rät und Lias von S.W. Brandenburg. *Jb. Geol.*, **1**, 491-531.

[41] Espitalié, J. and Sigal, J. 1960. Microfaunes du Domérien du Jura Méridional et du detroit de Rodez. *Rev. de Micropaléont.*, **3**, 52-59.

[42] Exton, J. 1979. Pliensbachian and Toarcian microfauna of Zambujal, Portugal: Systematic paleontology. *Carleton University Geological Paper*, **79-1**, i-viii, 1-104.

[43] Franke, A. 1936. Die foraminiferen des deutschen Lias. *Abh. Preuss. Geol. Landesant; N.F.*, (169), 1-140.

[44] Frentzen, K. 1941. Die foraminiferenfaunen des Lias, Doggers und unteren Malms des Umgeburg von Blumberg (Oberes Wutach-Gebiet). *Bietr. Naturkd. Forsch. Oberrheingeb.*, **6**, 125–402.

[45] Fuchs, W. 1970. Eine Alpine, Tiefliassische foraminiferenfauna von Hernstein in Niederosterreich. *Verh. Geol. B.A.* (1), 66-145.

[46] Gazdzicki, A. 1975. Lower Liassic ('Gresten Beds') microfacies and foraminifers from the Tatra Mts. *Acta Geol Polonika*, **25**, 385-398.

[47] Hallam, A. 1958. The concept of Jurassic axes of uplift. *Sci. Prog.*, **46**, 441–449.

[48] Henderson, I. J. 1934. The Lower Lias at Hock Cliff, Fretherne. *Proc. Bristol Nat. Soc.*, **4**, 549-564.

[49] Hofker, J. 1952. The Jurassic genus *Reinholdella* Brotzen (1948) (Foram.) *Paläont. Z.*, **26**, 15-29.

[50] Horton, A. and Coleman, B. E. 1978. The lithostratigraphy and micropalaeontology of the Upper Lias at Empingham, Rutland. *Bull. Geol. Surv. G.B.*, **62**, 1-12.

[51] Issler, A. 1908. Bietrage zur Stratigraphie und Mikrofauna des Lias in Schwaben. *Palaeontographica*, **55**, 1-103.

[52] Johnson, B. 1977. Ecological ranges of selected Toarcian and Domerian (Jurassic) foraminiferal species from Wales. *In:* Schafer, C. T. and Pelletier, B. R. (Eds.) First International Symposium on Benthonic Foraminifera of Continental Margins – Part B, 545-556. *Maritime Sediments Special Publication No. 1.*

[53] Jones, T. R. 1853. On the Lias at Fretherne, near Newham and Purton, near Sharpness, with an account of some new foraminifera discovered there. *Ann. Mag. Nat. Hist. Ser. 2,* **12**, 272-277.

[54] Jones, T. R. and Parker, W. K. 1860. On some fossil foraminifera from Chellaston near Derby. *Quart. Jl. geol. Soc. Lond.*, **16**, 452–456.

[55] Klingler, W. 1962. Lias Deutschlands. *In: Leitfossilien der Mikropalaontologie*, 73-122.

[56] Kopik, J. 1960. Micropalaeontological characteristic of Lias and Lower Dogger in Poland (in Polish). *Kuratalnik Geol.*, **4**, 921-935.

[57] Kristan-Tollmann, E. 1964. Die Foraminiferen aus den Rhatischen Zlambachmergeln der Fischerwiese bei Aussee in Jalzkammergut. *Jahrb. Geol. B.A.*, **10**, 1-189.

[58] Larsen, G., Buch, A., Christensen, O. B. and Bang, I. 1968. Oresund. Helsingor-Halsingborgslinien. Geologisk Rapport. *Dan. Geol. Unders. Rapp.*, **1**, 1-90.

[59] Lloyd, A. J., Savage, R. J. G., Stride, A. A. and Donovan, D. T. 1973. The geology of the British Channel floor. *Phil. Trans. R. Soc.*, **A274**, 595-626.

[60] Lord, A. 1978. The Jurassic Part 1 (Hettangian-Toarcian). *In:* Bate, R. H. and Robinson, E. (Eds.) *A Stratigraphical index of British Ostracoda. Geol. Journ. Special Issue No. 8,* 189-212.

[61] Macfadyen, W. A. 1936. D'Orbigny's Lias Foraminifera. *Jl. R. microsc. Soc.*, **56**, 147-153.

[62] Macfadyen, W. A. 1941. Foraminifera from the Green Ammonite Beds, Lower Lias, of Dorset. *Phil Trans. R. Soc.*, **B231**, 1-173.

[63] Magné, J., Seronié-Vivien, R. M. and Malmoustier, J. 1961. Le Toarcian de Thouars (Deuxsèvres). *In: Colloque sur le Lias Francais*, 357-370.

[64] Norling, E. 1972. Jurassic stratigraphy and foraminifera of Western Scania, Southern Sweden. *Sver. geol. Unders. Afh. Ser. Ca (47)*, 1-120.

[65] Nørvang, A. 1957. The Foraminifera of the Lias series in Jutland, Denmark. *Mede. Dansk. geol. Foren.*, **13**, 1-135.

[66] Paalzow, R. 1917. Bietrage zur Kenntnis der Foraminiferen fauna der Schwammerger des Unteren Weissen Jura Suddeutschland. *Abh. Naturhist. Ges. Nurnberg.*, **19**, 203-248.

[67] Pietrzenuk, E. 1961. Zur Mikrofauna einiger Liasvorkommen in der Deutschen Demokratischen Republik. *Freiberg. Forschungsh.*, C 113, Pal., 1-129.

[68] Richardson, L. 1908. On the section of Lower Lias at Hock Cliff, Fretherne, Gloucestershire. *Proc. Cotteswold Nat. Fld. Cl.*, **16**, 135-142.

[69] Ruget, C. H. and Sigal, J. 1967. Les Foraminifères du Sondage de Laneuveville-devant – Nancy (Lotharingien de la region type). *Est. Sc. de la Terre*, **12**, 33-70.

[70] Ruget, C. H. and Sigal, J. 1970. Le Lias moyen de Sao Pedro de Muel II. Les Foraminifères. *Communic. dos Serv. Geolog. de Portugal*, **54**, 79-108.

[71] Seronié-Vivien, R. M., Magné, J. and Malmoustier, J. 1961. Le Lias des Bordures Septrionale et

Orientale du Bassin d'Aquitaine. *In: Colloque sur le Lias Francais,* 757–791.

[72] Strickland, H. E. 1846. On two species of microscopic shells found in the Lias. *Quart. Jl. geol. Soc. Lond.,* **2,** 30–31.

[73] Tappan, H. 1955. Foraminifera from the Arctic slope of Alaska. Part II. Jurassic Foraminifera. *Prof. Pap. U.S. Geol. Surv.,* **236-B,** 21–90.

[74] Tate, R. and Blake, F. J. 1876. *The Yorkshire Lias.* London, J. van Voorst.

[75] Ten Dam, A. 1947. A new species of *Astergerina* from the Upper Liassic of England. *J. Paleont.,* **21,** 396–397.

[76] Terquem, O. 1858. Première mémoire sur les Foraminifères du Lias du Department de la Moselle. *Mém. Acad. Imper. Metz.,* **39,** 563–654.

[77] Terquem, O. 1862.. Rechèrches sur les Foraminifères de L'Etage Moyen et de L'Etage inférieur du Lias, Mémoire 2. *Mém. Acad. Imper. Metz.,* **42,** (ser. 2, 9), 415–466.

[78] Terquem, O. 1863. Troisième mémoire sur les Foraminifères du Lias des departements de la Moselle, de la Cote d'Or, du Rhone de la Vienne et du Calvados. *Mém. Acad. Imper. Metz.,* **44,** (2) (ser. 2, 11), 361–438.

[79] Terquem, O. 1864. Quatrième mémoire sur les Foraminifères du Lias comprenant les polymorphines des Departements de la Cote d'Or et de l'Indre, 233–305. Metz.

[80] Terquem, O. 1866(a). Cinquième mémoire sur les Foraminifères du Lias des Departements de la Moselle, de la Cote d'Or et de l'Indre, 313–454. Metz.

[81] Terquem, O. 1866(b). Sixième mémoire sur les Forminifères du Lias des Departements d l'Indre et de la Moselle, 459–532. Metz.

[82] Terquem, O. 1868. Première mémoire sur les Foraminifères du Systeme Oolithique Etude, du Fullers-Earth de la Moselle. *Bull. Soc. Hist. Nat. Moselle Metz,* **11,** 1–138.

[83] Terquem, O. and Berthelin, G. 1875. Etude microscopique des marnes du Lias Moyen d'Essey-lés-Nancy, zone inférieure de l'assise à *Ammonites margaritatus. Mém. Soc. Géol. France,* Ser. 2, **10,** 1–126.

[84] Trueman, A. E. 1918. The Lias of South Lincolnshire. *Geol. Mag.,* **5,** 64–73, 101–111.

[85] Usbeck, I. 1952. Zur Kenntnis von Mikrofauna and Stratigraphie im unteren Lias alpha Schwabens. *Neues Jb. Geol. Paläont. Abh.,* **95,** 371–476.

[86] Warrington, G. *et al.* 1980. A correlation of Jurassic rocks in the British Isles. *Geol. Soc. Lond., Special Report,* **13,** 78 pp.

[87] Warrington, G. and Owens, B. (Compilers) 1977. Micropalaeontological biostratigraphy of offshore samples from south-west Britain. *Rep. Inst. Geol. Sci.,* 77/7, 1–49.

[88] Welzel, E. 1968. Foraminferen und Fazies des frankischen Domeriums. *Erlarger geol. Abh.,* **69,** 1–77.

[89] Wernli, R. 1971. Les Foraminifères du Dogger du Jura méridional (France). *Arch. Soc. Genève,* **24,** 261–368.

[90] Wernli, I. and Septfontaine, M. 1971. Micropaléontologie comparée du Dogger du Jura méridional (France) et des Préalpes Medianes Plastiques romandes (Suisse). *Ecolg. Geol. Helv.,* **64,** 437–458.

[91] Wicher, C. A. 1938. Mikrofaunen aus Jura und Kreide, insbesondere Nordwestdeutschlands. I. Teil: Lias Alpha bis epsilon. *Abh. Preuss. Geol. L.A.,* **193,** 1–16.

[92] Wood, A. and Barnard, T. 1946. *Ophthalmidium.* A study of nomenclature, variation and evolution in the foraminifera. *Quart. Jl. geol. Soc. Lond.,* **102,** 77–113.

6.2 THE BAJOCIAN TO CALLOVIAN

B. Coleman

6.2.1 Introduction

There have been very few publications on British Middle Jurassic foraminifera. The fauna from a Bajocian sequence from Lyme Bay, Dorset, has recently been described by Coleman in Dingwall and Penn (in press). References to Bathonian foraminifera during the last century include a short paper by Jones (1884) on the foraminifera and ostracods from the Deep Boring at Richmond. The first extensive detailed study of Bathonian faunas was made by Cifelli (1959), and more recently by the Institute of Geological Sciences (Coleman, 1979).

The only publications on foraminifera from the British Callovian are by Coleman *et al.* (1974) and Medd in Richardson (1979). Foraminifera from the *lamberti* Zone are included in a paper by Cordey (1963) on the Oxford Clay fauna of Skye while the Callovian fauna of Brora has been described by Gordon (1967).

Outside Britain there is an extensive Middle Jurassic literature, including the classical works of Terquem (1868–1886) and Bartenstein and Brand (1937). More recent publications include Ruget (1973) and Munk (1978).

6.2.2 Location of collections of importance

British Museum (Natural History):

Cifelli 1959 collection on permanent loan from Harvard Museum of Comparative Zoology.

Cordey 1962 Including unpublished material from Oxford Clay, Gordon 1967.

Institute of Geological Sciences, Leeds

Coleman 1974 and 1979. Including faunal slides of Bajocian-Bathonian material from boreholes in Bath and Dorset areas.

6.2.3 Stratigraphic divisions

The Middle Jurassic is considered here as comprising the Bajocian, Bathonian and Callovian Stages (Arkell 1956). The base of the Najocian has been taken at the base of the *opalinum* Zone, the Aalenian Stage of continental workers being disregarded. No foraminifera have been obtained from the *opalinum* Zone since the only complete Bajocian sequence examined is from Lyme Bay, Dorset, where the basal Inferior Oolite is a sand facies (Bridport Sands).

Although a complete ammonite zonation of the Middle Jurassic exists (Callomon, 1968; Torrens, 1969, 1971; Parsons, 1974, 1975, 1977), ammonites are rare in British Bajocian and Bathonian rocks. This, together with a complex lithofacies development, has made correlation of the various divisions extremely difficult. The recent borehole programme of the Institute of Geological Sciences and in particular, the work of Penn and Wyatt, have done much to elucidate this very complex problem. For details of lithostratigraphical correlation see Dingwall and Penn (in press), Penn and Wyatt (1979) and Penn, Merriman and Wyatt (1979). A foraminiferal zonation of the Bathonian is outlined below (see also Coleman in Penn and Wyatt, 1979).

The Bathonian/Callovian boundary has been taken at the junction between the Lower and Upper Cornbrash (*discus*/*macrocephalus* Zones) which also marks a distinct change in foraminiferal faunas. The top of the Callovian occurs at the Middle -Upper Oxford Clay boundary, at the top of the *lamberti* Zone.

6.2.4 Faunal associations

6.2.4.1 BAJOCIAN

The only English Bajocian material studied has been a cored sequence from Lyme Bay, Dorset. The borehole, 50/03/329 (IGS offshore register No; NGR 3351 8115) penetrated rocks between basal Bathonian (Zigzag Bed) and the *murchisonae* Zone which in this area is the basal zone of the Inferior Oolite.

The foraminiferal fauna is intermediate in character, containing both Toarcian and Bathonian elements. The most diverse assemblages are found in the *murchisonae-discites* Zones, in which occur the upper range limits of several Lower Jurassic species viz: *Lenticulina dorbignyi, Nodosaria regularis, N. tenera, N. pulchra* and *Vaginulina* cf. *listi.* The first significant occurrence of *Lenticulina quenstedti*, an important Middle and Upper Jurassic species, is in the upper part of the *concavum* Zone, although rare specimens have been found at lower levels. The distinctive species *L. dictyodes* also appears at this level. Its range in Bajocian seems to be restricted to the *concavum-discites* Zones although it does reappear in the Upper Fuller's Earth of the Bathonian. Few foraminifera were obtained from the *laeviuscula* and *sauzei* Zones and no samples were obtained from the conglomeritic *humphriesianum-subfurcatum* Zones. The typical Bathonian fauna including *L. tricarinella, Planularia beierana* and *P. eugenii,* is developed in the *garantiana* and *parkinsoni* Zones, a feature also seen in material onshore from boreholes at Winterborne Kingston and Seabarn Farm. A small number of Liassic species such as *Baginulina clathrata* sensu stricto and *Nodosaria liassica,* extend into the early Bathonian.

6.2.4.2 BATHONIAN

The Bathonian argillaceous facies of southern England was first studied in detail by Cifelli (1959) who described four foraminiferal faunules, A B C and D, which appeared to occur consistently in the same stratigraphical order. The recent availability of extensively cored material from the Dorset coast to the area north of Bath, has made possible a revaluation of this scheme which was originally based on samples from widely separated surface outcrops. The four faunules described by Cifelli were recognised and, in the Frome-Bath area at least, it was possible to subdivide faunules B and C into 6 units: B1, B2, C1, C2, C3 and C4 (Coleman in Penn and Wyatt, 1979). This zonation was then applied with limited success to cores from Winterborne Kingston and Seabarn Farm boreholes in Dorset. As a result of this study it was concluded that in fact five assemblages could be distinguished in the Bathonian of southern England.

Faunule A. Range: Topmost Upper Inferior Oolite and Lower Fuller's Earth to the top of the Knorri Clays, *zigzag* Zone.

Definition: Characteristic species are *Planularia eugenii, Saracenaria oxfordiana, Nodosaria ingens, N. liassica* and *Vaginulina clathrata* sensu stricto. The *macrescens* Subzone of the *zigzag* Zone in Dorset has yielded a very diverse fauna including *Frondicularia nympha* and *Lenticulina* cf. *volubilis*, species which have been described from the Kuiavian (Upper Bajocian–Lower Bathonian) of Poland (Kopik 1969; Bielecka and Styk 1969).

Remarks: Compares well with Cifelli's faunule A.

Faunule B1, Range: succeeding beds of Lower Fuller's Earth Clay up to the Echinata Bed, *progracilis* Zone.

Definition: Although no species are restricted to it, the faunule is recognisable through the absence of faunule A elements and the presence of *Lenticulina quenstedti.*

Remarks: In Dorset, horizons rich in arenaceous foraminifera occur within this interval. Not exposed on the coast.

Faunules B2–C3. Range: Uppermost part of Lower Fuller's Earth Clay to lowermost Frome Clay, including the Wattonensis Beds in Dorset and the Combe Down Oolite at Bath, *subcontractus* – basal *aspidoides* Zones.

Definition: Characteristic species are *L. tricarinella* (also present in A and B1), *Ophthalmidium carinatum, L. dictyodes, N.* aff. *issleri* and *Vaginulina legumen. Lenticulina galeata* is not seen above Fuller's Earth Rock and its equivalents.

Remarks: Four distinct assemblages can be recognised at Bath (Coleman in Penn *et al.*, 1979) but in Dorset, such differentiation cannot easily be detected. The faunule corresponds in part to Cifellis' faunule B. However, a repetition of B2 faunal elements in C3 north of Frome led Cifelli to miscorrelate the Fuller's Earth Rock and the Wattonensis Beds.

Faunule C4: Range: Frome Clay, *aspidoides* Zone.
Definition: Characterised at Bath by the presence of *Lenticulina subalata* and in Dorset by the predominance of *Epistomina stelligera.* Other species which commonly occur are *Tristix oolithica* and *Gaudryina?* sp. 2 Lutze.
Remarks: Comparable to Cifelli's faunule C.

Faunule D: Range: Boueti Bed and Forest Marble, *discus* Zone. Locally it may include the topmost Beds of the Frome Clay.
Definition: presence of *Massilina dorsetensis.* Other characteristic species include *N. pectinata, N.* aff. *issleri, V. legumen* and *Cyclogyra liasina.*
Remarks: Comparabe to Cifelli's faunule D.

In the more calcareous facies north of Bath, faunules A, B1, B2, C1, C2 and D have been recognised in borehole material as far north as Carterton, Oxfordshire. The faunas of the Stonesfield slate and White Limestone are similar to that of the Forest Marble, containing species such as *Frondicularia oolithica, V. legumen* and *M.* cf. *dorsetensis.* In addition there are present large numbers of *Trocholina sp.* This is no doubt due to similar depositional environments.

6.2.4.3 CALLOVIAN

The English Callovian can be broadly divided into three foraminiferal assemblage Zones.

1. The *macrocephalus-calloviense* Zones are characterised by a chiefly arenaceous fauna which also contains a significant proportion of polymorphinids. Common species are *Ammobaculites agglutinans, Verneuilinoides tryphera, Gaudryina sp.* and *Eoguttulina liassica.*
2. This is replaced at or just below the base of the *jason* Zone by a dominant epistominid fauna which extends to the middle part of the *athleta* Zone. *Epistomina stelligera* is the most common species but *E.* cf. *nuda* is characteristic of the *jason-coraonatum* Zones.
3. The epistominid assemblage is succeeded just above the Lower/Middle Oxford Clay boundary by the influx of a highly diverse fauna dominated by nodosariids but also containing a strong miliolid element. This type of assemblage continues through the late Callovian into the Oxfordian.

In addition to these 3 main faunas, a number of species have proved to be good stratigraphical markers whenever they occur viz: *Miliammina jurassica* (*jason-obductum* Subzones), *Guttulina pera* (*jason* Zone) *Planularia eugenii* (*grossouvrei* Subzone-*athleta* Zone, middle part), *Citharinella moelleri* and *Citharinella sp.* A (*athleta* Zone, upper part).

The same distribution of foraminifera has been recorded in the Milton Keynes Boreholes (Coleman in Horton *et al.*, 1974), Worlaby E Borehole (Medd in Richardson 1979), Norman Cross Pit (Medd, in preparation) and a number of I.G.S. Boreholes from the Central and East Midlands (unpublished data).

The Callovian Brora was studied by Gordon (1967) who recorded a fauna thought to be more related to the English Oxfordian, perhaps due to similar lithofacies. The samples examined were chiefly from the Brora argillaceous series exposed in Brora Brick Pit and are representative of the *coronatum* and *lamberti* Zones only. Three types of assemblages were described (arenaceous, nodosariid and mixed), but these did not seem to have any stratigraphical or lithological significance. Only one sample (mixed assemblage) contained a high proportion of epistominids. Examination of the figured material has revealed that species attributed to *L. quenstedti, S, phaedra* and *F. moelleri* are comparable with *M. ectypa, S. oxfordiana* and *F. nikitini* (?) respectively in the present account. This brings at least 3 apparently anomolous ranges in line with those recorded from the English material.

6.2.5 Palaeoecology and Facies control

The predominant group of foraminifera in the Jurassic were the Nodosariidae which are believed to have occupied a far more diverse range of habitats than their living representatives. It is therefore more difficult to determine past environments from Mesozoic foraminiferal distributions. Nevertheless, some conclusions may be drawn by studying the foraminiferal assemblages in relation to the types of sediments in which they are found.

Nothing can be said at present regarding the lateral distribution of Bajocian foraminifera since very little material has been studied. In the Bathonian, a deeper water basin fauna is found in Dorset and a shallow water, more differentiated shelf fauna at Bath. In both areas there is a distinct shallowing upwards, the rich predominantly nodosariid fauna of the Lower Fuller's Earth becoming less diverse in the Upper Fuller's Earth and giving way to the predominantly miliolid assemblages of the Forest Marble. North of Bath, brackish and freshwater influences provide a gradually more unfavourable environment until at Oxford, only the lower part of the Lower Fuller's Earth contains assemblages comparable with those further south. Within this main distribution pattern small scale variations occur which seem to be directly related to the sedimentary cycles of the clay facies described by Penn and Wyatt (1979). One of these cycles in the Upper Fuller's Earth clay at Baggridge Borehole No. 3, was studied in some detail and shows the following sequence of events. The beginning of the cycle is marked by pale grey, silty, calcareous mudstone interbedded with finegrained argillaceous limestone with a predominant epifauna. These beds contain a rich nodosariid fauna together with miliolids, encrusting foraminifera, polymorphimids and large arenaceous species. The beds pass upward into somewhat darker, less silty, calcareous mudstone with a dominant infauna and subordinate epifauna. The miliolids become rare and large arenaceous forms are absent. In the succeeding unit of dark grey, smooth, fissile mudstone, there is a marked increase in the number of small arenaceous species, second in importance to a somewhat reduced nodosariid population. Polymorphinids are rare at this level. The dark grey calcareous, bedded clay towards the top of the cycle is marked by the return of diverse nodosariid assemblage, an increase in polymorphinids and the appearance of epistominids. Small arenaceous foraminifera become rare while miliolids and encrusting species are absent. It is clear from this that the distribution of foraminifera in the rocks is to some extent facies controlled which is not surprising when it is considered that all these forms had a benthonic mode of life. This should not however detract from their value as stratigraphic indices on a broader scale.

The lithology of the Oxford Clay is very uniform and accordingly the pattern of foraminiferal distribution appears to be fairly constant (but see Section 6.3.5). Although the fauna of the Lower Oxford Clay is not diverse, individual species are often present in very large numbers. All samples from the less bituminous Middle Oxford Clay contain rich foraminiferal assemblages. The Lower/Middle Oxford Clay boundary represents a major biostratigraphical boundary as well as a lithological one, marked by the first appearance of many Upper Jurassic species. There is no evidence of deposition under anaerobic conditions (Rutten, 1956).

6.2.6 Index Foraminifera

The foraminifera listed below are not all short range, stratigraphically diagnostic species. They are nevertheless included in order to give a true picture of the foraminiferal assemblages encountered in Middle Jurassic rocks. Those which have a more restricted range and which are considered to be of stratigraphical importance are included in the range chart (Fig. 6.2.2). A small number of species shown on the range chart are omitted here since they are figured in the previous section on Liassic foraminifera. Ranges are given for the Middle Jurassic only. The classification used is that adopted by Loeblich and Tappan (1964, 1974).

All specimens are in the collections of the Institute of Geological Sciences, Leeds.

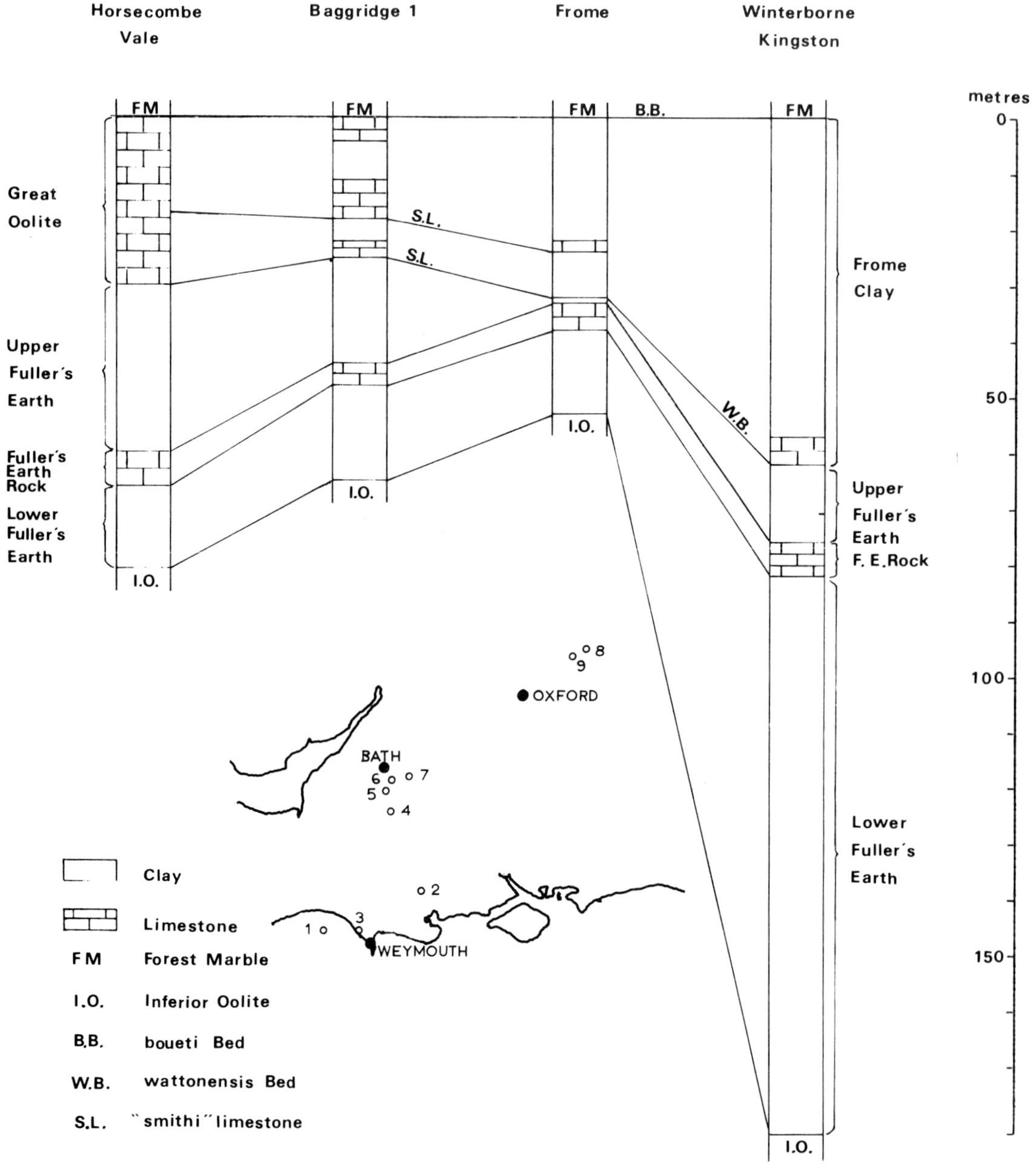

Fig. 6.2.1 – Correlation of Bathonian succession between Bath and Dorset (after Penn and Wyatt, 1979). Inset shows positions of boreholes: 1. SO/O3/329, Lyme Bay, 2. Winterborne Kingston. 3. Seabarn Farm, 4. Frome, 5. Naggridge No. 1, 6. Horsecombe Vale 15, 7. Atworth, 8. Milton Keynes 16, 9. Milton Keynes 24.

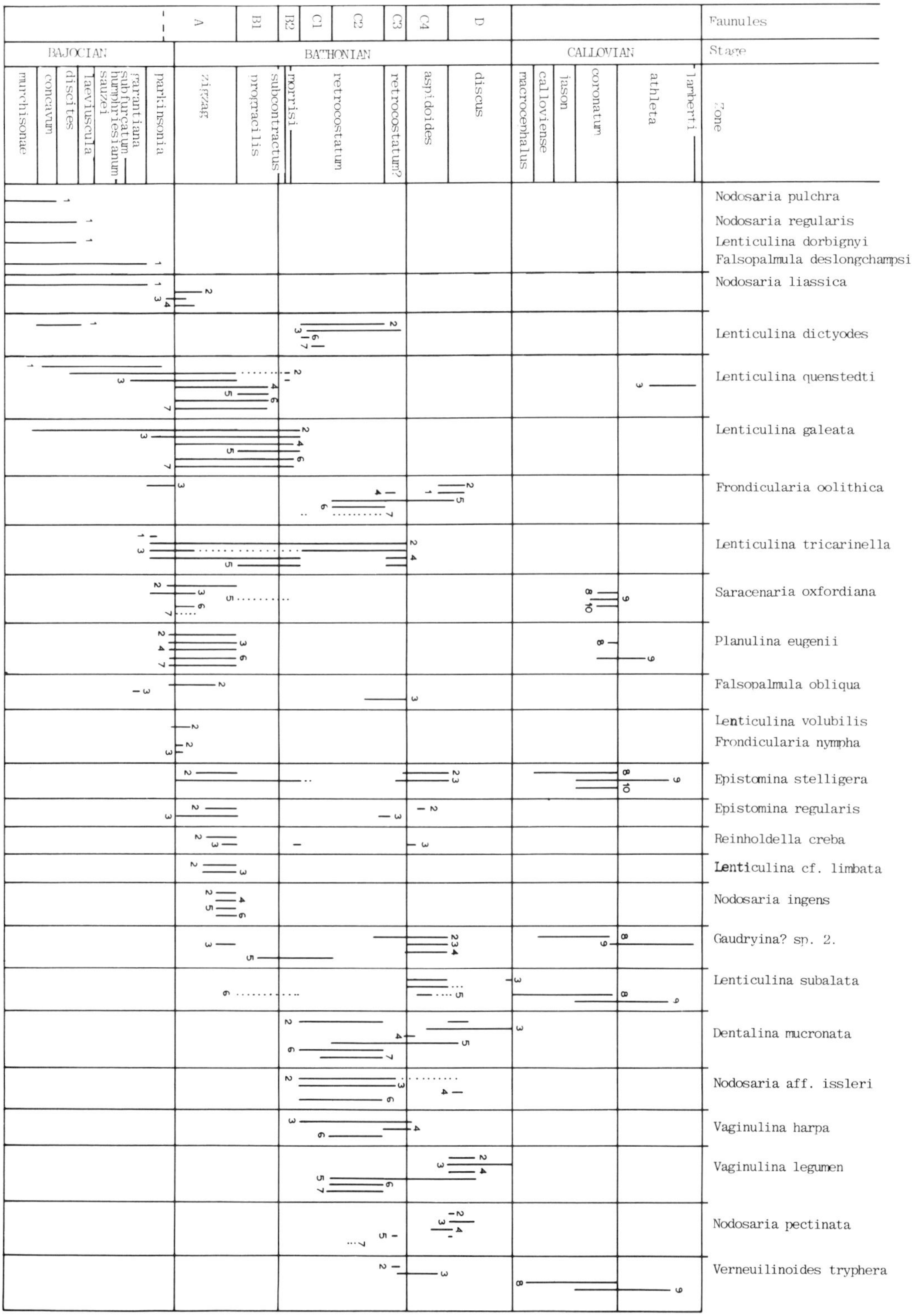

Fig. 6.2.2 (Part 1) – Range chart of Bojocian to Callovian foraminifera.

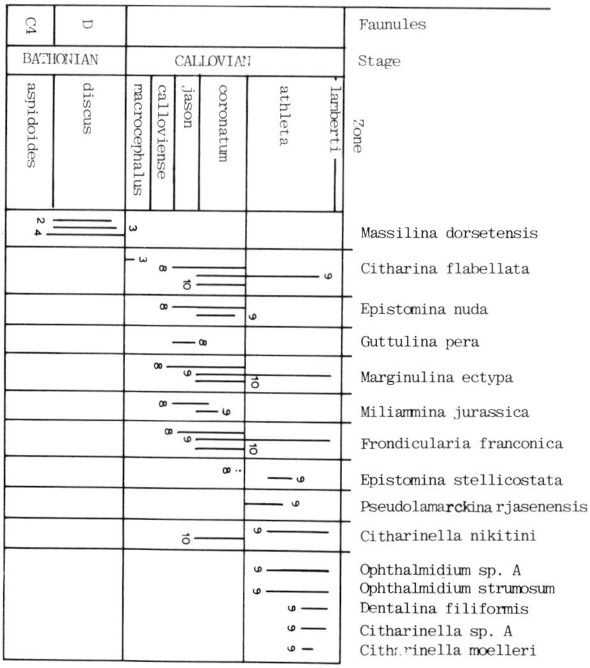

Fig. 6.2.2 (Part 2) – Range chart of Bojocian to Callovian foraminifera.

6.2.6.1 SUBORDER TEXTULARIINA

Ammobaculites agglutinans (d'Orbigny)
Plate 6.2.1, Fig. 1, MPK 2206, × 100, Upper Fuller's Earth, Bath = *Spirolina agglutinans* d'Orbigny, 1846. Description: Test small, finely agglutinated, early portion planispirally coiled, later rectilinear, circular in cross section; coil generally of 4 chambers with depressed umbilicus followed by well developed linear portion of 5–6 globular chambers which remain fairly constant in breadth, not exceeding the diameter of the coil; sutures distinct, depressed; aperture simple, terminal. Distribution: Common Jurassic species but does not become important component of the fauna until Late Bathonian: Upper Fuller's Earth Clay, *retrocostatum* Zone, Bath [6, 20], Frome Clay top *retrocostatum-aspidoides* Zones, Dorset [5]; Callovian: *macrocephalus-athleta* [11], *coronatum* [10] (as *Ammobaculites* sp. cf. *A. suprajurassica*) *athleta* [21].

Ammobaculites coprolithiformis (Schwager)
Plate 6.2.1, Fig. 2, MPK 22198, × 60, Lower Fuller's Earth, Bath = *Haplophragmium coprolithiformis* Schwager, 1867. Description: Test generally robust, coarsely agglutinated, early portion planispirally coiled, later rectilinear, circular in cross section; coil composed of 4–5 chambers followed by linear portion of 2–4, rarely 5 chambers; sutures usually indistinct, each chamber enveloping the previous one; aperture simple, terminal, produced on short neck. Distribution: Occurs throughout Bathonian and Callovian but is particularly common in the Lower Fuller's Earth Clay, *zigzag-progracilis* Zones [5, 20]; Callovian: *coronatum-lamberti* [10], *athleta* [11].

Ammobaculites fontinensis (Terquem)
Plate 6.2.1, Fig. 3, MPK 2392, × 75, Lower Fuller's Earth, Dorset = *Haplophragmium fontinense* Terquem, 1870. Description: Test variable in size, finely agglutinated, compressed, planispiral in early part; coil evolute, composed of numerous globular chambers gradually increasing in size and arranged in 2–3 whorls; linear portion composed of 3–4 broad chambers with depressed sutures; aperture simple, terminal. Distribution: Occurs sporadically throughout Bajocian [8] and Bathonian [5, 6, 20].

Gaudryina ? sp. 2 Lutze, 1960
Plate 6.2.1, Fig. 4, MPK 2375, × 166, Frome Clay, Dorset. Description: Test small, finely agglutinated; initial chambers globular, arranged in small trochospire with 4 chambers in the first whorl; final 4–5 chambers biserially arranged, increasing gradually in size, somewhat compressed; sutures depressed, particularly in later part of test to give slightly lobulate outline; aperture simple at base of final chamber. Distribution: Sporadic Bathonian–Callovian species. Common in Frome Clay, *aspidoides* Zone [5, 20]; Callovian: *athleta* [11]

Plate 6.2.1

Miliammina jurassica (Haeusler).
Plate 6.2.1, Fig. 5, SAB 515, FI × 100 Lower Oxford Clay, Milton Keynes = *Trochammina jurassica* Haeusler, 1882. Description: Test small, oval, finely agglutinated; composed of 5–9 elongated chambers arranged in quinqueloculine fashion; each chamber forms a half whorl and is slightly swollen at its initial end, tapering distally; sutures indistinct except in final chambers; aperture simple at end of final chamber. Distribution: Described originally from the Middle Oxfordian of the Swiss Jura, this species has proved to be a good marker for the English Middle Callovian, *jason-coronatum* Zones [11].

Suborder Rotaliina
Paalzowella feifeli (Paalzow). Plate 6.2.1, Fig. 6, MPK 2317, × 25, Middle Inferior Oolite, Lyme Bay = *Trocholina feifeli* Paalzow, 1932. Description: small calcareous, conical test with chambers arranged in 4–7 whorls visible on the spiral side; spiral suture generally elevated to give step-like appearance; umbilical side flat or concave; 3 chambers in final whorl. Distribution: Sporadic. Bajocian [8]–Bathonain [5, 20].

Suborder Textulariina
Tetrataxis ? sp.
Plate 6.2.1, Fig. 7, MPK 2368, × 250. Forest Marble, Dorset. Synonym: *Trochammina haeusleri* (Galloway) Cifelli, 1959. Description: Test calcareous, granular, conical, chambers arranged in trochoid spiral with 4 to a whorl in early portion, later 3; sutures flush, indistinct; aperture a slit at base of last chamber in umbilical region. Distribution: Bathonian [5, 6, 20].

Triplasia althoffi (Bartenstein)
Plate 6.2.1, Figs. 8–9, Fig. 8, MPK 2380, × 85, Frome Clay, Dorset; Fig. 9, SAB 471 Fl, × 65, Middle Oxford Clay, Milton Keynes, = *Flabellammina althoffi* Bartenstein, 1937. Common synonyms: *Triplasia variabilis* (Brady) Bartenstein and Brand, 1937; *Triplasia bartensteini* Loeblich and Tappan, 1952. Description: Test large, arenaceous, highly variable in outline; planispiral coil of 3–5 chambers followed by linear portion of 2–9 compressed, triangular or, rarely, quadrangular chambers with excavated sides; sutures straight initially, later strongly convex, slightly depressed; aperture oval, terminal, produced on a short neck. See Lindenberg (1967) for full description and synonymy. Distribution: Late Bajocian [5, 8]; Bathonian [5, 6, 20]; *coronatum* [10], *athleta* [11].

Trochammina globigeriniformis (Parker and Jones)
Plate 6.2.1, Fig. 10, MPK 2386, × 190, Lower Fuller's Earth, Dorset. = *Lituola nautiloidea* Lamarck var. *globigeriniformis* Parker and Jones, 1865. Description: Test small, agglutinated, trochoid; chambers arranged in 2–3 whorls, becoming increasingly bulbous with 3–4 chambers in final whorl; sutures indistinct initially but strongly depressed in final part of test; margin rounded, lobulate, aperture loop-shaped at base of last chamber. Distribution: Sporadic. Bathonian [5, 6, 20]; Callovian: *jason-athleta* [11].

Verneuilinoides tryphera Loeblich and Tappan, 1950
Plate 6.2.1, Fig. 17, MPK 2261, × 150, Frome Clay, Faulkland. Description: Test small, finely agglutinated, triserial, flaring distally; chambers subglobular, increasing very gradually in size as added; sutures distinct, depressed; aperture

crescent-shaped at base of final chamber. Distribution: Late Bathonian: *retrocostatum-discus* [5, 20]; Callovian: *macrophalus-athleta* [11].

Suborder Miliolina

Cyclogyra liasina (Terquem)

Plate 6.2.1, Fig. 11, MPK 2198, × 150, Forest Marble, Dorset = *Cornuspira liasina* Terquem, 1866. Common synonym: *Cornuspira orbicula* Terquem and Berthelin, 1875. Description: Test porcellaneous, imperforate, planispiral; large spherical proloculus followed by regular, undivided tube of 5–8 whorls, increasing very gradually in width; aperture simple, open end of tube. Distribution: Common Bajocian–Callovian [5, 8, 20, 11].

Massilina dorsetensis Cifelli, 1959

Plate 6.2.1, Fig. 12, MPK 2264, × 130, Forest Marble, Bath. Description: Test suboval, slightly biconvex; composed of 8 chambers, each one half whorl in length increasing in size as added, the first 6 arranged in a quinqueloculine series, the last 2 opposite; sutures flush, producing a smooth surface on both sides of test; apertue simple, without tooth. Distribution: Late Bathonian, Forest Marble, *discus* Zone [5, 6, 20].

Ophthalmidium carinatum Kübler and Zwingli, 1870

Plate 6.2.1, Fig. 13, MPK 2759, × 100, Wattonensis Beds., Lyme Bay. Description: Test compressed, planispiral, biconvex in cross section; only later chambers are clearly visible each one ½ whorl in length; initial chamber arrangement appears to be typical for the genus (See *Ophthalmidium strumosum*), periphery may be keeled although this is a variable feature; aperture simple, produced on short neck. Distribution: Bajocian ? [8], Bathonian [5, 20].

Ophthalmidium strumosum (Gümbel)

Plate 6.2.1, Fig. 14, SAB 461 F2 × 200, Middle Oxford Clay, Milton Keynes = *Guttulina strumosa* Gümbel, 1862, Description: Test smooth, highly compressed, planispiral, generally an elongate oval in outline; periphery rounded; 2–5 chambers visible in reflected light, each ½ whorl in length and swollen at its proximal end; initial chamber arrangement typically consists of proloculus followed by tube-like second chamber approximately ½ whorl in length and third chamber one complete turn in length; occasionally a chamber appears perpendicular to main planispiral; aperture simple, at end of long tapering neck. Distribution: Callovian: *athleta-lamberti* Zones [11].

Ophthalmidium sp. A. Coleman, 1974

Plate 6.2.1, Fig. 15, SAB 463 Fl, × 150, Middle Oxford Clay, Milton Keynes. Description: Test compressed, elliptical, with all chamber visible and separated by depressed sutures; proloculus gives rise to tube-like second chamber which forms approximately ½ whorl; third chambers form ¾ whorl about the proloculus; this and subsequent chambers are widest at their proximal ends, tapering distally, each forming ½ whorl; there are generally 6–8 chambers present; aperture simple, at end of long tapering neck. Distribution: Late Callovian, *athleta* Zone [11] *athleta-lamberti* [21].

Spirophthalmidium concentricum (Terquem and Berthelin)

Plate 6.2.1, Fig. 16, MPK 2395, × 150, Lower Fuller's Earth, Dorset = *Spiroloculina concentricum* Terquem and Berthelin, 1875. Description: Test compressed, planispiral; outline variable, generally elongate oval; initial chamber arrangement not clearly visible but appears to be small planispiral; later chambers ½ whorl in length with 2 final chambers distinctly larger than preceding ones and separated from them by depressed sutures; aperture simple at end of long neck. Distribution: Bajocian: *garantiana-parkinsoni* [5, 8]; Bathonian *zigzag--discus* [5, 6, 20].

6.6.2 SUBORDER ROTALIINA

Citharina flabellata (Gümbel)
Plate 6.2.2, Fig. 1, MPK 2760, × 70, Callovian, Lyme Bay = *Marginulina flabellata* Gümbel, 1862. Description: Test compressed, uniserial, composed of 5–10 chambers gradually increasing in size to give triangular outline; sutures generally depressed; ornamented by variable numbers of narrow, continuous ribs which sometimes bifurcate; aperture simple, terminal, produced on a prominent neck surrounded by a notched collar. Distribution: Callovian: *jason--athleta* [11], *coronatum* [10], *athleta* [21].

Citharinella nikitini (Uhlig)
Plate 6.2.2, Fig. 2, SAB 443 F1, × 40, Middle Oxford Clay, Milton Keynes = *Frondicularia nikitini* Uhlig, 1883. Description: Test large, highly compressed, quadrate; proloculus followed by asymmetrical or citharinid-type chamber, subsequent chambers symmetrical, chevron-shaped, uniserial, 7–12 in number; central depression gives slightly bilobed cross section; margin rounded, slightly lobulate; sutures distinct, depressed; surface ornamented by numerous, discontinuous arcuate costae; aperture simple, terminal, situated on short neck surrounded by notched collar.
Remarks: Differs from *Falsopalmula anceps* (Terquem) in having only one citharinid-type chamber and a larger number of chevron chambers, giving a more symmetrical outline. The extent of the ornament is highly variable, some specimens being almost smooth. Distribution: Callovian: *athleta* [11].

Citharinella moelleri (Uhlig)
Plate 6.2.2, Fig. 3, MPK 2761, × 50, Middle Oxford Clay, Milton Keynes = *Frondicularia moelleri* Uhlig, 1883. Description: Test large, highly compressed, quadrate; second chamber citharinid-type, remaining chambers chevron shaped, uniserially arranged to give a symmetrical outline; sutures indistinct; surface ornamented by variable number of strong, continuous costae; aperture simple, produced on short neck surrounded by notched collar. Distribution: Late Callovian, *athleta* Zone [11].

Citharinella sp. A Coleman, 1974
Plate 6.2.2, Fig. 4, SAB 443 F2, × 50. Middle Oxford Clay, Milton Keynes. Description: *Citharinella* with later chambers strongly overlapping earlier ones to give broad, almost semicircular outline to test; sutures limbate, depressed in later part of test; surface ornamented by variable number of discontinuous costae, generally absent from final one or two chambers;aperture simple, produced on short neck, surrounded by notched collar. Distribution: Late Callovian, *athleta* Zone [11,21].

Dentalina communis d'Orbigny, 1826
Plate 6.2.2, Fig. 5, MPK 2242, × 76, Lower Fuller's Earth, Bath. Description: Test uniserial, chambers increasing gradually in size giving slender, gently flaring, outline; suture oblique, generally indistinct at first, later depressed; aperture radiate, terminal. Distribution: Common Bajocian–Bathonian form [5, 6, 8, 20].

Dentalina filiformis (d'Orbigny)
Plate 6.2.2, Fig. 6, MPK 2762, × 100, Middle Oxford Clay, Milton Keynes = *Nodosaria filiformis* d'Orbigny, 1826. Description: Test slender, fragile, uniserial, composed of elongated chambers constricted at sutures; sutures horizontal but aperture, when present, is marginal rather than central: the species is therefore referred to the genus *Dentalina*. Distribution: Callovian: *athleta* [11].

Dentalina intorta Terquem, 1870
Plate 6.2.2, Fig. 7, MPK 2214, × 100, Forest Marble, Bath. Description: Test uniserial, composed of 4–9 chambers, all visible; sutures oblique, distinct, generally depressed; chambers fairly constant in breadth but increase gradually in height, later chambers strongly overlapping previous ones; aperture radiate, terminal. Distribution: Bajocian–Bathonian [5, 6, 8, 20].

Dentalina mucronata Neugeboren, 1856
Plate 6.2.2, Fig. 8, MPK 2215, × 100, Forest Marble, Bath. Description: Variable species of *Dentalina* with chambers broader than high and strongly oblique, depressed sutures; initial part of test usually slightly curved; aperture radiate, terminal.
Remarks: Sometimes difficult to distinguish from *Vaginulina legumen* (Linné). See Cifelli 1959 for full variation. Distribution: Upper Fuller's Earth Clay – Forest Marble, *retrocostatum-discus* [5, 6, 20].

Epistomina cf. *nuda* Terquem, 1883
Plate 6.2.2, Figs 9–10. Fig. 9, Dorsal view, SAB 519 Fl, × 150; Fig. 10, ventral view, SAB 519 F2, × 150. Lower Oxford Clay, Milton Keynes. Description: Test small, trochoid, smooth, biconvex; peripheral margin rounded, slightly lobate; chambers poorly defined on spiral side with 6 chambers in final whorl visible on umbilical side; sutures occassionally depressed, otherwise flush; secondary apertures narrow, crescent-shaped, visible close to and parallel with margin; primary aperture at base of final chamber.
Remarks: This species is distinct from that attributed to *B. nuda* by Pazdro (1969), the latter being a stratigraphically older form of *E. stelligera*. Distribution: Middle Callovain: *jason–coronatum* Zones [11].

Epistomina regularis Terquem, 1883
Plate 6.2.2, Figs. 11–12, Fig. 11, MPK 2763, × 130, Fig. 12, MPK 2896, × 130, Lower Fuller's Earth, Stowell. Synonym:

Plate 6.2.2

Epistomina mosquensis Uhlig, Bartenstein and Brand, 1937 pars. Description: Test trochoid, biconvex; peripheral margin acute, keeled; all chambers visible on the spiral side with generally 6–7 in final whorl; sutures raised, frequently ornamented with bosses and pits; on umbilical side only chambers of last whorl visible, sutures radial, usually masked by variable reticulate ornament; secondary apertures broad, crescent-shaped, parallel with margin; primary aperture low arch at base of final chamber. Distribution: Bajocian [8], Bathonian [5, 20].

Epistomina stellicostata Bielecka and Pozaryski, 1954
Plate 6.2.2, Fig. 13, SAB 473 F5, × 100, Middle Oxford Clay, Milton Keynes. Description: Test trochoid, biconvex; peripheral margin acute, keeled; on spiral side all chambers are visible with 7–8 chambers in final whorl; sutures raised, the costae in the central part being very broad, the chambers being reduced to round depressions; costae often bear small pits. On umbilical side the sutures are radial with pitted costae forming an irregular network covering a central umbilical disc; secondary apertures crescent-shaped, parallel with margin; primary aperture indistinct at base of final chamber. Distribution: Late Callovian, *athleta* Zone [11].

Epistomina stelligera (Reuss)
Plate 6.2.2, Figs. 14–15, Fig. 14, MPK 2377, × 160, Frome Clay, Dorset; Fig. 15, SAB 467 F2, × 100, Middle Oxford Clay, Milton Keynes = *Rotalia stelligera* Reuss, 1854. Common synonym: *Brotzenia parastelligera* Hofker, 1954. Description: Test trochoid, biconvex; peripheral margin acute, occasionally keeled; on spiral side early chambers are not visible; sutures limbate or flush, 5–8 chambers in final whorl; on umbilical side chambers of last whorl are marked by small triangular areas between broad sutural ribs which converge to form an umbilical boss; secondary apertures elongate, parallel with test margin; primary aperture at base of final chamber. Distribution: Bathonian [5, 6, 20]; Callovian: *macrocephalus–athleta* [11]. *coronatum* [10], *athleta* [21].

Falsopalmula deslongchampsi Terquem, 1863
Plate 6.2.3, Fig. 1, MPK 2285, × 75, Lower inferior oolite, Lyme Bay. Description: Test generally large, highly compressed, quadrate; initial planispiral portion enveloped by later equitant, chevron-shaped chambers; sutures distinct, often marked by low ribs, aperture radiate, terminal, central. Distribution: Bajocian [8].

Falsopalmula obliqua (Terquem, 1863)
Plate 6.2.3, Fig. 2, MPK 2764, × 100, Frome Clay, Frome. Description: Test highly compressed, early chambers arranged in loose planispiral followed by variable number of chevron-shaped chambers which do not extend to the coiled portion; aperture radiate, terminal, central.
Remarks: Treated here as distinct species since it tends to be characteristic of particular stratigraphical horizons. However, it is probably a morphological variant of *Planularia beierana* Gümbel. Distribution: Bajocian–Bathonian. Occurs most commonly at the top of Bajocian–basal Bathonian, *parkinsoni–zigzag* Zones, and Wattonensis Beds, top *retrocostatum?* Zone [5, 20].

Frondicularia franconica Gümbel, 1862
Plate 6.2.3, Fig. 3, SAB 452 F1, × 110, Middle Oxford Clay, Milton Keynes. Description: Test compressed, uniserial, smooth, oval or bilobate in cross section; margins rounded, unkeeled; chambers broad exhibiting variable growth rate to give irregular outline; sutures depressed or flush, usually chevron-shaped but may be transverse in later chambers; aperture radiate, terminal, central, elevated on a short neck. Distribution: Callovian: *jason–athleta* [11], *coronatum* [10], *athleta–lamberti* [21].

Frondicularia nympha Kopik, 1969
Plate 6.2.3, Fig. 4, MPK 2408, × 115, Lower Fuller's Earth, Dorset. Description: Test compressed, uniserial with variable outline; margins occasionally rounded but generally keeled; composed of 2–8 chevron-shaped chambers with thickened, raised sutures; aperture radiate, terminal, central, elevated on short neck. Distribution: Late Bajocian–Early Bathonian, *parkinsoni--zigzag* Zones [5].

Frondicularia oolithica Terquem, 1870
Plate 6.2.3, Fig. 5, MPK 2217, × 90. Upper Fuller's Earth, Bath. Description: Test compressed, uniserial, elongate to quadrate; spherical proloculus followed by 6–10 chevron-shaped chambers, sutures distinct, depressed, often marked by low ribs; aperture oval, terminal, central. Distribution: Bajocian: *parkinsoni* Zone; Bathonian: *retrcostatum–discus* [5, 6, 12, 20].

Guttulina pera Lalicker, 1950
Plate 6.2.3, Fig. 6, SAB 394 F1, × 150, Lower Oxford Clay, Milton Keynes. Description: Test small, globular in outline, early chambers strongly embraced by later ones added in planes 144° apart; sutures distinct, depressed; aperture radiate, terminal. Distribution: Callovian: *jason* Zone [11].

Lenticulina dictyodes (Deecke)
Plate 6.2.3, Fig. 7, MPK 2381, × 114, Upper Fuller's Earth, Dorset = *Cristellaria dictyodes* Deecke, 1884. Description: Test somewhat compressed, elongate, with initial 4–6 chambers arranged in small coil, later 2–4 chambers uncoiling; sutures may be slightly depressed but generally concealed by distinctive, fine, mesh-like ornament which covers entire surface of test; aperture radiate, terminal. Distribution: Bajocian: *concavum–discites* [8]; Bathonian *retrocostatum* [5, 20].

Plate 6.2.3

Lenticulina galeata (Terquem)
Plate 6.2.3, Figs. 8–9. Fig. 8, MPK 2402, × 70, Lower Fuller's Earth, Dorset; Fig. 9, MPK 2897, × 130, Lower Fuller's Earth, Stowell, Somerset = *Cristellaria galeata* Terquem, 1870. Description: Chambers arranged in slightly involute, biconvex planispiral, most close coiled but occasionally uncoiled in later portion; sutures deeply depressed, generally with strong ribs along the margins extending to the keel; development of both ribs and keel variable.
Remarks: A tendency towards uncoiling and a triangular, *Saracenaria*-like cross section, is prevalent in the Bathonian examples. Distribution: Bajocian–Bathonian. Most common in the Lower Fuller's Earth Clay and Fuller's Earth Rock, *zigzag*-basal *retrocostatum* Zones [5, 20].

Lenticulina cf. *limata* Schwager, 1867
Plate 6.2.3, Fig. 10, MPK 2404, × 78, Lower Fuller's Earth, Dorset. Description: Test compressed, smooth; initial planispiral coil of 3–5 chambers followed by slightly arcuate, uncoiled portion; sutures indistinct, flush; later chambers broader than high, fairly uniform in size, resulting in a parallel sided outline; final chamber occasionally smaller than preceding one; aperture radiate, terminal, peripheral. Distribution: Early Bathonian, *zigzag* Zone [5].

Lenticulina major (Bornemann)
Plate 6.2.3, Fig. 11, SAB 437 F2, × 90, Middle Oxford Clay, Milton Keynes = *Cristellaria major* Bronemann, 1854. Description: Test generally robust, somewhat compressed; initial coil of 3–5 chambers followed by rectilinear portion of up to 7 chambers, each one broader than high, increasing very little in size to give a parallel-sided outline; sutures marked by strongly limbate ribs. Distribution: Callovian: *jason–athleta* [11].

Lenticulina quenstedti (Gümbel)
Plate 6.2.3, Fig. 12, MPK 2247, × 130, Lower Fuller's Earth, Bath = *Cristellaria quenstedti* Gümbel, 1862. Common synonym: *Cristellaria polonica* Wisniowski, 1890. Description: Test biconvex, involute planispiral; chambers increasing gradually in size; sutures marked by distinct sharp ribs which generally converge to form a circular umbilical rib; keel well developed especially in older chambers; aperture radiate, terminal, peripheral.
Remarks: Some Lower Bathonian specimens are difficult to distinguish from *L. galeata* and *L. tricarinella*. Distribution: Bajocian: *concavum-parkinsoni* [5, 8]; Bathonian: chiefly *zigzag–progracilis* [5, 6, 20]; Callovian: *athleta* [11], *athleta-lamberti* [21].

Lenticulina subalata (Reuss)
Plate 6.2.3, Fig. 13, MPK 2246, × 67, Twinhoe Beds, Bath = *Cristellaria subalata* Reuss, 1854. Description: Test robust, biconvex, involute, planispiral; periphery acute, sometimes angular, often keeled; sutures marked by rounded ribs which converge to form an umbilical boss; aperture radiate, terminal, peripheral. Distribution: Common Jurassic species. Within the Bathonian is particularly common in the Twinhoe Beds and equivalent strata, *aspidoides* Zone [5, 6, 20].

Lenticulina tricarinella (Reuss)
Plate 6.2.3, Figs. 14–15; Fig. 14 MPK 2208, × 100, Upper Fuller's Earth, Bath; Fig. 15, MPK 2226, × 100, Lower Fuller's Earth, Bath = *Cristellaria tricarinella* Reuss, 1863. Common synonym: *Cristellaria polymorpha* Terquem, 1870. Description: Test compressed; chambers arranged in loose planispiral with parallel sides; peripheral keel and distinctive ornament of sharp sutural ribs and lateral keels; degree of coiling variable, occasionally the lateral keels are replaced by impersistent oblique costae along margins of test; aperture radiate, terminal peripheral. Distribution: Late Bajocian: *garantiana-parkinsoni* [5, 8]; Bathonian *zigzag-retrocostatum* [5, 6, 20].

Lenticulina volubilis Dain, 1958
Plate 6.2.3, Fig. 16, MPK 2410, × 112, Lower Fuller's Earth, Dorset. Description: Test compressed, involute planispiral with tendency for later chambers to uncoil; distinct keel; sutures slightly depressed, bordered by costae bearing pits or depressions and short secondary costae which branch off upwards; aperture radiate, terminal, peripheral, situated on short neck. Distribution: Late Bajocian–Early Bathonian, *parkinsoni–zigzag* Zones [5].

Lingulina longiscata (Terquem)
Plate 6.2.4, Fig. 1, MPK 2220, × 60, Upper Fuller's Earth, Bath = *Frondicularia longiscata* Terquem, 1870. Common synonym: *Frondicularia nodosaria* Terquem, 1870. Description: Test highly compressed, uniserial, lanceolate; oval proloculus followed by up to 10 chambers which increase gradually in size; sutures usually strongly arched, depressed; margins rounded, lobulate; surface ornamented by large number of fine striae which do not continue from one chamber to the next; aperture elongate slit, terminal, central, produced on short neck. Distribution: Rare in Bajocian strata [8]; common Bathonian species [5, 6, 20], Callovian: *coronatum* [10], *athleta* [11].

Marginulina ectypa (Loeblich and Tappan)
Plate 6.2.4, Fig. 2, SAB 376 F1, × 150, Lower Oxford Clay, Milton Keynes = *Astacolus ectypus* Loeblich and Tappan, 1950. Description: Test compressed, involute, planispiral; later chambers uncoiling but generally maintaining contact with coil on ventral margin; sutures deeply depressed, exaggerated by inflation of chambers just anterior to sutures and presence of sharp sutural ribs; aperture radiate, terminal, peripheral, produced on a short neck. Distribution: Callovian: *calloviense–athleta* [11], *coronatum* [10], (as *L. quenstedti*).

Nodosaria hortensis Terquem, 1866
Plate 6.2.4, Fig. 3, MPK 2212, × 144, Upper Fuller's Earth, Bath. Description: Test uniserial, highly variable, orna-

Plate 6.2.4

mented by 8–10 continuous costae; aperture radiate, terminal, central.
Remarks: Similar to *N.fontinensis* Terquem which occurs commonly in the Bajocian and Lower Bathonian; differs in more robust nature of test, stronger costae which are generally fewer in number and chambers which are less spherical. Distribution: Common Bathonian species [5, 6, 20].

Nodosaria ingens (Terquem)
Plate 6.2.4, Fig. 4, MPK 2253, × 80, Forest Marble, Frome = *Dentalina ingens* Terquem, 1870. Common synonym: *Nodosaria guttifera* Bartenstein and Brand non d'Orbigny, 1937. Description: Test uniserial composed of spherical proloculus followed by 2–3 inflated chambers deeply constricted at the sutures; aperture radiate, terminal, central. Distribution: Cheifly Early Bathonian, *zigzag* Zone [5, 6, 20].

Nodosaria aff. *issleri* Franke *sensu* Cifelli, 1959
Plate 6.2.4, Fig. 5, MPK 2239, × 60, Upper Fuller's Earth, Bath. Description: Test robust, uniserial, tapered, slightly arcuate, composed of 4–6 chambers, increasing in breadth and height as added; sutures depressed, transverse, ornamented by 8–12 ribs which extend from base of proloculus to aperture situated on a short neck. Distribution: Late Bathonian, *retrocostatum-discus* Zones [5, 6, 20].

Nodosaria opalini Bartenstein, 1937
Plate 6.2.4, Fig. 6, MPK 2213, × 142, Forest Marble, Bath. Description: Test generally small, uniserial, tapered; sutures depressed or flush; ornamented by large number of fine ribs which are sometimes difficult to distinguish; aperture radiate, terminal, central. Distribution: common Middle Jurassic species [5, 6, 8, 20].

Nodosaria pectinata (Terquem)
Plate 6.2.4, Fig. 7, MPK 2254, × 100, Forest Marble, Frome = *Dentalina pectinata* Terquem, 1870. Description: Test uniserial, nearly always arched, composed of slightly elongate chambers constricted at sutures and ornamented by large number of costae; aperture not seen as specimens are invariably broken. Distribution: Late Bathonian, *retrocostatum-discus* zones with a distinctive occurrence in the Boueti Bd at the base of the Forest Marble [5, 6, 20].

Paalzowella feifeli (Paalzow), see Plate 6.2.1, Fig. 6.

Planularia eugenii (Terquem)
Plate 6.2.4, Figs. 8–9, Fig. 8 SAB 477 F1, × 85, Lower Oxford Clay, Milton Keynes; Fig. 9, MPK 2765, × 114, Lower Fuller's Earth, Dorset = *Cristellaria eugenii* Terquem, 1864. Description: Test compressed, variable in outline but generally elongate with parallel margins; initial 4–5 chambers arranged in loose planispiral, later chambers uncoiling; entire surface ornamented by variable number of ribs which may or may not be continuous: aperture radiate, terminal, peripheral.
Remarks: Some specimens resemble *P. beierana* (Gümbel) but are distinguished by presence of ornament. Distribution: Characteristic of top Bajocian and early Bathonian *parkinsoni-zigzag* Zones [5, 6, 20]. Reappers in the Callovian (Lower Oxford Clay) although this may prove to be a separate species. *coronatum-athleta* [11].

Planularia beierana (Gümbel)
Plate 6.2.4, Fig. 10, MPK 2249, × 150, Lower Fuller's Earth, Bath = *Marginulina beierana* Gümbel, 1862. Description: Test highly compressed; initial chambers arranged in loose planispiral; later uncoiled chambers may or may not retain contact with the spiral portion; sutures generally flush but may be depressed or marked by low ribs; aperture radiate, terminal, peripheral.
Remarks: *Falsopalmula obliqua* (Terquem) is probably a flabelline variant of this species. Distribution: Common Middle Jurassic species.

Pseudolamarckina rjasanensis (Uhlig)
Plate 6.2.4, Figs. 11–13. Fig. 11, SAB 473 F3, × 150; Fig. 12, SAB 473 F4, × 150, Middle Oxford Clay, Milton Keynes = *Pulvinulina rjasanensis* Uhlig, 1883. Description: Test trochoid, plano-convex or concavo-convex; chambers semicircular, arranged in 2–3 whorls with 5 chambers in final whorl; coiling generally dextral; initial chambers not usually visible; sutures on spiral side marked by low ribs; margin rounded – acute; on umbilical side sutures are depressed and form characteristic branchings near umbilicus; aperture small, loop-shaped, umbilical. Distribution: late Callovian, *athleta* Zone [11].

Reinholdella crebra Pazdro, 1969
Plate 6.2.4, Figs. 13–14, Fig. 13, MPK 2387, × 110, Lower Fuller's Earth, Dorset; Fig. 14, MPK 2766, × 150, Lower Fuller's Earth, Lyme Bay. Description: Test trochoid, plano-convex, smooth; 12–16 semicircular chambers arranged in 2–3 whorls, coiling generally sinistral; margin acute; sutures on umbilical side radial, slightly depressed; umbilicus closed; aperture is deep, loop-shaped indentation on last septal suture, extending almost to middle of final chamber; previous apertures closed becoming less visible in earlier chambers, all perpendicular to suture and almost perpendicular to one another. Distribution: Bathonian [5, 20].

Saracenaria oxfordiana Tappan, 1958
Plate 6.2.4, Fig. 15, MPK 2767, × 120, Lower Oxford Clay, Milton Keynes. Common synonym: *Saracenaria triquetra* Gümbel, 1862. Description: Initial 4–7 chambers arranged in small planispiral, later 4–5 chambers uncoiling, triangular

in cross section, periphery keeled, margins of apertural face acute, elevated; sutures distinct, depressed; aperture radiate, terminal, peripheral.
Remarks: Differs from *S. cornucopiae* (Schwager) in larger size of coil, broader apertural face and smaller number of chambers in linear portion. Distribution: Late Bajocian: *garantiana-parkinsoni* [5]: Bathonian particularly *zigzag* Zone; Callovian: *coronatum* [10], *coronatum-athleta* [11], *athleta* [21].

Tristix oolithica (Terquem)
Plate 6.2.4, Figs. 16–17. Fig. 16, MPK 2453, × 125; Fig. 17, MPK 2269, × 90, Frome Clay, Lyme Bay = *Tritaxia oolithica* Terquem, 1856. Common synonym: *Tristix suprajurassica* (Paalzow) 1932. Description: Test uniserial, composed of spherical proloculus followed by 5–8 triangular or rarely quadrangular chambers which gradually increase in size to give flared outline; final chamber sometimes smaller than preceding one; sutures strongly depressed; margins of test rounded or keeled; aperture radiate, terminal central, on short neck. Distribution: Bathonian: *retrocostatum-aspidoides* [5, 20]; Callovian: *jason* [11], *coronatum* [10].

Vaginulina clathrata (Terquem) *eypensa* Cifelli, 1959
Plate 6.2.4, Fig. 18, MPK 2209 × 75, Upper Fuller's Earth, Bath. Description: Test large, robust, uniserial, composed of 5–7 chambers increasing rapidly in breadth as added, to give triangular outline; sutures oblique, slightly depressed; ornamented by numerous strong costae which are generally continuous over the surface of the test; aperture radiate, terminal, peripheral.
Remarks: As stated by Cifelli, this subspecies succeeds *V. clathrata* (Terquem) *sensu stricto* in the Bathonian. It is distinguished by a broader, more divergent test and more irregular development of the ribs. Distribution: Bathonian [5, 6, 20].

Vaginulina harpa (Roemer) *sensu* Cifelli, 1959
Plate 6.2.4, Fig. 19, MPK 2244, × 40, Frome Clay, Bath. Description: Test large, compressed, uniserial, composed of 8–9 chambers which at first gradually increase in breadth but then remain fairly constant to give parallel sided outline; sutures oblique, slightly depressed; ornamented by large number of fine ribs which are continuous over several chambers; aperture radiate, terminal, peripheral. Distribution: Late Bathonian: *retrocostatum-aspidoides* [5, 6, 20].

Vaginulina legumen (Linné)
Plate 6.2.4, Fig. 20, MPK 2238, × 45, Upper Fuller's Earth, Bath = *Nautilus legumen* Linné, 1758. Description: Test large, robust, uniserial, oval in cross section, curved or straight; chambers broad, increasing very gradually in size giving smooth outline to test; sutures oblique, generally flush although may be depressed in later part of test; aperture radiate, terminal peripheral. Distribution: Late Bathonian: *retrocostatum-discus* [5, 6, 12, 20].

6.2.7 REFERENCES

[1] Arkell, W. J. 1956. *Jurassic Geology of the World,* Oliver and Boyd, Edinburgh, 806 pp.

[2] Bartenstein, H. and Brand, E. 1937. Mikro-paläontologische Untersuchungen zur Stratigraphie des nordwest-deutschen Lias und Doggers. *Abh. Senckenb. Naturf. Ges,* **439**, 1–224.

[3] Bielecka, W. and Styk, O. 1969. Some stratigraphically important Kuiavian and Bathonian Foraminifera of the Polish Lowlands. *Rocznik Pol. Tow. Geol.,* **39**, 515–531.

[4] Callomon, J. H. 1968. The Kellaways Beds and the Oxford Clay, *In* Sylvester-Bradley, P. C. and Ford, T. D., Eds. *The Geology of the East Midlands,* Leicester University Press, 264–290.

[5] Calver, M. A. and Rhys, G. H. (Eds.) 1981. The Winterborne Kingston Borehole, Dorset, England. *Rep. Inst. Geol. Sci.,* 81/3.

[6] Cifelli, R. 1959. Bathonian foraminifera of England. *Bull. Comp. Zool. Harvard,* **121**, 265–368.

[7] Cordey, W. G. 1962. Foraminifera from the Oxford Clay of Staffin Bay, Isle of Skye, Scotland. *Senck. leth.* **43**, 375–409.

[8] Dingwall *et al.* (in press). *Bajocian stratigraphy of Borehole 74/48, Lyme Bay.*

[9] Gordon, W. A. 1967. Foraminifera from the Callovain of Brora, Scotland. *Micropaleontology,* **13**, 445–464.

[10] Gordon, W. A. 1970. Biogeography of Jurassic Foraminifera. *Bull Geol. Soc. Am.,* **81**, 1689–1704.

[11] Horton, A. *et al.* 1974. The geology of the new town of Milton Keynes. *I.G.S. Report* No. 74/16, 90–102.

[12] Jones, T. R. 1884. Notes on the Foraminifera and Ostracoda from the Deep Boring at Richmond. *Jl. geol. Soc. Lond.,* **11**, 765–777.

[13] Kopik, J. 1969. On some representatives of the family Nodosariidae (Foraminiferida) from the Middle Jurassic of Poland. *Rocznik. Pol. Tow. Geol,* **39**, 533–552.

[14] Lindenberg, H. G. 1967. Unterscuchungen an lituoliden Foraminiferen aus dem SW deutschen Dogger 2: Die Arten von *Haplophragmium* und *Triplasia.* Eine Bearbeitung auf biometrischer und paläokologischer Grundlage. *Abh. senckenb. naturf. Ges.,* **514**, 1–74.

[15] Loeblich, A. R. and Tappan, H. 1964. *Sarcodina. Chiefly 'Thecamoebians' and Foraminiferida. Treatise on Invertebrate Paleontology, Part C, Protista 2,* **1** and **2**, 900 pp.
[16] Loeblich, A. R. and Tappan, H. 1974. Recent advances in the classification of the Foraminiferida. *In* Hedley R. H. and Adams, C. G. (eds.) *Foraminifera.* Vol. 1, 1–53.
[17] Munk, C. 1978. Feinstratigraphische und mikropaläontologische untersuchungen an Foraminiferen-Faunen im Mittlersen und Oberen Dogger (Bajocien–Callovien) der Frankenalb. *Erlanger geol. Abh.,* **105**, 1-72.
[18] Parsons, C. F. 1974. The *sauzei* and 'so-called' *sowerbyi* Zones of the Lower Bajocian. *Newsletters on Stratigraphy,* Leiden, **3**, 153–180.
[19] Parsons, C. F. 1975. Ammonites from the Doulting Conglomerate Bed (Upper Bajocian, Jurassic) of Somerset. *Palaeontology,* **18**, 191–205.
[20] Penn, I. E., Merriman, R. J. and Wyatt, R. J. 1979. A proposed type section for the Fuller's Earth (Bathonian) based on the Horsecombe Vale No. 15 Borehole, near Bath, with details of contiguous strata, Part 1 in The Bathonian Strata of Bath–Frome areal. *Rep. Inst. Geol. Sci.* No. 78/22, pp. 1–22.
[21] Richardson, G. 1979. The Mesozoic stratigraphy of two boreholes near Worlaby, Humberside. *Bull. Geol. Surv. G.B.,* **58**, 24 pp.
[22] Ruget, C. 1973. Inventaire des microfaunes de Bathonien moyen de l'Algarve (Portugal). *Rev. Faculd. Ciencias Lisboa,* **17**, 515–542.
[23] Rutten, M. G. 1956. Depositional environment of Oxford Clay at Woodham Claypit. *Geol. en Mijrib.,* **18**, 344.
[24] Terquem, O. 1868. Premier Mémoire sur les Foraminifères du Système Oolithique Etude du Fullers-Earth de la Moselle. *Bull. Soc. Hist. Nat. Moselle,* **11**, 1–138.
[25] Terquem, O. 1870a. Deuxième Mémoire sur les Foraminifères du Système Oolithique. Zone à Ammonites parkinsoni de la Moselle. *Mém. Imp. Acad. Metz.,* **50**, 403–456.
[26] Terquem, O. 1870b. Troisième Mémoire sur les Foraminifères de Système Oolithique comprenant les genres *Frondicularia, Flabellina, Nodosaria,* etc. de la Zone *Ammonites parkinsoni* de Fontoy (Moselle) *Mem. Imp. Acad. Metz.,* **51**, 299–380.
[27] Terquem, O. 1883. Cinquième Mémoire sur les Foraminifères du Système Oolithique de la zone à Ammonites parkinsoni de Fontoy (Moselle).
[28] Terquem, O. 1886. Les Foraminifères et les Ostracodes de Fuller's Earth des environs de Varsovie. *Mém. Soc. Géol. France.* ser. **3, 4, 112** pp.
[29] Torrens, H. S. 1969. The stratigraphical distribution of Bathonian ammonites in central England. *Geol. Mag.,* **106**, pp. 63–76.
[30] Torrens, H. S. 1971. Standard zones of the Bathonian *in* Colloque du Jurassique, Luxembourge, 1967. *Mem. Bur. Rech. Geol. Min. Fr.,* **75**, 582–604.

6.3 THE CALLOVIAN TO PORTLANDIAN

D. Shipp, J. W. Murray

6.3.1 Introduction

The pioneer work on the study of Jurassic foraminifera was carried out in the latter half of the nineteenth century mainly by continental workers who initially concentrated their attention on the Lias. Some of the more important papers in this category include those by Bornemann (1854), Terquem (1855 onwards), Berthelin (1879) and Häusler (1881). English workers of this time also dealt mainly with the Lias and their papers include those of Strickland (1846), Jones and Parker (1860), Brady (1867) and Crick and Sherborn (1891).

Interest in Upper Jurassic foraminifera generally developed later than in Lias forms, although two of the earlier papers (those of Gümbel (1862) and Schwager (1865)) are both important and deal with the Oxford Clay. Other contributions from this period include those of Häusler (1883), Deeke (1886) and Wisniowski (1890 and 1891).

In Britain, Whitaker (1886) and Crick (1887) published unillustrated lists of Oxford Clay foraminifera. A short paper on adherent foraminifera from the Oxford Clay was produced by Sherborn (1888). Early papers on the Kimmeridge Clay include those by Blake (1875), Chapman (1897) and Woodward (1895).

Although these early workers laid the foundations of Jurassic micropalaeontology they are to some extent responsible for the confusion that now exists in Jurassic nomenclature. This is partly due to the fact that they often erected a number of separate species names for what were probably variations of the same species and partly because their work was often poorly illustrated.

After the war, Barnard initiated further studies on Late Jurassic foraminifera with his papers on the Oxford Clay (1952 and 1953). Both are useful and well illustrated. More recent papers on the British Late Jurassic have included those on the Oxford Clay by Cordey (1962, 1963), Gordon (1967) and Coleman (1974); the Corallian by Gordon (1965); the Ampthill Clay by Gordon (1962), and the Kimmeridgian by Lloyd (1959, 1962). These are generally stratigraphic studies. Palaeoecology and palaeogeography have mainly been neglected, except by Gordon (1970).

Continental workers have not neglected the Late Jurassic in recent times with useful papers being produced by Lutze (1960, Germany); Bizon (1958) and Guyader (1968), both from France; Norling (1972, Sweden); Bielecka and Pozaryski (1954) and Bielecka (1960, Poland).

Three useful papers of a different kind where the authors revise the works of 'classic' micropalaeontologists are Seibold and Seibold (1955), revising the work of Gümbel (1862); Seibold and Seibold (1956), revising the work of Schwager (1865) and Bizon (1960), revising some of Terquem's type species.

Apart from papers on micropalaeontology, mention must be made of ammonite workers who produced the zonal scheme used in the Jurassic. Foremost are the works of Arkell (1933-1956) which provide the basis of our present zones. Mention must also be made of Brinkman (1929) and Spath (1939) who have both contributed to our knowledge of Jurassic ammonites. More recently Callomon (1955, 1964, 1968 a, b), Callomon and Cope, (1971) and Cope (1967, 1978) have given much attention to refining the details of the Upper Jurassic zonation.

6.3.2 Collections of Upper Jurassic Foraminifera

British Museum (Natural History)

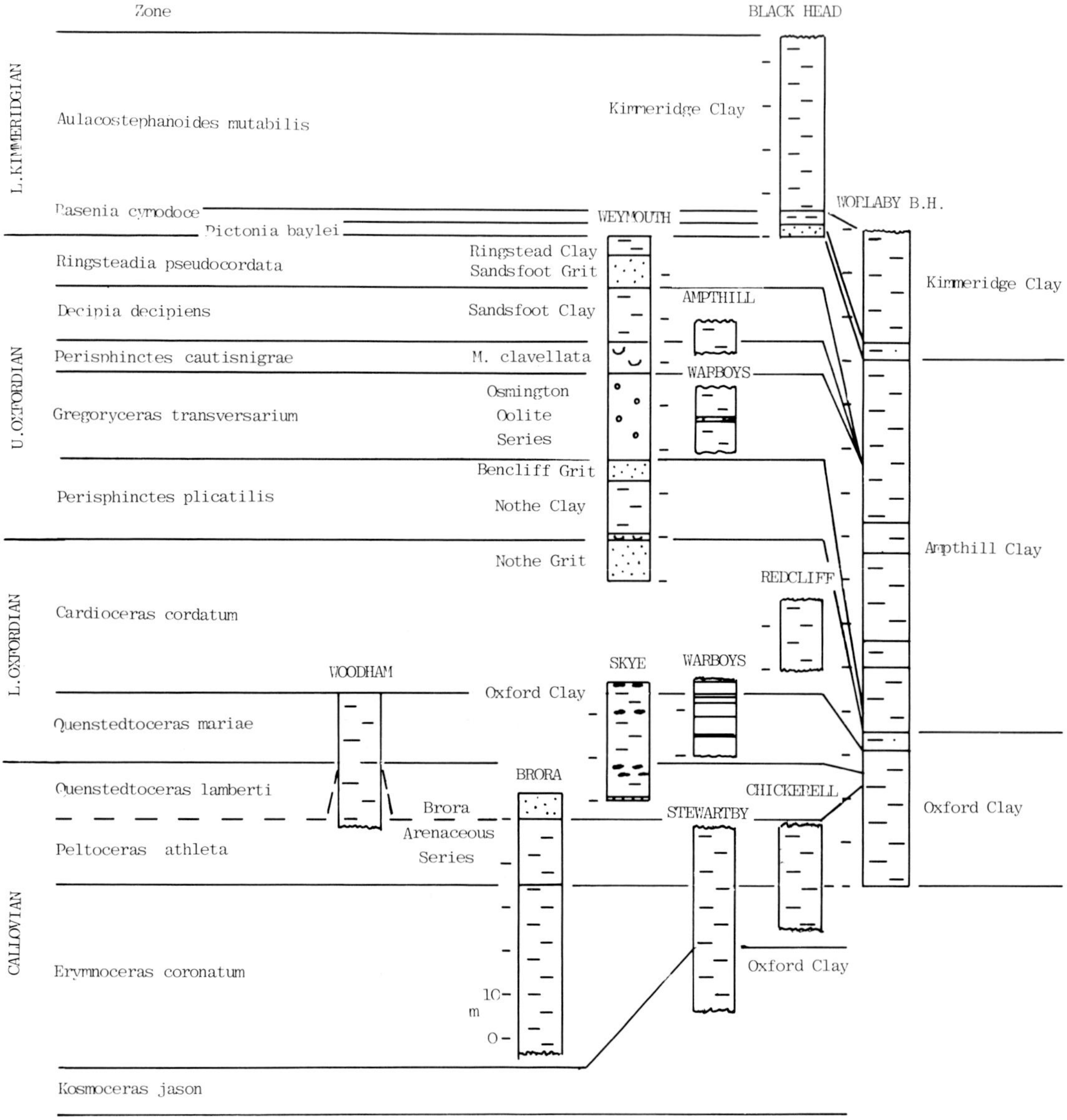

Fig. 6.3.1 – Stages, ammonite zones and lithostratigraphic successions. Brora (Gordon, 1967), Skye (Cordey, 1962), Woodham, Stewartby and Chickerell (Shipp, 1978), Warboys (Barnard, 1952; Shipp, 1978), Redcliff (Barnard, 1963; Shipp, 1978), Weymouth area (Gordon, 1965; Shipp, 1978), Ampthill (Gordon, 1962), Kimmeridge, Black Head (Lloyd, 1959), Worlaby, B. H. (Medd, 1979). For details of Upper Kimmeridge Clay see Table 6.3.1.

Barnard: Oxford Clay (Barnard, 1953)
Cordey: Oxford Clay (Gordon, 1962, 1963)
Gordon: Oxford Clay and Corallian (Gordon, 1963)
Lloyd: Kimmeridge Clay (Lloyd, 1959, 1962).
Institute of Geological Sciences
Medd: Oxford, Ampthill and Kimmeridge Clays (Medd, 1979)

A large collection of Upper Jurassic Foraminifera is also present in the Micropalaeontology Department of University College, London.

6.3.3 Stratigraphic Divisions

Although the Lower Oxford Clay has been briefly discussed in the Middle Jurassic section, it has also been included here because the faunas are of Upper Jurassic type. A summary of the stages, ammonite zones and lithological successions is given in Fig. 6.3.1 and Table 6.3.1. The range chart, Fig. 6.3.2, summarises the ranges from published and unpublished records.

Table 6.3.1

Summary of the stratigraphic units and the ammonite zones of the upper part of the Kimmeridge Clay (Based on data in Arkell, 1947; Cope, 1967, 1978; and Lloyd, 1959).

Zone	Bed	
Virgatopavlovia fittoni	Hounstout Marl Hounstout Clay Rhynchonella and Lingula Beds	
Pavlovia rotunda	Rotunda Shales Rotunda Nodule Bed Shales	Crushed ammonoid shales
Pavlovia pallasioides	Shales and Clay	
Pectinatites pectinatus	Shales and Clays Freshwater Steps Stone Band Shales with Middle White Band White Stone Band	
Pectinatites (Arkellites) hudlestoni	Dicey clays and shales Basal stone Shales Rope Lake Head Stone Band	
Pectinatites (Virgatosphinctoides) wheatleyensis	Shales Blackstone/Kimmeridge Oil Shale Shales and siltstones Grey Ledge Stone Band	
Pectinatites (Virgatosphinctoides) scitulus	Cattle Ledge Shales Yellow Ledge Stone Band	
Pectinatites (Virgatosphinctoides) elegans	Hen Cliff Shales	
Aulacostephanus autissiodorensis	Kimmeridge Bay Shales	
Aulacostephanus eudoxus	The Flats Stone Band	

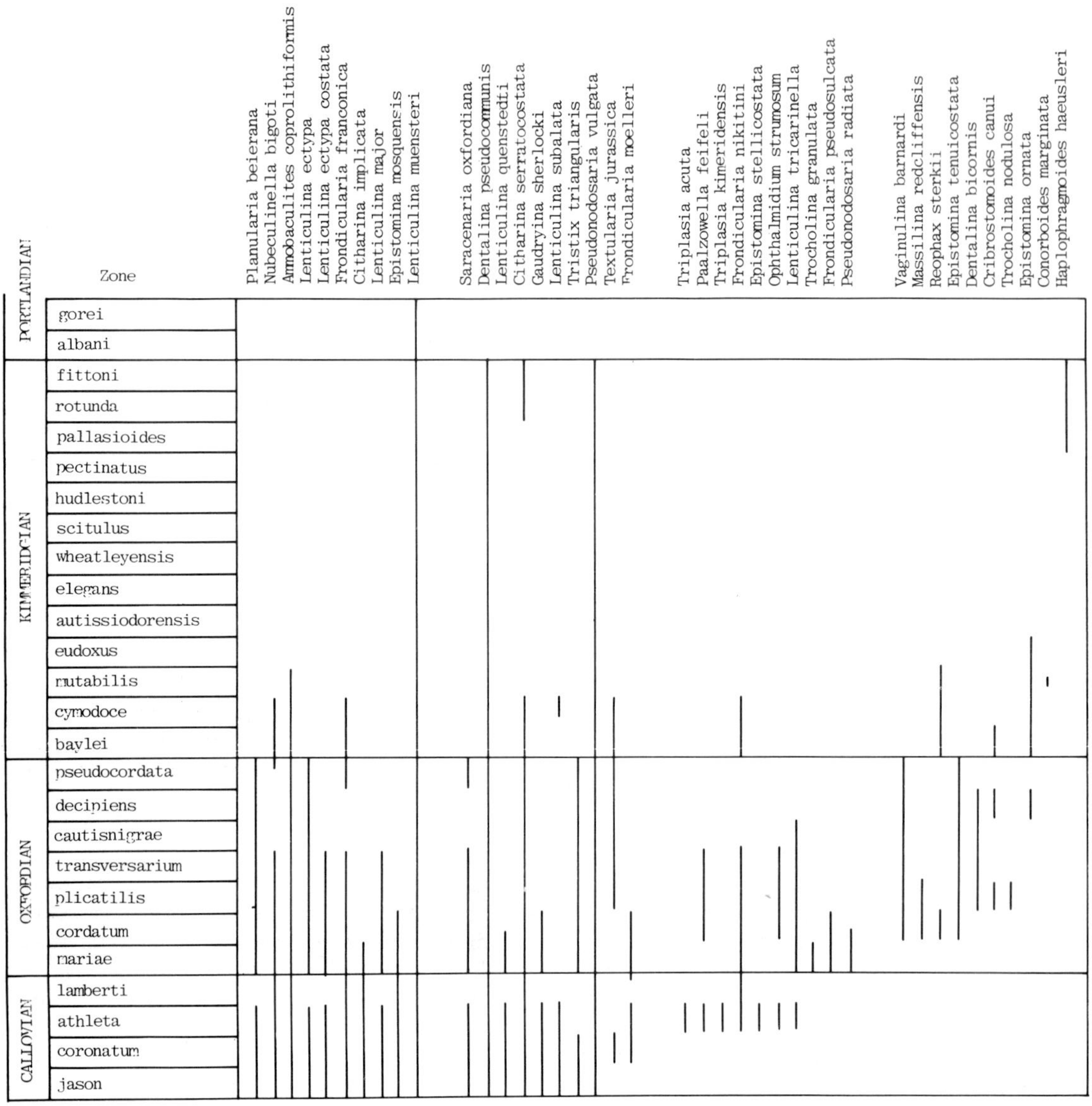

Fig. 6.3.2 – Summary range chart. For lithostratigraphy see Table 6.3.1 and Text-Fig. 6.3.1.

6.3.4 Foraminiferal Assemblages

6.3.4.1 *Oxfordian*

The argillaceous Brora Callovian succession was found to be characterised by three types of assemblage by Gordon (1967) and similar assemblages are found also in the Oxfordian:

(a) Those dominated by Textulariina, notably *Lagenammina difflugiformis, Reophax horridus, Ammobaculites suprajurassicus, Reophax sterkii* and *Ammobaculites coprolithiformis.* Such assemblages are characterised by their low diversity. It is possible that calcareous forms have been removed by solution during diagenesis.

(b) Those dominated by Rotaliina, especially nodosariaceans (60%) with some Textulariina (40%). The dominant species of *Lenticulina muensteri.* Other common forms are *Citharina flabellata, Dentalina guembeli, Lenticulina quenstedti, Marginulina scapha, Ammobaculites corprolithiformis, Eoguttulina liassica* and *Epistomina (Brotzenia) parastelligera.* Diversity is moderate.

(c) One assemblage with an even mixture of Rotaliina and Textulariina, characterised particularly by the dominance of *Epistomina (Brotzenia) parastelligera.*

The Oxford Clay of Skye (Cordey, 1962) extends through the *mariae* into the *cordatum* Zone (Fig. 6.3.1). Rotaliina, especially nodosariaceans, are dominant. *Lenticulina muensteri* is common to abundant in every sample. Other species common in one or two samples include *Planularia beierana, Epistomina (Brotzenia) mosquensis, Trocholina granulata* and *Conorboides pygmaea.* These assemblages are of (b) type above. In the Callovian *lamberti* Zone *Epistomina (Brotzenia) parastelligera* is common to abundant but is not associated with abundant Textulariina which are poorly preserved in this succession.

The Upper Oxford Clay of Warboys Pit, Huntingdonshire, spans the *mariae* and *cordatum* Zones. The asemblages are dominated by *Lenticulina quenstedti* and *L. muensteri* together with *Bullpora rostrata* (Barnard, 1952). These assemblages again are of type (b) above. The *coronatum* Zone at Redcliff Point, Dorset, yielded assemblages dominated by *Ammobaculites suprajurassica* and *Lenticulina muensteri* (Barnard, 1953). These may be type (c) assemblages. Barnard noted changes in overall abundance and in size in the foraminifera of this section although the ostracods showed no similar variation.

In addition Shipp (unpublished thesis 1978) has obtained a large and varied fauna from the English Oxford Clay from both the Callovian and Oxfordian. Preservation of the foraminifera is generally good. In the total number of species and individuals, the Lower Oxford Clay (Callovian) is poorer than the Middle (Callovian) or Upper (Oxfordian). He recorded fifty-four species from the Upper Oxford Clay and fifty-seven from the Middle, but only thirty-five from the Lower. The reason for this paucity of foraminifera in the Lower Oxford Clay is not apparent.

The fauna is dominated by the superfamily Nodosariacea, which constitutes over 75% of the species and a much greater percentage of the total population. The genus *Lenticulina* occurs in virtually every sample. Agglutinated foraminifera form the second most important element of the fauna with the species *Ammobaculites coprolithiformis* being especially prominent. Agglutinated forms often dominate individual samples in which calcareous forms are few or absent.

Species belonging to the Epistomininae and Ophthalmidiinae, although less significant in the total fauna, often swamp individual samples with genera such as *Ophthalmidium* and *Epistomina* frequently occurring in large numbers.

The Corallian of the Dorset coast east of Weymouth (Arkell, 1947) has been examined by Gordon (1965). Two species dominate the assemblages at most levels: *Ammobaculites coprolithiformis* and *Lenticulina muensteri.* Gordon commented on the similarity of the assemblages throughout this sequence. This seems surprising considering the varied environments of deposition inferred by Talbot (1973). However, examination of the details of Gordon's sampling shows that he restricted himself to clays and sands. Thus, not all the depositional environments were sampled for foraminifera. The Corallian assemblages are mainly of type (b).

Two levels within the Ampthill Clay, thought to be within the *cautisnigrae* and *decipiens* Zones, yielded assemblages which were considered to be similar to those of the Dorset Corallian (Gordon, 1962).

Shipp (1978, unpublished thesis) also studied the Corallian and recorded a microfauna, which although it does not contain as many species as seen in the Oxford Clay, is nevertheless numerous and diverse. Forty-eight species were identified of which eight are agglutinating, while of the remaining forty species the vast majority (26) are representatives of the Nodosariidae.

The fauna is dominated by four species; *Lenticulina muensteri, Ammobaculites coprolithiformis, Citharina serratocostata* and *Spirillina tenuissima. Lenticulina muensteri* constitutes almost 50% of the specimens recovered from the Dorset Coast and Warboys section and a quarter of the specimens from Millbrook. *Ammobaculites coprolithiformis* forms 23% of the fauna from Dorset and 14% of the fauna from Millbrook although it is much less numerous at Warboys.

Citharina serratocostata and *Spirillina tenuissima* are most common at Millbrook and Warboys but occur in much fewer numbers on the Dorset Coast. Ophthalmidiinae, Polymorphinidae and Epistomininae are generally less common than in the Oxford Clay.

6.3.4.2 *Kimmeridgian*

The type Kimmeridge Clay faunas were studied by Lloyd (1959, 1962) but the Nodosariacea were not discussed. Exton (personal communication) has provided the following information:

In the upper part of the Kimmeridge Clay (*scitulus* to *hudlestoni* Zones) the assemblages are dominated by small, flattened agglutinated forms of *Ammobaculites, Trochammina, Textularia* and *Haplophragmoides.* Nodosariids (mostly *Lenticulina muensteri* and *Marginulina* sp.) and other calcareous groups are rare. In the interval from the *pectinatus* to the *rotunda* Zone, nodosariids are almost as abundant as agglutinated genera. The abundance and size of the faunas also shows change. From the *scitulus* to lower *pectinatus* Zone the agglutinating faunas are rich in numbers of individuals but low in species diversity (bituminous shales are abundant at this level). From the upper *pectinatus* Zone to the *fittoni* Zone there is an increase in the size of the individuals associated with an increase in the proportion of quartz silt and calcareous content of the sediment. There may have been increased oxygen availability in the environment.

The Kimmeridge Clay of Humberside has been investigated in the Worlaby G borehole. Only the *baylei* and *cymodoce* Zones are adequately represented. The assemblages are similar to those of the underlying Ampthill Clay, being dominated by *Trochammina squamata, Ammobaculites* sp., *Epistomina porcellanea* (given as *Brotzenia*) and *Lenticulina muensteri* (Medd, 1979).

6.3.4.3 *Portlandian*

The only section examined from the Portlandian is from the *albani* Zone of the Portland Sand of Ringstead Bay, Shipp (1978, unpublished thesis).

The foraminifera seen in the Portland Sand are very sparse. Only a few specimens of *Lenticulina muensteri* have been recovered from the samples immediately below the Exogyra Bed.

6.3.5 Palaeoecology of the Foraminifera

The Oxford Clay has a rich benthic fauna of epifaunal and infaunal elements and this demonstrates that the sea was oxygenated down to the bottom (Hudson and Palframan, 1969; Duff, 1975). The presence of a benthic foraminiferal fauna supports this interpretation.

The environmental significance of the various groups of foraminifera found in the Oxford Clay is tentatively inferred by Shipp (1978, unpublished thesis) as follows:

(a) Small agglutinated specimens (*Trochammina, Gaudryina,* small *Ammobaculites,* etc.). These probably represent shallow water with reduced salinity or less oxygenated bottom conditions, especially when they occur alone.

(b) Large agglutinated specimens (*Ammobaculites coprolithiformis* mainly, but also *Triplasia*). These appear to represent generally deeper water conditions.

(c) Species of *Ophthalmidium* are thought to represent a shallow water environment.

(d) *Epistomina* appears to reflect normal marine and hence possibly deeper water conditions.

(e) Adherent species are thought to reflect breaks in the succession which may be due to a slow rate of deposition and/or to increased bottom currents.

(f) Nodosariacea, (excluding the Polymorphinidae) primarily indicate deeper water, but are not restricted to it, occurring frequently with the other groups.

(g) Polymorphinidae (mainly *Eoguttulina*). As yet the significance of this group is not clear, but as they often occur in large numbers, there may be some distinct factor, such as salinity or clarity of water, affecting their distribution.

The possible ecological significance of the various groups is based on the proposals of previous workers, together with information derived from Shipp's study. The suggestions are, however, tentative and open to modification in the light of subsequent work.

As stated by Fürsich (1976) the Corallian forms a shallow water deposit representing a variety of environments ranging from offshore shelf to intertidal between the deeper water Oxford and Kimmeridge Clays.

Several authors have worked on the environment of the Dorset Corallian; Arkell, 1933, 1935; Wilson 1968a, b; Talbot, 1973, 1974; Brookfield, 1973 and Fürisch (op. cit.). Fürisch also worked on the Corallian of Yorkshire and

Normandy and considered that Talbot's reconstruction was the most satisfactory. Shipp (1978, unpublished thesis) agreed with this conclusion and used Talbot's model as a basis for his investigation of the palaeoecology of the Corallian foraminifera.

In comparison with those of the Dorset coast the Corallian microfaunas at Millbrook and Warboys contain a more varied nodosariid assemblage together with a greater number of forms such as *Ophthalmidium*, *Quinqueloculina* and *Spirillina*, thought to indicate shallow water conditions. The greater variety of nodosariids suggests a more stable environment. *Epistomina*, thought to represent open marine shelf conditions, is absent from the Millbrook and Warboys sections although it occurs in parts of the Dorset section. Thus in the central area of England the depth of water appears to have been less than in the deeper part of the Dorset section although still in the general offshore shelf category. Conditions were apparently also more restricted here with less open marine influence. The main connection with the open sea at this time is thought to have been to the south with more normal marine conditions more likely in the Dorset area.

Townson (1975) has suggested that the Portland Group was deposited under generally shallow conditions with periods when low oxygen levels prevailed on the sea bed. This may account for the low numbers of foraminifera present and provides further proof of the wide environmental tolerance of *L. muensteri*, (Shipp 1978, unpublished thesis).

Ostracods recorded from this section occur in a greater number of samples than the foraminifera and appear to have been more tolerant of the rather harsh conditions that existed at this time.

The Oxfordian and Kimmeridgian represent the periods of maximum eustatic sea level rise according to Hallam (1969, 1978). In addition, regional subsidence in the Kimmeridgian is thought to have led to further deepening (Hallam and Sellwood, 1976). Bituminous shales developed at certain times (Morris, 1980).

Most of the Upper Jurassic foraminiferal faunas described from Britain are from muddy sediments. Many of the assemblages are dominated by *Lenticulina* and *Ammobaculites*. Other relatively common genera are *Epistomina*, *Citharina*, *Marginulina*, *Dentalina* and *Eoguttulina*. None of these genera is common in any modern shelf environment, but *Ammobaculites* is common in some brackish areas (see Murray, 1973, for a summary). *Lenticulina*, *Marginulina* and *Dentalina* are all known from modern shelf, slope and deep-sea, cold-water, normal marine environments.

6.3.6 Comparisons with other Faunas – Oxford Clay and Corallian

The faunas of the Oxford Clay and Corallian are very similar and comparable to faunas which have been described by various authors from sediments of a similar age from other localities. Cordey (1962) has reported a similar fauna from the Oxford Clay of Skye as has Coleman (1974) from the Kellaways beds and Oxford Clay of Milton Keynes. Gordon has described similar assemblages from the English Corallian (1965) and Ampthill Clay (1962) as well as from the Callovian of Brora (1967).

On the continent similar assemblages to those seen in the English Oxford Clay and Corallian have been recorded by Lutze from northeast Germany (1960), Norling from south Sweden (1972), Bielecka and Pozaryski from central Poland (1954) and Kapterenko-Chernousova *et al.* (1963) from the Ukraine. The fauna described by Guyader (1968) from the Seine Estuary is also similar, but includes some specimens of *Globigerina oxfordiana*. Planktonic foraminifera have also been recorded by Bignot and Guyader (1966) from Normandy, Grigelis (1958) from Lithuania and Seibold and Seibold (1960) from southern Germany. All these records come from deposits of comparable age. This species has not been found in the English Oxford Clay or in any fauna further north than those mentioned above. It is therefore suggested that the main influences on the faunas of the English Oxford Clay and Corallian came from the north and east. The main connection with the east appears to have been to the north of the Anglo-Belgian Island, across the Market Weighton axis. There appears to have been little, if any, influence from France to the south.

6.3.7 North Sea

Late Jurassic microfaunas from the North Sea are generally much less varied than those onshore. Anaerobic bottom conditions appear to have occurred frequently and foraminifera are often rare or absent. The assemblages are usually dominated by agglutinating forms, principally species of *Haplophragmoides* which are usually quite large but poorly preserved. Members of the Nodosariidae are sometimes present but are usually less numerous and varied than onshore. *Lenticulina* is the commonest genus. Radiolarians, such as *Lithostrobus* spp., are frequently the commonest microfaunal representatives although these are rarely recorded in onshore sections.

6.3.8 Index Species

The species selected include both short-ranging and common long-ranging forms. The classification is that of Loeblich and Tappan (1964). The references have been numbered to save space in the text; Shipp, refers to an unpublished thesis (Foraminfera from the Oxford Clay and Corallian of England and the Kimmeridgian of the Boulonnais, France. University of London, 1978). All the stratigraphic ranges are shown in Fig. 6.3.2.

6.3.8.1 Suborder Textulariina

Ammobaculites coprolithiformis (Schwager)
Plate 6.3.1, Figs. 1, 2, × 19, Osmington Oolite Series, *transversarium* Zone, Osmington, Dorset. = *Haplophramium coprolithiforme* Schwager, 1867. Description: test commences with a planispiral whorl of 2–6 chambers, followed by an uncoiled unserial portion of up to 6 chambers; these form approximately one third and two thirds the length of the test respectively; initial part compressed, later circular in cross section; final chamber often inflated, conical or pyridorm. Some specimens remain close-coiled; length 0.8–1.5 mm occasionally up to 3.0 mm.
Remarks: this form is common throughout the Late Jurassic, especially so in the Oxfordian sections on the Dorset Coast. Distribution: Callovian *coronatum-lamberti* [15], *jason-athleta* [Shipp], [9], Oxfordian *mariae-pseudocordata* [Shipp], [26], *transversarium* [22], Kimmeridgian (as cf.) *baylei-mutabilis* [17]. Europe: Oxfordian and Kimmerdigian of Poland [7], Callovian and Oxfordian of Germany (as *Haplophragmium aequale*) [20] and Middle Jurassic of the French Jura under the same name [27].

Cribrostomoides canui (Cushman)
Plate 6.3.1, Figs. 3, 4, × 76. Nothe Clay, *plicatilis* Zone, Bowleaze Cove, Dorset. = *Haplophragmoides canui* Cushman, 1910. Description: planispiral, umbilicate, involute, sometimes evolute in the last half whorl. 9 or 10 chambers in the final whorl; aperture circular or oval, centrally placed on apertural face; average diameter 0.55 mm.
Remarks: this species assigned to *Ammobaculites laevigatus* Lozo by Lloyd (1959). Distribution: Oxfordian *plicatilis* [14], [Shipp], Kimmeridgian *baylei* [17], Ampthill Clay, *?decipiens* [13]. Europe: Oxfordian of Normandy [12].

Gaudryina sherlocki (Bettenstaedt)
Plate 6.3.1, Figs. 5, 6, × 95. Oxford Clay, *jason* Zone, Calvert. = *Gaudryinella sherlocki* Bettenstaedt, 1952. Description: initial trochoid portion of 3–5 chambers (often difficult to see), later reducing to 3 chambers per whorl and finally 2; aperture varies from simple, central terminal to crescentric at base of final chamber; average length 0.35 mm.
Remarks: variation shows a tendency towards a uniserial arrangement but it is never fully developed and the final chamber is always in contact with the antepenultimate one. Distribution: Callovian *calloviense-athleta* (as *Gaudryina?* sp.) [2, 9], *jason-athleta* [Shipp], Oxfordian *mariae* [10], *mariae-cordatum* [Shipp] also recorded as *Gaudryina* sp. 2 from late Jurassic of Germany [20] and France [16]. First described from German Early Cretaceous by Bettenstaedt. Specimens from the Speeton Clay, Lincolnshire are also identical.

Haplophragmoides haeusleri Lloyd, 1959
Plate 6.3.1, Figs. 7, 8, × 70, Holotype, *Rhynchonella* Marls, Kimmeridge Clay, Hounstout, Dorset (redrawn from Lloyd, 1959). Description: planispiral, involute, with 5–6 chambers in the last whorl; early chambers have rounded periphery, later chambers compressed with a subrounded periphery; aperture an interiomarginal arch with peripheral lip; diameter approx. 0.7 mm.
Remarks: recorded only from Upper Kimmeridge Clay [17]. Distribution: Kimmeridgian *pallasoinides-fittoni* [17].

Reophax sterkii Haeusler, 1880
Plate 6.3.1, Figs. 9, 10, × 30. Oxford Clay, Brora, Scotland (redrawn from Gordon, 1967). Description: test uniserial, straight or slightly curved, with 3 to 4 (rarely 6) chambers which increase rapidly in size, the pyriform final chamber forming about half the length of the test; length 0.9–1.2 mm.
Remarks: this is a large form, the pyriform final chamber is characteristic. Distribution: Callovian *coronatum* [15], Kimmeridgian *baylei-mutabilis* [17].

Textularia jurassica (Guembel)
Plate 6.3.1, Figs. 11, 12, Fig. 11 × 190, Fig. 12 × 80. Fig. 11 Ampthill Clay, *transversarium* Zone, Millbrook; Fig. 12 *Myophorella clavellata* Beds, Weymouth (redrawn from Gordon, 1965). = *Textilaria jurassica* Guembel, 1862. Description: test only slightly flaring, initial planispiral whorl of 4–5 chambers seen in some specimens; test usually consists of 18–25 biserially arranged chambers; aperture an interiomarginal arcuate slit; average length 0.4 mm.
Remarks: *Textularia pugiunculus* (Schwager) recorded by Gordon (1965) is thought to be a variant of this species. Distribution: Oxfordian *cordatum-cymodoce* [22], *plicatilis-cautisnigrae* and as *T. pugiunculus cautisnigrae-pseudocordata* [14], also *plicatilis* and *parandieri* [Shipp].

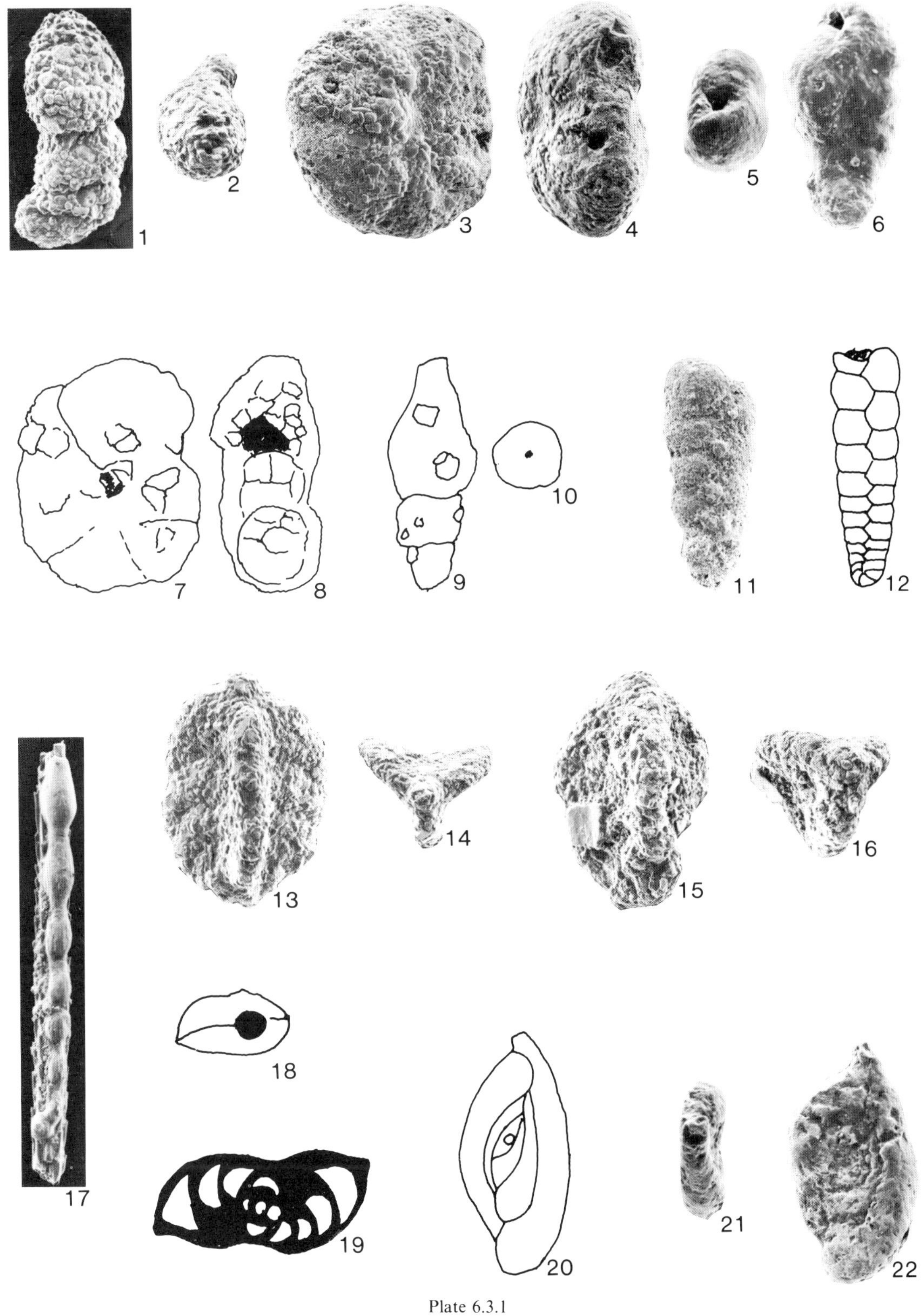

Plate 6.3.1

Triplasia acuta Bartenstein and Brand, 1951
Plate 6.3.1, Figs. 13, 14, × 38. Oxford Clay, *athleta* Zone, Woodham. Description: initial portion of 2–3 chambers planispiral, later uniserial portion of approximately 7 chambers; triangular in cross section, projections narrow and parallel-sided; aperture circular, terminal, possibly on a short neck; average length 1.00 mm, average greatest width 0.6 mm. Remarks: see remarks of *T. kimeridensis.* Distribution: Callovian *athleta* [Shipp].

Triplasia kimeridensis (Bielecka and Pozaryski)
Plate 6.3.1, Figs. 15, 16, × 50. Oxford Clay, *athleta* Zone, Woodham. = *Frankeina kimeridensis* Bielecka and Pozaryski, 1954. Description: at *T. acuta* except initial coil of 3–5 chambers, may be uniserial throughout; uniserial portion of 5–7 chambers.
Remarks: some specimens are quadrate in cross section. Occurs with *T. acuta* at Woodham but can be easily distinguished as *T. acuta* is smaller, has a more reduced initial coil while in cross section *T. kimeridensis* has less concave sides and the projections are wider and less parallel-sided. Distribution: Only found in Callovian *athleta* in Gt. Britain [Shipp]. Europe: Lower Kimmeridgian of Central Poland [7].

6.3.8.2 Suborder Miliolina
Massilina redcliffensis Gordon, 1965
Plate 6.3.1, Figs. 18–20, Fig. 18, and Fig. 20 × 100, Fig. 19 × 160). Holotype, Nothe Clay, near Weymouth (redrawn from Gordon, 1965). Description: test initially quinqueloculine, becoming massiline, approximately 12 chambers, periphery angular; sutures gently depressed to flush; aperture semicircular, with a broad, simple tooth; length 0.24–0.32 mm, up to 0.39 mm.
Remarks: this appears to be the only species of *Massilina* recorded from the Upper Jurassic. Distribution: Oxfordian *cordatum-plicatilis* [14].

Nubeculinella bigoti Cushman, 1930
Plate 6.3.1, Fig. 17, × 32. Nothe Clay, *plicatilis* Zone, Bowleaze Cove, Dorset. Description: Adherent test with initial coil of globose proloculus with a small tube of 2–7 chambers wound planispirally around it, followed by up to 12 flask-shaped hemispherical chambers; length of individual chamber 0.2–0.35 mm.
Remarks: usually attached to bivalve shells but also to echinoid spines, large foraminifera tests, etc. Often damaged with initial coil missing. Its presence may indicate a slow rate of sediment deposition. Common in Middle and Upper Oxford Clay, *athleta--cordatum* [Shipp]. Distribution: Callovian *jason–athleta* [Shipp], *athleta* [9], *athleta-lamberti* [22], Oxfordian *mariae-transversarium* [2, Shipp], *cordatum* [5], *pseudocordata* [22], Kimmeridgian *baylei-cymodoce* [22]. Originally recorded from *cordatum* Zone of Normandy [12], Adams [1] gives its range as Lias–Kimmeridgian.

Ophthalmidium strumosum (Guembel)
Plate 6.3.1, Figs. 21, 22, × 84, Nothe Clay, *plicatilis* Zone, Bowleaze Cove, Dorset. = *Guttulina strumosum* Guembel, 1862. Description: chambers half whorl in length tapering slightly from the base towards the oral end, sutures depressed; aperture terminal circular on short neck; length 0.3 mm–0.67 mm.
Remarks: the distinctive feature of this species is its inflated later chambers. Distribution: Oxfordian *cordatum-plicatilis* [Shipp], [14], *cordatum-transversarium* [22], also upper Oxfordian of Germany [20].

6.3.8.2 Suborder Rotaliina
Citharina implicata (Schwager)
Plate 6.3.2, Figs. 1, 2, × 80, Oxford Clay, *athleta* Zone, Bletchley. = *Cristellaria implicata* Schwager, 1865. Description: test oval in cross section, triangular in outline, average 7–8 chambers; sutures flush initially becoming depressed later; aperture marginal, terminal, simple; ornament of coarse ribs; average length 1.0 mm, average width 0.25 mm.
Remarks: *C. implicata* differs from *C. serratocostata* (Guembel), (see below) in having a narrower test and more steeply inclined sutures, it may however be an extreme variant. Distribution: Callovian *jason–mariae* [Shipp]. It has also been recorded from the Callovian of Germany [20] as *Citharina macilenta* (Terquem).

Citharina serratocostata (Guembel)
Plate 6.3.2, Figs. 3, 4, × 50. Oxford Clay, *athleta* Zone, Bletchley. = *Marginulina serratocostata* Guembel, 1862. Common synonym: *Citharina flabellata* (Guembel). Description: test compressed, triangular in outline, 7–16 chambers, sutures usually flush except at inner margin where they may be depressed; aperture marginal, terminal, radiate or simple; ornament of coarse, sometimes branching ribs, length 0.5 mm–1.5 mm, width 0.2–0.45 mm.
Remarks: see *C. implicata; C. serratocostata* is thought to be synonymous with *C. flabellata* [10], [13], [27], [Shipp]. This is a highly variable species and continuous variation to forms identical to *Falsopalmula anceps* (Terquem) and approaching *Frondicularia nikitini* Uhlig in appearance has been described [Shipp]. Distribution: common throughout the Callovian and Oxfordian [5 as cf], [9, 10 as *C. flabellata*], [13], [14], and [Shipp]. Kimmeridgian of Worlaby, *baylei-cymodoce* [22]. European occurrences include the Upper Jurassic of the Seine Estuary [16] and the Middle Jurassic of the French Jura [27].

Conorboides marginata Lloyd, 1962
Plate 6.3.2, Figs. 5, 9, 10, × 86. Holotype Kimmeridge Clay, Black Head, Dorset (redrawn from Lloyd, 1962). Des-

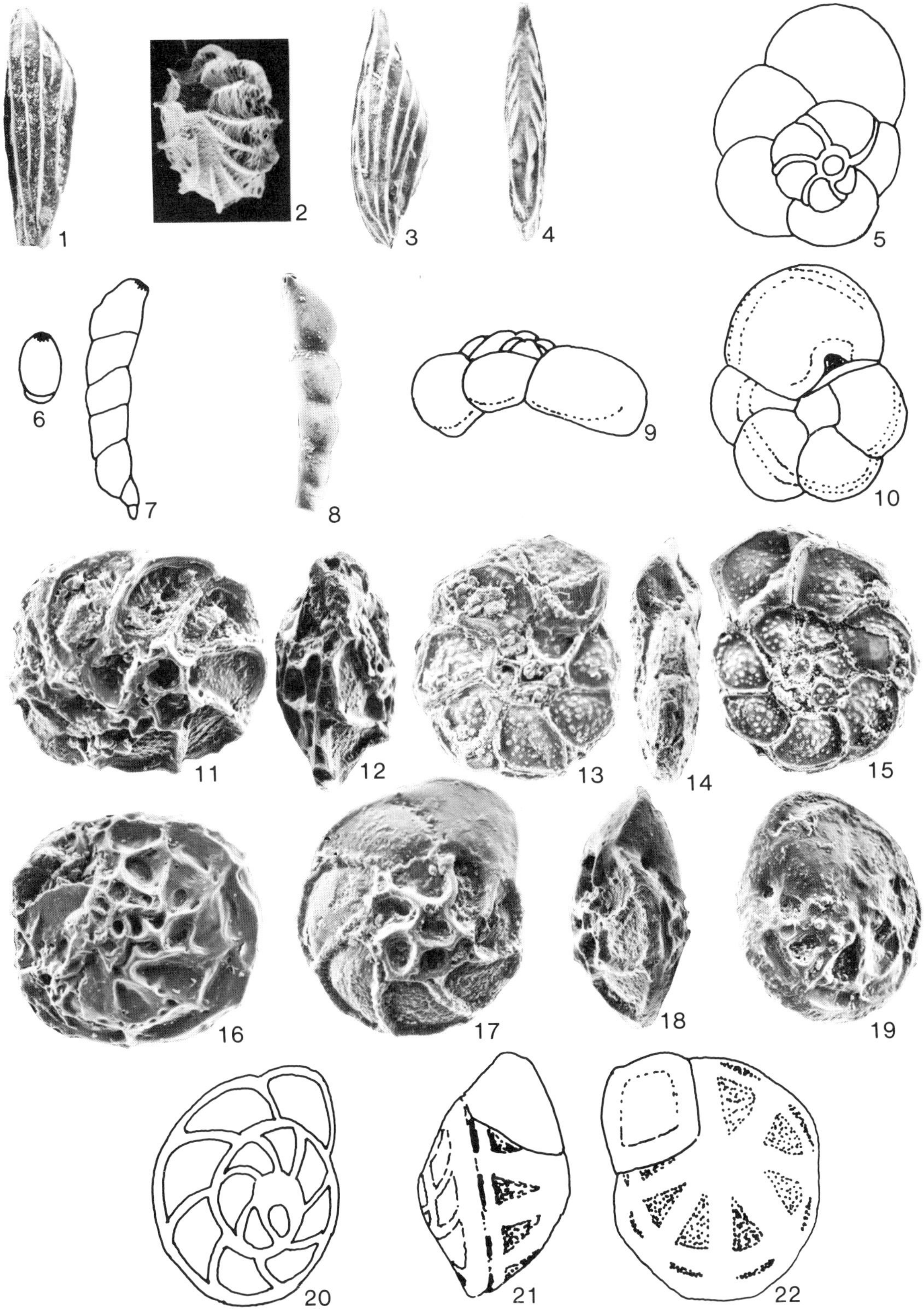

Plate 6.3.2

cription: test concavo-convex, trochospirally coiled with 4-6 inflated chambers in outer, whorl; periphery rounded; sutures initially limbate on spiral side but otherwise depressed; a submarginal thickening is visible on the umbilical side; aperture interiomarginal, loop-shaped; average diam. 0.4 mm.
Remarks: number of chambers and apical angle variable, a distinctive species with limited stratigraphic range. Distribution: Kimmeridgian middle of *mutabilis* where abundant [18].

Dentalina bicornis Terquem, 1870
Plate 6.3.2, Figs. 6, 7, × 32. Osmington Oolite Series, Osmington (redrawn from Gordon, 1965). Description: test uniserial, curved or rarely straight, about 8 chambers which increase in size away from the initial end; sutures depressed; aperture terminal radiate; length 0.75 mm–1.35 mm occasionally greater. Distribution: Oxfordian *athleta-plicatilis* [22], *plicatilis-decipiens* [14].

Dentalina pseudocommunis Franke, 1936
Plate 6.3.2, Fig. 8, × 120. Oxford Clay, *coronatum* Zone, Calvert. Description: test uniserial, arcuate consisting of 6–10 chambers becoming progressively larger, proloculum often apiculate; sutures oblique, slightly depressed; aperture terminal radial; length 0.4 mm–1.0 mm occasionally longer.
Remarks: *D. guembeli* Schwager appears to be very similar but the original specimens have more depressed sutures producing a lobate outline. Many later records of *D. guembeli* are thought to belong to *D. pseudocommunis* although with this rather variable foraminifer the latter may in fact be a junior synonym. Distribution: This is a common form throughout the Upper Jurassic, [10], as cf. *communis*), [Shipp]; [14, 9] (both as *D. guembeli*). Identical forms have also been recorded from the Lias [2].

Epistomina mosquensis Uhlig, 1883
Plate 6.3.2, Figs. 11, 12, 16, × 95. Oxford Clay, *jason* Zone, Stewartby. Common synonym: *Brotzenia mosquensis* (Uhlig). Description: test biconvex, with umbilical side more convex than dorsal, trochospiral; 6–8 chambers in outer whorl; intercameral and spiral sutures thickened and raised, those on spiral side curved, those on umbilical side radial about an umbilical ring; a weakly developed peripheral double keel, between which are the occluded peripheral apertures; diameter 0-35 mm–0.82 mm.
Remarks: close to *E. ornata* but the latter is more flattened with strongly developed peripheral keels. Distribution: Callovian *lamberti* [10], *jason-athleta* [Shipp], Oxfordian *mariae* [10], *mariae-cordatum* [Shipp], *cordatum* [11]. European records include, Lower Malm Poland [6], Late Jurassic Germany [24], [20], France, Seine Estuary [16], Wernli has recorded it from the Middle Jurassic of the French Jura [27].

Epistomina ornata (Roemer)
Plate 6.3.2, Figs. 13–15, × 95. Sandsfoot Clay, *decipiens* Zone, Shortlake, Dorset. = *Planulina ornata* Roemer, 1841. Description: test a flattened biconvex with 2 peripheral keels, trochospiral; 6–7 chambers in outer whorl, intercameral and spiral sutures thickened, raised and tuberculate; on umbilical side umbo is filled with a lattice of raised tuberculate ribs; on both surfaces the exposed parts of the chamber walls are finely punctate and covered with low, hemispherical tubercles; peripheral apertures between keels closed by a porous plate except in last chamber; septal foramen round and areal; diameter 0.3 mm–0.52 mm.
Remarks: distinguished by its distinctive ornament. Distribution: Oxfordian *decipiens* [Shipp], (as *E. stellicostata* var. *granulosa*), Kimmeridgian *baylei-euxodus* [18]. Europe: Upper Kimmeridgian of Poland [7], (as *E. stellicostata* var. *granulosa*).

Epistomina stellicostata Bielecka and Pozaryski, 1954
Plate 6.3.2, Figs. 17–19, × 47. Oxford Clay, *coronatum* Zone, Stewartby. Common synonym: *Brotzenia stellicostata* (Bielecka and Pozaryski). Description: test trochoidal, convex with 7–8 chambers in final whorl; sutures slightly raised into flat sutures; sutures on ventral surface are radially arranged and bear costae which form an irregular network over the central boss where they are covered in little pits; on dorsal surface costae often very broad and pitted; periphery bears two keels; peripheral apertures lie between the keels; final chamber has intermarginal septal aperture; diameter 0.22 mm–0.56 mm.
Remarks: distinguished by the pitted costae and central boss. Distribution: Callovian *athleta* [9], [Shipp]. Also Kimmeridgian of Poland [7].

Epistomina tenuicostata Bartenstein and Brand, 1951
Plate 6.3.2, Figs. 20–23, × 82. Ringstead Waxy Clay, Ringstead, Dorset (redrawn from Gordon, 1965). Common synonyms: *Brotzenia tenuicostata, Voorthuysenia tenuicostata*. Description: test biconvex, trochospiral with 7–8 chambers in final whorl; spiral side less convex than umbilical side; sutures limbate and curved backwards on spiral side, limbate and radial on umbilical side; apertures are present close to the periphery on the umbilical side; diameter 0.2 mm–0.4 mm, rarely larger. Distribution: Oxfordian *cordatum* [5], (as *E.* cf. *elegans*), *plicatilis-pseudocordata* [13], [15], [Shipp].

Frondicularia franconica Guembel, 1862
Plate 6.3.3, Figs. 1, 2, × 47. Oxford Clay, *mariae* Zone, Millbrook. Common synonym: *Frondicularia irregularis*, Terquem. Description: test compressed, uniserial, composed of 4–12 chambers, each a broad chevron shape, sutures generally flush; aperture central, terminal, radiate, length 0.25 mm–1.0 mm.

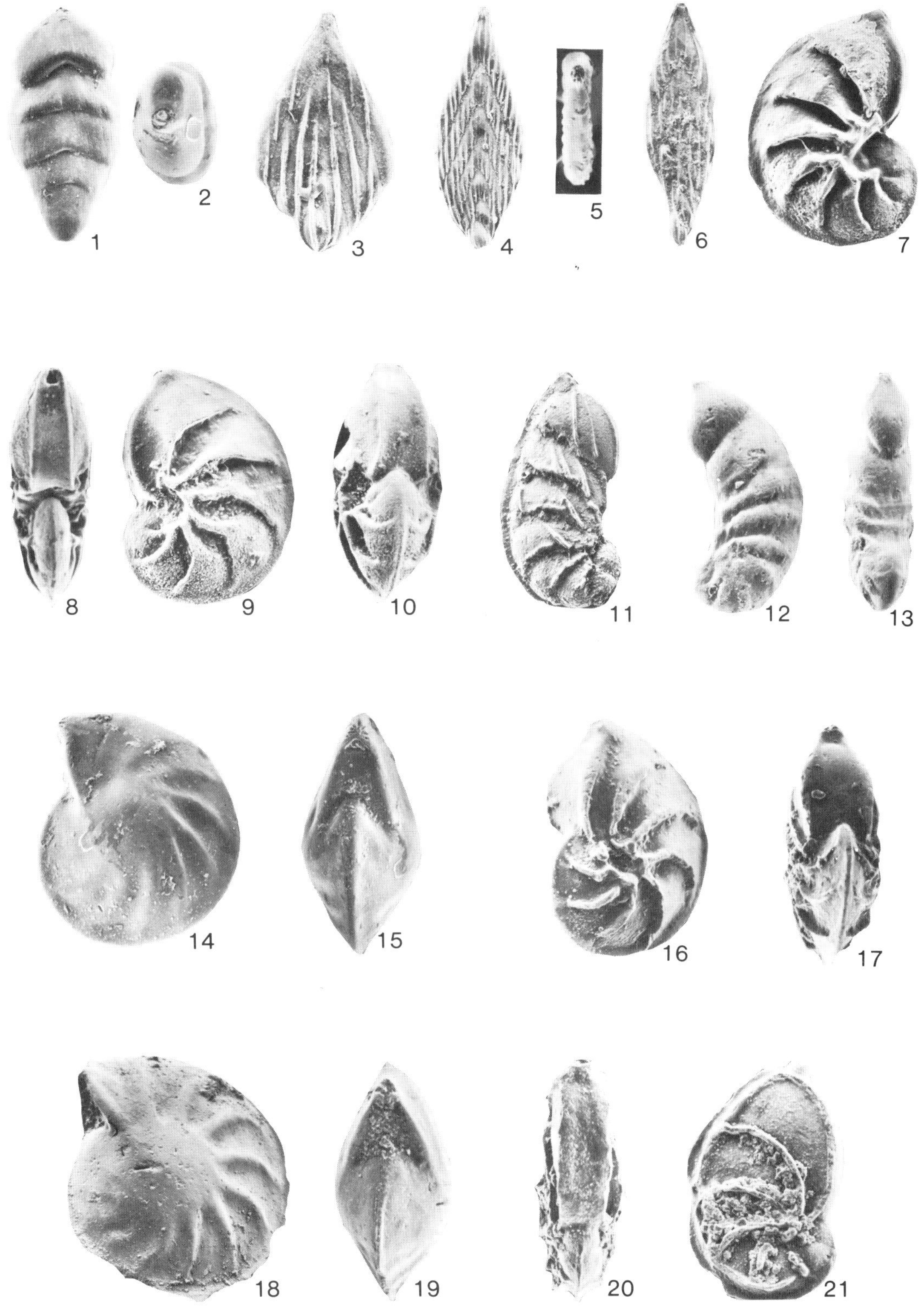

Plate 6.3.3

Remarks: this form is very variable. Chambers are added somewhat irregularly so the periphery is sometimes stepped Especially common in Oxford Clay. Distribution: Callovian *jason-athleta* [Shipp], [9], *athleta-lamberti* [22], *coronatum* [13], *lamberti* [10], Oxfordian *mariae-cordatum* [4], [Shipp], *mariae-transversarium, pseudocordata* [22] *cordatum* [5], Kimmeridgian *baylei-cymodoce* [21]. Europe: Bathonian to Lower Oxfordian of S. Sweden [23] Callovian of Germany [20].

Frondicularia moelleri Uhlig, 1883
Plate 6.3.3, Fig. 3, × 47. Oxford Clay, *athleta* Zone, Weymouth, Dorset. Description: test quadrate compressed, consisting of 3–11, chevron-shaped, uniserially arranged chambers, sutures usually indistinct and flush; ornament of 4–9 costae often extending into the proloculus; aperture circular central terminal on a short neck surrounded by a notched collar; length 0.4 mm–2.5 mm.
Remarks: the ornament of this species is its distinctive feature. It appears to be closely related to *Frondicularia nikitini* Uhlig but has stronger more continuous costae. The specimen recorded by Gordon (1965) is not typical and may not belong to this species. Distribution: Callovian *coronatum* [15], *athleta* [Shipp], [9] (as *Citharinella moelleri*). Germany, *athleta-mariae* [20].

Frondicularia nikitini Uhlig, 1883
Plate 6.3.3, Figs. 4, 5, × 35. Oxford Clay, *athleta* Zone, Woodham. Description: test usually large, compressed, of 5–10 chevron-shaped chambers, periphery smooth to lobate; sutures distinct, depressed; aperture circular terminal usually radiate; ornament varied, typically consists of approximately 15–20 discontinuous arcuate costae interrupted by the sutures, length 0.8 mm–1.8 mm.
Remarks: variation in the ornament has been described by Cordey, 1962. Chevron-shaped chambers sometimes do not appear until the fourth or fifth chamber after the proloculus giving forms identical to *Citharinella exarata* Loeblich and Tappan. See also remarks of *Citharina serratocostata* and *Frondicularia nikitini.* Distribution: Callovian *athleta* [Shipp], [9] (as *Citharinella nikitini*), *athleta-lamberti* [22], Oxfordian *mariae* to lower *cordatum* [Shipp], *mariae-transversarium* [22], Kimmeridgian *baylei-cymodoce* [22]. European records include the Callovian and Oxfordian of Germany [20], and the Bathonian and Callovian of Sweden [23].

Frondicularia pseudosulcata Barnard, 1952, emend. Barnard, 1957
Plate 6.3.3, Fig. 6, × 41. Oxford Clay, *mariae* Zone, Warboys. Description: Test compressed, rhomboid to elongate-rhomboid of 6–10 chevron-shaped chambers, sutures flush to slightly compressed, proloculus often apiculate; aperture central, terminal, radiate, ornament of approximately 10 costae more continuous in the central region of the test; length 0.9 mm–1.2 mm.
Remarks: The costae are more continuous than in *Frondicularia nikitini* especially on the early part of the test where the apertures are less distinct. The Lias species *Frondicularia sulcata* (see Barnard, 1957) differs from *F. pseudosulcata* in having more strongly developed costae. Distribution: Oxfordian *mariae* to lower *cordatum* of Warboys [4], [Skipp].

Lenticulina ectypa (Loeblich and Tappan)
Plate 6.3.3, Figs. 7–10, × 95. Oxford Clay, *coronatum* Zone, Stewartby. = *Astacolus ectypus* Loeblich and Tappan, 1950. Description: test planispiral, coiled to partly uncoiled, keeled, 8–12 chambers in the final whorl. Sutures deeply depressed on proximal side of each suture line, often ribbed on distal side; aperture simple, circular, peripheral; length 0.25 mm–0.7 mm.
Remarks: can be distinguished from *Lenticulina quenstedti* by its longitudinal section [20, 10] but also by the deep recess on the proximal side of each rib. Rare in Corallian [Shipp]. Distribution: Callovian *coronatum-athleta* [9], and *athleta* [22] (both as *Marginulina ectypa*), *jason-athleta* [Shipp], Oxfordian *mariae-pseudocordata* [Shipp], *transversarium* [24] (as *Marginulina ectypa*). Originally described from the *cordatum* Zone from U.S.A. [19].

Lenticulina ectypa (Loeblich and Tappan) *costata* Cordey, 1962
Plate 6.3.3, Fig. 11, × 95. Ampthill Clay, *transversarium* Zone, Warboys. Description: as *L. ectypa* but ribs more prominent and an additional ornament of 2–10 costae arising from each of these ribs; keel more pronounced; length 0.25 mm–0.9 mm. Distribution: Callovian *jason-athleta* [9], Oxfordian *mariae* [10], *plicatilis-transversarium* [Shipp, 24, – as *Marginulina*].

Lenticulina major (Bornemann)
Plate 6.3.3, Figs. 12, 13, × 40. Oxford Clay, *mariae* Zone, Weymouth. = *Cristellaria major* Bornemann, 1854. Common synonym: *Lenticulina suprajurassica* Gordon. Description: test loosely coiled, planispiral, compressed with 4–8 chambers in the uniserial portion, sutures flush or marked by low ribs, aperture terminal, marginal, radiate; length 0.5 mm–1.8 mm.
Remarks: the variation of this species includes forms described by Gordon [13, 14] under *Lenticulina suprajurassica.* This species is distinguished by its short evolute coil. Some specimens lack the depressions in the test noted by Gordon [13] and have a smooth outline approaching *Vaginulina barnardi* in general appearance but lack the faint striations discernible on this latter species. Distribution: Callovian *jason–athleta* [9], [Shipp], Oxfordian *mariae* [10], *mariae-transversarium* [Shipp], *cordatum-plicatilis* [13, 14] (as *Lenticulina suprajurassica*), Kimmeridgian *cymodoce* [22]. Europe: Callovian and Oxfordian of Germany [20]. Originally described from the Lias by Bornemann.

Lenticulina muensteri (Roemer)
Plate 6.3.3, Figs. 14, 15, × 53. Oxford Clay, *mariae* Zone, Millbrook = *Robulina muensteri* Roemer, 1839. Description: test biconvex, planispiral involute, rarely becoming evolute with uniserial portion, periphery subangular to carinate; sutures limbate, usually flush or slightly raised, sigmoid; umbilici with clear bosses showing earlier chambers underneath, flush or rarely raised; aperture marginal, radiate; diameter 0.3 mm–1.0 mm, uncoiled forms length 0.8 mm–1.1 mm.
Remarks: this is possibly the commonest species in the Late Jurassic. It is closely related to *Lenticulina subalata* (Reuss). Distribution: This species has been recorded from virtually every horizon and locality where nodosariids have been recorded in the Late Jurassic. This includes the Portlandian [Shipp].

Lenticulina quenstedti (Guembel)
Plate 6.3.3, Figs. 16, 17, × 50. Oxford Clay, *mariae* Zone, Warboys. = *Cristellaria quenstedti* Guembel, 1862. Description: test planispiral, biconvex involute with 8–10 chambers in final whorl, sometimes a later uncoiled portion of 2-3 chambers; periphery keeled, less marked on uncoiled portion; strong ribs on sutures meet at umbilicus which is bounded by a rib connecting the others at right angles; diameter 0.25 mm–0.85 mm, max. length uncoiled 1.0 mm.
Remarks: the distinctive feature of this species is its ornament. It is closely related to *Lenticulina brueckmanni* (Mjatliuk). Distribution: Callovian *jason–athleta* [Shipp], *athleta* [9], Oxfordian *mariae* and lower *cordatum* [4], *mariae* [10], *mariae-parandieri* [Shipp]. Europe: Bathonian–Oxfordian of Sweden [23], Callovian and Oxfordian of Germany (as cf. [20]) and Middle Jurassic of the French Jura [27].

Lenticulina subalata (Reuss)
Plate 6.3.3, Figs. 18, 19, × 62. Oxford Clay, *athleta* Zone, Woodham. = *Cristellaria subalata* Reuss, 1854. Description: test planispiral, involute, 8–10 chambers in last whorl, occasionally uncoiled, margin sharp with broad keel, aperture terminal, marginal, radiate; raised boss and thick radial ribs.
Remarks: similar to *Lenticulina muensteri* (Roemer) to which it appears to be related. It can be distinguished, however, by the raised opaque boss and sutural ribs. Similar forms exist from the Lias–Early Cretaceous. Distribution: Callovian *macrocephalus-athleta* [9], *jason–athleta* [Shipp], *athleta* [22], Kimmeridgian *cymodoce* [22]. Europe: Oxfordian and Callovian of Germany [20], Kimmeridgian of Poland [7] and the Boulonnais [Shipp].

Lenticulina tricarinella (Reuss)
Plate 6.3.3, Figs. 20, 21, × 100. Oxford Clay *athleta* Zone, Woodham. = *Cristellaria tricarinella* Reuss, 1863. Description: test planispiral, evolute, flaring, laterally compressed with flattened sides; final whorl of 5–6 chambers; periphery with angular keel and angled shoulders on each side; aperture terminal, marginal radiate; length 0.3 mm–0.6 mm, rarely up to 0.72 mm.
Remarks: a distinctive but long ranging species being found in the Middle Jurassic [8]; it was originally described from the Lower Cretaceous. Distribution: Callovian *athleta* [24, Shipp], Oxfordian *plicatilis-cautisnigrae* [14], *mariae-transversarium* [Shipp]. European records include Callovian and Oxfordian, Germany [24], [20], Bathonian–Callovian Sweden [23], Middle Jurassic, French Jura [27].

Paalzowella feifeli (Paalzow)
Plate 6.3.4, Figs. 1 and 3, × 160. Ampthill Clay, *transversarium* Zone, Millbrook; Fig. 2, × 175, Fig. 4, × 215, vertical sections of high and low spired forms, Nothe Grit, *cordatum* Zone, Ham Cliff. = *Trocholina feifeli* Paalzow, 1932. Description: test conical, of single tubular chamber, spirally enrolled, almost involute on umbilical side, evolute on spiral side; radial striations present on umbilical surface of some specimens; aperture at open end of tube, simple; diameter 0.2 mm–0.3 mm.
Remarks: the height of the spire is quite variable. Distribution: Callovian *athleta* [22], *cordatum-transversarium* [22, Shipp]. Europe: Oxfordian Germany [25].

Planularia beierana (Guembel)
Plate 6.3.4, Fig. 5, × 54. Oxford Clay, *mariae* Zone, Millbrook. = *Marginulina beierana* Guembel, 1862. Description: test smooth, compressed, of 4–10 chambers, 3–6 of which are in contact with proloculus; in microspheric generation proloculus completely enclosed, sutures district, usually depressed, poorly developed keel present on early chambers of some specimens; aperture simple, circular, terminal on short neck; some specimens bear ornament of 1 or 2 fine striae on apertural margin of last few chambers; length 0.2 mm–0.75 mm.
Remarks: a highly variable species, the shape depends on the varied growth rate. The variation was studied by Cordey [10]. A common Late Jurassic species most prominent in the Oxford Clay. Distribution: Callovian *jason-athleta* [9], [Shipp], *coronatum* (as *Marginulina scapha* Lalicker [15]), *lamberti* [22], Oxfordian *mariae* [10], *mariae-cordatum* (as *P. protracta* (Bornemann), [4], *mariae-transversarium* [22], *mariae-pseudocordata* [Shipp]. European records include the Lower Malm of Poland [6], Upper Jurassic of the Seine Estuary, [16] and Kimmeridgian of the Boulonnais [Shipp].

Pseudonodosaria radiata (Barnard)
Plate 6.3.4, Figs. 6, 7, × 80. Oxford Clay, *mariae* Zone, Millbrook. = *Pseudoglandulina radiata* Barnard, 1952. Description: test a uniserial series of 5–10 chambers; sutures initially flush but may be constricted between later chambers; central radiate aperture; distinctive ornament of 20–24 longitudinal costae which are slightly twisted with respect to the growth axis; average length 0.5 mm.
Remarks: a rare but stratigraphically restricted species, its distinctive feature is its spirally arranged costae. Distribution: Oxfordian *mariae* [4], *mariae* to lower *cordatum* [Shipp].

Pseudonodosaria vulgata (Bornemann)
Plate 6.3.4, Fig. 8, × 50. Ampthill Clay, *transversarium* Zone, Millbrook. = *Glandulina vulgata* Bornemann, 1854. Common synonyms: *P. humilis* (Roemer), *P. oviformis* (Terquem), *P. tenuis* Bornemann. Description: test uniserial of 4–10 chambers, chambers strongly overlapping, especially earlier ones; outline smooth initially becoming lobate as later chambers may be inflated; aperture central, terminal, radiate, length 0.25 mm–0.65 mm occasionally larger.
Remarks: this is a highly variable species common throughout the Jurassic and Cretaceous. It includes many forms previously described as separate species [23]. Distribution: British records (under various names) include Lias [3], Bathonian [8], Callovian [9], [Shipp], [15], Oxfordian [14], [10], [Shipp]. It has also been widely recorded on the continent.

Saracenaria oxfordiana Tappan, 1955
Plate 6.3.4, Fig. 9, × 38. Oxford Clay, *mariae* Zone, Millbrook. Common synonym: *Saracenaria triquetra* (Guembel). Description: test planispiral with small initial coil followed by uncoiled portion of 2–6 chambers, keeled; sutures marked by low ribs; aperture radiate, terminal, sometimes on short neck; apertural face flattened; length 0.2 mm–0.45 mm.
Remarks: can be distinguished from other *Saracenaria* by presence of ribs. Guembel first used the specific name *triquetra* for an Eocene form [26]. Distribution: Callovian *coronatum-athleta* [9], *jason-athleta* [21], *athleta* [22], Oxfordian *mariae* [10], *mariae-transversarium* [Shipp], *pseudocordata* [14]. Europe: Callovian Germany [20] and Lower Oxfordian of Sweden [23]. Tappan proposed the new name for forms from the Oxfordian and Lower Kimmeridgian of N. Alaska [26].

Tristix triangularis Barnard, 1953
Plate 6.3.4, Figs. 10, 11, × 44. Oxford Clay, *mariae* Zone, Millbrook. Common synonyms: *T. acutangulus* (Reuss), *T. oolithica* (Terquem). Description: test a rectilinear series of 4–8 chambers, triangular in cross-section; sutures slightly depressed, arched towards the aperture; edges of the test often keeled; aperture central, terminal, simple or radiate; length 0.3 mm–0.9 mm.
Remarks: according to Ellis and Messina the types of *T. acutangulus* (Reuss) and *T. oolithica* Terquem should be referred to genera other than *Tristix*. Gordon [14] has described the variation of this species and included forms with rounded and carinate margins. Rare specimens are quadrate. Distribution: Callovian *jason* [9], *coronatum* [15]; Oxfordian *mariae-transversarium* [Shipp], *?cautisnigrae* [13], *plicatilis-pseudocordata* [14]. European records include the Callovian and Oxfordian of Germany [20].

Trocholina granulata Cordey, 1962
Plate 6.3.4, Figs. 12–14, × 100. Holotype, Oxford Clay, Skye, Scotland (redrawn from Cordey, 1962). Description: test plano-convex, consisting of proloculus and an undivided second chamber of 4–5 whorls; spiral side convex with a

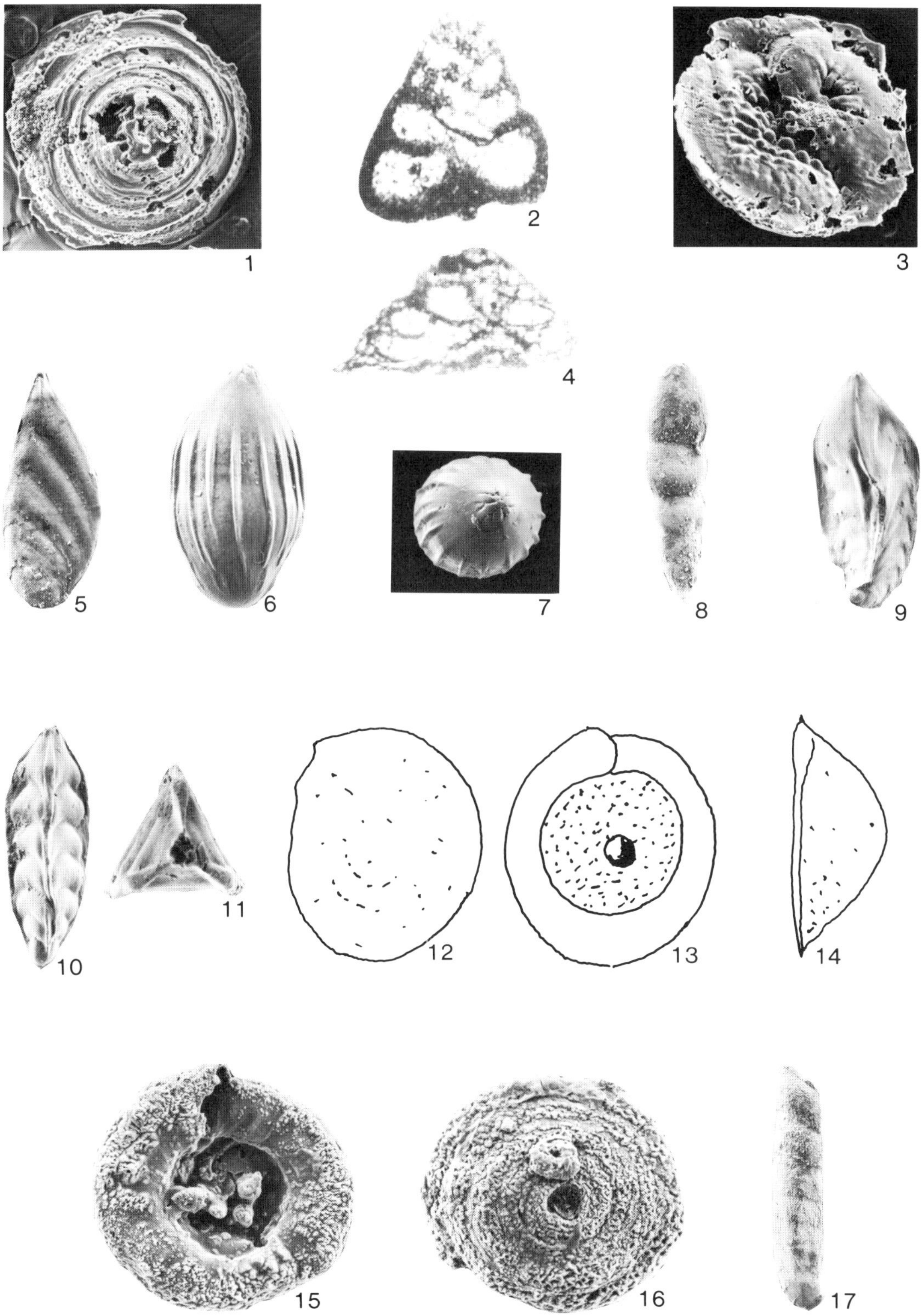

Plate 6.3.4

granular texture; umbilical side flat, infilled with calcite and may have a central pustule; aperture simple, at end of chamber; diameter 0.2 mm–0.3 mm. Distribution: Oxfordian *mariae* [10].

Trocholina nodulosa Seibold and Seibold, 1960
Plate 6.3.4, Figs. 15, 16, × 163. Elsworth Rock Series, *plicatilis* Zone, Millbrook. Description: as *T. granulata* except dorsal surface smooth not granular, ventral surface pustulate with several pustules, greater apical angle (approx. 120°); average diameter 0. 3 mm.
Remarks: Probably closely related to *T. granulata*. Distribution: Oxfordian *plicatilis-parandieri* [Shipp]. Recorded from European Oxfordian [24], [20].

Vaginulina barnardi Gordon, 1965
Plate 6.3.4, Fig. 17, × 30. Ampthill Clay, *transversarium* Zone, Warboys. Description: test uniserial, straight or curved, slightly compressed; 9–12 chambers, rarely up to 16; sutures oblique flush; aperture terminal marginal radiate; ornament of fine, longitudinal striations, length 0.3 mm–1.0 mm occasionally longer.
Remarks: only recorded from the Corallian, striations are always present but may be very faint. Distribution: Oxfordian: *cordatum-decipiens* [14], *plicatilis-pseudocordata* [Shipp], *transversarium* [22].

Acknowledgements

Due to pressure of work the author originally scheduled to write the chapter on the Late Jurassic was unable to carry out the study. Consequently, an original draft manuscript on the Late Jurassic was written at short notice by J. W. Murray. It was read by E. J. Coleman, J. Exton, B. Johnson and D. Shipp, all of whom made useful comments and suggestions for revision. Following this D. J. Shipp and J. W. Murray jointly produced the final manuscript. The limited amount of time spent preparing this chapter compared to the others should, however, be borne in mind.

6.3.9 References

The numbers refer to publications used to compile the distribution data.

[1] Adams, C. G. 1962. Calcareous Adherent Foraminifera from the British Jurassic and Cretaceous. *Micropaleontology*, **5**, 149–170.
Arkell, W. J. 1933. *The Jurassic System in Great Britain*, Oxford: Clarendon Press, 681 pp.
Arkell, W. J. 1935. The Corallian Beds of Dorset. *Proc. Dorset Nat. Hist. Archaeol. Soc.*, **57**, 59–93.
Arkell, W. J. 1947. The Geology of the country around Weymouth, Swanage, Corfe and Lulworth. *Mem. geol. Surv. Gt. Br.*, 1–386.
Arkell, W. J. 1956. *Jurassic Geology of the World.* Oliver and Boyd, Edinburgh and London.
[2] Barnard, T. 1950a. Foraminifera from the Lower Lias of the Dorset Coast. *Quart. J. geol. Soc. Lond.*, **105**, 341–391.
[3] Barnard, T. 1950b. Foraminifera from the Upper Lias of the Dorset Coast. *Quart. J. geol. Soc. Lond.* **106**, 1–36.
[4] Barnard, T. 1952. Foraminifera from the Upper Oxford Clay (Jurassic) of Warboys, Huntingdonshire. *Proc. Geol. Ass.*, **63**, 336–350.
[5] Barnard, T. 1953. Foraminifera from the Upper Oxford Clay (Jurassic) of Recliff Point, near Weymouth, England. *Proc. Geol. Ass.* **64**, 183–197.
Berthelin, G. 1879. Foraminifères du lias moyen de la Vendée. *Rev. mag. Zool.*, (3), 7, 24–41.
[6] Bielecka, W. 1960. Micropalaeontological Stratigraphy of the Lower Malm in the vicinity of Chrzanowa (southern Poland), *Inst. Geol. Prace.*, **33**, 1–155.
[7] Bielecka, W. and Pozaryski, W. 1954. Micropalaeontological Stratigraphy of the Upper Malm in Central Poland. *Inst. Geol. Prace.* **12**, 143–206.
Bignot, G. and Guyader, J. 1966. Découverte de Foraminifères planctoniques dans l'Oxfordien du Havre (Seine-Maritime). *Rev. Micropal.*, **9**, 104–110.
Bizon, J. J. 1958. Foraminifères et Ostracodes de l'Oxfordien de Villers sur mer (Calvados). *Rev. Inst. Fr. Pét.*, **13**, 1–145.
Bizon, G. 1960. Revision de quelques espèces-types de Foraminifères du Lias du Bassin Parisien de la collection Terquem. *Rev. Micropal.*, **3**, 3–18.
Blake, J. F. 1875. On the Kimmeridge Clay of England. *Quart. J. geol. Soc. Lond.*, **31**, 196–237.
Bornemann, J. G. 1854. *Uber die Liasformation in der Umgegund von Göttingen und ihre organischen Einschlusse.* A. W. Schade, Berlin.
Brady, H. B. 1867. Synopsis of the middle and upper Lias of Somerset. *Proc. Somerset Arch. and Nat. Soc.*, **13**.
Brinkmann, R. 1929. Statistisch-biostratigraphische Untersuchungen an mittel-jurassichen Ammoniten über Artbegriff und Stammesentwicklung. *Abhandl. Ges. Wiss. Göttingen, Math-phys.*, Kl., N.F. **13**, 1–249.
Brookfield, M. 1973. Palaeogeography of the Upper Oxfordian and Lower Kimmeridgian (Jurassic) in Britain. *Palaeogeogr. Palaeoclimatol., Palaeoecol.*, **14**, 137–167.
Callomon, J. H. 1955. The Ammonite Succession in the Lower Oxford Clay and Kellaways Beds at Kidlington, and the Zones of the Callovian Stage. *Phil. Trans. R. Soc.*, **B. 239**, 215–264.
Callomon, J. H. 1964. Notes on the Callovian and Oxfordian Stages. *Compte rendu et Mem. Coll. Jur. Luxembourg,* 1962. *Inst. Gr.-duc Sect. sci. nat. phys. math.*, 269–291.
Callomon, J. H. 1968a. The Kellaways Beds and Oxford Clay. Chapter 14, pp. 264–80. In: *The Geology of the East Midlands* (ed. Sylvester-Bradley, P. C. and Ford, T. D.), Leicester Uni-

versity Press.
Callomon, J. H. 1968b. The Corallian Beds, the Ampthill Clay and the Kimmeridge Clay. Chapter 15, pp. 291-299. Ibid.
Callomon, J. H. and Cope, J. C. W. 1971. The stratigraphy and ammonite succession of the Oxford and Kimmeridge Clays in the Warlingham Borehole. *Bull. geol. Surv. Gt. Br.*, **36**, 147-176.
Chapman, F. 1897. Notes on the Microzoa from the Jurassic beds at Hartwell. *Proc. Geol. Ass.*, **14**, 96-97.
[8] Cifelli, R. 1959. Bathonian Foraminifera of England. *Bull. Mus. Comp. Zool.*, **121**, 265-368.
[9] Coleman, B. E. 1974. Foraminifera of the Oxford Clay and Kellaways Beds. Appendix 3 of The geology of the new town of Milton Keynes. *Rep. Inst. Geol. Sci.*, No. 74/1b.
Cope, J. W. 1967. The palaeontology and stratigraphy of the lower part of the Upper Kimmeridge Clay of Dorset. *Bull. Brit. Mus. (Nat. Hist.) Geol.* **15**, (1), 1-79.
Cope, J. W. 1978. The ammonite faunas and stratigraphy of the upper part of the Upper Kimmeridge Clay of Dorset. *Palaeontology*, **21**, 469-533.
[10] Cordey, W. G. 1962. Foraminifera from the Oxford Clay of Staffin Bay, Isle of Skye, Scotland. *Senckenbergiana leth.*, **43**, 375-409.
[11] Cordey, W. G. 1963. The genera *Brotzenia* Hofker 1954, and *Voorthuysenia* Hofker 1954 and Hofker's Classification of the *Epistomariidae*. *Palaeontology*, **6**, 653-657.
Crick, W. D. 1887. Note on some foraminifera from the Oxford Clay at Keystone near Thrapston. *Northants. Nat. Hist. Soc.*, **4**, 233.
Crick, W. D. and Sherborn, C. D. 1891. On some Liassic foraminifera from Northamptonshire. *J. Northants Nat. Hist. Soc.*, **6**.
[12] Cushman, J. A. 1930. Note sur quelques foraminifères Jurassique de Auberville (Calvados) *Bull. Soc. Linn. Normandie* (8), **2**, (1929), 134.
Deeke, W. 1886. Les Foraminifères de l'Oxfordien des environs de Montbeliard. *Mèm. Soc. emulation* (3), **16**, 283-335.
Duff, K. L. 1975. Palaeoecology of a Bituminous Shale – The Lower Oxford Clay of Central England. *Palaeontology*, **18**, 443-482.
Fürisch, F. T. 1976. The use of microinvertebrate associations in interpreting Corallian (Upper Jurassic) environments *Palaeogeogr., Palaeoclimatol., Palaeocol.*, **20**, 235-256.
[13] Gordon, W. A. 1962. Some Foraminifera from the Ampthill Clay, Upper Jurassic of Cambridgeshire. *Palaeontology*, **4**, 520-537.
[14] Gordon, W. A. 1965. Foraminifera from the Corallian Beds Upper Jurassic of Dorset, England. *J. Paleont.*, **39**, 838-863.
Gordon, W. A. 1966. Variation and its significance in Classification of some English Middle and Upper Jurassic Nodosariid Foraminifera. *Micropaleontology*, **12**, 325-333.
[15] Gordon, W. A. 1967. Foraminifera from the Callovian (Middle Jurassic) of Brora, Scotland. *Micropaleontology*, **13**, 445-464.
Gordon, W. A. 1970. Biogeography of Jurassic Foraminifera. *Bull. geol. Soc. Am.*, **81**, 1689-1704.
Grigelis, A. A. 1958. (*Globigerina Oxfordiana* sp. – occurrence of Globigerines in the Upper Jurassic deposits of Lithuania) (Russian) *Nauchnye Doklady Vysshey Shkoly, Geol.-Geogr. Nauki, Moscow*, 1958, No. 3, 110, 111.
Gümbel, C. W. 1862. Die Streitbergerer Schwammergel und ihre Foraminferen-Einschlüsse. *Jahres. Ver. Vaterl. Naturk. Wurttemberg*, Jarg. **18**, 192-238.
[16] Guyader, J. 1968. Le Jurassique Supèrieur de la Baie de la Seine-Etude Stratigraphique et Micropaléontologique. Thèse de Docteur, University of Paris.
Hallam, A. 1969. Tectonism and Eustasy in the Jurassic. *Earth Sci. Rev.*, **5**, 45-68.
Hallam, A. 1978. Eustatic Cycles in the Jurassic. *Palaeogeogr. Palaeoclimatol., Palaeoecol.*, **23**, 1-32.
Hallam, A. and Sellwood, B. W. 1976. Middle Mesozoic sedimentation in relation to tectonics in the British area. *J. Geol.*, **84**, 301-321.
Häusler, R. 1881. *Unterschungen über die mikroskopischen Strukturverhältniss der Aargauer Jurakalke mit besonderer Berücksichtigung ihre Foraminiferen fauna.* Diss Univ. Zurich, 47 pp.
Hudson, J. D. and Palframan, D. F. B. 1969. The Ecology and Preservation of the Oxford Clay Fauna at Woodham, Buckinghamshire. *Quart. Jl. geol. Soc. Lond.*, **124**, 387-418.
Jones, T. R. and Parker, W. K. 1860. One some fossil foraminifera from Chellaston near Derby. *Quart. Jl. geol. Soc. Lond.*, **16**, 452-458.
Kapterenko-Chernousova, O. K., Goljak, L. M., Zerneckij,, B. F., Krajeva, E. J. and Lipnik, E. S. 1963. Atlas of Characteristic foraminifera from the Jurassic, Cretaceous and the Palaeogene of the Ukrainian Platform. (In Russian). *Trudu Akad, Nauk USSR, Ser. strat. paleont.* **V. 45**, 1-200.
[17] Lloyd, A. J. 1959. Arenaceous Foraminiferal Faunas from the Type Kimmeridgian. *Palaeontology* **1**, 298-320.
[18] Lloyd, A. J. 1962. Polymorphinid, Miliolid and Rotaliform Foraminifera from the Type Kimmeridgian. *Micropaleontology*, **8**, 369-383.
[19] Loeblich, A. R. and Tappan, H. 1950. North American Jurassic Foraminifera I. The Type Redwater Shale (Oxfordian) of South Dakota. *J. Paleont.*, **24**, 39-60.
Loeblich, A. R. and Tappan, H. 1964. Sarcodina. Chiefly "Thecameobians" and Foraminiferida, Treatise on Invertebrate Paleontology (ed. Moore, R. C.) Part C, Protista 2, **1 and 2**, *Geol. Soc. Am. University of Kansas Press.*
[20] Lutze, G. F. 1960. Zur Stratigraphie und Paläontologie des Callovian und Oxfordian in Nordwest-Deutchland. *Geol. Jahrb.*, **77**, 391-532.
[21] Macfadyen, W. A. 1935. Jurassic Foraminifera. The Mesozoic palaeontology of British Somaliland. Pt. II, pp. 7-20.
[22] Medd, A. W. 1979. In Richardson, G. The Meso-

zoic Stratigraphy of two boreholes near Worlaby, Humberside. *Bull. geol. Surv. G.B.*, **58**, 1–24.

Morris, K. A. 1980. Comparison of major sequences of organic-rich mud deposition in the British Isles. *J. geol. Soc. Lond.*, **137**, 157–170.

Murray, J. W. 1973. *Distribution and ecology of living benthic foraminiferids.* London, Heinemann, 288 pp.

Neaverson, E. 1921. The Foraminifera of the Hartwell Clay and subjacent beds. *Geol. Mag.*, **58**, 454–473.

[23] Norling, E. 1972. Jurassic Stratigraphy and Foraminifera. *Sver. Geol. Unders.* Ser. Ca, **47**, 120 pp.

Rutten, M. G. 1956. Depositional Environment of Oxford Clay at Woodham Clay Pit. *Geol Mijnb. N.S.*, **18**, 344–347.

Schwager, C. 1865. Beitrag zur Kenntnis der mikroskopischen Fauna Jurassicher Schichten. *Jb. Ver. Vaterl. Naturkde. Wurtt, Jahrg.* **21**, 82–151.

Seibold, E. and Seibold, I. 1955. Revision der Foraminiferen-Bearbeitung G. W. Gümbel's (1862) aus den Streitberger Schwammumergeln (Oberfranken), Unterer Malm. *Neues Jahrb. Geol. Paläont. Abh.*, **101**, 91–134.

Seibold, E. and Seibold, I. 1956. Revision der Foraminiferen-Bearbeitung C. Schwager's (1865) aus den Impressaschichten (Unterer Malm) Suddeutschland. *Neues. Jahrb. Geol. Paläont. Abh.*, **103**, 91–154.

[24] Seibold, E. and Seibold, I. 1960. Foraminiferen der Bank – und Schwamm-Fazies im unteren Malm Suddeutschland. *Neues. Jahrb. Geol. Paläont. Abh.*, **109**, 309–438.

Sherborn, C. D. 1888. Notes on *Webbina irregularis* (d'Orbigny) from the Oxford Clay at Weymouth. *Proc. Bath Nat. Hist. Antiq. Fld. Cl.*, **6**, 322–333.

[25] Spath, L. F. 1939. The Ammonite Zones of the Upper Oxford Clay of Warboys, Huntingdonshire. *Bull. Geol. Surv. Gt. Br.*, **1**, 82–98.

Strickland, H. E. 1846. On two species of microscopic shells found in the Lias, *Quart. J. Geol. Soc. Lond.*, **2**, 31.

Sykes, R. M. and Callomon, J. H. 1979. The *Amoeboceras* Zonation of the Boreal Upper Oxfordian. *Palaeontology*, **22**, 893–903.

Talbot, M. R. 1973. Major sedimentary cycles in the Corallian Beds. *Palaeogeog., Palaeoclimatol., Palaeocol.*, **14**, 293–317.

Talbot, M. R. 1974. Ironstones in the Upper Oxfordian of southern England. *Sedimentology*, **21**, 433–450.

[26] Tappan, H. 1955. Foraminifera from the Arctic Slope of Alaska; Part 2. Jurassic Foraminifera. *Prof. Pap. U.S. geol. Surv.*, 236–B.

Terquem, O. 1855. Paléontologie de l'étage infèrieur de la Formation Liassique de la province de Luxembourg, grandduché (Hollande) et de Hettange du départment de la Moselle. *Mém. Soc. géol. France* (2), **5**, (1854), pt. 2, mém. 3.

Townson, W. G. 1975. Lithostratigraphy and Deposition of the type Portlandian. *Jl. geol. Soc. Lond.*, **131**, 619–638.

[27] Wernli, R. 1971. Les Foraminiféres du Dogger du Jura Méridoinal (France). *Arch. des Science.*, **24**, 305–364.

Whittaker, W. 1866. On some borings in Kent. A contribution to the deep-seated geology of the London Basin. *Quart. J. geol. Soc. Lond.*, **42**, 26–48.

Wilson, R. C. C. 1968a. Upper Oxfordian Palaeogeography of southern England. *Palaeogeog., Palaeoclimatol., Palaeoecol.*, **4**, 5–28.

Wilson, R. C. C. 1968b. Carbonate facies variation within the Osmington Oolite Series in southern England. *Palaeogeog., Palaeoclimatol., Palaeoecol.*, **4**, 89–123.

Wimbledon, W. A. and Cope, J. C. W. 1978. The ammonite faunas of the English Portlandian Beds and the zones of the Portlandian Stage. *Jl. geol. Soc. Lond.*, **135**, 183–190.

Wisniowski, T. 1890. Mikrofauna i ow ornatowych okolicy Krakowa. 1. – Otwornice gornego Kelloeayu w Grojco. *Pam. Akad. Um. Krakowie, Wydz. Mat. – Przyr.*, **17**, 181–242.

Woodward, H. B. 1895. The Jurassic rocks of Britain (Yorkshire excepted) Vol. V, Middle and Upper Oolithic Rocks. *Mem. geol. Surv. G.B.*

6.4 SUMMARY

B. Coleman, P. Copestake, B. Johnson
J. W. Murray, D. Shipp

The original intention was to have a single chapter on the Jurassic in which all the results were integrated but this objective has not been achieved owing to changes of authorship. Instead the results have been presented as three sections: Hettangian to Toarcian, Bajocian to Callovian, and Callovian to Portlandian. The duplication of the Callovian was difficult to avoid because the Callovian–Oxfordian boundary falls within the Oxford Clay.

Table 6.4.1 is a compilation by all the contributors to this chapter to show the ranges of the species throughout the Jurassic of Britain. It is apparent that the majority of the species are long ranging and are not truly index forms. They have been included because they form a major part of the fauna. The few records of foraminifera in the Rhaetian have also been included as they herald the incoming of the Jurassic faunas. (See Chapter 5). Species which continue into the Lower Cretaceous are discussed in Chapter 7.

Table 6.4.1
Table showing ranges of species throughout the Jurassic of Britain.

Species, genus	Rhaetian	Hettangian	Early Sinemurian	Late Sinemurian	Early Pliensbachian	Late Pliensbachian	Toarcian	Bajocian	Bathonian	Callovian	Oxfordian	Kimmeridgian	Portlandian
acuta, Triplasia										X			
agglutinans, Ammobaculites									X	X	X	X	
althoffi, Triplasia								X	X	X			
anceps, Palmula (see F. nikitini)										X	X		
barnardi, Vaginulina											X		
beierana, Planularia									X	X		X	
bicornis, Dentalina				X	X			X	X	X			
bigoti, Nubeculinella				X	X			X	X	X	X		
byfieldensis, Nodosaria							X						
canui, Cribrostomoides											X	X	
carinatum, Ophthalmidium								X	X				
clathrata, Vaginulina/Citharina							X	X	X				
colliezi, Citharina							X	X	X	X	X		
communis, Dentalina = D. pseudocommunis		X	X	X	X	X	X	X	X	X	X	X	
concentricum, Spirophthalmidium				X	X			X	X				
coprolithiformis, Ammobaculites									X	X	X	X	
creba, Reinholdella									X				
curva, Vaginulina		X	X										

Species, genus	Rhaetian	Hettangian	Early Sinemurian	Late Sinemurian	Early Pliensbachian	Late Pliensbachian	Toarcian	Bajocian	Bathonian	Callovian	Oxfordian	Kimmeridgian	Portlandian
deslongchampsi, Falsopalmula — — —							x	x					
dictyodes, Lenticulina — — — — —								x	x				
dorbignyi, Lenticulina — — — — —							x	x					
dorsetensis, Massilina — — — — —									x				
ectypa, Lenticulina/Marginulina — — —										x	x		
ectypa costata, Lenticulina — — — —										x	x		
eugenii, Planularia — — — — — —				x	x	x		x	x	x			
exarata, Vaginulinopsis — — — — —		?	x	?									
feifeli, Paalzowella — — — — — —								x	x		x		
filiformis, Dentalina — — — — — —										x			
flabellata, Citharina (see C. serratocostata) —										x	x	x	
fontinensis, Ammobacultes — — — —		x	x	x	x	x	x	x	x				
franconica, Frondicularia — — — — —										x	x	?	
galeata, Lenticulina — — — — — —								x	x				
globigeriniformis, Trochammina — — —			x	x	x	x	x		x	x	?	x	
globulata, Nodosaria — — — — — —						x							
granulata, Trocholina — — — — —											x		
gryci, Trochammina — — — — — —			x	x									
guembeli, Dentalina (see D. pseudocommunis)										x			
haeusleri, Haplophragmoides — — — —												x	
hamus, Marginulina/Saracenaria — — —			x										
harpa, Vaginulina — — — — — —									x				
hortensis, Nodosaria — — — — — —			x	x	x				x	x			
implicata, Citharina — — — — — —								x		x	?	x	
inaequistriata, Planularia — — — — —		x	x	x									
ingens, Nodosaria — — — — — —									x				
intorta, Dentalina — — — — — —								x	x	x			
irregularis, Frondicularia (see F. franconica)										x	x	?	
issleri, Nodosaria — — — — — —				x					x				
jurassica, Miliammina — — — — —										x			
jurassica, Textularia — — — — — —									x	x	x		
jurensis, Thurammina — — — — —							x						
kimeridensis, Tripasia — — — — —										x			
langi, Carixia — — — — — — —					x	x							
legumen, Vaginulina — — — — — —									x				
liasina, Cyclogyra — — — — — —		x	x	x	x	x	x	x	x	x			
liasina, Tristix — — — — — — —					x	x	x						
liasica, Brizalina — — — — — — —				x	x	x	x						
liassica, Involutina — — — — — —		x	x	x	x	x	x						
liassica, Nodosaria — — — — — —							x	x	x				

Species, genus	Rhaetian	Hettangian	Early Sinemurian	Late Sinemurian	Early Pliensbachian	Late Pliensbachian	Toarcian	Bajocian	Bathonian	Callovian	Oxfordian	Kimmeridgian	Portlandian
longiscata, Lingulina		x	x	x	x	x	x	x	x	x			
macfadyeni, Reinholdella						x	x						
major, Lenticulina		x	x	x	x	x	x			x	x		
magarita, Reinholdella			x	x									
marginata, Conorboides												x	
matutina, Dentalina			x	x	x	x							
mauritii, Verneuilinoides				x	x								
mediomatricorum, Vaginulinopsis				x									
moelleri, Citharinella/Frondicularia										x			
mosquensis, Epistomina									x	x	x		
mucronata, Dentalina		x	x	x	x	x			x				
muensteri, Lenticulina	x	x	x	x	x	x	x	x	x	x	x	x	x
nikitini, Citharinella/Frondicularia										x	x		
nodulosa, Trocholina											x		
nuda, Epistomina										x			
nympha, Frondicularia								x	x				
obliqua, Falsopalmula							x	x	x				
obliquecostulata, Marginulina			x										
oolithica, Frondicularia								x	x				
oolithica, Tristix (see T. triangularis)									x	x	x	x	
opalini, Nodosaria								x	x	x			
ornata, Epistomina											x	x	
oxfordiana, Saracenaria								x	x	x	x		
pachyderma subsp. A., Reinholdella					x								
pera, Guttulina										x			
pictonica, "Placentula"				x	x	x							
planiconvexa, Reinholdella		x	x	x	x								
prima interrupta, Marginulina				x	x	x	x						
prima spinata, Marginulina				x	x	x	x						
pseudocommunis, Dentalina (see D. communis)		x	x	x	x	x	x	x	x	x	x	x	
pseudosulcata, Frondicularia											x		
pulchra, Nodosaria								x					
quenstedti, Lenticulina							x	x	x	x	x		
radiata, Pseudonodosaria											x		
redcliffensis, Massilina											x		
regularis, Epistomina								x	x				
regularis regularis, Nodosaria			x	x	x	x		x					
regularis subsp. A, Nodosaria							x						
rjasanensis, Pseudolamarckina										x			

Species, genus	Rhaetian	Hettangian	Early Sinemurian	Late Sinemurian	Early Pliensbachian	Late Pliensbachian	Toarcian	Bajocian	Bathonian	Callovian	Oxfordian	Kimmeridgian	Portlandian
scapha, Marginulina (see P. beierana)									x	x		x	
semireticulata, Astacolus			x										
serratocostata, Citharina										x	x	x	
sherlocki, Gaudryina										x	x		
stellicostata, Epistomina										x			
stelligera, Epistomina									x	x			
sterkii, Reophax										x	?	x	
strumosum, Ophthalmidium										x	x		
subalata, Lenticulina							x	x	x	x			
subfavosa, Thurammina							x						
sublaevis sublaevis, Saracenaria						x	x						
sp., Tubinella				x	x	x							
sp. A, Citharina							x						
sp. A, Marginulina			x										
sp. A, Saracenella							x						
sp. A, 'Spiroloculina'			x	x									
sp. B, 'Spiroloculina'		x	x	?									
sp. 2, Neobulimina			x	x									
sp. 2, Ophthalmidium							x						
tenera collenoti, Lingulina	x	x											
tenera occidentalis, Lingulina				x	x	x	x						
tenara subprismatica, Lingulina			x	x	x	x							
tenera subprismatica, Lingulina		x	x										
tenuicostata, Epitomina											x		
tenuistriata, Palmula							x						
terquemi subsp. A, Frondulularia		x											
terquemi subsp. B, Frondicularia						x	x						
terquemi muelensis, Frondicularia				x	x	x	x						
testudinaria, Lingulina					x	x							
tibia, Nubeculinella							x				x		
triangularis, Tristix (see T. oolithica)									x	x	x	x	
tricarinella, Lenticulina								x	x	x	x		
trochoides, Conicospirillina					x	x							
tryphaera, Verneuilinoides									x	x			
varians subsp. D, Lenticulina						x	x						
varians haeusleri, Dentalina			x	x	x								
volubilis, Lenticulina								x	x				
vulgata, Pseudonodosaria		x	x	x	x	x	x	x	x	x	x	x	

7
Cretaceous

M. B. Hart, H. W. Bailey, B. Fletcher, R. Price, A. Sweicicki

7.1 INTRODUCTION

The Cretaceous System, proposed by d'Omalius d'Halloy in 1822, takes its name from *Creta* – the Latin for chalk. Chalk is the most distinctive rock type of the System, not only in Britain (Fig. 7.1), but over a considerable area of the globe. Upper Cretaceous sediments at one time covered the whole of north west Europe, save for a few isolated areas of ancient massifs like the Baltic Shield. Today, sediments of Cretaceous age are limited (onshore) to a narrow band down eastern England, together with a wide tract of country across southern England. Offshore, Cretaceous strata are more extensive, covering, or known from, large areas of the North Sea Basin, the English Channel, the Celtic Sea and the South West Approaches Basin.

The final text was prepared by MBH; the SEM photography was by MBH and AS (assisted by B. Lakey and C. Jocelyn – Plymouth Polytechnic Electron Microscopy Unit); plates were prepared by MBH (assisted by Media Services, Plymouth Polytechnic); and the diagrams and charts were prepared by MBH. Data were prepared by all contributors to the Cretaceous section, but thanks are also extended to D. J. Carter (Imperial College), S. Crittenden (Plymouth Polytechnic), C. S. Harris (Robertson Research International Ltd), P. J. Bigg (Institute of Geological Sciences), C. J. Wood (Institute of Geological Sciences), and F. Robaszynski (Mons Polytechnic, Belgium), all of whom assisted in various ways during preparation. B. N. Fletcher's contribution is published with the approval of the Director, Institute of Geological Sciences.

7.2 COLLECTIONS OF CRETACEOUS FORAMINIFERA

British Museum (Natural History)

This museum contains the largest, and most important, collections of Cretaceous foraminifera. Many of the suites of specimens contain material from specific publications, although there is also a large amount of reference material from all over the British Isles. The most important collections are those of Barnard (1962, 1963), Barr (1962), Carter and Hart (1977), Chapman (1891-98),

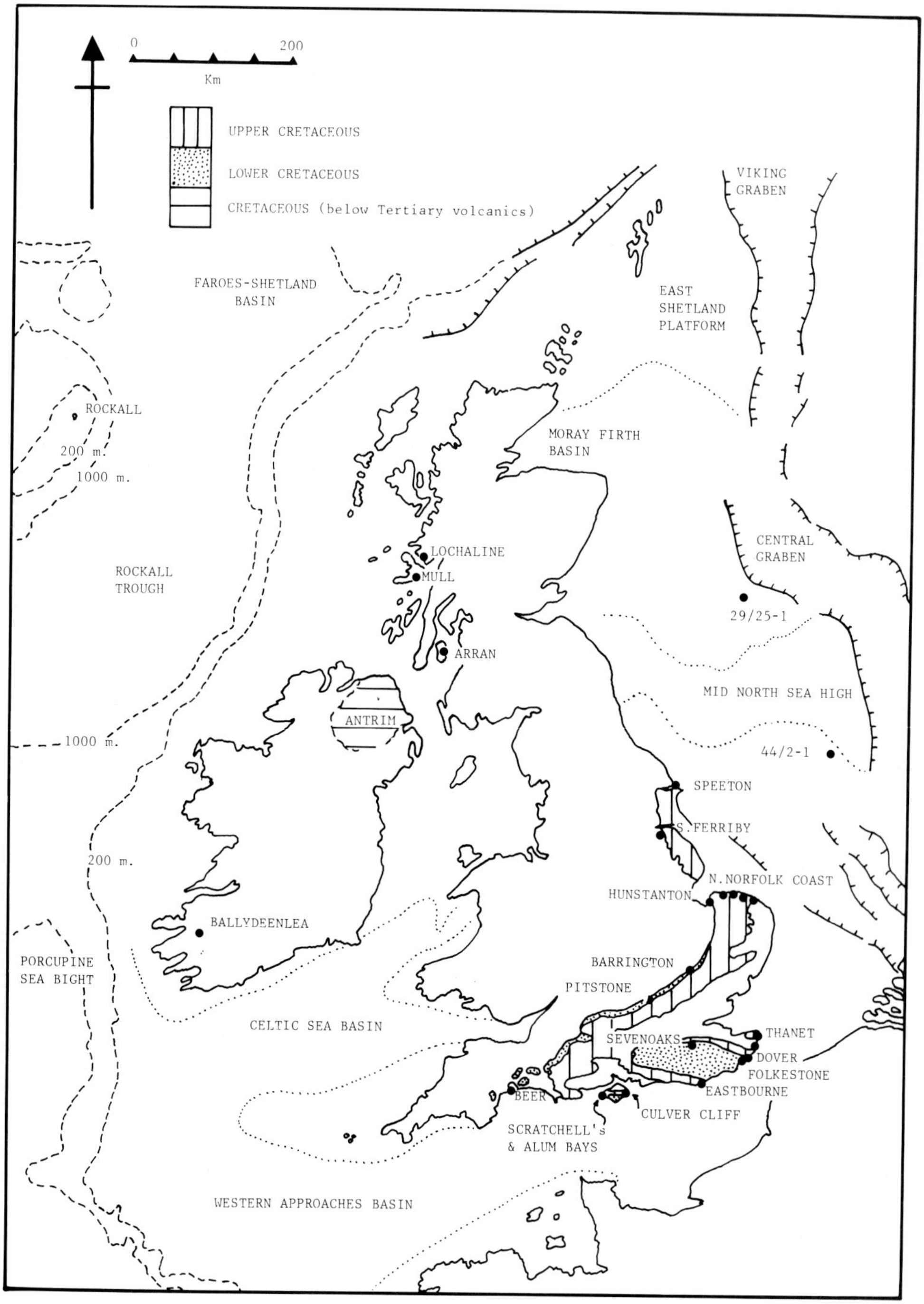

Fig. 7.1 – Locality map for the British Isles and adjacent off-shore areas.

Jukes-Browne (1898) and Williams-Mitchell (1948). Important samples of material, or selected foraminifera, have been donated to the collections by Adams, Blow, Brady, Curry, Davis, Elliot, Heron-Allen, Hodgkinson, Jones, Khan, Owen and Rowe.

Institute of Geological Sciences (London and Leeds)
The collections of the Institute include material from shallow and deep boreholes, from both on-shore and off-shore. The micropalaeontological collections are currently being concentrated in the Leeds office.

Department of Geology, Imperial College, London
The material from the Channel Tunnel Site Investigation (Albian–Turonian) is housed in this department under the supervision of Mr. D. J. Carter.

There are other, smaller, collections at many universities and colleges through the British Isles, including University College London (Prof. T. Barnard), Hull University (Prof. J. W. Neale), University College of Wales, Aberystwyth (Dr. J. Haynes), and Plymouth Polytechnic (Dr. M. B. Hart).

7.3 STRATIGRAPHIC DIVISIONS

The Cretaceous Period lasted approximately from 135 Ma–65 Ma. In recent years several attempts have been made to produce an acceptable time scale for the various stages within the Cretaceous (Casey, 1964; Bandy, 1967; Obradovich and Cobban, 1975; Kauffman, 1970; van Hinte, 1976) but there has been little, if any, agreement. In Table 7.1 the stages used in this account are listed, together with the radiometric dates proposed by van Hinte (1976).

Table 7.1

Stage	Boundary age (Ma)	
PALAEOCENE		
	65	
MAASTRICHTIAN		
	70	
CAMPANIAN		
	78	
SANTONIAN		LATE CRETACEOUS
	82	
CONIACIAN		
	86	
TURONIAN		
	92	
CENOMANIAN		
	100	
ALBIAN		
	107	
APTIAN		
	115	
BARREMIAN		EARLY CRETACEOUS
	121	
HAUTERIVIAN		
	126	
VALANGINIAN		
	131	
RYAZANIAN		
	136	
VOLGIAN		

Ryazanian – Based on the Ryazan Beds of the Moscow Platform (Sazonov, 1951), this stage was applied to the British succession by Casey (1973) and Rawson *et al.* (1978). In both accounts the problems of relating Ryazanian to the Berriasian stage are discussed.

Valanginian – The type section in the Seyon Gorge, Valangin, Neuchatel, Switzerland, contains a succession of shallow-water, calcareous sediments. The fauna is rather limited, the foraminifera having little in common with the faunas recorded from Britain and the rest of N.W. Europe.

Hauterivian – Renévier (1874) established this stage using a succession of shallow water carbonate sediments at Hauterive, Neuchatel, Switzerland. The microfauna, as above, is rather poor, and Kemper (1973) has provided some important data concerning the correlation of this section with that in N.W. Europe.

Barremian – Following work by Busnardo (1965) the type succession is based on the Angles (Basse-Alpes, France) road section. The limestones contain a good ammonite fauna but correlation of the microfauna again relies on data from N. Germany (Neale, 1974; Fletcher, 1973).

Aptian – The succession of Apt (Basse-Alpes, France) has been fully described by Breistroffer (1947). The sub-stages proposed for that section are not used in Britain, Casey (1961) having proposed a simple grouping of zones into lower and upper sub-stages. The Aptian microfauna of the British Isles is poorly known.

Albian – This stage, proposed by d'Orbigny (1842) for the Aptian–Cenomanian interval, takes its name from the Aube region of Eastern France. The foraminifera of the Aube have been exceptionally well described by Magniez-Jannin (1975) and this has allowed detailed correlation with the British Isles (Price, 1977a).

Cenomanian – Introduced by d'Orbigny for the marginal sands, gravels, marls, and chalks of the Sarthe area of France, the boundaries of the stage have posed many problems (Rawson *et al.*, 1978). The microfauna of the type area has been discussed by Marks (1967) and Carter and Hart (1977).

Turonian – The type area, from Saumur to Montrichard (along the Loire–Cher Valley), has been fully investigated in recent years (e.g. Lecointre, 1959). The foraminifera of the Cher Valley (just outside the designated type area) have been described by Butt (1966), Bellier (1971) and Carter and Hart (1977).

Coniacian – Introduced by Coquand (1857), and based on the succession near Cognac (Aquitaine); a detailed correlation with the British Isles is not yet possible (Rawson *et. al.*, 1978).

Santonian – Introduced by Coquand (1857), and based on the succession near Saintes (Aquitaine), the Santonian, like the Coniacian, has proved difficult to correlate precisely with the British succession by means of the macrofauna. The microfauna of the British Isles is quite diagnostic, with many zonally important species of planktonic Foraminifera being recorded (Bailey, 1978, unpublished thesis), despite statements to the contrary in Rawson *et al.* (1978).

Campanian – The Campanian stage was introduced by Coquand in 1857, using sections described in 1856, notably those at Aubeterre-sur-Dronne in Charente. The position of the Campanian (and Maastrichtian) in S.W. France has recently been completely reviewed (Séronie-Vivien, 1972). The microfauna of the type sections is rather poor and correlation with other regions is difficult.

Maastrichtian – This stage was introduced by Dumont (1850), and based on sections near the town of Maastricht, Holland. The early definitions were rather unsatisfactory and recently (Voigt, 1956; Deroo, 1966; Schmid, 1959, 1967) the situation has been clarified. The type sections are, however, in a rather different facies to those normally encountered elsewhere in N.W. Europe and the British Isles.

7.4 HISTORY OF INVESTIGATIONS

British Cretaceous foraminifera were little used for stratigraphic, or other purposes, before 1948. All these early publications were monographic in nature, typified by the publications of Burrows, Sherborn and Bailey (1890) on the microfauna of the Red Chalk of Yorkshire, and Chapman (1891–98) on the foraminifera of the Gault Clay.

Williams-Mitchells' (1948) account of the Chalk foraminifera from the Portsdown borehole was the first to suggest that a series of microfaunal zones could be established in the Upper Cretaceous. This stratigraphic approach was extended to the Upper Cretaceous of Northern Ireland by the work of McGugan (1957). There then followed a series of publications (Barnard, 1958, 1963a, b; Barnard and Banner, 1953) on selected taxa. The works of Burnaby (1962) and Jefferies (1962, 1963) showed how statistical methods could help in palaeoecological interpretation, but more than that they showed (particularly Jefferies (1962)) how a small part of the succession *should* be investigated in detail. From then on parts of the Cretaceous succession began to receive more detailed treatment, and this work is continuing.

The Speeton Clay of Yorkshire was fully investigated by Fletcher (1966, Lower Cretaceous foraminifera from the Speeton Clay, Yorkshire; unpublished thesis, University of Hull, 1973), and the foraminifera of the Red Chalk were described by Dilley (1969). In the south, the Gault Clay was investigated by Hart (1970, The distribution of the Foraminiferida in the Albian and Cenomanian of S.W. England, unpublished thesis, University of London; 1973), Carter and Hart (1977) and Price (1975, Biostratigraphy of the Albian Foraminifera of North-West Europe, unpublished thesis, University of London, 1976, 1977a, b). The Upper Cretaceous succession has been the subject of numerous major projects during the last twenty years (Barr, 1962, 1966a; Bailey, 1975, 1978, A foraminiferal biostratigraphy of the Lower Senonian of Southern England, unpublished thesis, (CNAA) Plymouth Polytechnic; Hart, 1970, unpublished thesis; Carter and Hart, 1977; Swiecicki, 1980, A foraminiferal biostratigraphy of the Campanian and Maastrichtian Chalks of the United Kingdom, unpublished thesis (CNAA) Plymouth Polytechnic; Hart and Bailey, 1979; and Bailey and Hart, 1979) although views on the nature of the 'chalk palaeoenvironment' are as varied as ever.

Elsewhere in Europe, Cretaceous foraminifera have been investigated by many workers and some of their results are applicable to Britain. Of particular note are the publications of Robaszynski *et al.* (1980) in Belgium and northern France; Stenestad (1969) in Denmark; Berthelin (1880). Marie (1938, 1941), Goel (1965), and Magniez-Jannin (1975) in France; Bartenstein (1952, 1974, 1976a, 1976b, 1977, 1978a, 1978b, 1979), Bartenstein and Bettenstaedt (1962), Bartenstein, Bettenstaedt and Bolli (1957), Bartenstein and Brand (1951), Bartenstein and Kaever (1973), Bettenstaedt (1952), and Aubert and Bartenstein (1976) in Germany; Fuchs (1967), and Ten Dam (1947, 1948a, b, 1950) in Holland; Gawor-Biedowa (1969, 1972) in Poland; Brotzen (1934a, b, 1936, 1942, 1945, 1948) in Sweden; Caron (1966) and Klaus (1960a, b, c) in Switzerland; and Neagu (1965, 1969) in Rumania. Recently many European micropalaeontologists have collaborated (under the auspices of IGGP No. 58 'Mid-Cretaceous Events') to produce a monograph of the mid-Cretaceous planktonic foraminifera (Robaszynski and Caron, 1979a, b).

In America the Cretaceous foraminifera have been investigated by many specialists, including J. A. Cushman. Nearly every publication on Cretaceous foraminifera will include an extensive list of his references and none have been singled out for special mention here. Other workers in the Cretaceous include Loeblich and Tappan (1961, 1964), Eicher (1965, 1966, 1967, 1969) Eicher and Worstell (1970), Douglas (1969a, b), Sliter (1968), Pessagno (1967), Frizzell (1954), Plummer (1931), Olsson (1964), Petters (1977a, b), Ascoli (1976), and Masters (1977).

7.5 DEPOSITIONAL HISTORY

The Late Cimmerian movements at the end of the Jurassic (P. A. Ziegler, 1975) effectively terminated active sedimentation over large areas of Europe, and, in particular, revived a series of positive areas between Britain and Poland (the London-Brabant, Rhenish, and Bohemian massifs). This formed a highly effective barrier between the 'Boreal' realm of the North Sea Basin, Danish-Polish Furrow, and Russian Platform and the 'Tethyan' area of Alpine Europe, and the evolving North Atlantic Ocean. This interconnected chain of massifs controlled surface water movements and faunal distributions well into the Late Cretaceous, even though they were probably inundated by the Early Turonian.

The British Isles, sitting astride the London-Brabant Massif, was effectively part of two depositional basins, and these were only occasionally interconnected in the pre-Aptian interval. During and after the Aptian the connection was almost continuous. Apart from a short-lived marine phase near the Jurassic-Cretaceous boundary (Casey, 1973) the southern basin remained essentially non-marine until the early Aptian (Fig. 7.2). There followed a marine or near-marine phase during the Aptian-Albian interval, during which time the Atherfield Clay, Lower Greensand, Gault Clay and Upper Greensand were deposited

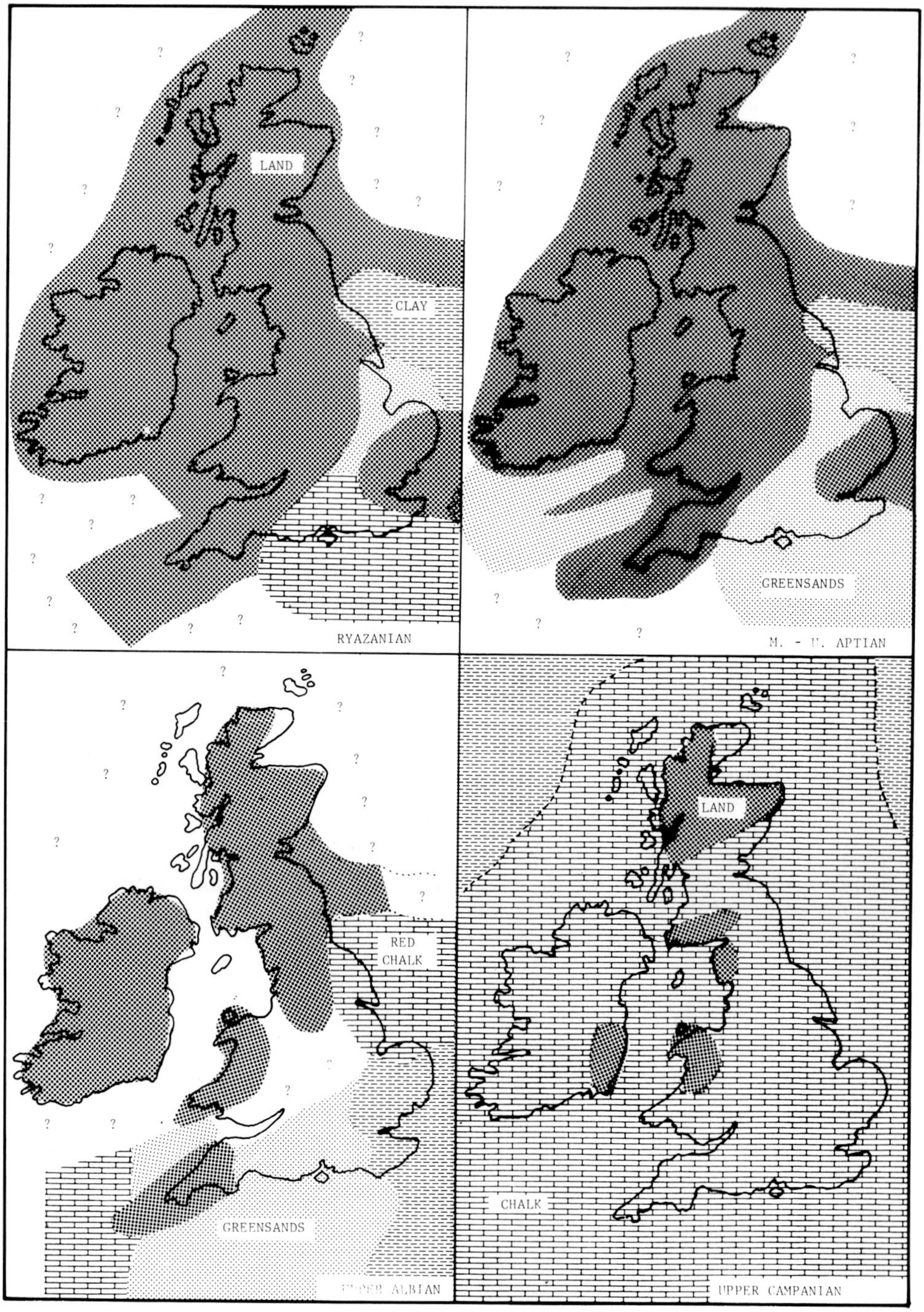

Fig. 7.2 – Cretaceous palaeogeography of the British Isles (Ryazanian, M.-Aptian, Upper Albian and Upper Campanian).

over southern England and the English Channel Basin. This interval also marks the first appearance of the foraminifera in the Cretaceous of southern England, although the fauna is very restricted in parts of the Lower Greensand. In the North Sea Basin (which effectively includes parts of Yorkshire, Humberside and Lincolnshire) the pre-Albian interval is characterised by the Speeton Clay Formation of the Cromer Knoll Group (Rhys, 1975). This marine clay succession is fossiliferous throughout and shows a marked similarity to faunas from N. Germany. The uppermost Lower Cretaceous in the North Sea Basin is characterised by the Red Chalk Formation. This thin, red-coloured, limestone succession contains a rich foraminiferal fauna, that contains large numbers of small, simple, planktonic foraminifera. The same planktonic fauna can also be recognised in the uppermost part of the Gault Clay of southern England. The Upper Albian (Fig. 7.2), therefore, marks the point in the succession when the two basins of deposition became faunally united. From the Cenomanian onwards the British fauna is, in most respects, uniform, although a few regional trends can be identified. As early as the Late Cenomanian the fauna becomes recognisable as typical of the whole of northern Europe, with many species being quite widespread in occurrence. At several levels (notably un the Upper Cenomanian, Turonian, Santonian, and Maastrichtian (especially in the North Sea Basin)), planktonic foraminifera of international biostratigraphic, importance are found, despite statement to the contrary (Rawson *et al.*, 1978, p. 25). The apparently uniform chalk succession of the Cenomanian–Maastrichtian interval means that the evolution of such benthic genera as *Gavelinella, Bolivinoides,* and *Stensioina* is of great value in inter-regional correlation (Bailey and Hart, 1979).

Since all the British Cretaceous sediments were being deposited in an extensive shelf sea, changes in sea level were clearly of fundamental importance. The eustatic changes now known from the Cretaceous (Hancock, 1976; Hancock and Kauffman, 1979; Cooper, 1977) have had a profound effect on the distribution, and evolution, of both the benthic and planktonic foraminifera (Hart and Bailey, 1979; Swiecicki, 1980, unpublished thesis; Hart, 1980a, b).

7.6 STRATIGRAPHIC DISTRIBUTION OF CRETACEOUS FORAMINIFERA

Many sections of Cretaceous strata are available for investigation in the British Isles, and those chosen for inclusion in this account are probably some of the best exposed, and at present yield the best, most typical, biostratigraphic data. This has inevitably meant that successions in marginal areas such as Devon, Northern Ireland, N.W. Scotland, and to some extent Central England, have been omitted from the discussion. In Devon and Dorset the faunas of the Albian to Coniacian interval have been thoroughly investigated by Hart (1970, unpublished thesis, 1973, 1975), Hart and Weaver (1977), Hart, Manley and Weaver (1979), and Bailey (1975, 1978, unpublished thesis). In Northern Ireland the Cretaceous succession is poorly exposed, and much of the Chalk is very hard, having been affected by Tertiary volcanism. The work of McGugan (1957) is the most complete account of the foraminifera of that area, although Barr (1966b) has described a Cretaceous fauna from a cave infill in Eire. The Scottish sections have been considered occasionally over the years, but unpublished work by one of us (MBH) on material from Mull and Lochaline has failed to provide data worthy of inclusion in this account.

The successions considered in this account, therefore, are those at Speeton, Folkestone, Dover, S.E. Kent, Scratchells Bay/Alum Bay (Isle of Wight), the north Norfolk coast, and the Southern North Sea Basin (Well 44/2-1). These are shown in Fig. 7.3, together with the appropriate lithostratigraphic terminology.

Throughout the Cretaceous succession the fauna changed in response to the environmental changes brought about by marked changes in sea level (see Fig. 7.4). These fluctuations drastically changed the nature of the fauna from a Lituolacea/Nodosariacea/Robertinacea dominated fauna in the Early Cretaceous to a Lituolacea/Cassidulinacea/Globigerinacea/Nodosariacea fauna in the mid-Cretaceous and Lituolacea/Cassidulinacea/Buliminacea/Globigerinacea fauna in the Late Cretaceous. This is summarised in Fig. 7.4, which gives an indication of the relative dominance of all the superfamilies throughout the succession. The use made of these various superfamilies is also interesting. The Nodosariacea are important for biostratigraphic work in the Early Cretaceous as

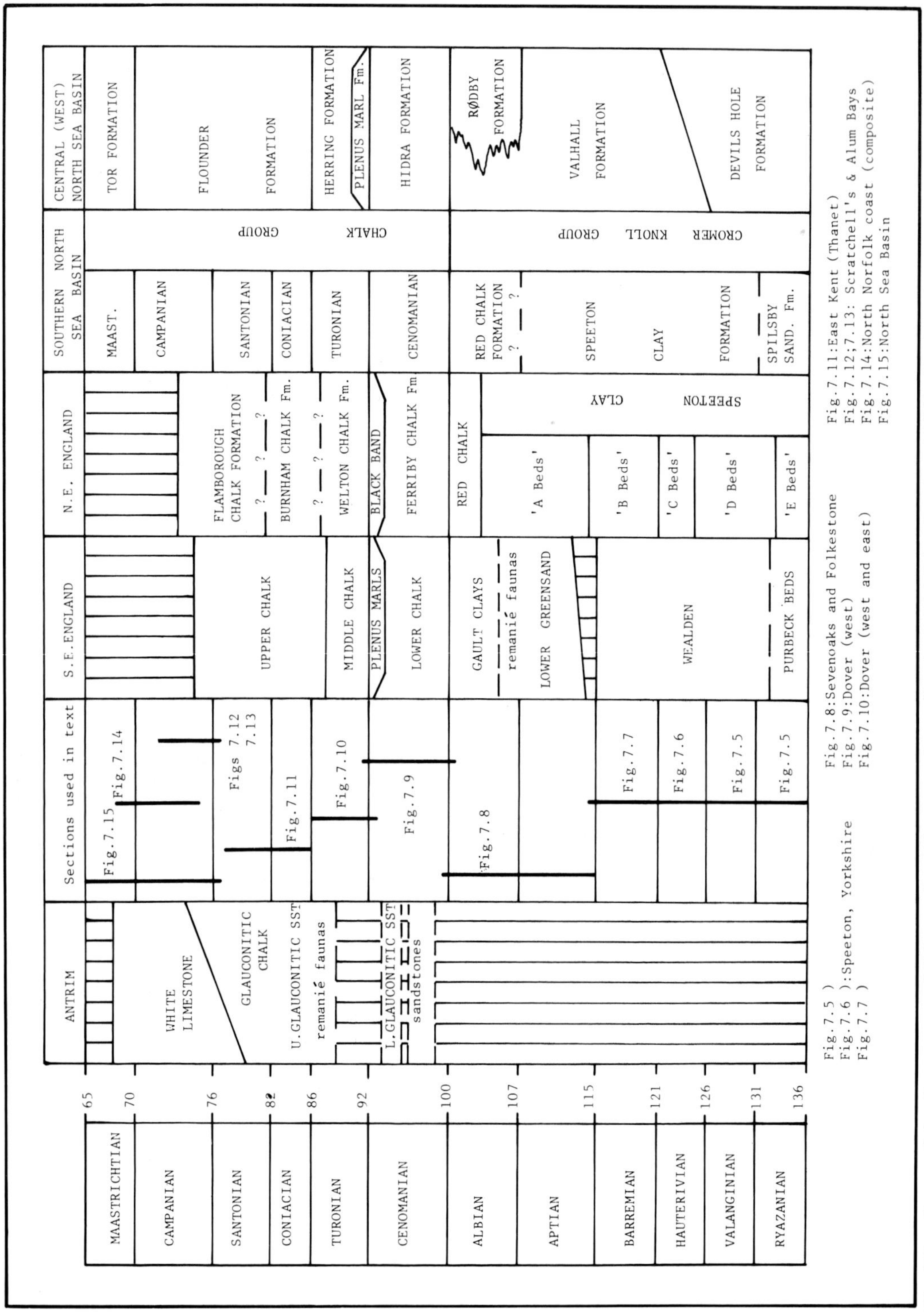

Fig. 7.3 – Stages of the Cretaceous and lithostratigraphic nomenclature for the British Isles. Stratigraphic distribution of sections discussed in the text is indicated in the central column.

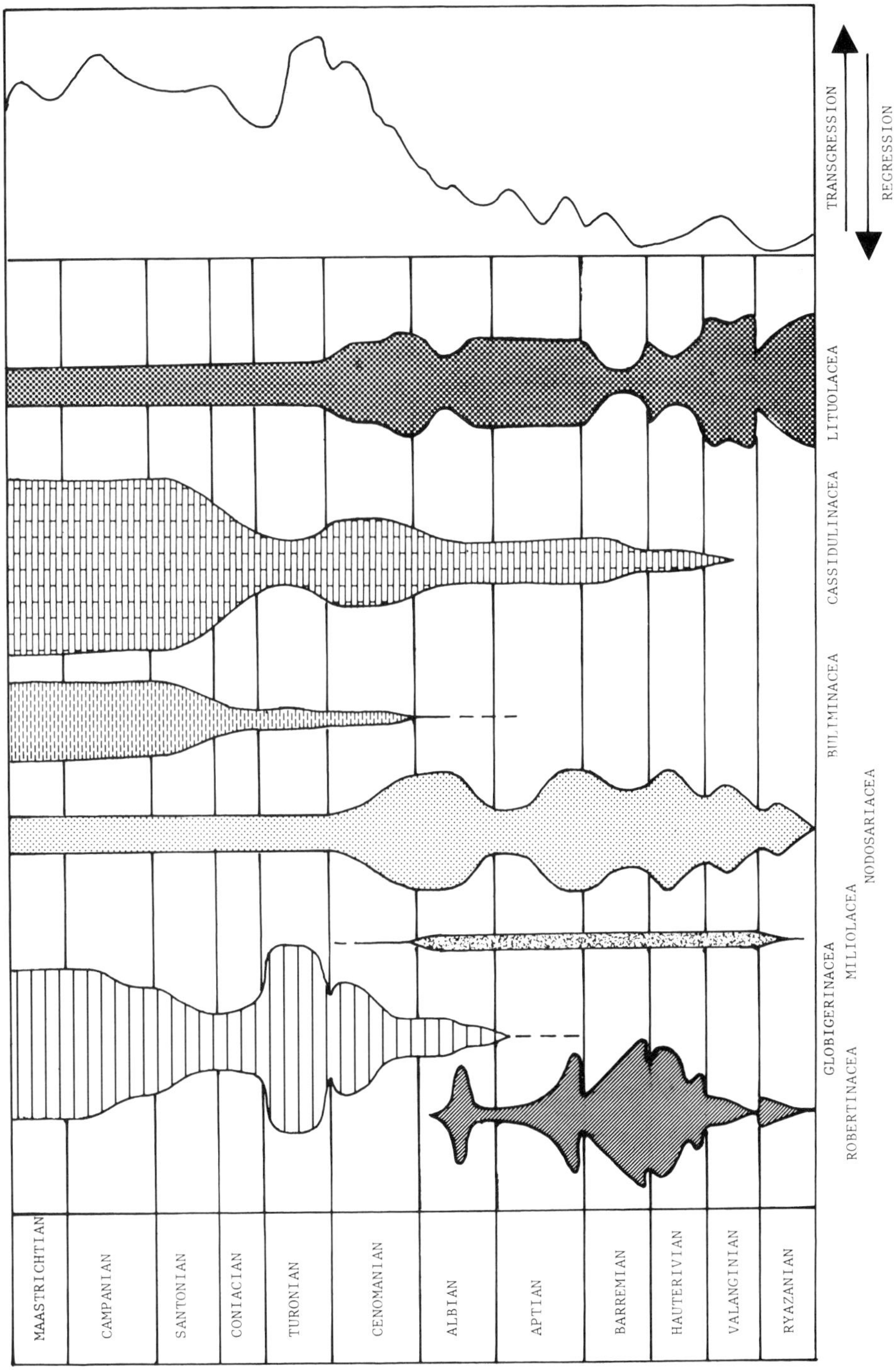

Fig. 7.4 – Schematic distribution of the Major superfamilies of Foraminifera throughout the Cretaceous.

there is little else available, while in the Late Cretaceous (though still present in reasonable numbers) they are little used, as other groups appear more attractive. The Lituolacea are mainly used for zonation in the mid-Cretaceous where the *Arenobulimina, Plectina,* and *Flourensina* lineages are well-developed. In the early Cretaceous and late Cretaceous the Lituolacea are rarely used, even though some genera and species appear to be quite distinctive at certain levels in the succession (Adams *et al.,* 1973). In the marginal facies of the mid-Cretaceous, *Orbitolina* has been recorded widely (Hart, Manley and Weaver, 1979) and work in progress may improve the stratigraphic value of the genus in local stratigraphy.

The Globigerinacea are perhaps the most useful for international correlation (as distinct from inter-regional) and many key species are now being recorded from the British succession for the first time. Recently Magniez-Jannin (personal communication, 1981) has found specimens of *Rotalipora appenninica* (Renz) and *Planomalina buxtorfi* (Gandolfi) in the uppermost Albian of Folkestone. Other distinctive species such as *Rotalipora reicheli* (Mornod), *R. greenhornensis* (Morrow), *Praeglobotruncana helvetica* (Bolli), *Marginotruncana coronata* (Bolli), *M. renzi* (Gandolfi), *M. sigali* (Reichel), *Dicarinella concavata* (Brotzen), *Globotruncana fornicata* (Plummer), *G. contusa* (Cushman), and *Globotruncanella havanensis* (Voorwijk) are also now being recorded, and while their stratigraphic distribution must be viewed after consideration of the palaeogeography and palaeocirculation patterns, many do appear to have real value.

In the mid-late Cretaceous, apart from the Turonian (Hart and Bailey, 1979) the fauna is dominated by the benthic foraminifera, and an accurate inter-regional correlation (Bailey and Hart, 1979) can be effected using members of the *Gavelinella, Stensioina,* and *Bolivinoides* lineages. These have been well documented by Koch (1977), Bailey (1978, unpublished thesis) and Swiecicki (1980, unpublished thesis). The prolonged period of chalk sedimentation seems to have allowed their gradual evolution in the N.W. European area, and a very refined stratigraphy is now possible.

The palaeoenvironments represented by these faunas are all very similar. Throughout the Cretaceous, marine sedimentation was typical of a shallow-water to deeper-water shelf area. In no case can one recognise the passage from marine to non-marine deposition by means of foraminifera. Most workers have therefore concentrated on estimating or guessing the depth of water over the shelf at any particular time. Various estimates range from 9.0 m to 600 m in the mid-late Cretaceous (Burnaby, 1962; Hart and Bailey, 1979; Kennedy, 1970; Hancock, 1976; Hancock and Kauffman, 1979). Recently the problem of anoxic events in the Cretaceous succession has been considered by Schlanger and Jenkyns (1976) and Hart and Bigg (in press), especially with reference to the Black Band in Yorkshire. Such palaeoenvironmental problems will probably be further considered in the future, especially now that there is an accurate biostratigraphy for the British Cretaceous succession.

7.7 INDEX SPECIES

The classification used in this account is that proposed by Loeblich and Tappan (1964, Treatise; 1974), but also includes genera incorporated into the literature since that time. In the following section the species are grouped by sub-order (Textulariina, Miliolina and Rotaliina), and then discussed in alphabetical order.

The species selected for inclusion are, in most cases, the most stratigraphically useful, and those which would probably be encountered in any work on the Cretaceous succession. However, in an attempt to provide a balanced distribution throughout the succession, some stratigraphically useful (e.g. *Marssonella kummi* Zedler, *Orbitollina* sp., *Plectina mariae* (Franke), *Dicarinella calnaliculata* (Reuss), *D. imbricata* (Mornod), *Globotruncana arca* (Cushman), *Lingulina denticulcocarinata* (Chapman), *Marginotruncana renzi* (Gandolfi), *Racemiguembelina fructicosa* (Egger)) species have had to be ommitted.

The range charts (Figs. 7.5–7.15) present the stratigraphic distribution of foraminifera as seen in several key sections in southern and eastern England, including one composite section from the Southern North Sea Basin (with the approval of Shell U.K. Exploration and Production Ltd. and ESSO Exploration and Production U.K. Ltd.). Summary charts (Fig. 7.16) are also included for the full Cretaceous succession.

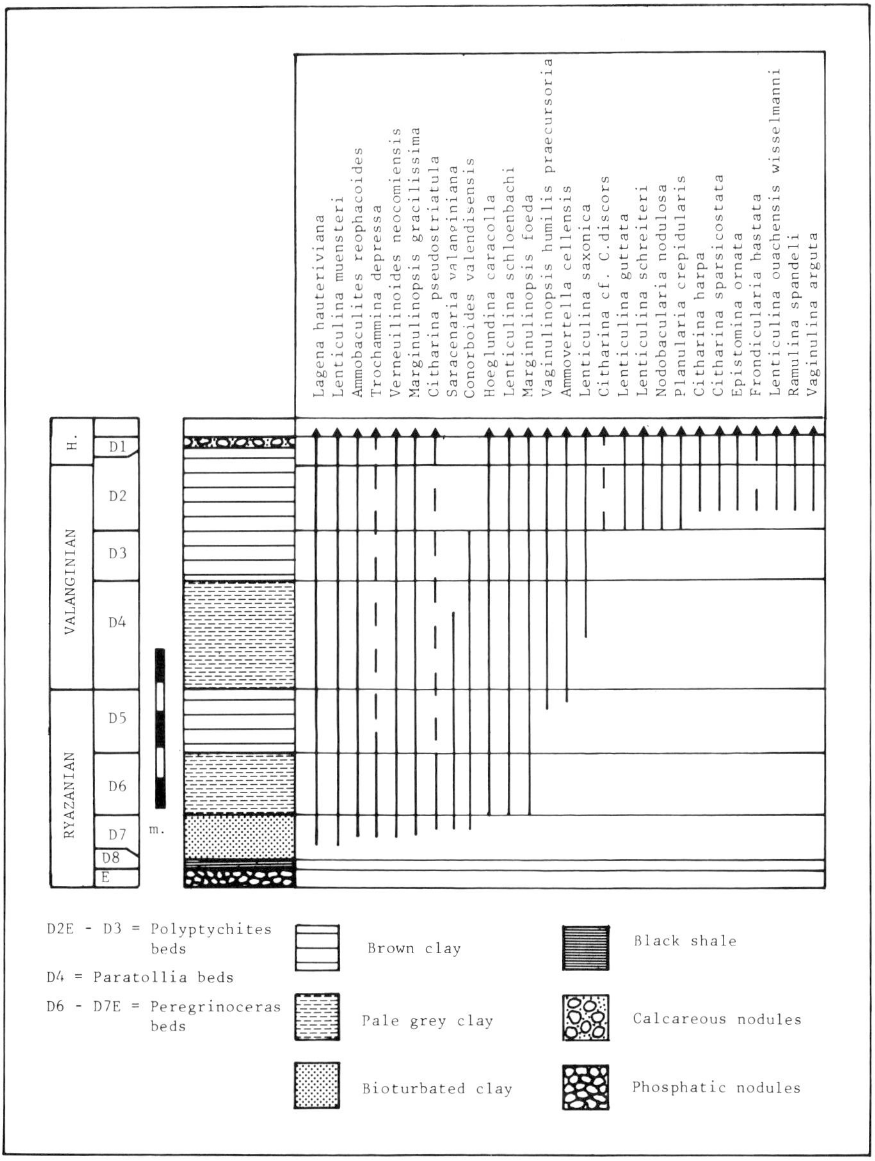

Fig. 7.5 – The Speeton Clay (Ryazanian and Valanginian) of Speeton, Yorkshire.

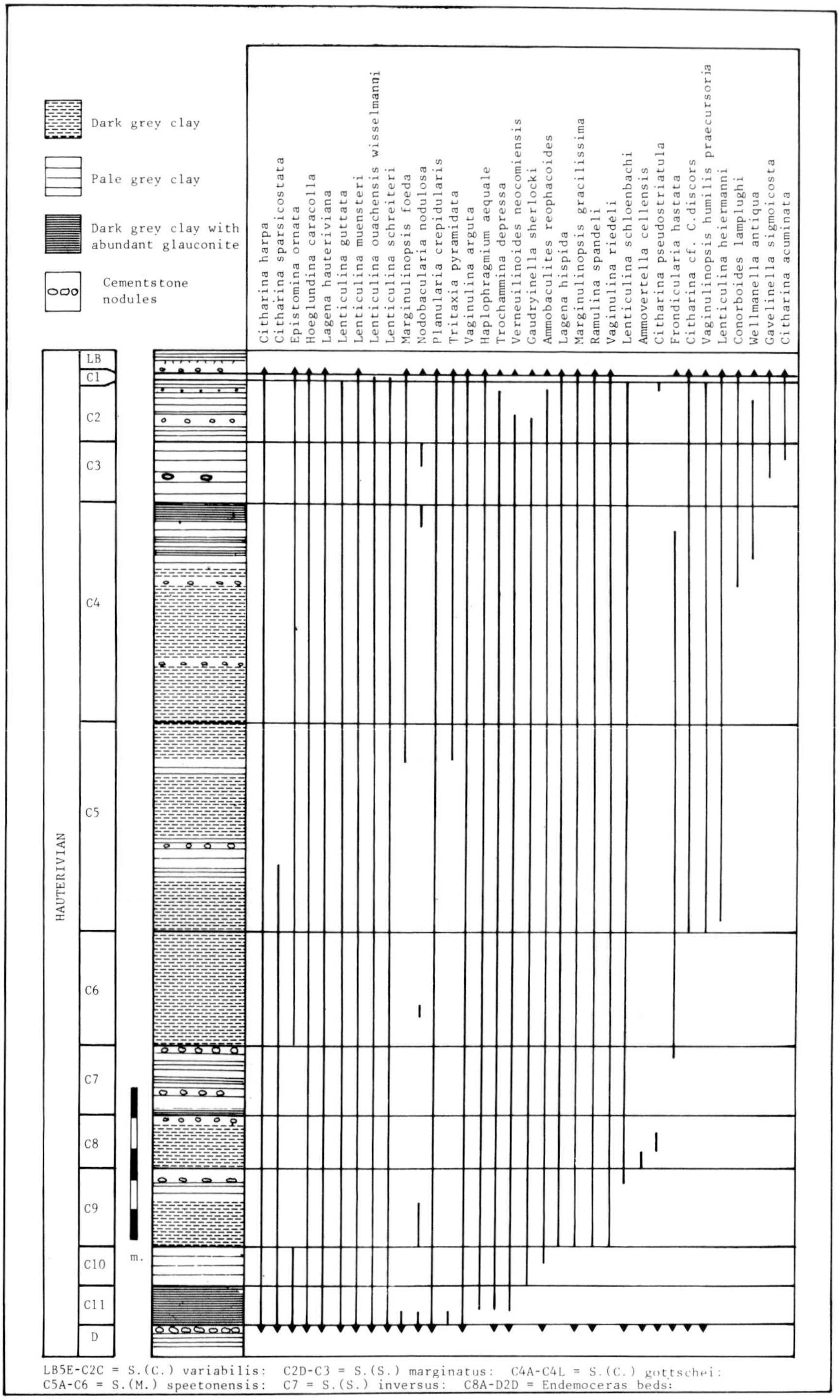

Fig. 7.6 – The Speeton Clay (Hauterivian) of Speeton, Yorkshire.

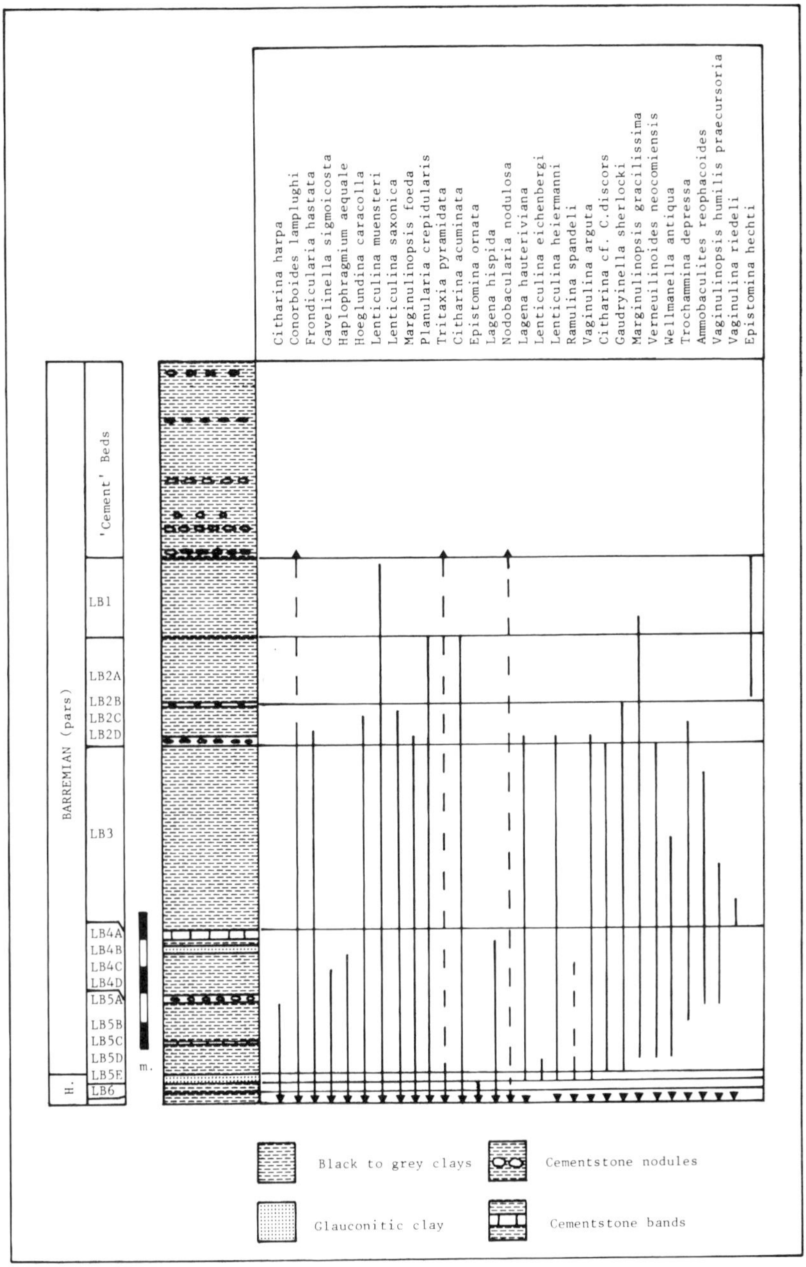

Fig. 7.7 – The Speeton Clay (Barremian) of Speeton, Yorkshire.

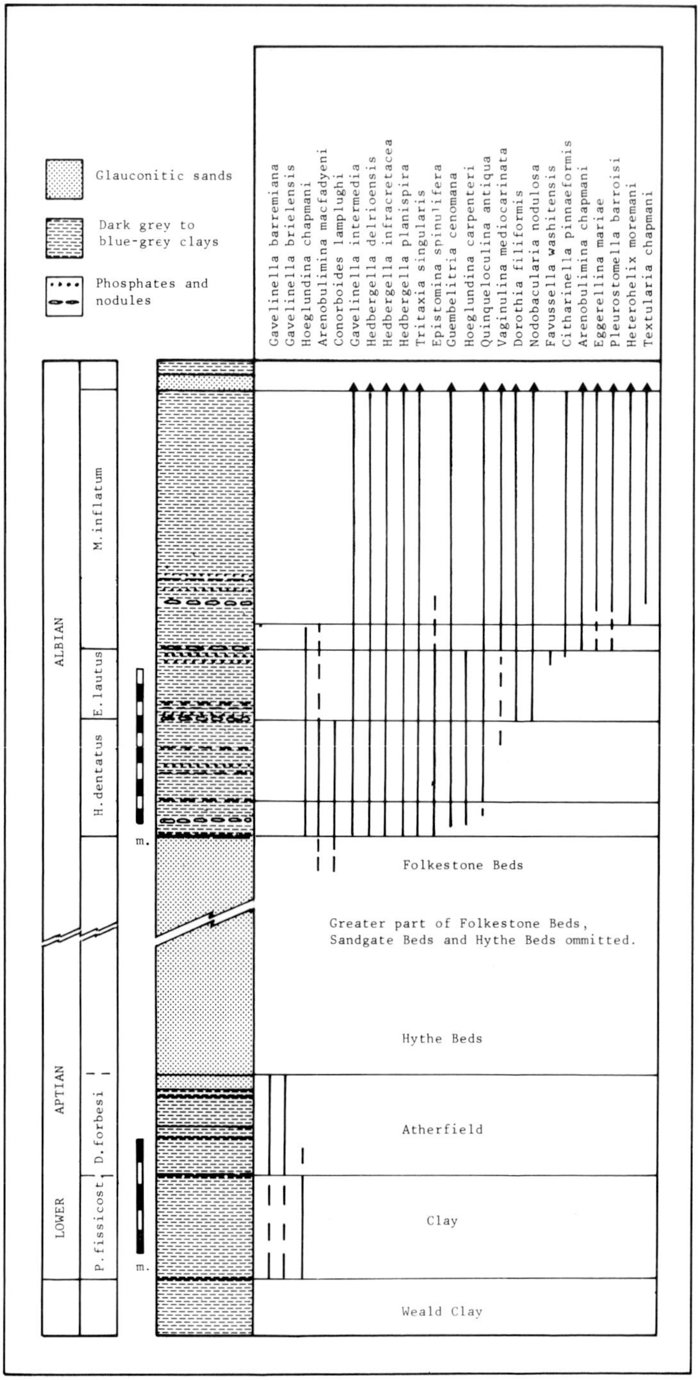

Fig. 7.8 – The Atherfield Clay (Lower Aptian) and Gault Clay (Middle and Upper Albian) of Sevenoaks and Folkestone, Kent.

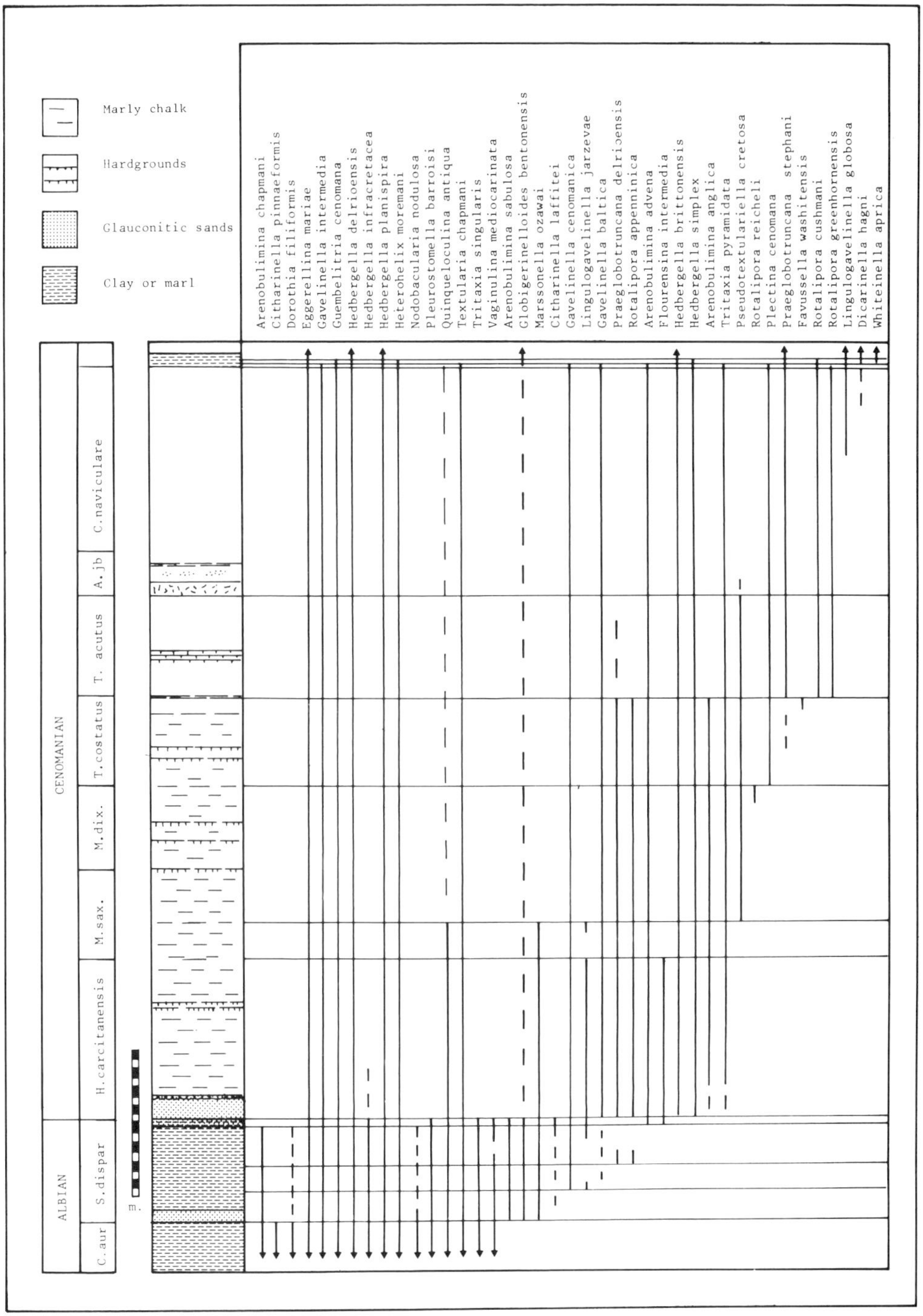

Fig. 7.9 – The uppermost Gault Clay (Upper Albian) and Lower Chalk (Cenomanian) of Folkestone and Dover, Kent.

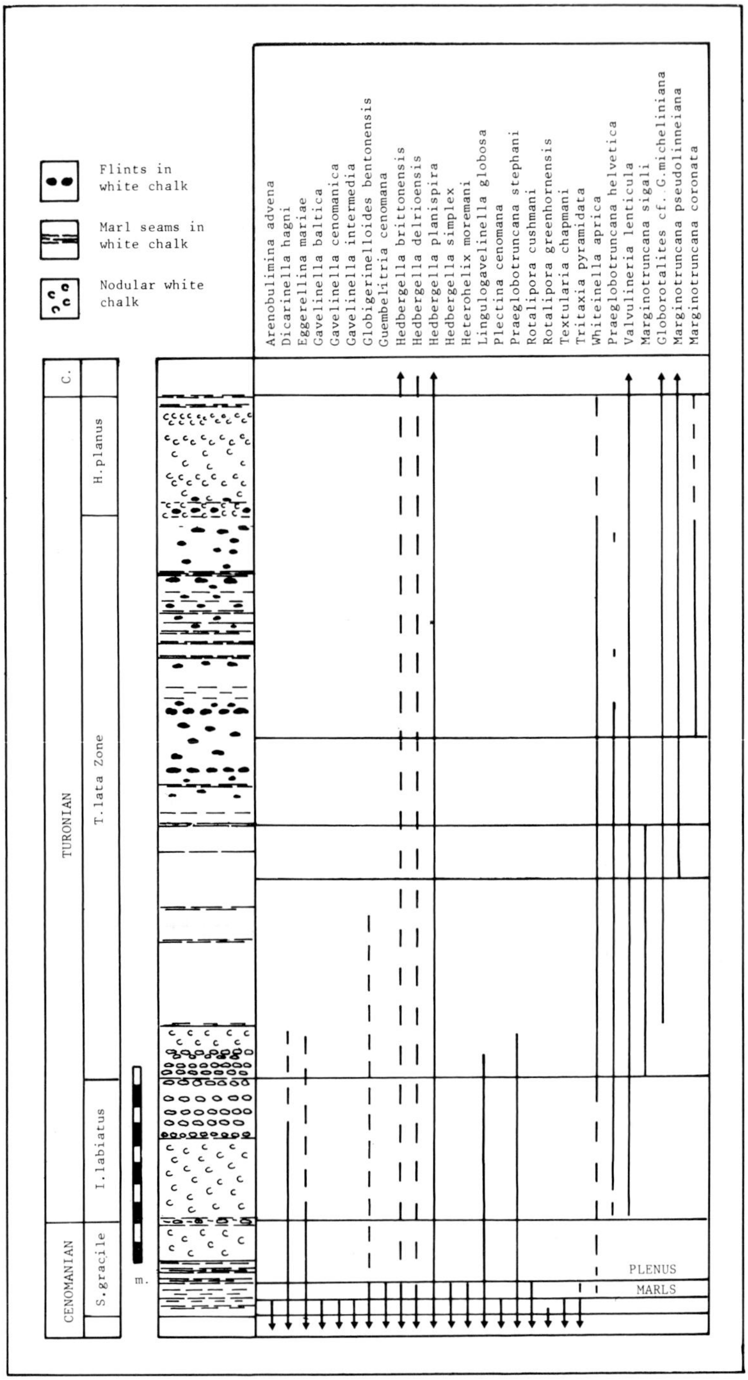

Fig. 7.10 – The Middle Chalk (Turonian) of Dover, Kent.

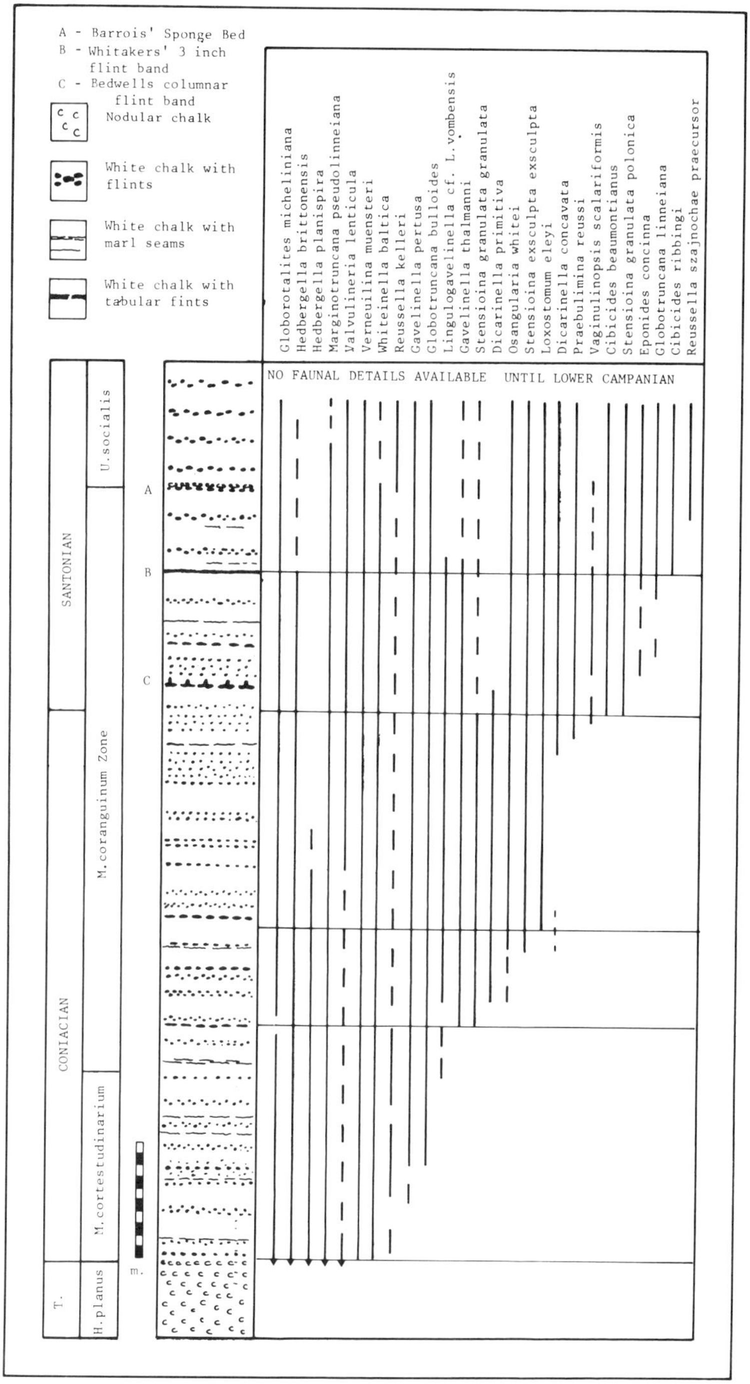

Fig. 7.11 – The lower part of the Upper Chalk (Turonian to Santonian) of South-East Kent.

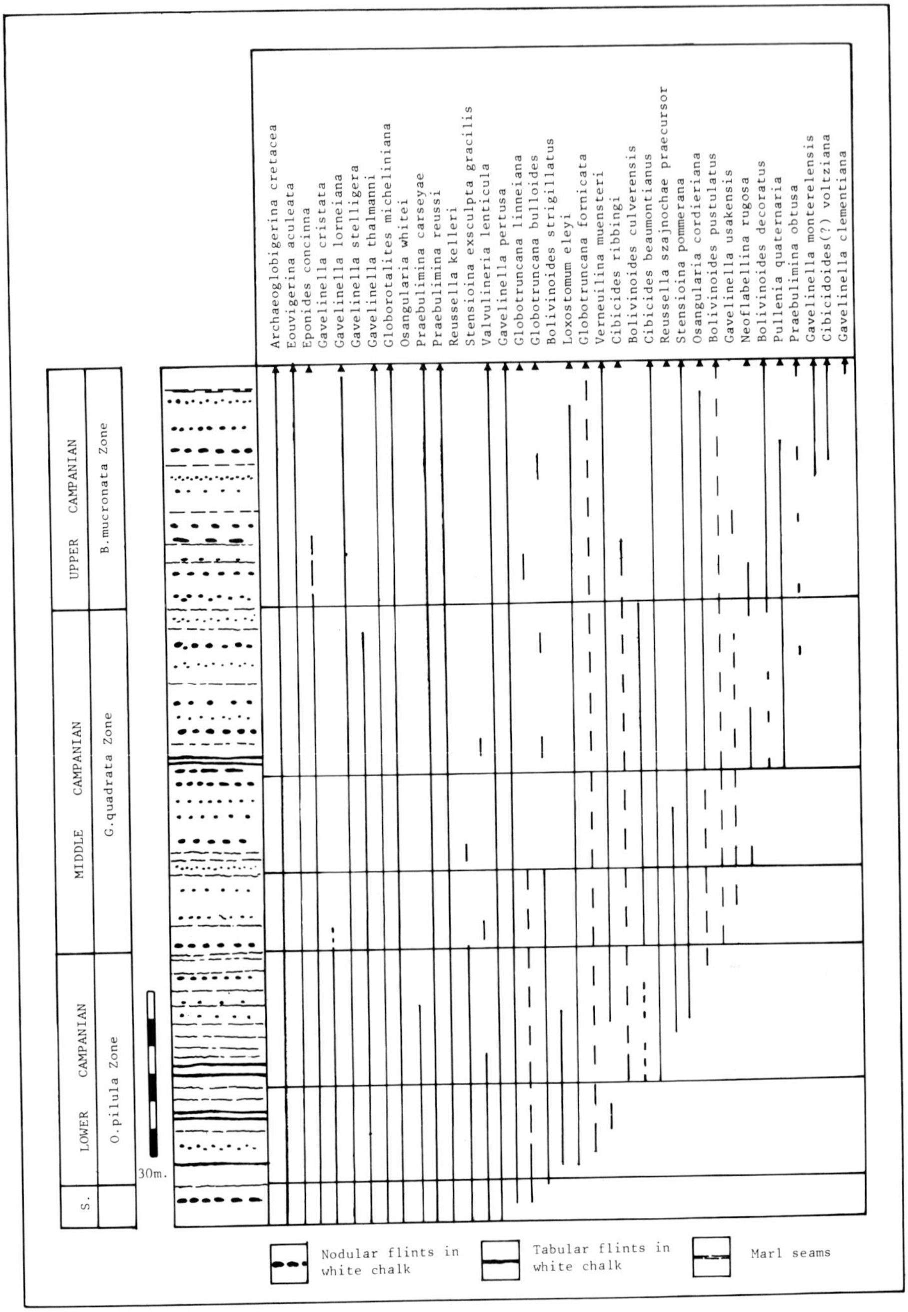

Fig. 7.12 – The middle of the Upper Chalk (Santonian to Upper Campanian) of Scratchell's Bay, Isle of Wight. (N.B. this section is accessible only by boat, and then in favourable weather conditions).

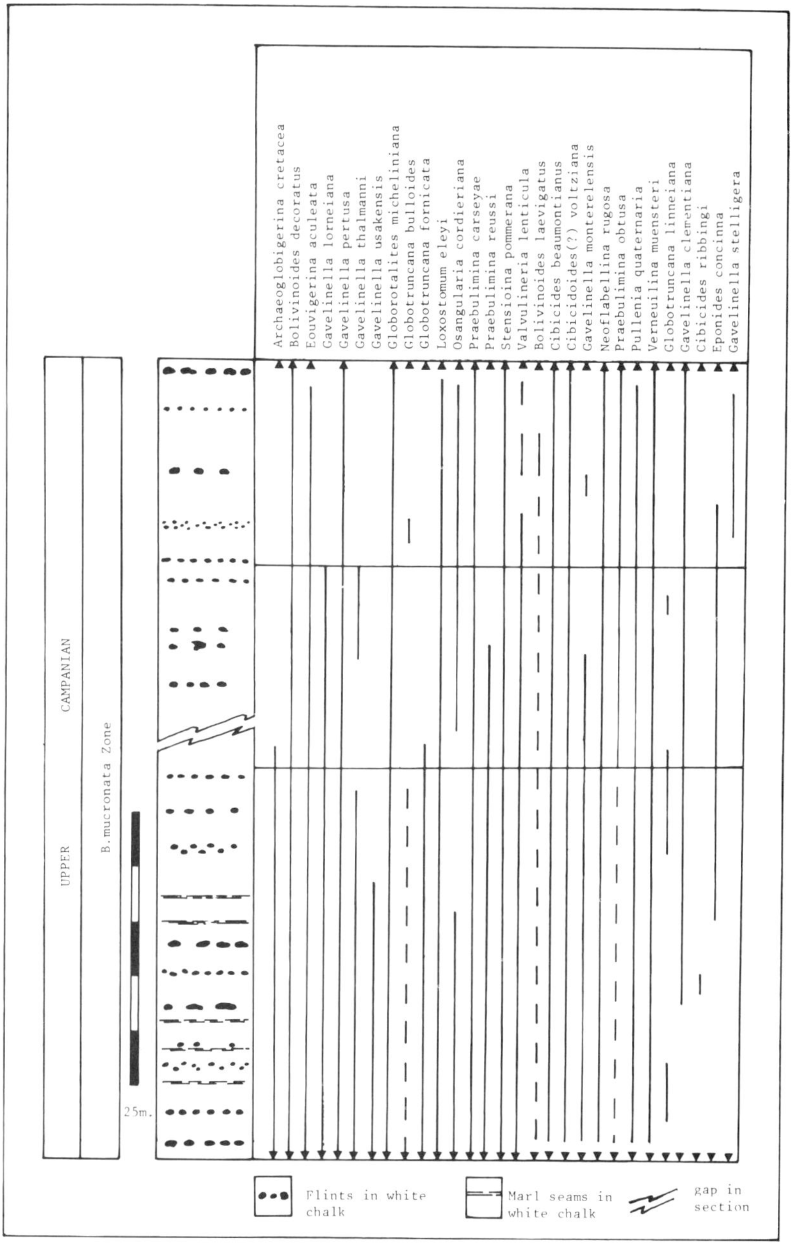

Fig. 7.13 – The middle of the Upper Chalk (Upper Campanian) of Alum Bay, Isle of Wight.

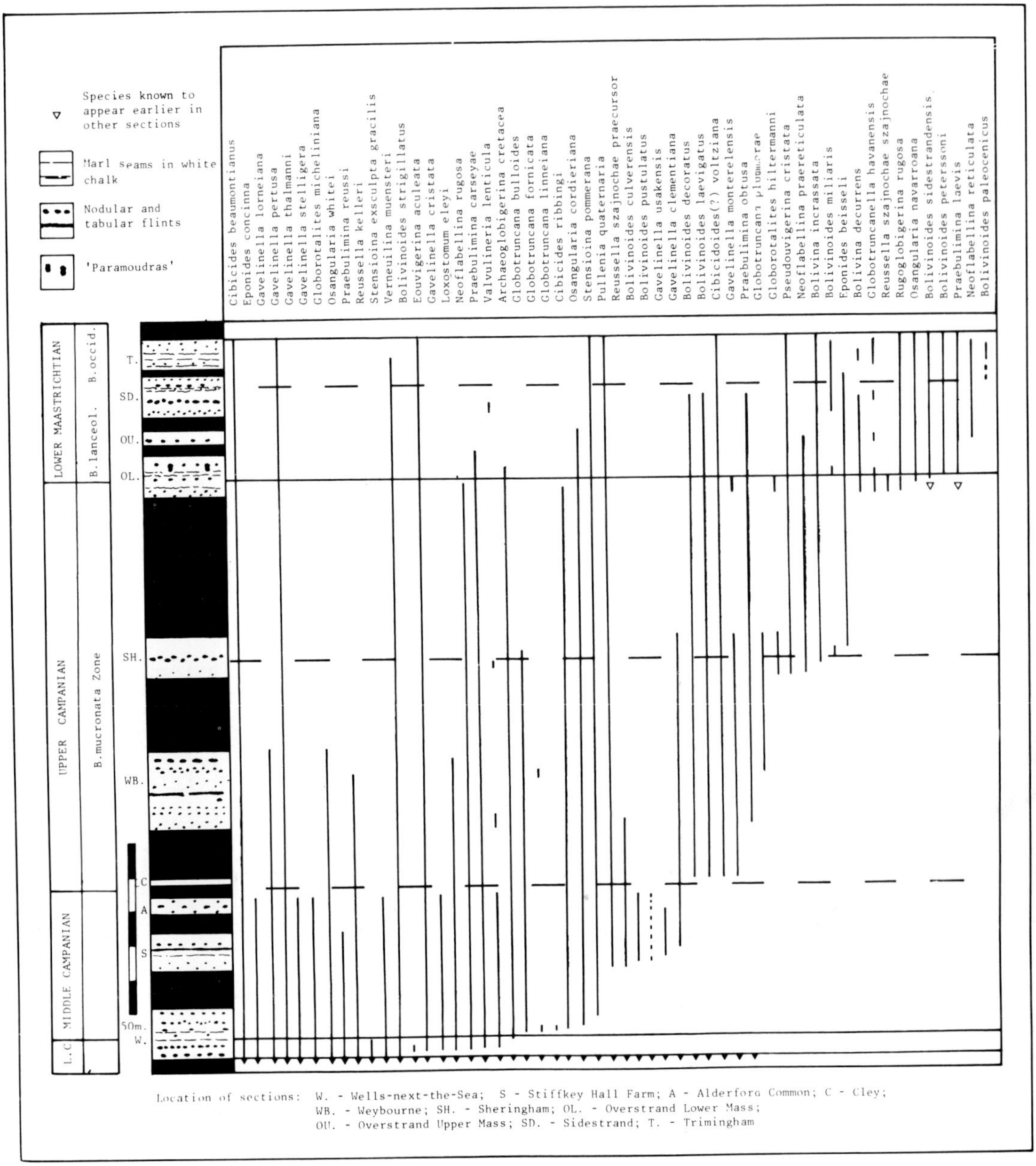

Fig. 7.14 – The upper part of the Upper Chalk (Lower Campanian to Lower Maastrichtian) of the North Norfolk coast. This is a composite section based on several localities; for detailed description of the section see Swiecicki (Thesis, 1980).

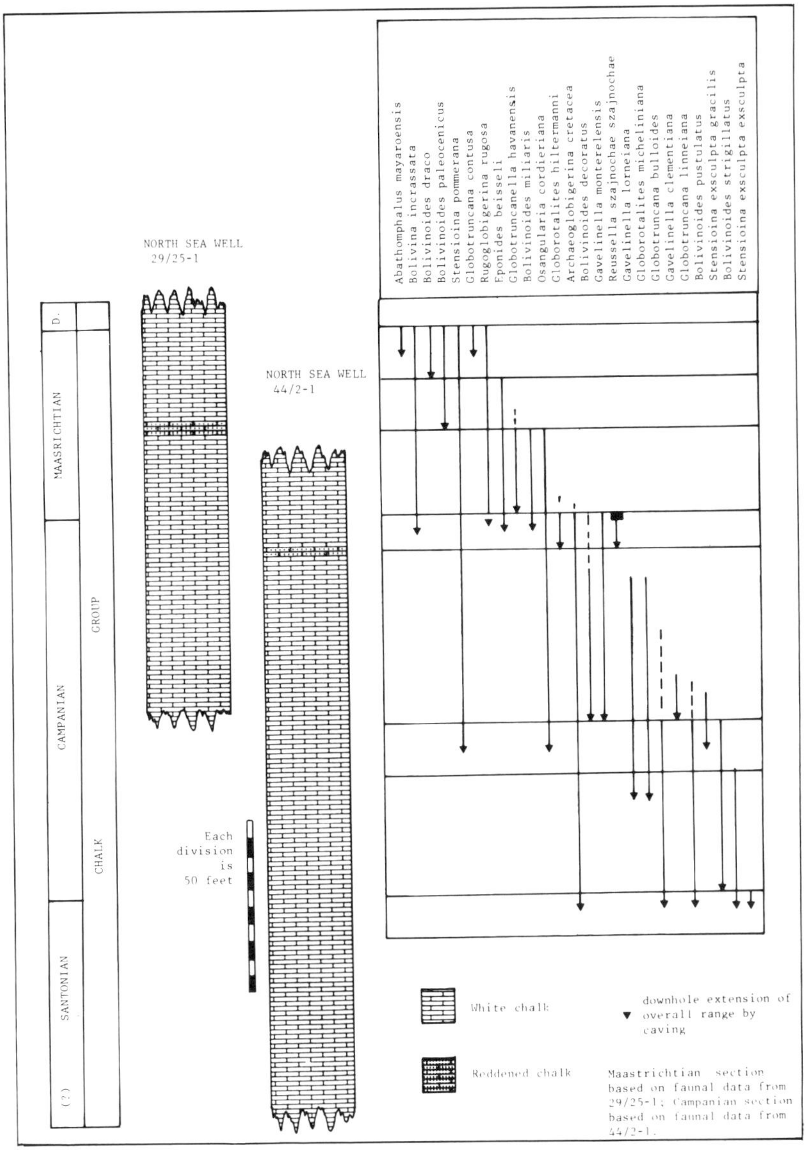

Fig. 7.15 – The upper part of the Chalk Group in the Southern North Sea Basin (Maastrichtian to Campanian).

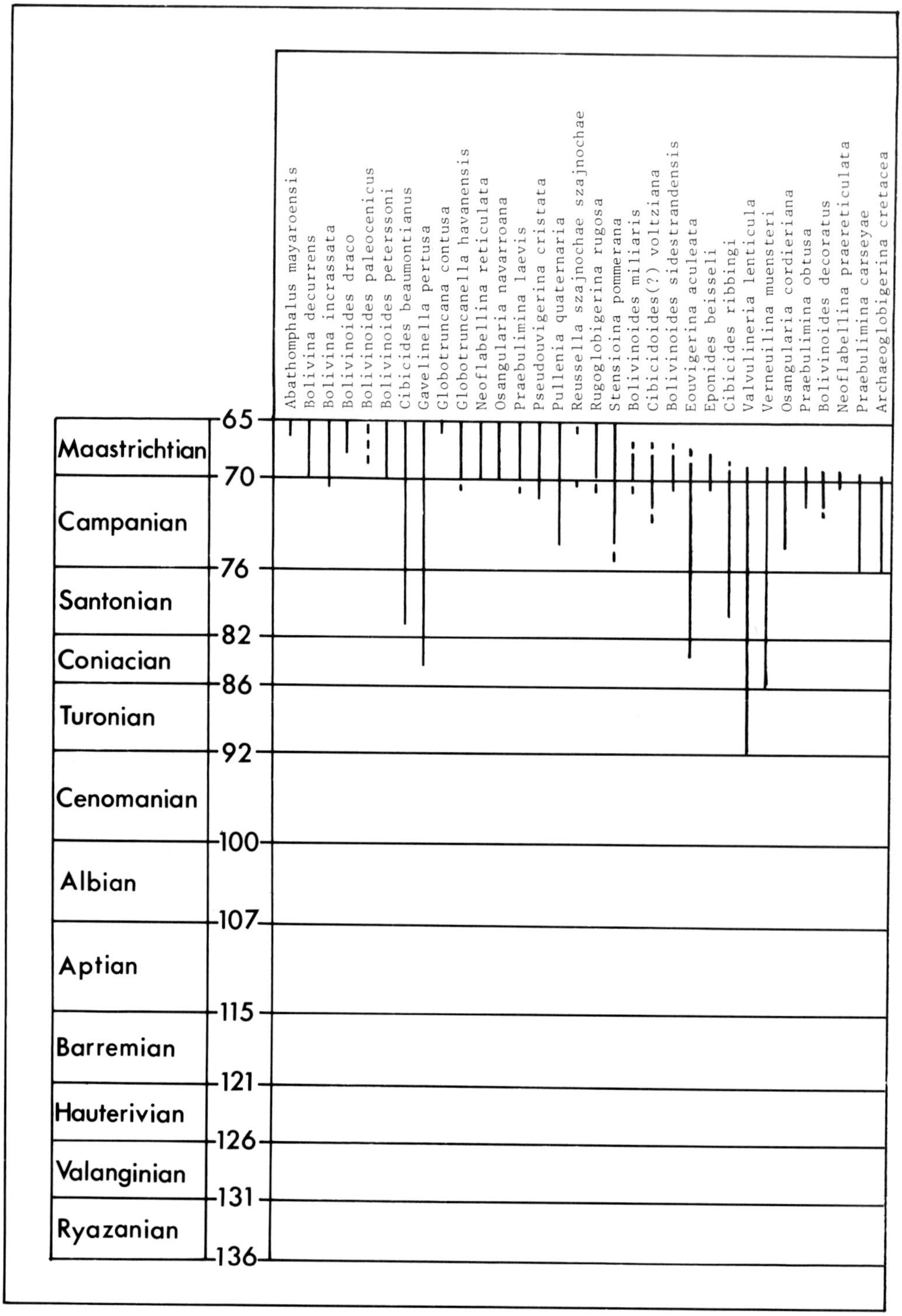

Fig. 7.16 – Summary of the distribution of Cretaceous Foraminifera.

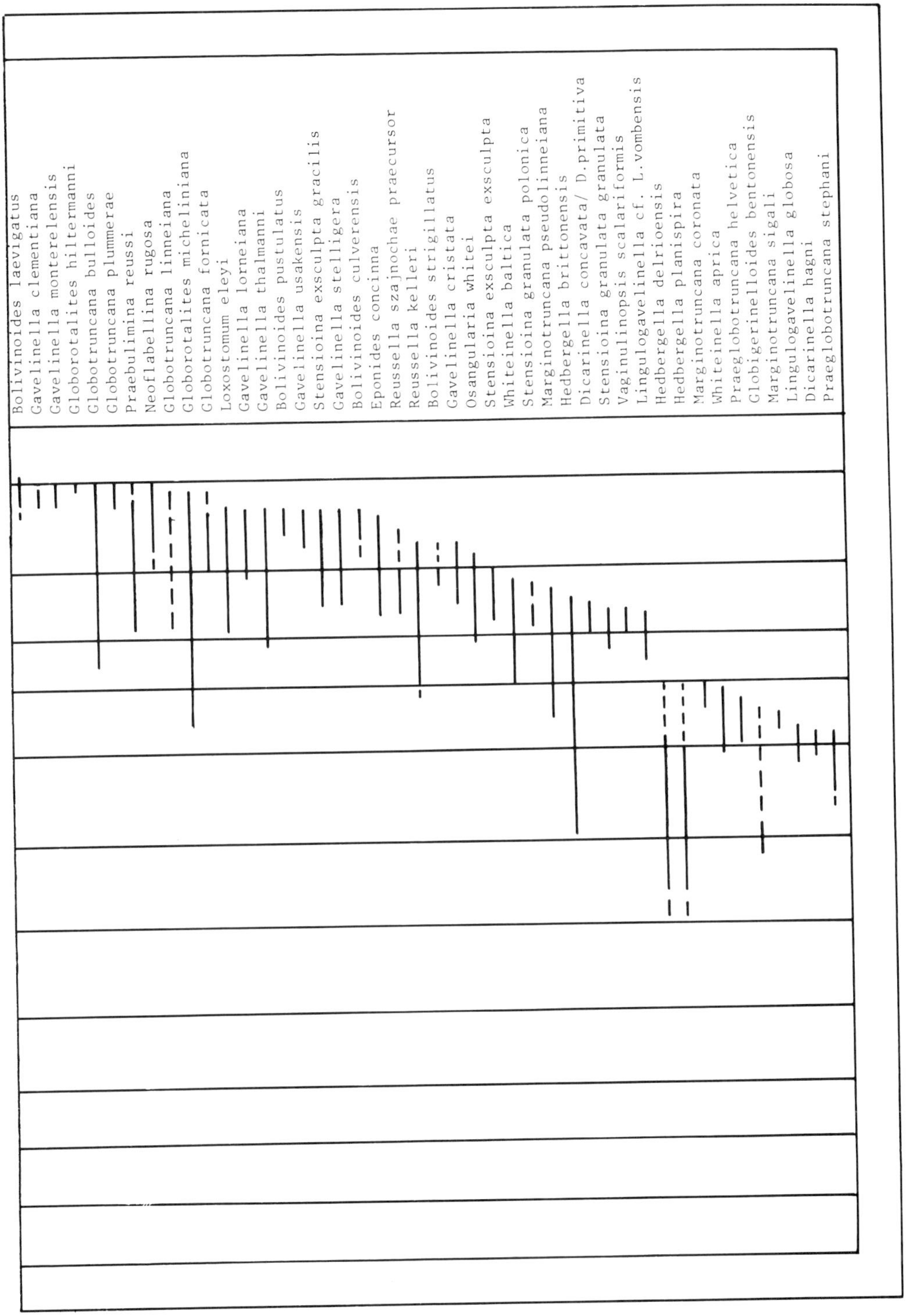
Bolivinoides laevigatus
Gavelinella clementiana
Gavelinella monterelensis
Globorotalites hiltermanni
Globotruncana bulloides
Globotruncana plummerae
Praebulimina reussi
Neoflabellina rugosa
Globotruncana linneiana
Globorotalites micheliniana
Globotruncana fornicata
Loxostomum eleyi
Gavelinella lorneiana
Gavelinella thalmanni
Bolivinoides pustulatus
Gavelinella usakensis
Stensioina exsculpta gracilis
Gavelinella stelligera
Bolivinoides culverensis
Eponides concinna
Reussella szajnochae praecursor
Reussella kelleri
Bolivinoides strigillatus
Gavelinella cristata
Osangularia whitei
Stensioina exsculpta exsculpta
Whiteinella baltica
Stensioina granulata polonica
Marginotruncana pseudolinneiana
Hedbergella brittonensis
Dicarinella concavata/ D.primitiva
Stensioina granulata granulata
Vaginulinopsis scalariformis
Lingulogavelinella cf. L.vombensis
Hedbergella delrioensis
Hedbergella planispira
Marginotruncana coronata
Whiteinella aprica
Praeglobotruncana helvetica
Globigerinelloides bentonensis
Marginotruncana sigali
Lingulogavelinella globosa
Dicarinella hagni
Praeglobotruncana stephani

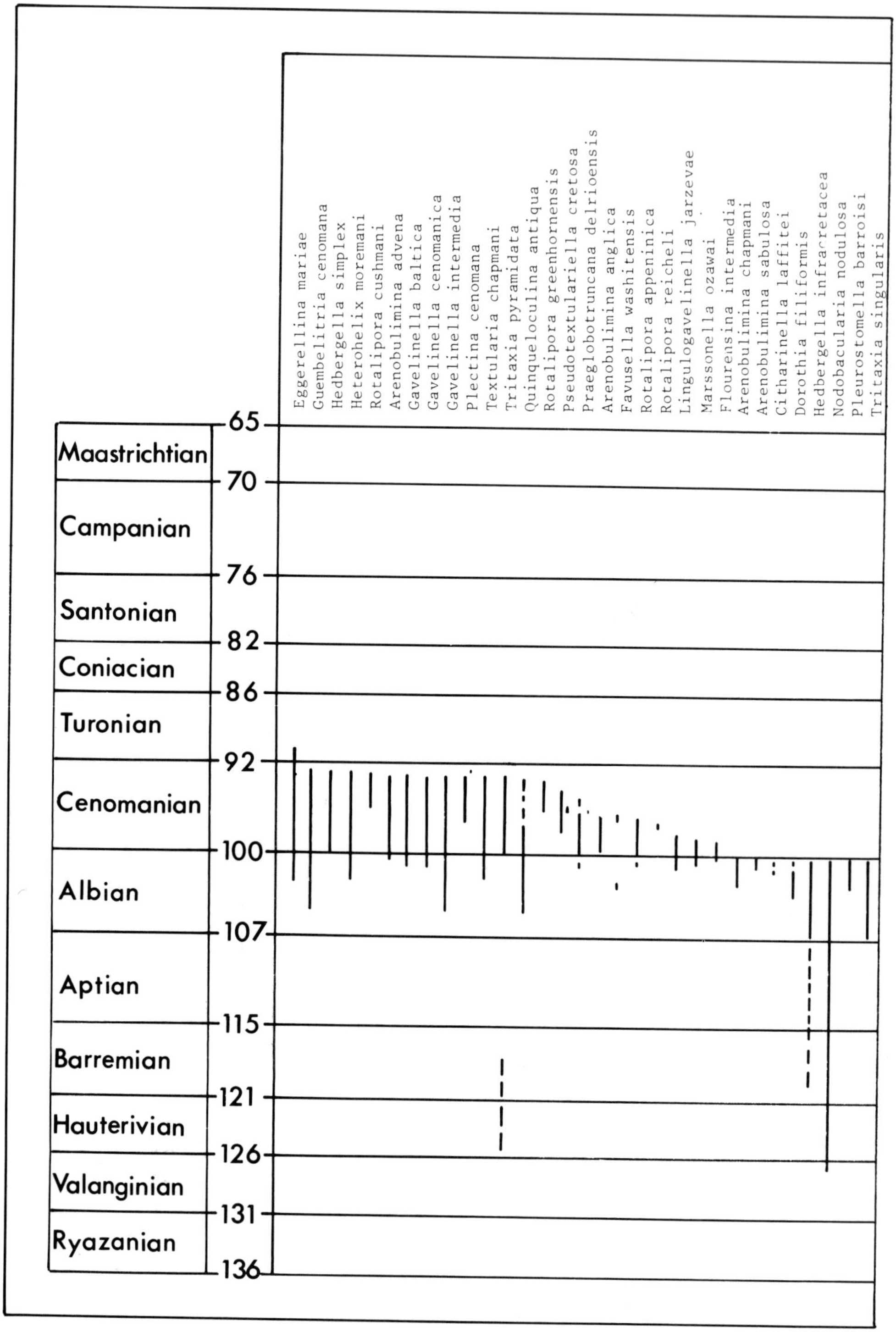

Fig. 7.16 – Summary of the distribution of Cretaceous Foraminifera.

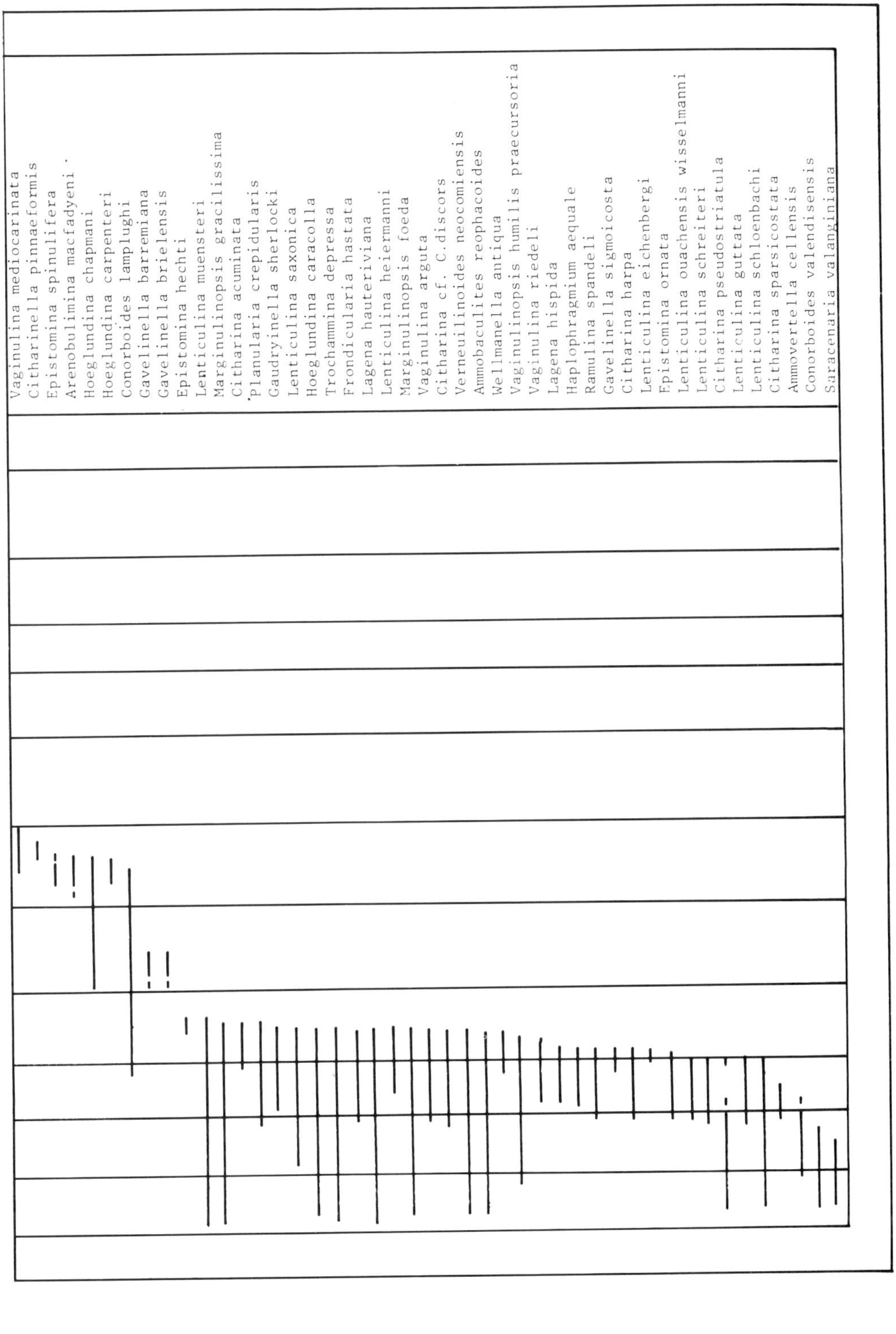

Vaginulina mediocarinata
Citharinella pinnaeformis
Epistomina spinulifera
Arenobulimina macfadyeni
Hoeglundina chapmani
Hoeglundina carpenteri
Conorboides lamplughi
Gavelinella barremiana
Gavelinella brielensis
Epistomina hechti
Lenticulina muensteri
Marginulinopsis gracilissima
Citharina acuminata
Planularia crepidularis
Gaudryinella sherlocki
Lenticulina saxonica
Hoeglundina caracolla
Trochammina depressa
Frondicularia hastata
Lagena hauteriviana
Lenticulina heiermanni
Marginulinopsis foeda
Vaginulina arguta
Citharina cf. C.discors
Verneuilinoides neocomiensis
Ammobaculites reophacoides
Wellmanella antiqua
Vaginulinopsis humilis praecursoria
Vaginulina riedeli
Lagena hispida
Haplophragmium aequale
Ramulina spandeli
Gavelinella sigmoicosta
Citharina harpa
Lenticulina eichenbergi
Epistomina ornata
Lenticulina ouachensis wisselmanni
Lenticulina schreiteri
Citharina pseudostriatula
Lenticulina guttata
Lenticulina schloenbachi
Citharina sparsicostata
Ammovertella cellensis
Conorboides valendisensis
Saracenaria valanginiana

SUB-ORDER TEXTULARIINA

Ammobaculites reophacoides Bartenstein, 1952
Plate 7.1, Figs. 1, 2 (× 50), Speeton Clay, Speeton, Yorkshire, Barremian (LB3). Description: test free, small, initial coil followed by 2–4 uniserial chambers; uniserial chambers broader than high, elliptical in cross section, sutures radial and indistinct in coiled portion, distinct, horizontal, depressed in uniserial part; aperture terminal rounded.
Remarks: commonly crushed, thereby giving a variable outline. When crushed it is very similar to *A. subcretacea* Cushman and Alexander. Range: Ryazanian to lowermost Albian.

Ammovertella cellensis Bartenstein and Brand, 1951
Plate 7.1, Figs. 3, 4 (× 20), Speeton Clay, Speeton, Yorkshire, Valanginian (D2E). Description: test free or attached, long, tubular with constrictions and bends; test may or may not branch in an irregular manner; large amount of cement giving a smooth test; aperture at the end of tube, or branching tubes.
Remarks: this distinctive species shows great variation in external shape. Range: Middle and Upper Valanginian, rare in Hauterivian; branching forms commoner in Upper Valanginian.

Arenobulimina advena (Cushman)
Plate 7.1, Fig. 5 (× 60), Lower Chalk, Eastbourne, Sussex, Cenomanian. = *Hagenowell advena* Cushman, 1936. Description: test free, trochospiral, with last three chambers occupying over half of test; chambers slightly inflated, sutures distinct, depressed; interior of test divided by complex partitions; aperture an interiomarginal loop, set in a hollow in face of last chamber.
Remarks: the complexity of the internal partitions varies throughout the Cenomanian, being relatively simple in the Lower Cenomanian and becoming progressively more complex in the Upper Cenomanian. Range: Very uppermost Albian to Cenomanian.

Arenobulimina anglica Cushman, 1936
Plate 7.1, Fig. 6 (× 70), Lower Chalk, Eastbourne, Sussex, Middle Cenomanian. Description: test free, trochospiral; last whorl occupies over half of test; chambers rounded, slightly inflated, sutures distinct, depressed; last chamber rounded, almost terminal; agglutinated material fine grained, giving a saccharoidal appearance to the test; aperture an interiomarginal loop, set in hollow in last chamber.
Remarks: the sugary appearance is characteristic. In the upper Middle Cenomanian forms with slightly crenulated chamber margins have been recorded. Range: Lower to Middle Cenomanian.

Arenobulimina chapmani Cushman, 1936
Plate 7.1, Fig. 7 (× 70), Bed X111, Gault Clay, Folkestone, Kent, Upper Albian. Description: test free, trochospiral, proximal end pointed, widening rapidly distally, rounded in cross section; adult test consists of five whorls, four chambers in each except the last which has five; sutures depressed; aperture a loop shaped in a trough like depression, running from top of test to edge of last chamber.
Remarks: characterised by its rapidly tapering test and large last whorl. It is easily distinguished from its predecessor, *A. mafadyeni* Cushman, being larger, more inflated, and more coarsely agglutinated. Range: Upper Albian.

Arenobulimina macfadyeni Cushman, 1936
Plate 7.1, Figs. 8, 9 (× 45), Bed IV, Gault Clay, Folkestone, Kent, Lower Albian. Description: test free, small, trochospiral, proximal end sharply pointed, gradually widening distally; longitudinal outline triangular, rounded in cross section; test surface finely agglutinated; adult test comprises 4–5 whorls with three chambers in each; sutures depressed; aperture a loop-shaped opening, surrounded by a raised lip.
Remarks: the smallest and most finely agglutinated of all the Albian arenobuliminids. Range: Lower and Middle Albian, with occasional, atypical and/or reworked specimens in the lowermost Upper Albian (Carter and Hart, 1977; Price, 1977a, b).

Arenobulimina sabulosa (Chapman)
Plate 7.1, Fig. 10 (× 55), Bed XIII, Gault Clay, Folkestone, Kent, Upper Albian. = *Bulimina preslii* Reuss var. *sabulosa* Chapman, 1892. Description: test free, stout, trochospiral, subtriangulate in longitudinal outline, subquadrate in cross section; adult test comprises 4–5 whorls, quadriserial, the last chamber covering the complete upper surface of the test; sutures depressed, but often obscured by the very coarse agglutinated material; aperture loop-shaped.
Remarks: the sub-quadrate cross section and coarsely agglutinated test are distinctive. Range: Uppermost Albian.

Dorothia filiformis (Berthelin)
Plate 7.1, Figs. 11, 12 (× 115), Bed XI, Gault Clay, Folkestone, Kent, Upper Albian. = *Gaudryina filiformis* Berthelin, 1880. Description: test free, narrow and elongate, proximal portion trochospiral with four chambers per whorl; distal portion biserial, thin and elongate, with almost sub-parallel sides; aperture an interiomarginal slit developed on the last chamber, rarely visible.
Remarks: easily recognised by its small size and almost parallel sides. Range: Lower to Upper Albian.

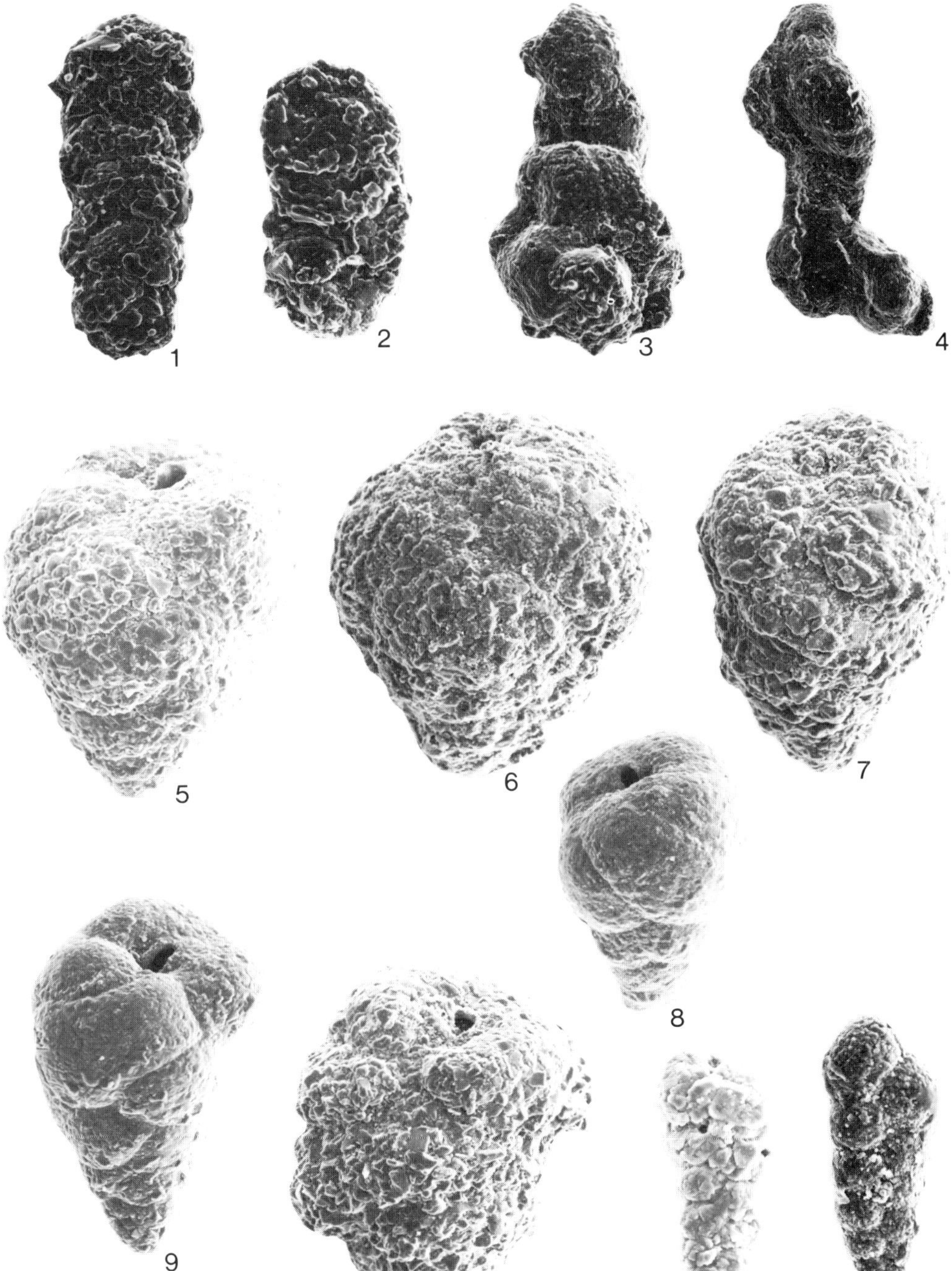

Plate 7.1

Eggerellina mariae Ten Dam, 1950
Plate 7.2, Figs. 1, 2 (× 70), Cambridge Greensand, Barrington, Cambridgeshire, Lower Cenomanian. Description: test free, varying from short pyramidal to long and narrow, triserial, conical to ovoid in longitudinal outline; chambers very often very inflated and embracing; wall very finely agglutinated; aperture narrow, hook-shaped, interiomarginal, and extending about half way up chamber face.
Remarks: very variable in outline within the Cenomanian and Turonian, as noted by Carter and Hart (1977). Range: Upper Albian to Lower Turonian.

Flourensina intermedia Ten Dam, 1950
Plate 7.2, Fig. 3 (× 60), Cambridge Greensand, Barrington, Cambridgeshire, Lower Cenomanian. Description: test free, trochospiral, coarsely agglutinated; chamber margins irregular, with depressed sutures; distinctly triserial, giving markedly triangular cross-section; aperture loop-shaped on face of final chamber.
Remarks: can be confused with the quadriserial *Arenobulimina sabulosa* (Chapman), although there should be no problem if viewed from the proximal end. Range: very uppermost Albian to Lower Cenomanian.

Gaudryinella sherlocki Bettenstaedt, 1952
Plate 7.2, Figs. 4, 5 (× 65), Speeton Clay, Speeton, Yorkshire, Barremian (LB4B). Description: test free, elongate, subcylindrical; earlier portion of test triserial, followed by biserial that may tend to becom uniserial; chambers subglobular, giving lobate periphery; sutures distinct, depressed in biserial part of test; aperture terminal, central, flush, circular or ovate.
Remarks: there is little or no variation, any recorded normally being due to compression during preservation. Range: Barremian to Lower Albian, although it is most abundant in the Lower to mid-Barremian.

Haplophragmium aequale (Roemer)
Plate 7.2, Figs. 6, 7 (× 60), Speeton Clay, Speeton, Yorkshire, Hauterivian (C8). = *Spirolina aequalis* Roemer, 1841, Description: test free, elongate, slightly compressed, early stage close-coiled, planispiral, later portion uniserial, 3–5 chambers; chambers distinct, inflated, sutures distinct, depressed, straight; aperture rounded, terminal.
Remarks: there is considerable variation in test outline, particularly in smaller specimens in which the uncoiled portion has not developed fully. Range: Hautervarivian to lowermost Barremian.

Marssonella ozawai Cushman, 1936
Plate 7.2, Figs. 8, 9 (× 60), Cambridge Greensand, Barrington, Cambridgeshire, Lower Cenomanian. Description: test free, elongate trochospiral, conical; proximal portion triserial, later biserial; biserial portion of test with sub-parallel sides, rounded in cross-section; sutures flush to slightly depressed; aperture a small, low opening at the inner margin of the last chamber.
Remarks: Carter and Hart (1977) indicated that this species can be differentiated from the closely related *M. trochus* (d'Orbigny) by its coarser agglutination and almost sub-parallel sides. Rare transitional forms between the two species are found in the Upper Albian. Range: Uppermost Albian to Lower Cenomanian.

Plectina cenomana Carter and Hart, 1977
Plate 7.2, Fig. 10 (× 60), Lower Chalk, Eastbourne, Sussex, Middle Cenomanian. Description: test free, trochospiral; 2–3 whorls, with up to five chambers per whorl, reducing to three chambers per whorl in later growth stages; overall appearance triserial, rapidly tapering; sutures distinct, depressed; aperture rounded, occasionally oval, positioned in a slight depression in final chamber.
Remarks: the relationship between this species, *Plectina mariae* (Franke) and *Arenobulimina frankei* Cushman is rather unclear. Range: Middle to Upper Cenomanian.

Pseudotextulariella cretosa (Cushman)
Plate 7.2, Figs. 11, 12 (× 50), Lower Chalk, Eastbourne, Sussex, Middle Cenomanian. = *Textulariella cretosa* Cushman, 1932. Description: test free, very large; subconical, with an earlier triserial and later biserial growth form; chambers internally complex with horizontal and vertical marginal partitions; aperture interiomarginal.
Remarks: a very distinctive species, but its distribution is probably facies controlled. Range: in S.E. England the range is Lower to Middle Cenomanian, although rare, small, individuals have been found in the Red Chalk (Upper Albian) of Yorkshire and Humberside.

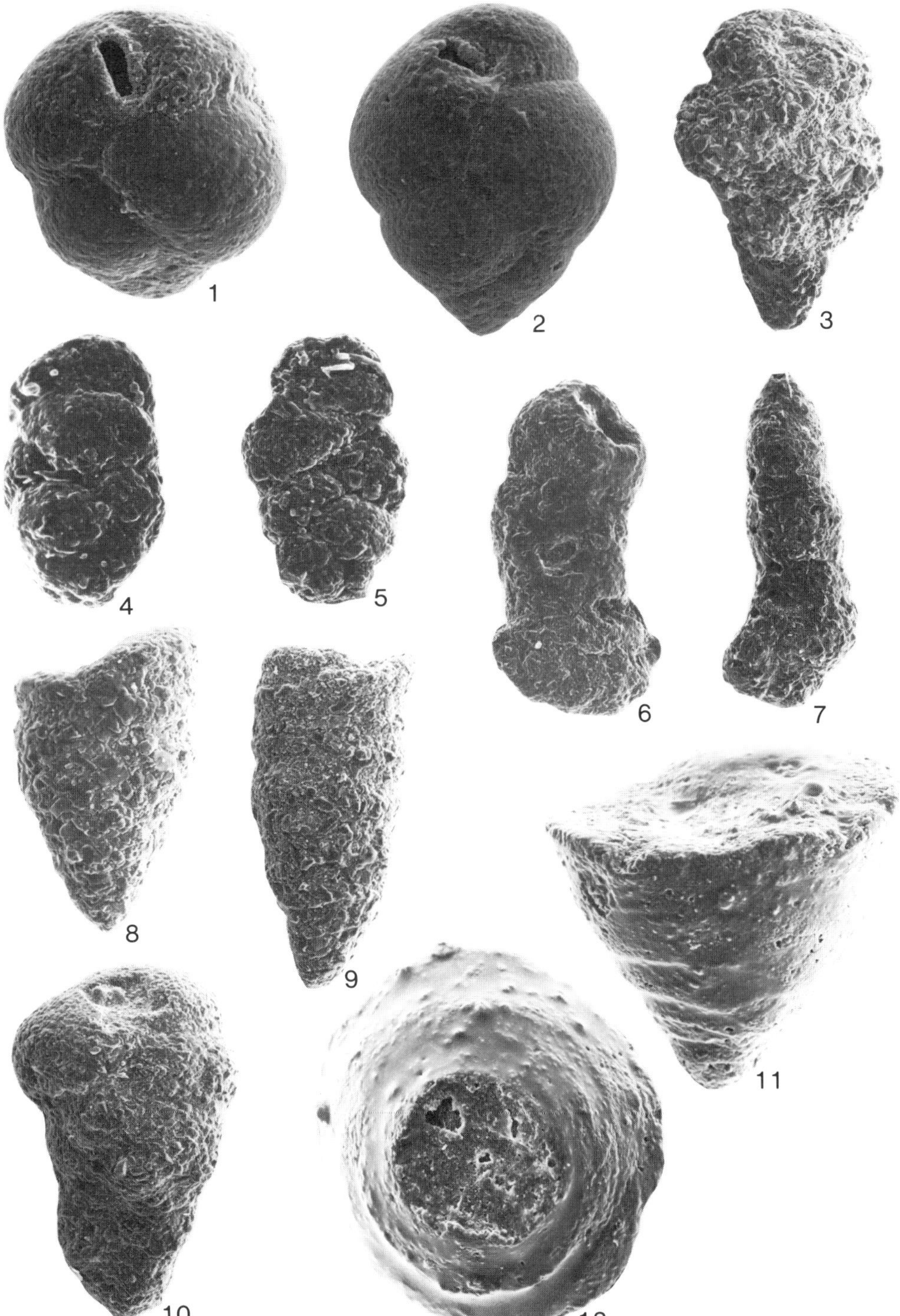

Plate 7.2

Textularia chapmani Lalicker, 1935
Plate 7.3, Fig. 1 (× 50), Lower Chalk, Eastbourne, Sussex, Lower Cenomanian. Description: test free, small, biserial, with 8–10 chambers; test widening rapidly, with broad distal portion, triangulate in longitudinal outline, sub-compressed; sutures indistinct, slightly depressed; aperture a low arch at base of last chamber.
Remarks: the rapidly widening growth form is distinctive. Range: Upper Albian to Upper Cenomanian.

Tritaxia pyramidata Reuss, 1862
Plate 7.3, Figs. 2, 3 (× 30), Lower Chalk, Folkestone, Kent, Lower Cenomanian (Fig. 2) and Upper Cenomanian (Fig. 3). Description: test free, triserial, triangular in cross-section; sides generally concave but may be straight; up to 9 triserially arranged chambers overlapping ¼–½ their widths; sutures indistinct in early stages becoming more depressed later; aperture circular, may become terminal.
Remarks: while this is a distinctive species in the upper part of its range it can be confused with *Tritaxia singularis* Magniez-Jannin in the Lower Cretaceous. The latter species differs from *T. pyramidata* mainly in having more excavated sides, and a more coarsely agglutinated test. *T. pyramidata* shows a wide range of variation in both shape and size. In the Upper Cenomanian some individuals tend to become uniserial. Range: typically Upper Albian and Cenomanian, but has been variously recorded from the Hauterivian onwards.

Tritaxia singularis Magniez-Jannin, 1975
Plate 7.3, Fig. 4 (× 50), Bed XIII, Gault Clay, Folkestone, Kent, Upper Albian. Description: test free, triserial, pyramidal, triangular in cross-section; sides characteristically excavated, giving sharp edges to the test; sutures distinct, depressed; aperture circular, normally terminal.
Remarks: can be confused with early forms of *T. pyramidata*. Range: Albian.

Trochammina depressa Lozo, 1944
Plate 7.3, Figs. 5, 6 (× 75), Speeton Clay, Speeton, Yorkshire, Hauterivian (C4K). Description: test free, trochoid, compressed, lobulate periphery; chambers distinct, 5–6 in final whorl; chambers very slightly inflated, sutures distinct, depressed; aperture a small opening at base of last chamber.
Remarks: any variation is usually caused by degree of compression and asymmetrical deformation during preservation. *T. murgeanui* Neagu from the Barremian of Rumania may be synonymous. Range: Ryazanian to Lower Barremian.

Verneuilina muensteri Reuss, 1854
Plate 7.3, Figs. 7, 8 (× 40), Upper Chalk, Redbournbury, Hertfordshire, Santonian. Description: test elongate, triangular in cross-section, increasing uniformly and rapidly in size; sides flat to slightly concave; chambers distinct; sutures flush to slightly raised, strongly curved; aperture loop-shaped opening towards the centre of the inner margin of the apertural face.
Remarks: specimens referred to his species are also close to forms described as *V. limbata* Cushman. This trivial name is already occupied (Terquem, 1883), but some authors have followed Cushman, and used *V. limbata* for this species. Range: Coniacian to Lower Maastrichtian.

Verneuilinoides neocomiensis (Mjatliuk)
Plate 7.3, Fig. 9 (× 40), Speeton Clay, Speeton, Yorkshire, Valanginian (D2E). = *Verneuilina neocomiensis* Mjatliuk, 1939. Description: test free, narrow, elongate, triangular in cross-section; chambers arranged triserially in 7–8 rows; chambers distinct, giving a lobulate outline in uncrushed specimens; sutures distinct, depressed; aperture a slit at base of last chamber.
Remarks: there is considerable variation in the degree of taper of the test, although most specimens are crushed. It is very similar to *V. subfiliformis* Bartenstein, from which it differs in being shorter and broader. Range: Ryazanian to Barremian.

SUB-ORDER MILIOLIINA

Nodobacularia nodulosa (Chapman)
Plate 7.3, Fig. 14 (× 46), Bed XIII, Gault Clay, Folkestone, Kent, Upper Albian. = *Nubecularia nodulosa* Chapman, 1891. Description: test free, porcelanous, always fragmentary, generally 1–3 chambers, joined by narrow tube; chambers vary from nodulose to pyriform; aperture simple, terminal, circular.
Remarks: Chapmans' specimens (BMNH No. P.4597) from the Albian are smaller and more delicate than the more robust forms normally encountered in the earlier Lower Cretaceous. There is a great deal of variation. Range: Ryazanian to Albian.

Quinqueloculina antiqua (Franke)
Plate 7.3, Figs. 12, 13 (× 40), Bed XIII, Gault Clay, Folkestone, Kent, Upper Albian. = *Miliolina (Quinqueloculina) antiqua* Franke, 1928. Description: test free, in typical 'quinqueloculine' coil; wall finely porcelanous; aperture on short neck, with tooth not normally visible.
Remarks: Price (1977a) has retained both *Q. antiqua* and *Q. ferussaci* (d'Orbigny) in his zonation of the Albian, although Carter and Hart (1977) used only the former name for all Albian quinqueloculines. The Polish form described by Gawor-Biedowa (1972) as *Q. kozlowski* may also be synonymous. Range: Middle Albian to mid-Cenomanian; rare atypical forms have been recorded in the Upper Cenomanian.

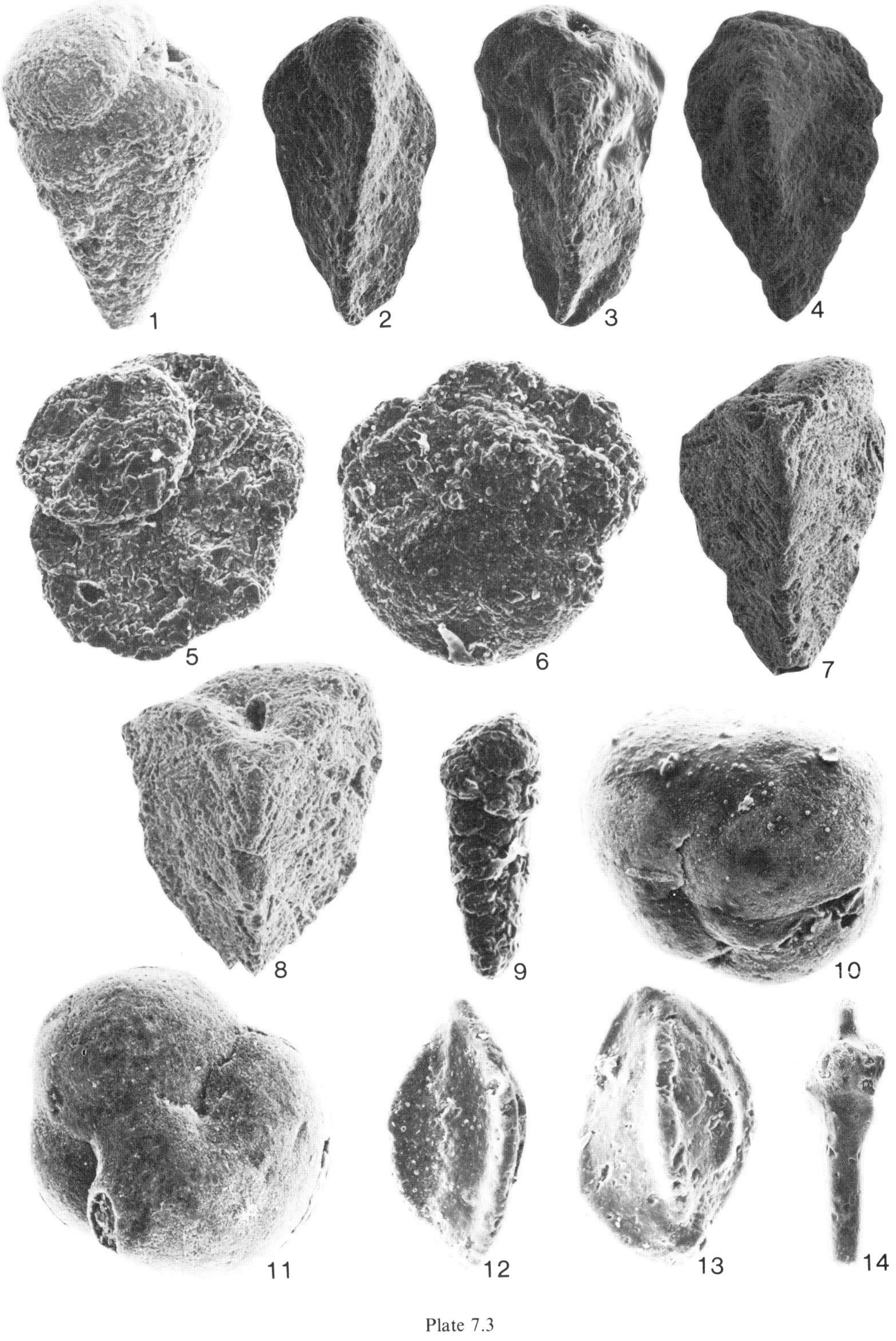

Plate 7.3

Wellmanella antiqua (Reuss)
Plate 7.3, Figs. 10, 11 (× 45), Speeton Clay, Speeton, Yorkshire, Barremian (LB5B). = *Hauerina antiqua* Reuss, 1863. Description: test free, broadly ovate, periphery rounded; chambers inflated, strongly curved, often indistinct, generally triloculine, but may have four chambers per whorl; sutures generally indistinct; aperture slit-like, at the base of the last chamber.
Remarks: there is considerable variation in the shape and arrangement of the chambers, and in the size and form of the aperture. Range: Upper Hauterivian to mid-Barremian.

SUB-ORDER ROTALIINA

Abathomphalus mayaroensis (Bolli)
Plate 7.4, Figs 1–3 (× 100), Shell/Esso 44/2–1, North Sea, Maastrichtian. = *Globotruncana mayaroensis* Bolli, 1951. Description: test free, low trochospiral coil, spiral side weakly convex, umbilical surface typically concave; outline sub-circular, lobate; periphery truncated, with widely spaced, beaded, double keel; chambers distinct, 2½–3 whorls, 5–6 chambers in final whorl; sutures straight, radial, depressed on umbilical side, but crescentric, raised, slightly beaded on spiral side; umbilicus shallow, covered by a tegilla. primary aperture interiomarginal, extraumbilical.
Remarks: distinguished by its shallow umbilicus, and apertural characteristics. Range: Uppermost Maastrichtian.

Archaeoglobigerina cretacea (d'Orbigny)
Plate 7.4, Figs. 4, 5 (× 100), Upper Chalk, Alderford, Norfolk, upper Middle Campanian. = *Globigerina cretacea* d'Orbigny, 1841. Description: test free, low trochospiral coil of 2½–3 whorls, outline subcircular, lobate; periphery weakly truncated, initially rounded; chambers distinct, subglobular, with weakly developed, faint keels bordering a raised, imperforate band; sutures radial, depressed; umbilicus broad, deep, covered by a tegilla with accessory apertures; primary aperture umbilical.
Remarks: *Globigerina cretacea* d'Orbigny is referred to the genus *Archaeoglobigerina* on the basis of its non-truncate, globular chambers, and near-radial, depressed sutures. Range: uppermost Santonian and Campanian.

Bolivina decurrens (Ehrenberg)
Plate 7.4, Figs. 6, 7 (× 100), Upper Chalk, Overstrand, Norfolk, Lower Maastrichtian. = *Grammostomum decurrens* Ehrenberg, 1854. Description: test free, elongate, slender, 2–3 times as long as wide, initially bluntly pointed, occasionally spinose; test compressed, margins sub-acute, spinose; chambers biserial throughout, initially indistinct, but becoming inflated; aperture elongate, elliptical slit, extending from basal suture of final chamber to occupy a sub-terminal position.
Remarks: easily distinguished by its spinose periphery and elongated, slender, form. Range: Maastrichtian.

Bolivina incrassata Reuss, 1851
Plate 7.4, Figs. 8, 9 (× 45), Upper Chalk, Trimingham (Fig. 8) and Overstrand (Fig. 9), Norfolk, Lower Maastrichtian. Description: test free, elongate, varying in length from a long slender form to one that is very stout and robust: test compressed, with sub-rounded margins, periphery entire, occasionally becoming slightly lobate; chambers numerous, sutures distinct, slightly depressed, steeply inclined; aperture elongate, wide, ovate, highly inclined subterminal opening.
Remarks: highly variable and many subspecies have been recognised (Koch, 1977). Range: uppermost Campanian and Maastrichtian.

Bolivinoides culverensis Barr, 1967
Plate 7.4, Figs. 10, 11 (× 100), Upper Chalk, Overstrand, Norfolk, Lower Maastrichtian. Description: test free, elongate, moderately flaring; proloculus sub-globular, followed by 7–8 pairs of biserial chambers; sutures indistinct, slightly curved; test surface possessing raised, broad, elongate lobes, usually three per chamber; initial portion of test possesses only faintly raised nodules.
Remarks: represents transitional forms between the ancestral *B. strigillatus* (Chapman) and the descendant *B. decoratus* (Jones). Range: Transitional forms between *B. strigillatus* and *B. culverensis* occur sporadically in the upper Lower Campanian, and the species occurs abundantly throughout the Middle Campanian.

Bolivinoides decoratus (Jones)
Plate 7.4, Fig. 12 (× 75), Upper Chalk, Caister St. Edmunds, Norfolk, Upper Campanian. = *Bolivina decorata* Jones, 1875. Description: test free, elongate, rhomboid outline; periphery sub-rounded; cross-section compressed, elliptical; globular proloculus, followed by 7–9 pairs of biserial chambers, sutures oblique, indistinct, obscured by strongly developed lobes; aperture narrow, loop-shaped.
Remarks: the erection of a lectotype (Barr, 1966a) has clarified the taxonomic position of this widely recorded species. Range: Upper Campanian and Lower Maastrichtian.

Bolivinoides draco (Marsson)
Plate 7.4, Fig. 13 (× 100), Shell/Esso 44/2–1, North Sea, Maastrichtian. = *Bolivina draco* Marsoon, 1878. Description: test free, rhomboidal, compressed; margins sub-acute to acute, often carinate; initial end bluntly rounded, followed by 6–7 pairs of biserially arranged chambers; test surface covered by strongly developed, longitudinally elongated lobes, four on each chamber, which coalesce to form longitudinal ribs; aperture wide, loop-shaped, bordered by thin lip, and possessing an internal tooth-plate.

Plate 7.4

Remarks: separated from *B. miliaris* Hiltermann and Koch by its more regular rhomboid test, and the fusion of the lobes into ridges. Range: uppermost Lower Maastrichtian and the Upper Maastrichtian.

Bolivinoides laevigatus Marie, 1941
Plate 7.5, Figs. 1, 2 (X 110), Upper Chalk, Eaton, Norfolk, Upper Campanian. Description: test free, elongate; cross-section compressed, periphery sub-acute; globular proloculus followed by 7–9 pairs of biserial chambers; chambers distinct, sutures slightly depressed; test surface possessing weakly developed circular to elongate nodes, generally 2–3 per chamber; aperture narrow, loop-shaped.
Remarks: a distinctive, weakly ornamented, species. Range: Upper Campanian and Lower Maastrichtian.

Bolivinoides miliaris Hiltermann and Koch, 1950
Plate 7.5, Fig. 3 (X 100), Upper Chalk, Trimingham, Norfolk, Lower Maastrichtian. Description: test free, kite-shaped, compressed, margins acute; proloculus globular, followed by 7–8 pairs of slightly inflated, biserial chambers; sutures indistinct, except at the periphery; test surface initially pustulose, later possessing elongate, narrow, lobes, generally three per chamber; aperture loop-shaped, occasionally bordered by an indistinct lip.
Remarks: *B. miliaris* is morphologically intermediate between *B. decoratus* (Jones) and *B. draco* (Marsson). Range: uppermost Upper Campanian and Lower Maastrichtian.

Bolivinoides paleocenicus (Brotzen)
Plate 7.5, Fig. 4 (X 200), Upper Chalk, Sidestrand, Norfolk, Lower Maastrichtian. = *Bolivina paleocenica* Brotzen, 1948. Description: test free, kite-shaped, compressed, periphery sub-acute; 6–7 pairs of uninflated, biserial chambers; sutures near periphery are distinct, depressed, but in centre of test are obscured by raised network of intersecting narrow ridges; aperture narrow, loop-shaped.
Remarks: a highly distinctive species, recognised by the compressed nature of the test and the surface sculpture. Range: moderately common in the upper Lower Maastrichtian, and rare in the Upper Maastrichtian.

Bolivinoides peterssoni Brotzen, 1945
Plate 7.5, Fig. 5 (X 100), Upper Chalk, Overstrand, Norfolk, Lower Maastrichtian. Description: test free, elongate, compressed, periphery slightly lobate, cross-section elliptical, periphery sub-acute; proloculus globular, followed by 7–8 pairs of slightly inflated, biserial chambers; sutures distinct, depressed; median part of test surface possessing well-developed, elongate, lobes which extend across the chamber faces perpendicular to the sutures; aperture narrow, loop-shaped, basal opening.
Remarks: this distinctive species may be distinguished from its probable ancestor (*B. laevigatus* Marie) by its broader, markedly compressed form and surface sculpture, which is even more restricted to the median region. Range: Maastrichtian.

Bolivinoides pustulatus Reiss, 1954
Plate 7.5, Fig. 6 (X 150), Upper Chalk, Alderford, Norfolk, upper Middle Campanian. Description: test free, cross-section elliptical, periphery sub-acute; globular proloculus followed by 6–8 pairs of slightly inflated, biserial chambers; sutures indistinct, depressed, obscured by numerous weakly raised, somewhat elongate, pustules that often merge with one another; aperture narrow, loop-shaped, slit.
Remarks: may be conspecific with *B. texana* Cushman. Range: Upper Middle and lower Upper Campanian.

Bolivinoides sidestrandensis Barr, 1966
Plate 7.5, Fig. 7 (X 100), Upper Chalk, Overstrand, Norfolk, Lower Maastrichtian. Description: test free, elongate, periphery rounded to sub-acute; globular proloculus followed by 7–8 pairs of inflated biserial chambers; sutures oblique, depressed, generally obscured by the well-defined, narrow, elongate, lobes that extend across the chambers; aperture narrow, loop-shaped, bordered by a faint lip.
Remarks: *B. delicatulus regularis* (sensu Koch, 1977) is a common synonym of this species. For full taxonomic discussion of this species see Swiecicki (1980, unpublished thesis). Range: Upper Campanian and Lower Maastrichtian.

Bolivinoides strigillatus (Chapman)
Plate 7.5, Fig. 8, (X 100), Upper Chalk, Scratchells Bay, Isle of Wight, Lower Campanian. = *Bolivina strigillata* Chapman, 1892. Description: test free, elongate, cross-section sub-circular, only slightly compressed; subglobular proloculus followed by 6–8 pairs of slightly inflated, biserial chambers; sutures indistinct, obscured by the 2–3 raised lobes, that extend the length of each chamber; aperture narrow, loop-shaped, with an indistinct lip.
Remarks: *B. strigillatus,* originally described from the Upper Santonian phosphatic chalk of Taplow, England, is the earliest known species of the genus. Range: uppermost Santonian and Lower Campanian.

Cibicides beaumontianus (d'Orbigny)
Plate 7.5, Figs. 9, 10 (X 85), Upper Chalk, Culver Cliff, Isle of Wight, Upper Santonian. = *Truncatulina beaumontiana* d'Orbigny, 1840. Description: test attached, plano-convex, shape variable, margins distinctly rounded; chambers distinct, sub-globular, 5–6 in final whorl; sutures distinct, depressed, straight; wall smooth, to very slightly rugose on the ventral side; aperture an equatorial slit to semi-circular hole, extending back along the dorsal side, following the spiral suture.
Remarks: the attached habit leads to considerable morphological variation, especially on the dorsal side, along which the attachment is made. Range: Santonian to Maastrichtian.

Plate 7.5

Cibicides ribbingi Brotzen, 1936

Plate 7.6, Figs. 1–3 (X 85), Upper Chalk, Scratchells Bay, Isle of Wight, Middle Campanian. Description: test attached, plano-convex, ventral side only slightly inflated; peripheral outline extremely variable, especially in later chambers, which show a tendency to become irregular; margin accutely angled; chambers distinct, sutures flush on dorsal side, slightly depressed on the ventral side, straight, radiate; aperture a slit on the inner margin of the final chamber, extending round towards the umbilicus.

Remarks: there is a possibility that this species is an ecophenotype of *C. beaumontianus* (d'Orbigny), as suggested by Barr (1966b), although this was not supported by Bailey (1978, unpublished thesis) nor Swiecicki (1980, unpublished thesis). Range: Santonian to Lower Maastrichtian.

Cibicidoides (?) voltziana (d'Orbigny)

Plate 7.6, Figs 4–6 (X 80), Upper Chalk, Alum Bay (Figs. 4, 6) and Claister St. Edmunds (Fig. 5), Upper Campanian. = *Rotalina voltziana* d'Orbigny, 1840. Description: test free, trochospiral, 3–4 whorls, plano-convex, involute, with moderate to large calcite boss; umbilical side flat, may have small, central calcite boss; periphery sub-acute; chambers indistinct, dorsally inflated in final whorl; sutures distinct, slightly depressed; aperture an interiomarginal, broad, low, arch, extending onto umbilical surface.

Remarks: a distinctive species, although its generic position is still under discussion. Range: Upper Campanian and Lower Maastrichtian.

Citharina acuminata (Reuss)

Plate 7.6, Figs. 7, 8 (X 30), Speeton Clay, Speeton, Yorkshire, Barremian (LB2 – Fig. 7 and LB4D – Fig. 8). = *Vaginulina acuminata* Reuss, 1863. Description: test free, compressed, elongate, slender, delicate, dorsal edge nearly straight; sutures strongly oblique, flush with surface, frequently obscured by ornament, which comprises 6–9 closely spaced, longitudinal, fine ribs; aperture radiate, on a sub-cylindrical neck.

Remarks: the only variation concerns the placing of the fine ribs on the surface of the test. Range: Upper Hauterivian to Upper Aptian but most abundant in the Barremian.

Citharina cf. *C. discors* (Koch)

Plate 7.6, Figs. 9, 10 (X 18), Speeton Clay, Speeton, Yorkshire, Lower Barremian (LB3) (Fig. 9) and Hauterivian (C4E) (Fig. 10). = *Vaginulina discors* Koch, 1851. Description: test free, large, compressed, test outline sub-triangular; chambers low, broad, gently curving, extending to base proximally; 12–15 chambers following an elliptical proloculus that may have a small spine; sutures distinct oblique, often obscured by the strong, longitudinal costae; aperture radiate, terminal, on a short neck.

Remarks: differs from *C. discors* var. *gracilis* Marie in that the proloculus is not so elongated and drawn out. *C. d'orbigny* Marie differs in the greater number of costae and their strong divergence. Range: Valanginian to Lower Barremian, but most common in the Upper Hauterivian.

Citharina harpa (Roemer)

Plate 7.6, Fig. 11 (X 15), Speeton Clay, Speeton, Yorkshire, Hauterivian (C4E). = *Vaginulina harpa* Roemer, 1841. Description: test free, robust, sub-triangular in outline, compressed, dorsal margin has prominent thin keel; chambers numerous, parallel sided, curved; proloculus elliptical, sometimes with a short spine; sutures limbate, oblique, flush with surface; ornament comprises 10–16 strong, thin ribs, parallel to the dorsal margin, with occasional short costae between the main ribs; aperture terminal, radiate, on a short neck.

Remarks: there is some variation in the overall shape and strength of the ornamentation. Range: Upper Valanginian to the Hauterivian/Barremian boundary.

Citharina pseudostriatula Bartenstein and Brand, 1951

Plate 7.6, Figs. 12, 13 (X 40), Speeton Clay, Speeton, Yorkshire, Hauterivian (C2A). Description: test free, small, triangular in outline, compressed; globular proloculus followed by 6–12 curving chambers, sutures inclined, limbate, flush; ornament comprises longitudinal costae, which bifurcate, as many as 18 being seen on the final chamber; aperture radiate, marginal, on a small tubular extension.

Remarks: the overall outline of the test is very variable. This species is very similar to *C. rusocostata* Bartenstein and Brand, but differs from it in having bifurcating ribs. Range: Upper Ryazanian and Lower Valanginian but with some rare occurrences in the Hauterivian.

Citharina sparsicostata (Reuss)

Plate 7.6, Fig. 14 (X 16), Speeton Clay, Speeton, Yorkshire, Hauterivian (D2C). = *Vaginulina sparsicostata* Reuss, 1863. Description: test free, large, compressed, triangular in outline, dorsal margin typically with strong keel extending from proloculus to aperture; oval proloculus followed by up to 16 chambers; sutures limbate, flush, sometimes obscured by ornament of short costae; aperture radiate, protruding at dorsal angle.

Remarks: there is a great variation in size, shape and ornamentation. In many specimens there may be a trend to longer, more continuous, costae. *C. cristellaroides* (Reuss) differs in having slightly inflated chambers, and much weaker, more regularly spaced ornament which tends to be confined to the chamer. Range: Lower Hauterivian.

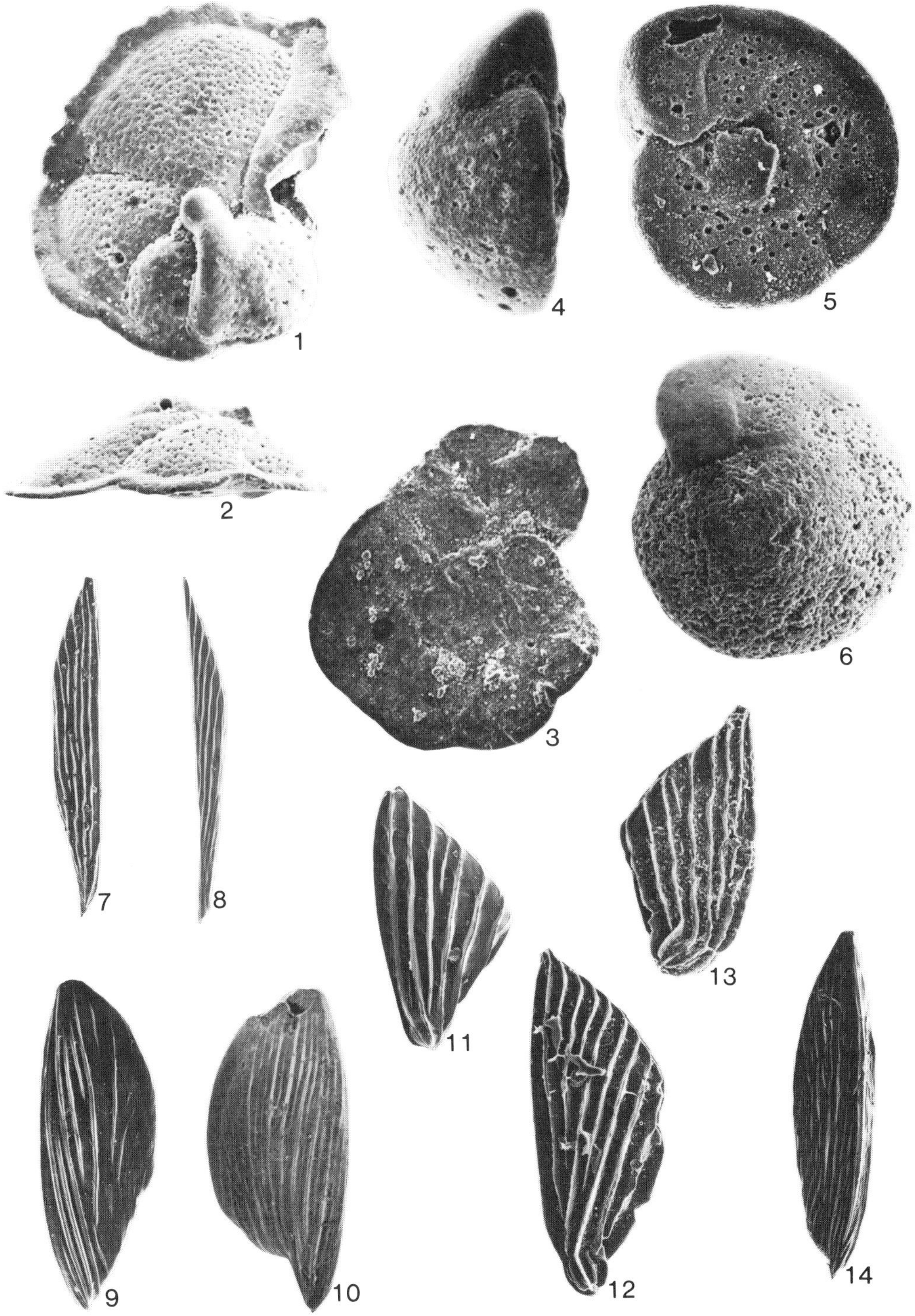

Plate 7.6

Citharinella laffittei Marie, 1938
Plate 7.7, Fig. 1 (X 25), uppermost Gault Clay, Arlesey, Bedfordshire, Upper Albian. Description: test free, lanceolate; early chambers indistinct, rapidly becoming chevron-shaped, symmetrical; 10–12 chambers, with the last one extending from a third to one half the length of the test; chambers ornamented by regular numerous, discontinuous, short, narrow striations; aperture terminal, radiate.
Remarks: very similar to both *C. karreri* (Berthelin) and *C. lemoinei* Marie. However it is much narrower than *C. karreri* and lacks the more continuous ribs at the proximal end. *C. lemoinei* is a much smoother species, lacking the prominent ribbing across the sutures. Range: Upper Albian, but rare except in the very uppermost levels.

Citharinella pinnaeformis (Chapman)
Plate 7.7, Fig. 2 (X 25), Bed XI, Gault Clay, Folkestone, Kent, Upper Albian. = *Frondicularia pinnaeformis* Chapman, 1893. Description: test free, lanceolate; early chambers indistinct, becoming chevron-shaped, symmetrical; 5–16 chambers with the last extending over half the length of the test; test surface smooth except for two prominent, sharply-edged, sub-parallel, longitudinal ridges which extend the entire length of the test; aperture terminal, radiate.
Remarks: the two prominent ribs are most distinctive and this species can usually be identified even from fragmentary material. Range: uppermost Middle and lower Upper Albian.

Conorboides lamplughi (Sherlock)
Plate 7.7, Figs 3–5 (X 90), Bed III, Gault Clay, Folkestone, Kent, Middle Albian. = *Pulvinuline lamplughi* Sherlock, 1914. Description: test free, biconvex to plano-convex, trochoid, 2–3 whorls, with 5–6 chambers in the final whorl; periphery acute, slightly lobulate; sutures on dorsal side distinct, slightly raised, curved, meeting the periphery tangentially; thickened sutures of early chambers fuse to form an umbonal boss; ventral sutures depressed, radial; aperture interiomarginal.

Remarks: *Lamarckina hemiglobosa* Ten Dam differs in having a low, very broadly rounded, dorsal surface. In *C. lamplughi* there is some variation in the height of the spire, and some forms have a more acute apical angle. Range: Upper Hauterivian to Middle Albian.

Conorboides valendisensis Bartenstein and Brand, 1951.
Plate 7.7, Figs. 6–8 (X 85), Speeton Clay, Speeton, Yorkshire, Valanginian (D4). Description: test free, small, trochoid, 2–3 whorls plano-convex or concavo-convex, periphery acute: initial chambers indistinct on dorsal side, sutures on dorsal side narrow, gently curved, flush or slightly raised; aperture a low, interiomarginal, umbilical arch with short flap.
Remarks: there is little variation and no difficulty in identification. Most specimens are poorly preserved, the majority being infilled with pyrite. Range: Ryazanian and Valanginian.

Dicarinella concavata (Brotzen)
Plate 7.7, Figs. 9, 10 (X 60), Upper Chalk, Kelveden, Essex, Santonian. = *Rotalia concavata* Brotzen, 1934. Description: test free, low trochospire, dorsal side flat to slightly concave, ventral side strongly convex; angular periphery with two distinct, closely set keels; chambers distinct, rapidly increasing in size in final whorl, of 5–6 chambers; sutures distinct, slightly elevated and curved on the spiral side, radial and deeply constricted on the umbilical side; aperture an interiomarginal low arch, umbilical to slightly extra-umbilical.
Remarks: this distinctive species occurs only sporadically in the U.K., but nevertheless is of great stratigraphic value. Range: uppermost Coniacian and Lower Santonian.

Dicarinella hagni (Scheibnerova)
Plate 7.7, Figs. 11, 12 (X 80), Middle Chalk, Shillingstone, Dorset, Lower Turonian. = *Praeglobotruncana hagni* Scheibnerova, 1962. Description: test free, trochospiral, symmetrical biconvex to plano-convex; 2–2½ whorls, 5–8 chambers in the final whorl; chambers on umbilical side triangular, inflated to globular, sutures, radial depressed; chambers on spiral side petaloid, sutures raised, curved forwards; peripheral margin with two closely-spaced, parallel, keels; aperture extraumbilical-umbilical.
Remarks: this species has had a confused taxonomic history, fully discussed in Robaszynski and Caron (1979). It appears to be restricted to more northerly latitudes. Range: uppermost Cenomanian and Lower Turonian.

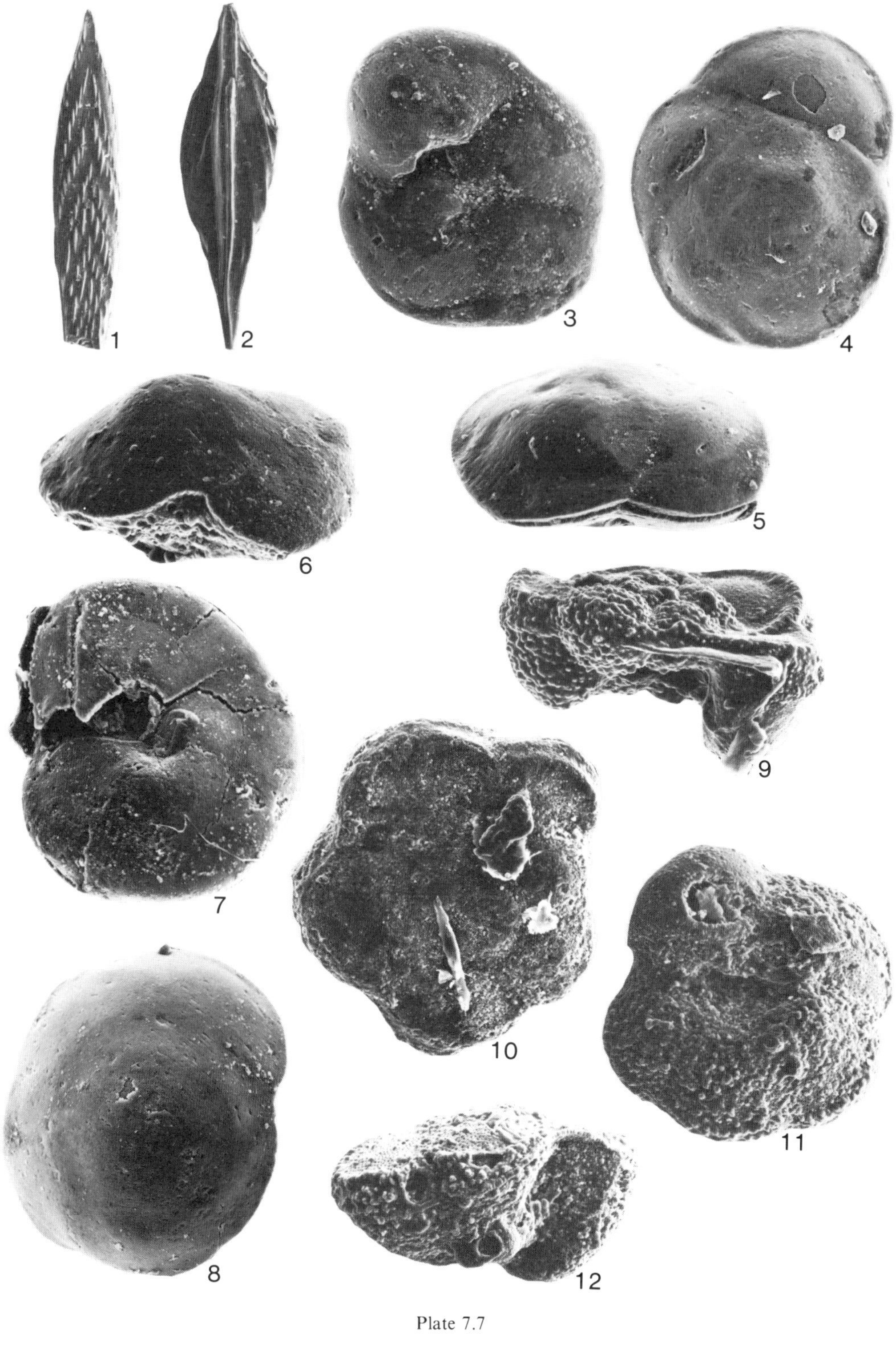

Plate 7.7

Dicarinella primitiva (Dalbiez)
Plate 7.8, Figs. 1, 2 (× 100), Upper Chalk, Borehole 25, Thames Barrier Site, Santonian. = *Globotruncana ventricosa primitiva* Dalbiez, 1955. Description: test free, trochospiral, biconvex to slightly umbilico-convex; 2–2¼ whorls, 5–6 chambers in final whorl; chambers on umbilical side inflated, sutures radial, depressed; chambers on spiral side petaloid, sutures oblique, slightly raised; equatorial periphery lobate, margin marked by two very close keels; primary aperture extraumbilical-umbilical.
Remarks: very close to *D. concavata* (Brotzen), which is more plano-convex, and has chambers that are more inflated ventrally. Range: Coniacian to early Santonian.

Eouvigerina aculeata (Ehrenberg)
Plate 7.8, Figs. 3, 4 (× 220), Upper Chalk, Redbournbury, Hertfordshire, Santonian. = *Loxostomum aculeata* Ehrenberg, 1854. Description: test free, small, biserial, margins rounded; chambers distinct, elongated, slightly pyriform, final chamber centrally positioned; sutures slightly depressed, oblique; terminal aperture bordered by a lip, usually at the end of a short neck.
Remarks: the stratigraphic value of this species has been largely devalued by its unclear taxonomic status. Specimens have been variously referred to *E. americana* Cushman, *E. cretacea* (Heron-Allen and Earland), and *E. serrata* (Chapman). Range: Coniacian to Lower Maastrichtian.

Epistomina hechti Bartenstein, Bettenstaedt and Bolli, 1957
Plate 7.8, Figs. 5–7 (× 70), Speeton Clay, Speeton, Yorkshire, Barremian (LB2). Description: test free, trochoid, biconvex, composed of 2½–3 whorls, the last having 9–10 chambers; periphery acute; sutures straight, radiating from central boss; dorsal structures curved; apertures of two types, a small oval hole on the terminal face, and a long, slit-like, latero-marginal depression parallel to the periphery on the ventral surface.
Remarks: very close to *Hoeglundina chapmani* Ten Dam, but differs from it in not having raised sutures and depressed chambers in the early whorls. Range: Lower and Middle Barremian.

Epistomina ornata (Roemer)
Plate 7.8, Figs. 8–10 (× 40), Speeton Clay, Speeton, Yorkshire, Hauterivian (C4). = *Planulina ornata* Roemer, 1841. Description: test free, lenticular, biconvex, trochoid; 1½–2 whorls, 10 chambers in final whorl; chambers on dorsal side curved, depressed, chambers on ventral side triangular in shape, depressed; dorsal sutures tuberculate, raised, ventral sutures gently curved, radiating from central umbilical pit; areal aperture rarely seen, latero-marginal apertures seen on ventral surface, near periphery.
Remarks: similar in general appearance to *E. spinulifera* (Reuss) but lacks the lobulate outline and the peripheral spines. The surface ornament is highly variable as seen in Pl. 7.8, Figs 8–10; when well-developed it can completely envelope the specimen. Range: occurs throughout the Hauterivian but is most abundant in the Upper Hauterivian.

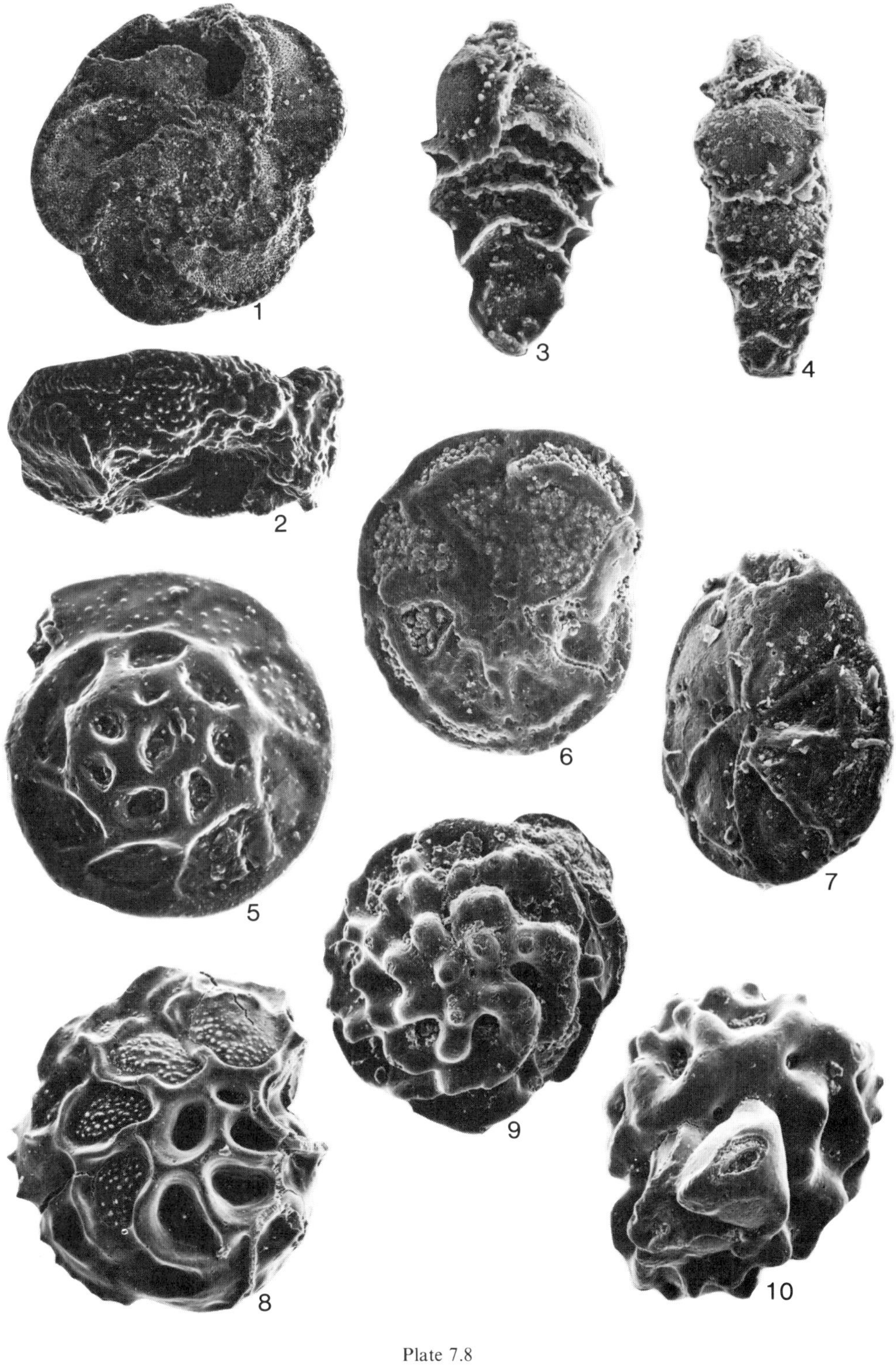

Plate 7.8

Epistomina spinulifera (Reuss, 1862)
Plate 7.9, Figs. 1-3 (X 45), Bed IV, Gault Clay, Folkestone, Kent, Middle Albian. = *Rotalia spinulifera* Reuss, 1862. Description: test free, unequally biconvex, lenticular, trochospiral; 2–2½ whorls, 7–10 chambers in the final whorl; periphery with sharp, wide, crenulate, spinose keel; all sutural ribs high and sharp, strongly acruate; chamber surfaces depressed; aperture often damaged but thin, fairly wide, tooth-plate seen running nearly perpendicular to vertical axis of test.
Remarks: the ornamentation is quite variable. In the Middle Albian the overall size and degree of ornamentation increases gradually up-section (Carter and Hart, 1977), reaching an acme just below the Middle–Upper Albian boundary. Range: Middle Albian.

Eponides beisseli Schijfsma, 1946
Plate 7.9, Figs. 4, 5 (X 65), Upper Chalk, Sidestrand, Norfolk, Lower Maastrichtian. Description: test free, asymmetrically biconvex, trochospiral; outline circular, occasionally lobate; chambers on dorsal side indistinct, only 6–7 chambers of final whorl visible; chambers triangular on ventral side, uninflated; sutures slightly depressed; aperture a distinct, narrow, slit along inner margin of final chamber, bordered by a thick, distinct, lip.
Range: Upper Campanian and Lower Maastrichtian.

Eponides concinna Brotzen, 1936
Plate 7.9, Figs. 6, 7 (X 100), Upper Chalk, Kelveden, Essex, Santonian. Description: test free, plano-convex to biconvex; periphery circular to slightly lobate, sub-acute; on dorsal side there are 3 whorls of 7–10 uninflated, crescentric, curved, chambers; sutures indistinct on dorsal side but slightly depressed on ventral side; umbilicus variable, filled with a boss of clear calcite; aperture a narrow slit along inner margin of final chamber, bordered ventrally by a lip.
Remarks: a very variable species both in the number of chambers in the final whorl and in the presence and size of the umbilical boss (see varieties erected by Vasilenko (1961)). Range: Santonian and Campanian.

Favusella washitensis (Carsey)
Plate 7.9, Figs. 8–10 (X 70), Lower Chalk, Dover, Kent (Fig. 8), Gault Clay, Folkestone, Kent (Figs. 9, 10), mid-Cenomanian (Fig. 8) and uppermost Middle Albian (Figs. 9, 10). = *Globigerina washitensis* Carsey, 1926. Description: test free, low to high trochospiral coil, periphery lobate; chambers subspherical to spherical; 2–3 whorls, 4–5 chambers in the final whorl; sutures radial to slighlty curved, depressed; chamber surfaces covered by a reticulate network of fine to coarse ridges, which produces a characteristic honeycomb pattern; aperture a low to moderate, interiomarginal umbilical–extraumbilical arch, bordered by a narrow lip.
Remarks: the surface ornamentation is quite diagnostic although there is great variation in the overall size and shape. The very strange stratigraphic distribution probably indicates a pronounced ecological control, and the validity of this species could, and has been, questioned by Masters (1977). Range: this species occurs at two levels; one near to the Middle–Upper Albian boundary, and the other in the mid-Cenomanian, immediately below the mid-Cenomanian non-sequence (Carter and Hart, 1977).

Frondicularia hastata Roemer, 1842
Plate 7.9, Fig. 11 (X 15), Speeton Clay, Speeton, Yorkshire, Lower Barremian (LB2). Description: test free, large, compressed, elongate, lanceolate; 8–16 chambers, increasing slowly in size as added; periphery smooth, rounded; sutures distinct, strongly angled at centre of the test, depressed; large proloculus, occasionally ornamented by 1 or 2 costae; aperture terminal, radiate, on a long neck.
Remarks: very similar to *F. midwayensis* Cushman, but the latter differs in having slightly raised sutures and a finely papillate surface. Range: uppermost Valanginian to mid-Barremian.

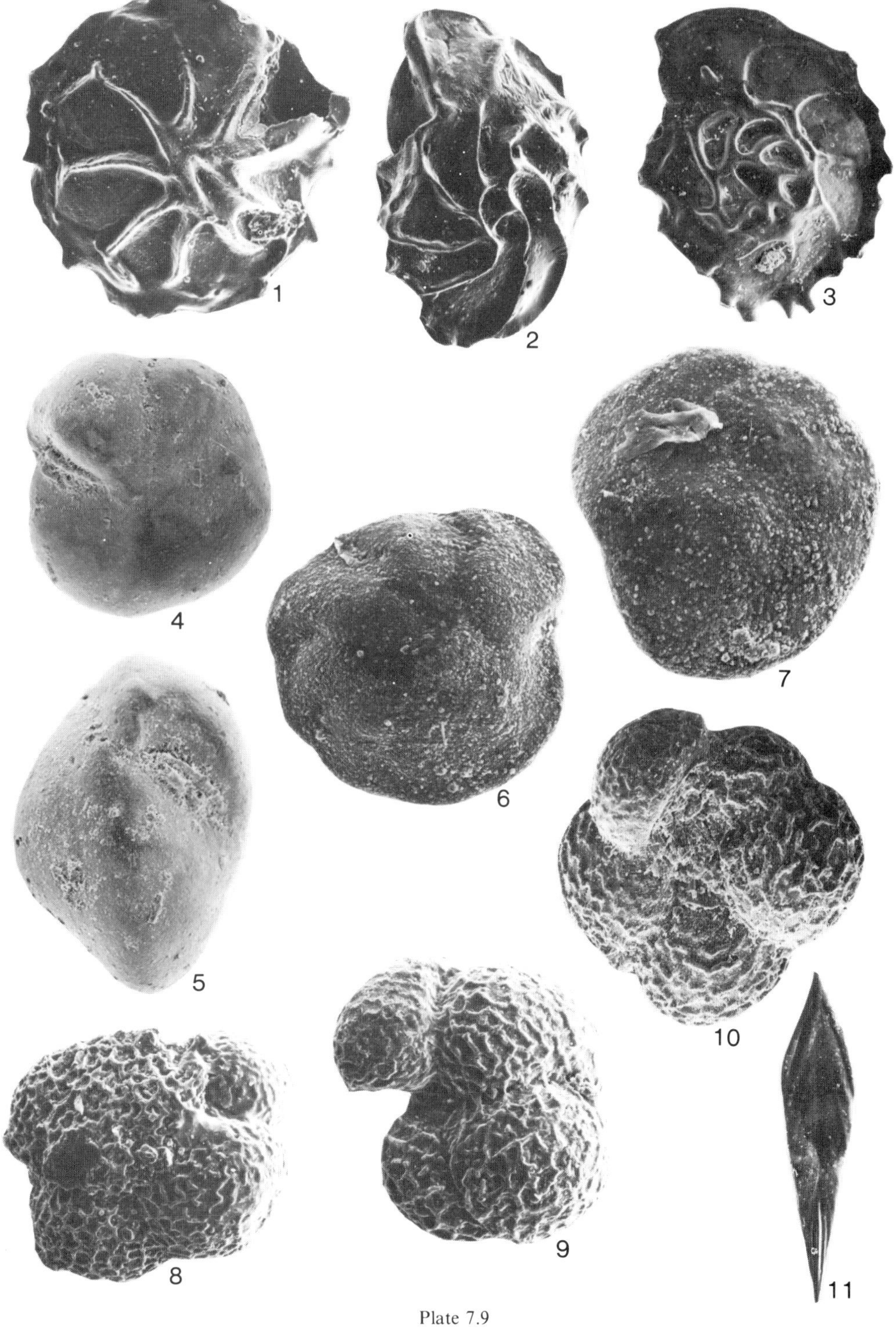

Plate 7.9

Gavelinella baltica Brotzen, 1942
Plate 7.10, Figs. 3–5 (× 50), Lower Chalk, Pitstone, Hertfordshire, Middle Cenomanian. Description: test free, trochospiral, biconvex, sides flattened, periphery rounded; 2–2½ whorls, 8–12 chambers in the final whorl; chambers all slightly inflated, increasing gradually in size as added except for the last 3–4 which expand more markedly; sutures distinct, depressed; aperture a low, interiomarginal slit, extending from near periphery to the umbilicus.
Remarks: can be confused with *G. intermedia* (Berthelin), the ancestral form, in the uppermost Albian and lowermost Cenomanian. Range: uppermost Albian to Upper Cenomanian.

Gavelinella barremiana Bettenstaedt, 1952
Plate 7.10, Figs. 1, 2 (× 60), Lower Atherfield Clay, Sevenoaks, Kent, Lower Aptian. Description: test free, trochospiral, biconvex, with sides flattened; periphery rounded; 2½–3½ whorls, 10–12 chambers in final whorl, exapnding uniformly in size; sutures distinct, flush; aperture a low interiomarginal slit extending from near the periphery to the umbilicus.
Remarks: there is very little variation in this species, which can usually be recognised by the large number of chambers in the final whorl. Range: Barremian to Aptian.

Gavelinella brielensis Malapris–Bizouard, 1974
Plate 7.10, Figs. 6–8 (× 60), Lower Atherfield Clay, Sevenoaks, Kent, Lower Aptian. Description: test free, trochospiral, flattened biconvex, periphery rounded; 2–2½ whorls, 9–10 chambers in final whorl, chambers only slightly inflated; sutures distinct, tending to become raised; aperture a low interiomargianl slit, extending from near periphery to umbilicus.
Remarks: *G. brielensis* seems to be (Malapris–Bizouard, 1974) the ancestral form of *G. intermedia* (Berthelin), which appears in the Albian. Range: Lower Aptian.

Gavelinella cenomanica (Brotzen)
Plate 7.10, Figs. 9–11 (× 50), Lower Chalk, Pitstone, Hertfordshire, Lower Cenomanian. = *Cibicidoides (Cibicides) cenomanica* Brotzen, 1945. Description: test free, trochospiral, biconvex, periphery rounded to slightly angled; 2–2½ whorls, 8–10 chambers in the fianl whorl, all expanding gradually in size; sutures distinct, slightly depressed; umbilicus is characterised by the presence of a more-or-less marked calcite rim, which forms a rim around the central depression; aperture a low, interiomarginal arch/slit, extending from near periphery to umbilicus.
Remarks: the prominence of the beaded, umbilical rim varies from very weak in the Upper Albian, where this species evolves from *G. intermedia* (Berthelin), to the highly diagnostic form, more typical of the Cenomanian. Range: uppermost Albian to Upper Cenomanian.

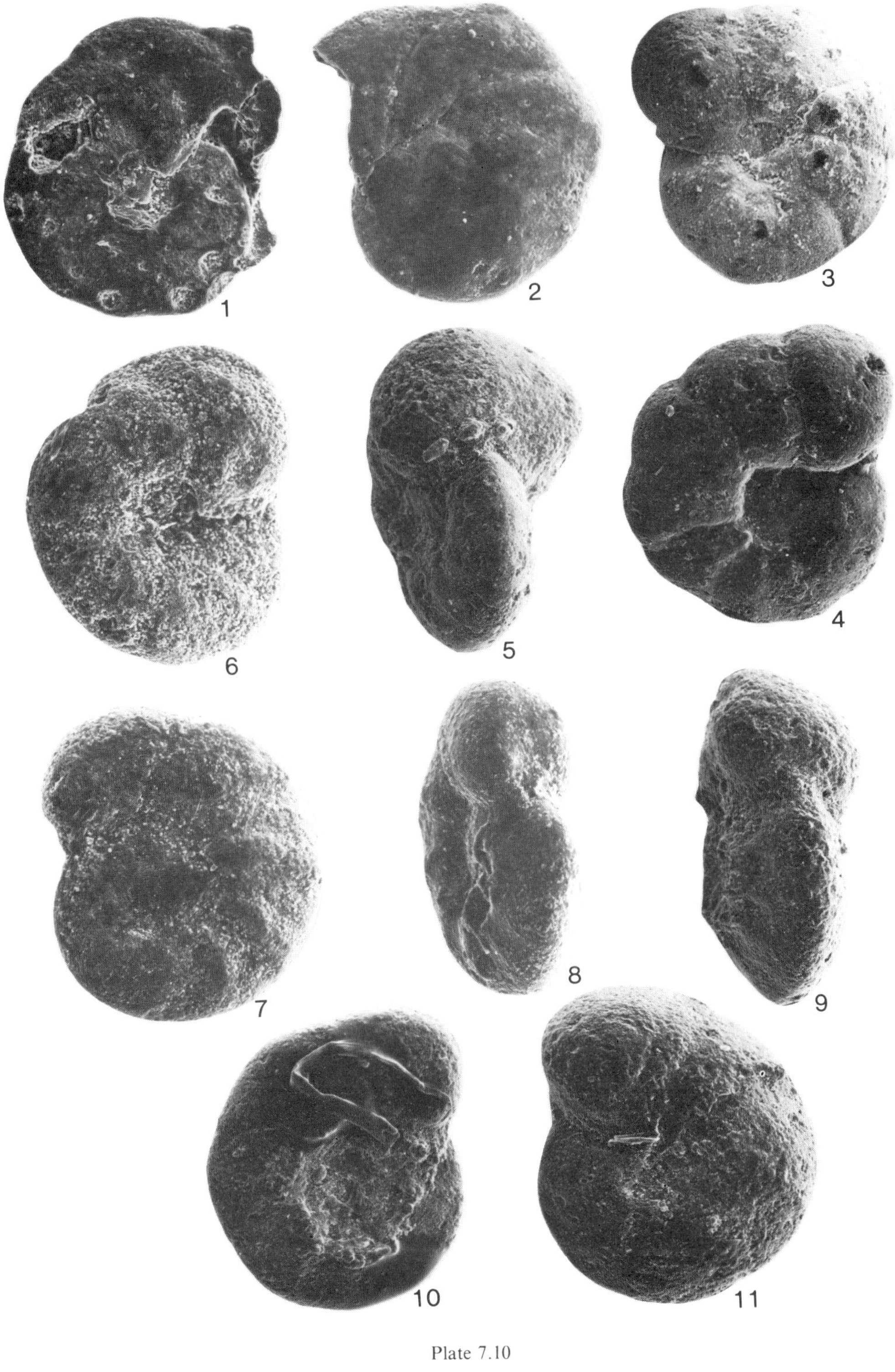

Plate 7.10

Gavelinella clementiana (d'Orbigny)
Plate 7.11, Figs. 1–3 (× 65), Upper Chalk, Alum Bay, Isle of Wight, Upper Campanian. = *Rosalina clementiana* d'Orbigny, 1840. Description: test free, low trochospiral coil, periphery rounded; spiral side flattened, evolute, early whorls obscured by ornament; umbilical side weakly convex, involute, with small calcite boss, chambers distinct, 8–10 in final whorl; sutures initially raised, tending to become depressed; aperture a low interiomarginal slit, bordered by a distinct lip, extending from periphery into umbilicus.
Remarks: a distinctive and characteristic species in the Upper Campanian. Range: Upper Campanian.

Gavelinella cristata (Geol)
Plate 7.11, Figs. 4–6 (× 100), Upper Chalk, Scratchells Bay, Isle of Wight, Lower Campanian. = *Pseudovalvulineria cristata* Goel, 1965. Description: test free, low trochospiral coil, periphery rounded; umbilicus filled with irregular calcite boss; 9–10 chambers in final whorl, distinct, weakly inflated; sutures distinct, raised, thickened, becoming flush or depressed; aperture a low interiomarginal slit, bordered by a distinct lip, extending from periphery towards umbilicus, covered by distinct, sub-triangular, chamber flaps.
Remarks: *G. cristata* may be distinguished from *G. clementiana* (d'Orbigny) by its convex spiral side and stronger ornamentation. Range: Upper Santonian and Lower Campanian.

Gavelinella intermedia (Berthelin)
Plate 7.11, Figs. 7–9 (× 100), Lower Chalk, Pitstone, Hertfordshire, Lower Cenomanian. = *Anomalina intermedia* Berthelin, 1880. Description: test free, periphery rounded, but may be slightly angled; dorsal side semi-involute with 1½–2 whorls visible, 10–12 chambers in final whorl; sutures distinct, depressed to slightly raised, slightly arcuate; aperture interiomarginal-equatorial, surrounded by a fairly wide lip.
Remarks: abundant in the Albian and Cenomanian of Britain. As used in this account, it does not include those forms with an umbilical boss (see Malapris, 1965; Michael, 1966). Range: Albian and Cenomanian.

Gavelinella lorneiana (d'Orbigny)
Plate 7.11, Figs. 10–12 (× 110), Upper Chalk, Alum Bay, Isle of Wight, Upper Campanian. = *Rosalina lorneiana* d'Orbigny, 1840. Description: test free, low trochospiral coil of 2½–3 whorls, outline circular, becoming lobate; periphery rounded to sub-acute with the occasional development of an imperforate peripheral band; chambers distinct, 8–10 in final whorl; sutures depressed; aperture a low interiomarginal slit bordered by an indistinct lip, extanding into umbilicus where it is covered by sub-triangular, imperforate, chamber flaps.
Remarks: the well-developed, imperforate chamber flaps and narrow umbilicus are highly distinctive. Range: Turonian to lower Upper Campanian.

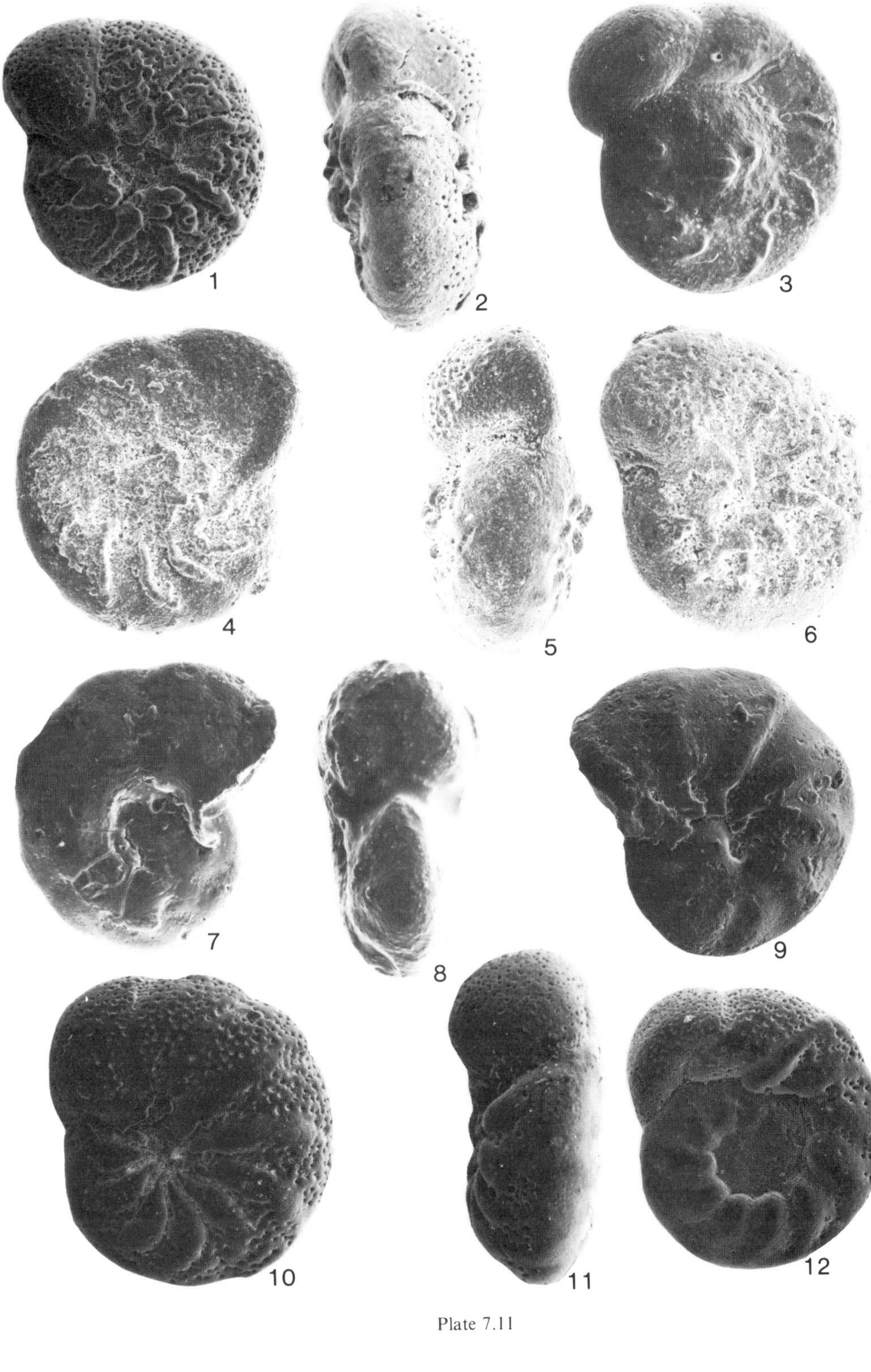

Plate 7.11

Gavelinella monterelensis (Marie)
Plate 7.12, Figs. 1–3 (× 90), Upper Chalk, Alum Bay, Isle of Wight, Upper Campanian. = *Anomalina monterelensis* Marie, 1941. Description: test free, biconvex, low trochospiral coil of 2½–3 whorls; biumbonate, umbilical boss distinct, raised; spiral side boss distinct, broad, low; chambers numerous, 12–14 in final whorl; periphery circular, becoming weakly lobate, possessing indistinct imperforate keel; sutures distinct, depressed; aperture a moderately high interio-marginal arch, bordered by a thin lip, and extending into the umbilicus along the spiral suture partly covered by backward pointing, sub-triangular, chamber flaps.
Remarks: a distinctive, stratigraphically useful species. Range: Upper Campanian.

Gavelinella pertusa (Marsson)
Plate 7.12, Figs. 4–6 (× 110), Upper Chalk, Overstrand, Norfolk, Upper Campanian. = *Discorbina pertusa* Marsson, 1878. Description: test free, low trochospiral coil, partially evolute, especially on spiral side, periphery rounded; chambers distinct, gradually increasing in size, 8–9 in final whorl; sutures flush to slightly depressed, indistinct, radial; test coarsely perforate; aperture an arcuate slit following the inner margin of the final chamber from the umbilicus round onto the periphery; umbilicus broad, deep.
Remarks: the broad, open umbilicus is characteristic of this species. Range: Coniacian to Maastrichtian.

Gavelinella sigmoicosta (Ten Dam)
Plate 7.12, Figs. 7–9 (× 55), Speeton Clay, Speeton, Yorkshire, Upper Hauterivian (C1). = *Anomalina sigmoicosta* Ten Dam, 1948. Description: test free, plano-convex, periphery rounded, to sub-acute; ventral side convex, elevated, with deep umbilicus; 7–9 narrow, sigmoidal, depressed, chambers in final whorl; dorsal sutures strongly curved, raised, thickened; ventral sutures less distinct, slightly curved; aperture a narrow slit at base of apertural face, extending from umbilicus to half way towards the periphery.
Remarks: often very poorly preserved, or even fragmentary in the Speeton Clay succession. Range: Upper Hauterivian to Lower Barremian.

Gavelinella stelligera (Marie)
Plate 7.12, Figs. 10–12 (× 100), Upper Chalk, Scratchells Bay, Isle of Wight, Middle Campanian. = *Planulina stelligera* Marie, 1941. Description: test free, low trochospiral coil of 2½–3 whorls, periphery rounded to sub-acute; umbilical side has a broad shallow depression surrounding umbilicus, which is almost completely filled with chamber flaps; chambers distinct, 12–13 in final whorl; sutures indistinct, curved, flush; on spiral side chamber margins imperforate, thickened anterior to sutures, giving the appearance of a series of oblique, curved, ribs; aperture a narrow interiomarginal slit, bordered by an indistinct lip.
Remarks: the broad, curved, imperforate ribs and compressed test are characteristic of this species. Range: atypical forms recorded from the Upper Santonian, Lower to Middle Campanaian, becoming very rare in the lower Upper Campanian.

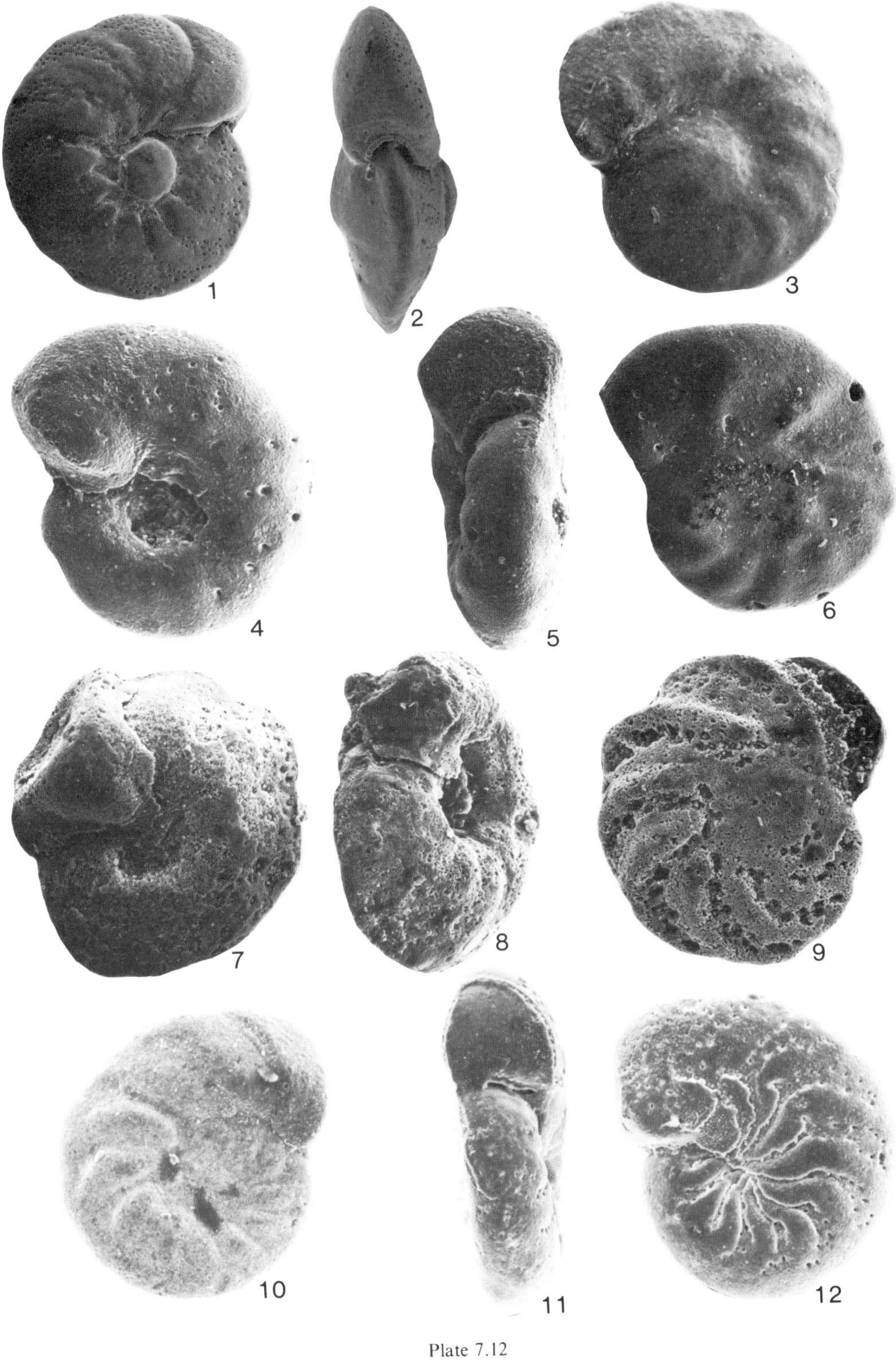

Plate 7.12

Gavelinella thalmanni (Brotzen)

Plate 7.13, Figs. 1–3 (× 140), Upper Chalk, Euston, Suffolk (Figs. 1, 2) and Scratchells Bay, Isle of Wight (Fig. 3), Upper Santonian (Figs. 1, 2) and Middle Campanian (Fig. 3). = *Cibicides thalmanni* Brotzen, 1936. Description: test free, trochospiral, but appearing almost planispiral; chambers indistinct, not inflated, increasing gradually in size as added, 11–12 in final whorl; sutures flush, indistinct, straight, radial; ventral surface smooth, but dorsal surface with raised, rugose ribs, producing an irregular, nodose ornament; aperture an interiomarginal slit, extending from the periphery into the umbilical area of the ventral side.

Remarks: this stratigraphically useful species is easily distinguished from other members of the genus by the flattened nature of the ventral side, and the wide, shallow, dorsal umbilicus. Range: Coniacian to lower Upper Campanian.

Gavelinella usakensis (Vasilenko)

Plate 7.13, Figs. 4–6 (× 100), Upper Chalk, Scratchells Bay, Isle of Wight, Middle Campanian. = *Anomalina (Pseudovalvulineria) clementiana* d'Orbigny var. *usakensis* Vasilenko, 1961. Description: test free, low to moderately high trochospiral coil, periphery rounded; spiral side convex, evolute, with earlier whorls obscured by thickened, smooth, calcite layer; umbilical side weakly convex with small umbilical boss; chambers distinct, 9–10 in final whorl, weakly inflated; sutures distinct, slightly depressed; aperture a low interiomarginal slit, bordered by a distinct lip, extending from periphery to umbilicus.

Remarks: this species is distinguished by its smooth, moderately convex, spiral side. Range: Middle Campanian to lower Upper Campanian.

Globigerinelloides bentonensis (Morrow)

Plate 7.13, Figs. 7–9 (× 100), Bed XIII, Gault Clay, Folkestone, Kent, Upper Albian. = *Anomalina bentonensis* Morrow, 1934. Description: test free, planispiral, bi-umbilicate, involute; 2 whorls, 6–9 chambers in the final whorl; sutures distinct, depressed, straight to slightly curved; aperture a low interiomarginal, umbilical–equatorial arch, with a prominent lip; relict apertural lips visible around the umbilicus on both sides.

Remarks: this is a characteristic species in the uppermost Albian of the British Isles. There is however a great deal of taxonomic confusion between this species and *G. cushmani* (Tappan), *G. caseyi* (Bolli, Loeblich and Tappan), and *G. eaglefordensis* (Moreman). After examination of topotype material Carter and Hart (1977) rejected *G. caseyi* and *G. eaglefordensis*, using *G. bentonensis* for the Upper Albian and Cenomanian forms found in the British succession. Masters (1977) has erroneously suggested that *G. bentonensis* should be restricted to the Albian, even though Morrow's type material was of mid-Cenomanian age. However Masters does not extend its geographic distribution to the U.K. Both Masters and Price would refer this species to *G. cushmani*. Range: uppermost Albian and Cenomanian, although it is most abundant immediately below the Albian–Cenomanian boundary.

Globorotalites hiltermanni Kaever, 1961

Plate 7.13, Figs. 10, 11 (× 90), Upper Chalk, Caister St. Edmunds, Norfolk, Upper Campanian. Description: test free, biconvex, trochospiral, 1½–2½ whorls, periphery acute, slightly keeled; chambers indistinct on spiral side; 7–9 chambers visible on umbilical side, indistinct, becoming slightly inflated; sutures indistinct, flush to weakly depressed; aperture narrow, elongate, interiomarginal slit, with distinct murus reflectus.

Remarks: the almost equally biconvex test, and closed umbilicus characterise this species. Range: only recorded from the upper Upper Campanian.

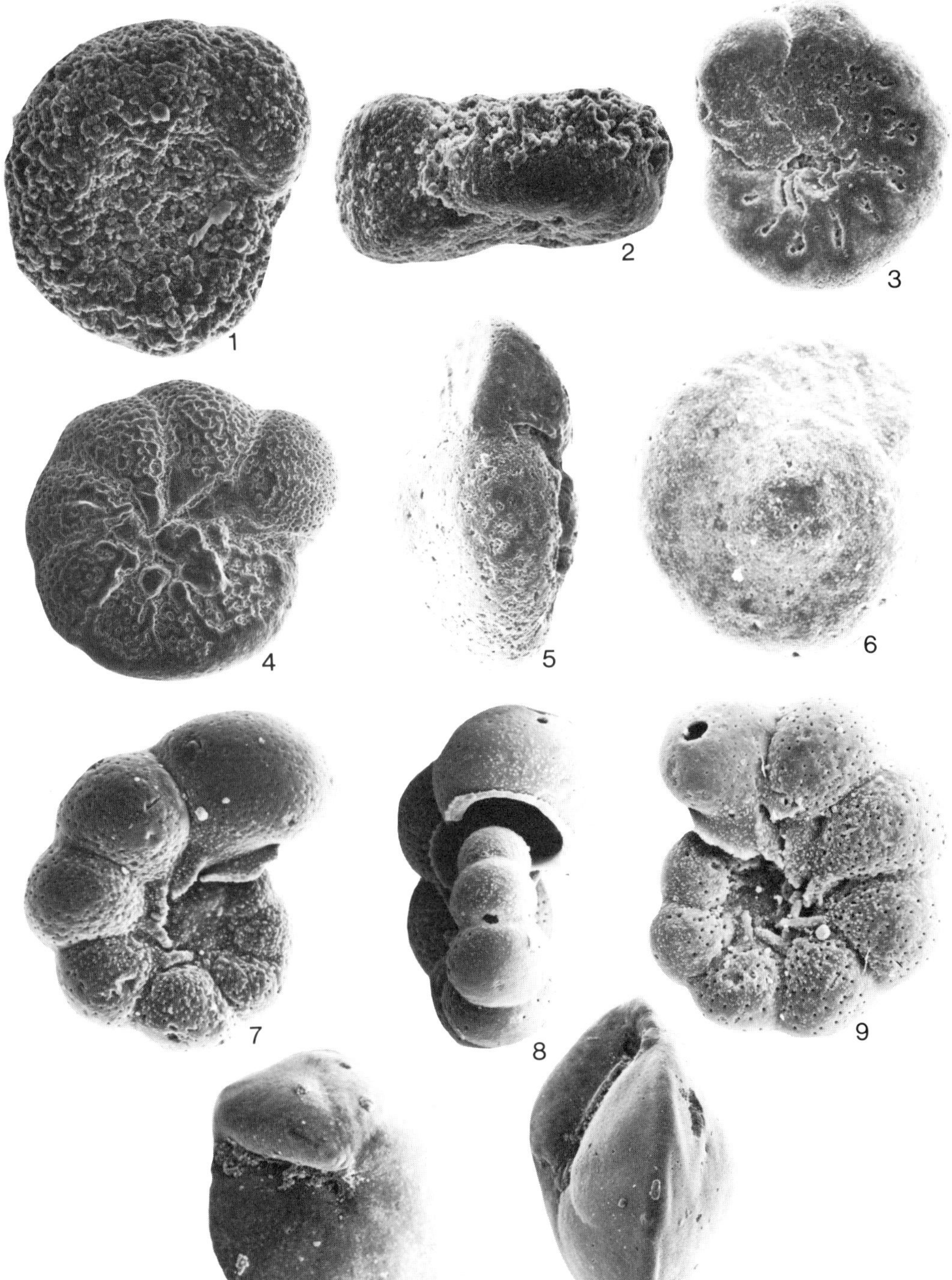

Plate 7.13

Globorotalites micheliniana (d'Orbigny, 1840)
Plate 7.14, Figs. 1, 2 (X 85 and X 70), Upper Chalk, Catton Grove, Norfolk, Upper Campanian. = *Rotalina micheliniana* d'Orbigny, 1840. Description: test free, plano-convex, trochospiral; umbilical side involute, steeply conical, normally showing no pseudo-umbilicus; outline circular, periphery acute to sub-acute, slightly keeled; 6–7 chambers in final whorl, indistinct, conical; spiral side sutures indistinct; umbilical side sutures distinct, slightly depressed; aperture a narrow, elongate,, interiomarginal slit, which may be a slight lip.
Remarks: *G. micheliniana* was the first Upper Cretaceous species of this genus to be described. Morphologically it shows a broad range of variation with respect to the height of the umbilical side and the width of the pseuo-umbilicus. Little of this variation (Goel, 1963) would appear to have any stratigraphic value. *G. cushmani* Goel – as used by Bailey (1978, unpublished thesis) – is probably an earlier member of the *G. micheliniana* plexus. Range: Turonian to Campanian.

Globotruncana bulloides Vogler, 1941
Plate 7.14, Figs. 3–5 (X 85), Upper Chalk, Sherringham (Fig. 3) and Keswick (Figs. 4, 5), Norfolk, Upper Campanian. Description: test free, low trochospire, slightly biconvex, margins truncated with two distinct, widely-spaced, beaded, keels; chambers distinct, inflated both dorsally and ventrally, 6–7 in final whorl; sutures distinct, depressed, curved, masked by a lobate keel on the umbilical surface; aperture umbilical; umbilicas small with evidence for the development of a tegilla, although this is rarely preserved except in the Campanian.
Remarks: this species is distinguished by its small umbilicus, and widely spaced keels. There is a strong possibility that *G. paraventricosa* (Hofker) should be regarded as synonymous. Range: Coniacian to Campanian.

Globotruncana contusa (Cushman)
Plate 7.14, Figs. 6, 7 (X 75), North Sea Basin, Maastrichtian. = *Pulvinulina arca contusa* Cushman, 1926. Description: test free, trochospiral, spiro-convex with a deep umbilicus, outline sub-circular, weakly lobate; periphery truncated by two closely spaced, double keels; chambers indistinct, 5–7 in the final whorl, crescentric on spiral side, elliptical on umbilical side; sutures distinct, strongly curved, weakly raised and beaded on spiral side, while depressed, radial and slightly curved on the umbilical side; primary aperture interiomarginal-umbilical.
Remarks: this distinctive species is stratigraphically important in the North Sea Basin. Range: Upper Maastrichtian.

Globotruncana fornicata Plummer, 1931
Plate 7.14, Figs. 8–10 (X 110), Upper Chalk, Alderford, Norfolk, Middle Campanian. Description: test free, trochospiral, asymmetrically biconvex; margins angular, truncated by two well-spaced keels; chambers distinct, 4–5 in final whorl; chambers elongate, crescentric on dorsal side, sub-rectangular on the umbilical side; umbilicus narrow, deep, covered by tegilla; primary aperture interiomarginal, umbilical.
Remarks: readily recognised by its moderately convex spiral side, and elongate, crescentric chambers. The initial chambers in the first whorl are also distinctly 'hedbergellid' in appearance. Range: uppermost Santonian and Campanian in the U.K., although the range in Southern Europe continues into the Maastrichtian.

Globotruncana linneiana (d'Orbigny)
Plate 7.14, Figs. 11, 12 (X 85), Upper Chalk, Scratchells Bay, Isle of Wight (Fig. 11) and Shell/Esso 44/2–1, North Sea Basin (Fig. 12), Campanian. = *Rosalina linneiana,* d'Orbigny, 1839. Description: test free, low trochospiral, both sides almost flat, margins distinctly angular and truncated, with two sutures curved, raised on spiral side, slightly lobate to radial on umbilical side; primary aperture interiomarginal but preservation too poor to record development of a tegilla.
Remarks: this species, first described from Recent beach sands on Cuba, has had a much confused taxonomic history (Swiecicki, 1980, unpublished thesis). It is never common in the British succession. Range: Santonian and Campanian.

Plate 7.14

Globotruncana plummerae Gandolfi, 1955.
Plate 7.15, Figs. 3–5 (× 110), Upper Chalk, Arminghall, Norfolk, Upper Campanian. Description: test free, low trochospiral coil, asymmetrically biconvex; periphery sub-rounded to sub-acute, truncated by prominent double keel; chambers distinct, 2–2½ whorls, 4–5 chambers in fianl whorl; chambers on dorsal side elongate, crescentric, inflated; chambers on umbilical side are elongate, sub-rectangular and weakly inflated; sutures depressed and curved; umbilicus narrow, deep; primary aperture interiomarginal, umbilical, covered by a tegilla.
Remarks: can be distinguished by its elongate, and inflated chambers, with those of the final whorl increasing rapidly in size as added. Range: Upper Campanian.

Globotruncanella havanensis (Voorwijk)
Plate 7.15, Figs. 1, 2 (× 150), Shell/Esso 44/2–1, North Sea Basin (Fig. 1) and Upper Chalk, Trimingham, Norfolk (Fig. 2), Maastrichtian. = *Globotruncana havanensis* Voorwijk, 1937. Description: test free, low trochospiral coil; periphery acute to sub-acute, may possess imperforate, peripheral band; 4–5 distinct chambers in final whorl; sutures distinct, depressed, radial, slightly curved on umbilical side and strongly curved on spiral side; umbilicus narrow, moderately deep; aperture high umbilical–extraumbilical arch, opening into an umbilicus which may be covered by a delicate tegilla.
Remarks: a very distinctive species which possibly evolved from 'whiteinellid' stock in the latest Campanian. *G. cita* Bolli is a common synonym of this species. Range: Maastrichtian.

Guembelitria cenomana (Keller)
Plate 7.15, Fig. 6 (× 225), Bed XIII, Gault Clay, Folkestone, Kent, Upper Albian. = *Guembelina cenomana* Keller, 1935. Description: test free, triserial throughout, extremely small; chambers inflated and globular; sutures distinct, depressed; aperture an interiomarginal arch at base of final chamber, variable in outline but normally low.
Remarks: Masters (1977) has concluded, after examination of the holotype of *G. harrisi* Tappan, that is a junior synonym of *G. cenomana.* Range: Middle Albian to lowermost Turonian.

Hedbergella brittonensis Loeblich and Tappan, 1961
Plate 7.15, Figs. 7–9 (× 150), Lower Chalk, Pitstone, Hertfordshire, Cenomanian. Description: test free, high asymmetrical trochospire; 2–2½ whorls, 5½–7 chambers in final whorl; chambers inflated, globular, sutures radial, depressed; primary aperture umbilical–extraumbilical, bordered by lip.
Remarks: there is a problem in separating this species from the lower-spired *H. delrioensis* (Carsey), the higher-spired *Whiteinella paradubia* (Sigal), and the ancestral *H. infracretacea* (Glaessner). Higher in the Cenomanian and Turonian the aperture tends to become more umbilical in position, thereby changing the generic position to that of a *Whiteinella.* Range: Cenomanian to Santonian.

Hedbergella delrioensis (Carsey)
Plate 7.15, Figs. 10–12 (× 200), Lower Chalk, Pitstone, Hertfordshire, Cenomanian. = *Globigerina cretacea* d'Orbigny var. *delrioensis* Carsey, 1926. Description: test free, trochospiral, biconvex, umbilicate, periphery rounded, gently lobate; chambers globular, sutures distinct, depressed; normally five chambers per whorl; aperture a simple interiomarginal, extraumbilical–umbilical arch, commonly bordered by a narrow lip.
Remarks: *H. delirioensis* is recognised by its low-medium spire, but in the Cenomanian its separation from the higher-spired *H. brittonensis* Loeblich and Tappan may sometimes be rather arbitrary. Range: Middle Albian to Upper Turonian.

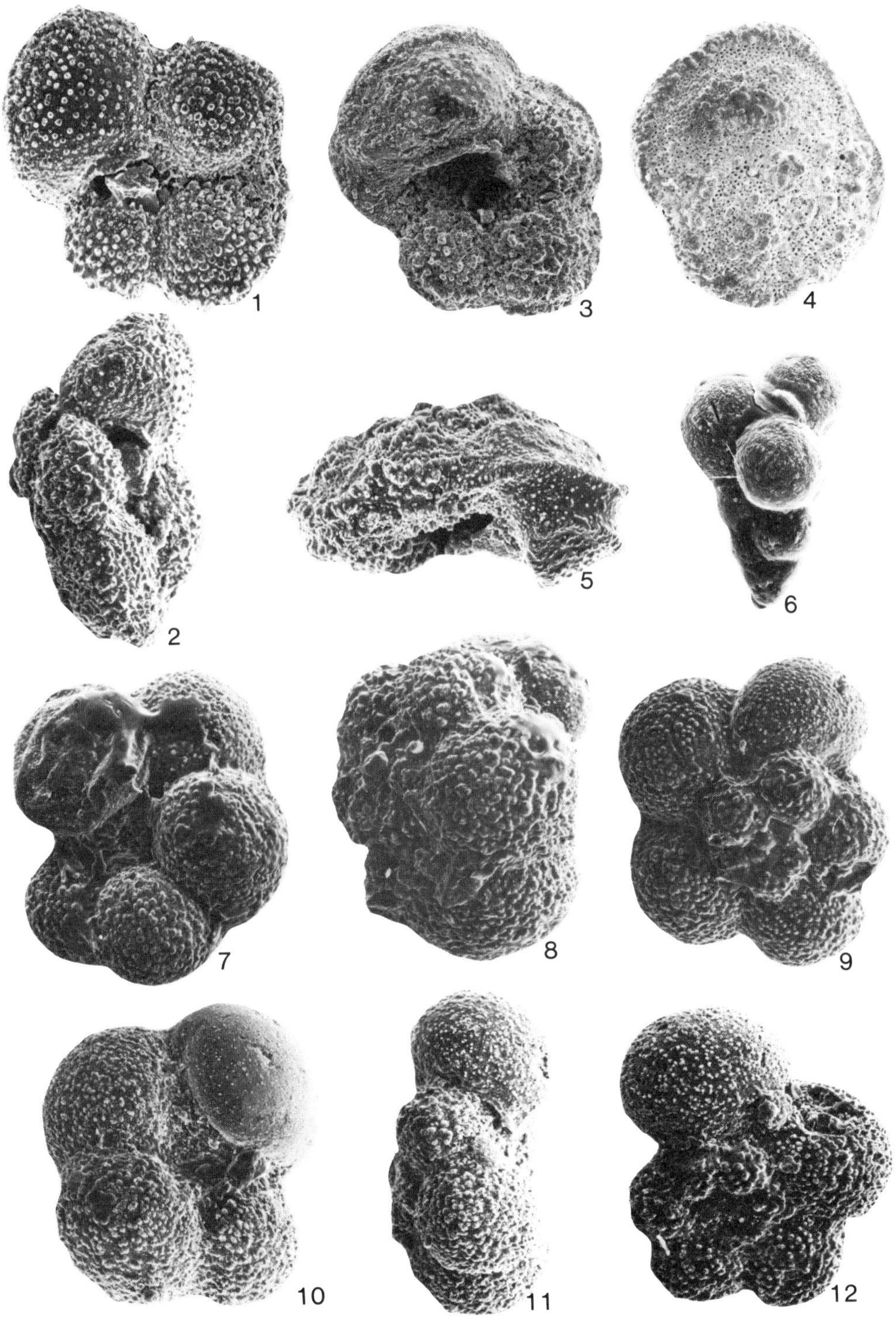

Plate 7.15

Hedbergella infracretacea (Glaessner)
Plate 7.16, Figs. 1-3 (× 200), Bed XIII, Gault Clay, Folkestone, Kent, Upper Albian. = *Globigerina infracretacea* Glaessner, 1937. Description: test free, trochospiral, with moderately high spire, periphery rounded; 5 sub-spherical chambers per whorl, sutures radial, depressed; aperture extraumbilical, with imperforate lip.
Remarks: although Masters (1977) placed this species in the synonym of *H. delrioensis* (Carsey), both Price (1977b) and Carter and Hart (1977) have upheld their separate identity, while still recognising that they are both ends of a complete range of variability. Range: Upper Albian to Barremian in the North Sea Basin, but onshore the range is Middle and Upper Albian.

Hedbergella planispira (Tappan)
Plate 7.16, Figs. 4, 5 (× 200), Bed XIII, Gault Clay, Folkestone, Kent, Upper Albian. = *Globigerina planispira* Tappan, 1940. Description: test free, small, very low trochospire, appearing almost planispiral in peripheral view; 2–2½ whorls, 6–7 chambers in final whorl, slowly increasing in size as added; chambers inflated, sutures distinct, depressed; aperture a low interiomarginal, extraumbilical-umbilical arch, bordered by an imperforate flap.
Remarks: small and very distinctive, although larger specimens are known from higher levels in the succession. Range: Middle Albian to Upper Cenomanian, with rare, atypical specimens, being found in the Turonian.

Hedbergella simplex (Morrow)
Plate 7.16, Figs. 6–8 (× 150), Lower Chalk, Maiden Newton, Dorset, Upper Cenomanian. = *Hastigerinella simplex* Morrow, 1934. Description: test free, trochospiral, slightly asymmetrical, concavo-convex; 2 whorls, 4–6 chambers in final whorl; chambers globular, becoming radially elongated, sutures radial, very depressed; periphery rounded, clavate; aperture extraumbilical-umbilical, bordered by a well-developed lip.
Remarks: a distinctive species which includes *H. amabilis* Loeblich and Tappan and *H. simplicissima* (Magné and Sigal) in its synonomy. Range: Cenomanian.

Heterohelix moremani (Cushman)
Plate 7.16, Fig. 9 (× 400), Bed XIII, Gault Clay, Folkestone, Kent, Upper Albian. = *Guembelina moremani* Cushman, 1938. Description: test free, slender, gradually tapering; biserial, 13–20 chambers, generally subequal in width, height, and thickness, slowly increasing in size; sutures depressed; aperture low-moderately high interiomarginal arch, bordered by narrow imperforate lip.
Remarks: no microspheric forms (i.e. those possessing an initial coil) have been recognised in the British succession. Range: Middle Albian to Upper Cenomanian (although Masters (1977) gives a Middle Albian to Maastrichtian range for this species).

Hoeglundina caracolla (Roemer)
Plate 7.16, Figs. 10–12 (× 50), Speeton Clay, Speeton, Yorkshire, Barremian (LB4D). = *Gyroidina caracolla* Roemer, 1841. Description: test free, biconvex, trochoid, periphery acute, slightly keeled, 1½–2½ whorls, 6–10 chambers in final whorl; chambers triangular, depressed on umbilical side, curved and depressed on the spiral side; sutures prominent, raised, limbate; aperture oval, areal, with conspicuous lateromarginal apertures along the ventral periphery.
Remarks: there is a great deal of variation both in size and shape, although poor preservation hinders detailed study. Range: uppermost Ryazanian to Lower Barremian; often occurs in flood abundance, particularly in the Upper Hauterivian.

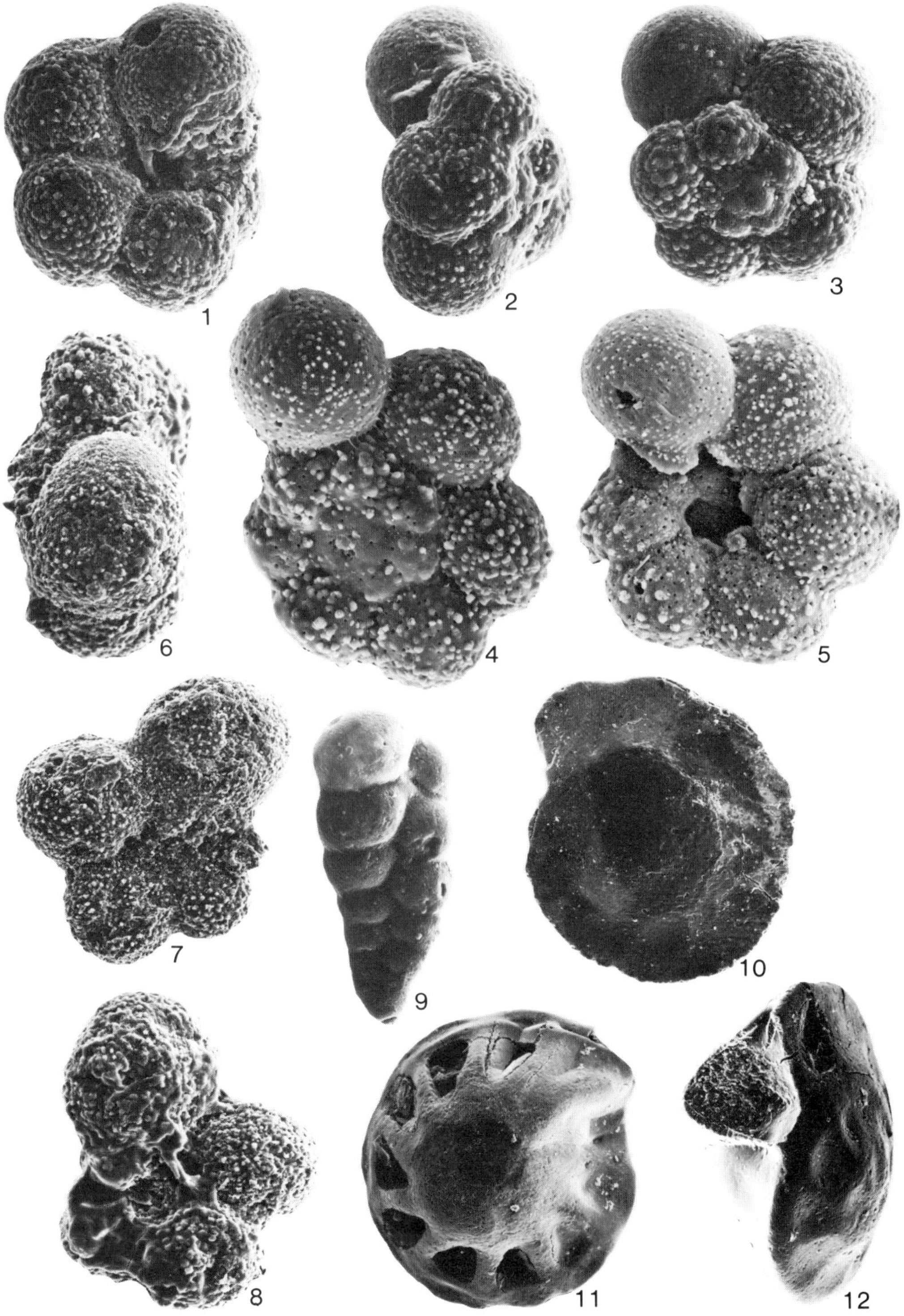

Plate 7.16

Hoeglundina carpenteri (Reuss)
Plate 7.17, Figs. 1, 2 (× 65), Bed III, Gault Clay, Folkestone, Kent, Middle Albian. = *Rotalia carpenteri* Reuss, 1861. Description: test free, biconvex, low trochospiral coil; periphery sharp, acute, crenulate but not developed into marked spines, 2–2½ whorls, 6–9 chambers in final whorl; chambers triangular on apertural side, curved on spiral side; all chambers slightly depressed between raised sutures; primary aperture a slit in the face of final chamber, with a slightly raised lip; lateromarginal chambers also present.
Remarks: the test ornament is highly variable ranging from smooth to more typically pustulose. It differs from *H. chapmani* (Ten Dam) in being larger and in having a crenulate periphery. Range: Lower and Middle Albian, with rare, (?) reworked, specimens being found in the lowermost Upper Albian.

Hoeglundina chapmani (Ten Dam)
Plate 7.17, Figs. 3–5 (× 65), Bed IV, Gault Clay, Folkestone, Kent, Middle Albian. = *Epistomina chapmani* Ten Dam, 1948. Description: test free, biconvex to rarely plano-convex, low trochospiral coil; periphery acute, carinate, non-crenulate; 2–2½ whorls, 8 or 9 chambers in the final whorl; chambers triangular on apertural side, curved on spiral side; last 1 or 2 chambers may be depressed, otherwise smooth; sutures distinct, but flush with surface; primary aperture a slit in face of final chamber; lateromarginal apertures may also be present.
Remarks: this species, from the Aptian–Middle Albian interval of the British succession is clearly closely related to *E. caracolla* (Roemer) and *E. hechti* Bartenstein, Bettenstaedt and Bolli, both recorded from the Lower Cretaceous of N.E. England. It is possible that they may all be part of the same group, but as they occur at different stratigraphic levels in two separate basins of deposition such a conclusion would be difficult to substantiate. Range: Lower Aptian to lowermost Upper Albian.

Lagena hauteriviana Bartenstein and Brand, 1951
Plate 7.17, Fig. 6 (× 65), Speeton Clay, Speeton, Yorkshire, Hauterivian (C6). Description: test free, small oval to spherical, circular in transverse section; unilocular; central, tapering initial spine; aperture simple, terminal, at end of tubular neck.
Remarks: a distinctive species, though variation makes it difficult to identify at the sub-specific level. The length-breadth ratio above shows a continuous variation with *L. hauteriviana hauteriviana* at one extreme, and *L. Hauteriviana cylindracea* Bartenstein and Brand at the other. Range: Upper Ryazanian to Lower Barremian, but most abundant in the Hauterivian.

Lagena hispida Reuss, 1863
Plate 7.17, Figs. 7, 8 (× 60), Speeton Clay, Speeton, Yorkshire, Barremian (LB4). Description: test free, small, with spherical, inflated chamber, and a long basal spine; surface is fine to coarsely hispid; aperture central, simple, on a long tubular neck.
Remarks: differs from *L. oxystoma* Reuss in having a coarsely hispid surface. Range: Hauterivian to Lower Barremian.

Lenticulina eichenbergi Bartenstein and Brand, 1951
Plate 7.17, Figs. 9, 10 (× 25), Speeton Clay, Speeton, Yorkshire, Hauterivian (C7). Description: test free, biconvex, periphery acute, with delicate keel; up to 10 sub-triangular chambers in final whorl; sutures curved, limbate, ornamented by elevated pustules, sutures of later chambers tend to be unornamented; aperture radiate at peripheral angle.
Remarks: in the British succession *L. eichenbergi* and *L. guttata* (Ten Dam) have the same range, and are regarded as one, varying, population by Fletcher (1966, unpublished thesis), but Bartenstein (1977) gives two distinct ranges for these species in Germany. Range: Valanginian to Upper Aptian.

Lenticulina guttata (Ten Dam)
Plate 7.17, Figs. 11, 12 (× 25), Speeton Clay, Speeton, Yorkshire, Hauterivian (C3). = *Planularia gutatta* Ten Dam, 1946. Description: test free, biconvex, periphery acute, keeled; close coiled but with a tendency for the later chambers to uncoil; chambers subtriangular. 10–13 in the final whorl; sutures curved, limbate, raised, ornamented with numerous small, guttiform, pustules; aperture at peripheral angle, terminal, radiate.
Remarks: there is a great deal of variation in the degree of coiling, and also in the strength of the ornamentation (Fletcher, 1966, unpublished thesis). Range: Upper Valanginian to Middle Barremian.

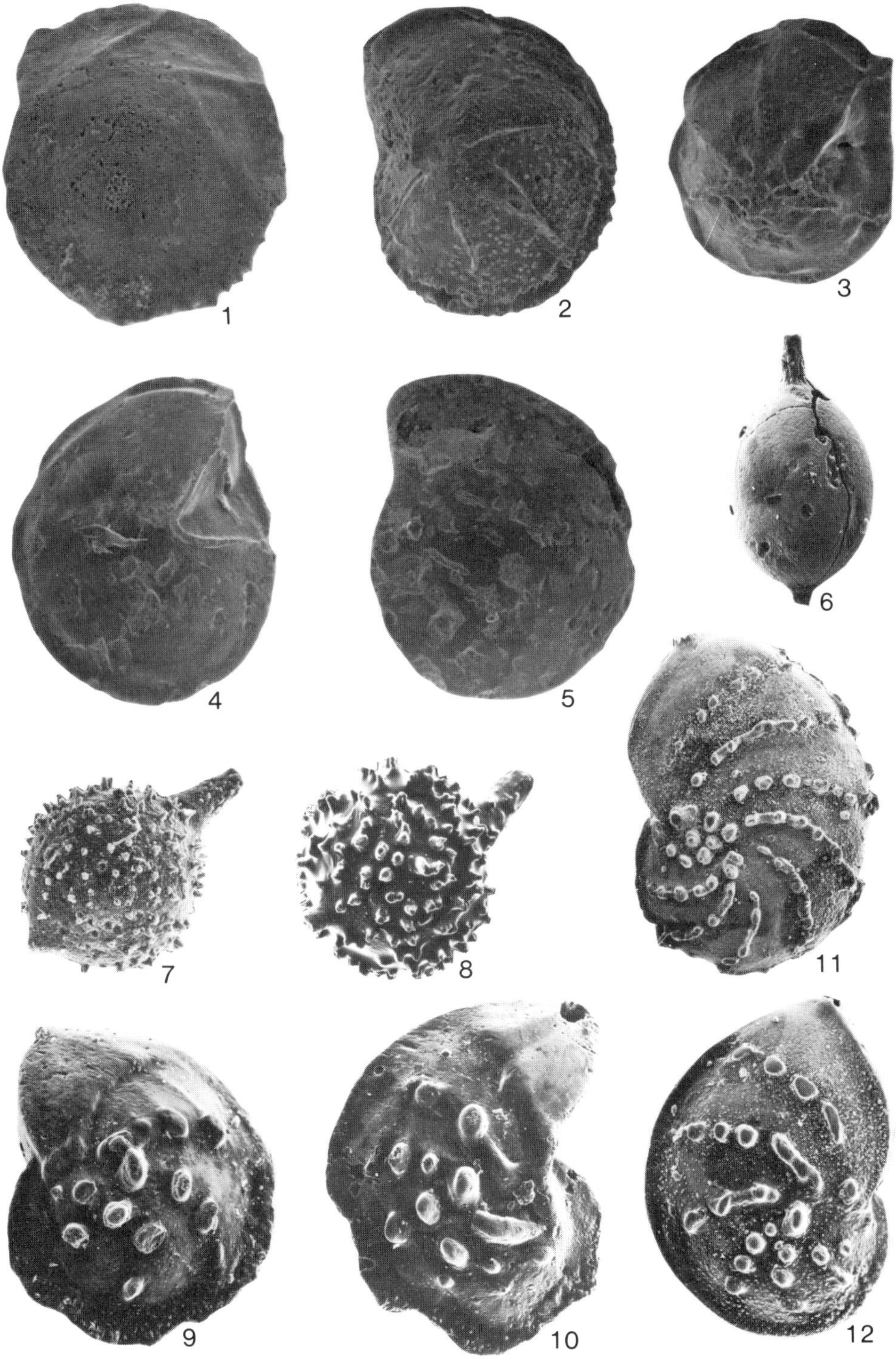

Plate 7.17

Lenticulina heiermanni Bettenstaedt, 1952
Plate 7.18, Fig. 1 (X 20), Speeton Clay, Speeton, Yorkshire, Barremian (LB5D). Description: test free, biconvex, periphery curved or slightly sub-angular, keeled; involute, 10–11 chambers in the final whorl; sutures limbate, raised, curving backwards to peripheral margin; clear, raised, calcite umbilical boss to which sutures are fused; aperture radiate, commonly on a small neck.
Remarks: there is very little variation, apart from the degree of elevation of the sutures above the surface of the test. The keel is easily broken giving a ragged outline. Range: Upper Hauterivian to Lower Aptian, but most abundant in the Lower Barremian.

Lenticulina muensteri (Roemer)
Plate 7.18, Fig. 2 (X 30), Speeton Clay, Speeton, Yorkshire, Hauterivian (C1B). = *Robulina munsteri* Roemer, 1839. Description: test free, biconvex, periphery smooth, acute, but not keeled; test involute 9–13 chambers in the final whorl; sutures limbate, flush with surface; large clear calcite boss, flush with surfaces; aperture radiate, normally at peripheral angle, but sometimes on a small extension.
Remarks: a distinctive species, but the range of variation gives rise to many morphotypes in the Jurassic and Lower Cretaceous. These have been well documented by Jendryka-Fuglewicz (1975). Range: This species ranges from the Ryazanian to Lower Barremian in the Lower Cretaceous.

Lenticulina ouachensis wisselmanni Bettenstaedt, 1952
Plate 7.18, Fig. 3 (X 25), Speeton Clay, Speeton, Yorkshire, Hauterivian (C7). Description: test free, biconvex, periphery acute, with well-developed keel; involute, but with a tendency to uncoil; 7–12 chambers in the final whorl; sutures limbate, raised, umbilical area surrounded by a raised rim; aperture radiate, on a small nect at the peripheral angle.
Remarks: extremely variable but can be distinguished by the characteristic high, sharp, rib, enclosing an oval umbilical area. Range: Hauterivian.

Lenticulina saxonica Bartenstein and Brand, 1951
Plate 7.18, Fig. 4 (X 45), Speeton Clay, Speeton, Yorkshire, Barremian (LB4D). Description: test free, biconvex, periphery acute, with sharp keel; involute, 7–10 chambers in final whorl. sutures limbate, raised, curving towards aperture and meetin periphery at a low angle; raised sutures meet in the centre of the specimen; aperture terminal, radiate, on a small extension.
Remarks: there is little variation, but in some cases the raised sutures may fail to reach the centre. The keel is fragile and easily broken, giving a ragged outline to the test. Range: Upper Valanginian to Lower Barremian.

'Lenticulina' schloenbachi (Reuss)
Plate 7.18, Fig. 5 (X 50), Speeton Clay, Speeton, Yorkshire, Hauterivian (C5K). = *Cristellaria schloenbachi* Reuss, 1863. Description: test free, elongate, compressed oval in cross-section; early portion close coiled, becoming uncoiled with 5–6 chambers in a rectalinear series; sutures indistinct in coiled portion, later distinct, oblique, and sometimes depressed; aperture radiate, at peripheral angle.
Remarks: similar to *Lenticulina (Astacolus) pachynota* (Ten Dam) but lacks the limbate, elevated sutures. The generic position of this species, in either *Astacolus* or *Lenticulina,* is a subject of debate. Range: Upper Ryazanian to Lower Barremian.

'Lenticulina' schreiteri (Eichenberg)
Plate 7.18, Figs. 6, 7 (X 40), Speeton Clay, Speeton, Yorkshire, Hauterivian (C2F). = *Elphidium schreiteri* Eichenberg, 1935. Description: test free, elongate, compressed; initially coiled, becoming uncoiled, with 3–8 low, broad, sub-triangular chambers in uncoiled part of test; surface of test with strong, reticulate ornament; aperture radiate, on a distinct neck.
Remarks: the ornament is quite variable, from strong and regular to very irregular and sometimes quite weak. To a lesser extent the degree of uncoiling is also quite variable causing problems of generic determination (*Lenticulina* vs. *Astacolus*). Taxonomically it is very close to *Vaginulinopsis reticulosa* (Ten Dam). Range: Upper Valanginian to Barremian.

Lingulogavelinella globosa (Brotzen)
Plate 7.18, Figs. 8–10 (X 75), Plenus Marls, Bed 4, Betchworth Quarry, Surrey, Upper Cenomainian. = *Anomalinoides globosa* Brotzen, 1945. Description: test free, trochospiral, equally biconvex; periphery broadly rounded, lobate, spiral side centrally depressed; 2–2½ whorls, 6–8 inflated chambers in final whorl; sutures distinct, depressed, curved; aperture interiomarginal, slit-like, bordered by a lip; remnant lips of previous chambers forming star-shaped pattern around the umbilicus.
Remarks: *L. globosa* is a distinctive species, which appears to form the end of the Albian–Cenomanian 'lingulogavelinellid' lineage. Range: Upper Cenomanian to Middle Turonian.

Lingulogavelinella jarzevae (Vasilenko)
Plate 7.18, Figs. 11–13 (X 75), Glauconitic Marl, Folkestone, Kent, Lower Cenomanian. = *Cibicides (Cibicides) jarvezae* Vasilenko, 1954. Description: test free, plano-convex, periphery rounded to sub-acute; apertural side flat, with a star-shaped pattern made up of the relict apertural flaps; spiral side elevated, with 5–7 highly inflated chambers; sutures very depressed on spiral side, almost straight; aperture an interiomarginal slit extending from the periphery to umbilicus.

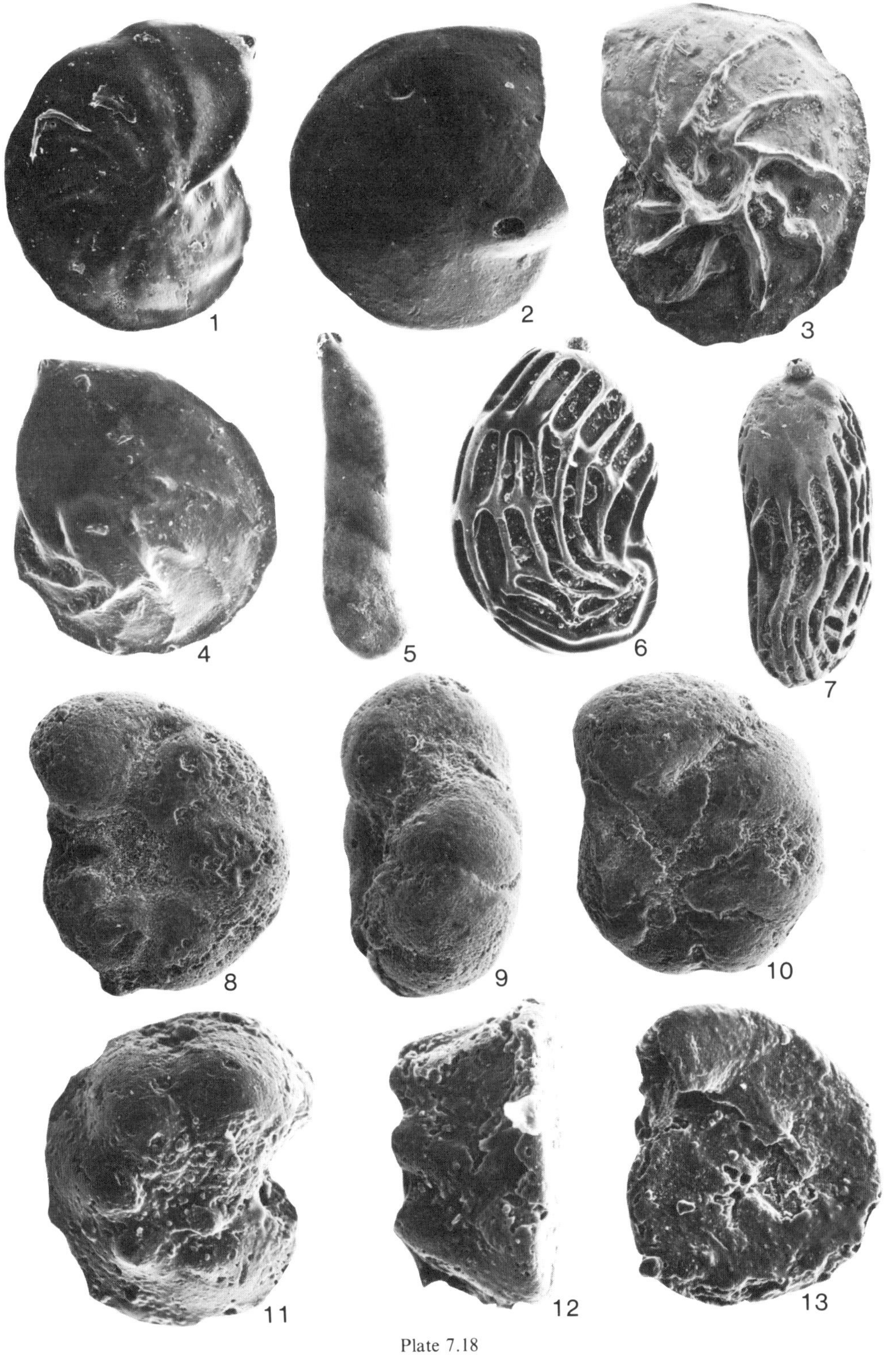

Plate 7.18

Remarks: this very distinctive species is morphologically very close to *Cibicides formosa* Brotzen, and is regarded as synonymous by some authors. The ranges of the two species do not, however, overlap and the use of a Campanian/ Maastrichtian specific name is rejected for the present. Range: uppermost Albian to Lower Cenomanian.

Lingulogavelinella sp. cf. *L. vombensis* (Brotzen)
Plate 7.19, Figs. 1-3 (X 95, Upper Chalk, Helhoughton, Norfolk, Santonian. = *Pseudovalvulineria vombensis* Brotzen, 1942. Description: test free, low trochospiral, periphery rounded to sub-angular, early whorls on dorsal surface masked by a large calcareous boss, ventral side involute; sutures flush to slightly depressed on dorsal side, raised on ventral side; aperture a narrow, interiomarginal slit, covered by an apertural flap on the ventral surface; umbilicus filled with remnant apertural flaps, forming a radiate pattern.
Remarks: a distinctive species of great stratigraphic value. *L. vombensis* (Brotzen) is a Maastrichtian species and does not appear to be an extension of the present species' range. Range: Coniacian to Santonian.

Loxostomum eleyi (Cushman)
Plate 7.19, Figs. 4, 5 (X 160), Upper Chalk, Helhoughton, Norfolk, Santonian. = *Bolivinita eleyi* Cushman, 1927. Description: test free, elongate, compressed, biserial, edge truncated, sub-angular; chambers flat, overlapping, reniform in outline; sutures distinct, flush to slightly raised, curved; aperture terminal, ovoid slit, surrounded by slightly raised lip.
Remarks: has a very useful range over a wide geographical area of Europe. Range: Santonian to Middle Campanian; very rare in the Upper Campanian.

Marginotruncana coronata (Bolli)
Plate 7.19, Figs. 6, 7 (X 50), Middle Chalk, Beer, Devon, Turonian. = *Globotruncana lapparenti* Brotzen subsp. *coronata* Bolli, 1945.Description: test free, low trochospiral coil; 2–2½ whorls, 6–8 chambers in final whorl; chambers on umbilical side reniform, chambers on spiral side petaloid, chambers not inflated; sutures raised; chamber margins truncated, with two keels separated by an imperforate band; primary aperture extraumbilical–umbilical, may be covered with apertural flaps or a tegilla (especially in later forms).
Remarks: differs from *M. pseudolinneiana* (Pessagno) in having a more compressed profile, petaloid chambers on the spiral side, and in keels that are closer together. Range: Upper Turonian.

Marginotruncana pseudolinneiana Pessagno, 1967
Plate 7.19, Figs. 8, 9 (X 75), Middle Chalk, Beer Devon, Middle Turonian. Description: test free, very low trochospiral, both sides flat to slightly convex; periphery truncated, with two closely spaced, beaded keels; chambers elongate petaloid on spiral side, lobate on umbilical side, flat, 5-7 in final whorl; sutures slightly depressed at the margins, otherwise raised; aperture extraumbilical–umbilical, low, interiomarginal arch, tegilla not recorded.
Remarks: there is a continuous morphological gradation from *M. pseudolinneiana* to *G. linneiana* (d'Orbigny). Range: Turonian to Santonian.

Marginotruncana sigali (Reichel)
Plate 7.19, Fig. 10 (X 75), Middle Chalk, Beer, Devon, Middle Turonian. = *Globotruncana sigali* Reichel, 1950. Description: test free, trochospiral, single peripheral keel, formed by a very close double row of pustules; 2-3 whorls, 5-7 chambers in final whorl; chambers on umbilical side sub-rectangular, not inflated; chambers on spiral side subtrapezoidal, flat; sutures raised on umbilical side, 'U'-shaped, continuing along umbilical periphery; primary aperture extraumbilical–umbilical, umbilicus surrounded or covered by portici.
Remarks: a very distinctive, biconvex species; the oldest single keeled marginotruncanid in the British succession. Range: Middle Turonian.

Marginulinopsis foeda (Reuss)
Plate 7. 19, Figs. 11, 12 (X 45), Speeton Clay, Speeton, Yorkshire, Hauterivian (C4C). = *Cristellaria foeda* Reuss, 1863. Description: test small, elongate, circular in cross-section; in microspheric form early chambers coiled, followed by 3–4 uncoiled chambers; chambers slightly inflated, sutures depressed; surface of test strongly hispid; aperture terminal, radiate, on a long tubular, lipped neck.
Remarks: there is considerable variation in the density and size of the hispid ornamentation; It is similar in shape to *M. gracilissima* (Reuss), but the latter species is completely smooth. Range: Upper Ryazanian to mid- Barremian.

Marginulinopsis gracilissima (Reuss)
Plate 7.19, Figs. 13, 14 (X 50), Speeton Clay, Speeton, Yorkshire, Barremian (LB3). = *Cristellaria gracillissima* Reuss, 1863. Description: test free, elongate, circular in cross-section; 4–5 coiled chambers followed by an uncoiled linear series of 3–4 chambers, the last of which is usually considerably inflated; sutures distinct, depressed, oblique in uncoiled portion of the test; aperture radiate, on a distinct neck.
Remarks: the overall shape of the test varies greatly. This species is very similar to *M. foeda* (Ruess) but difers in having a smooth test. Range: Ryazanian to Lower Barremian.

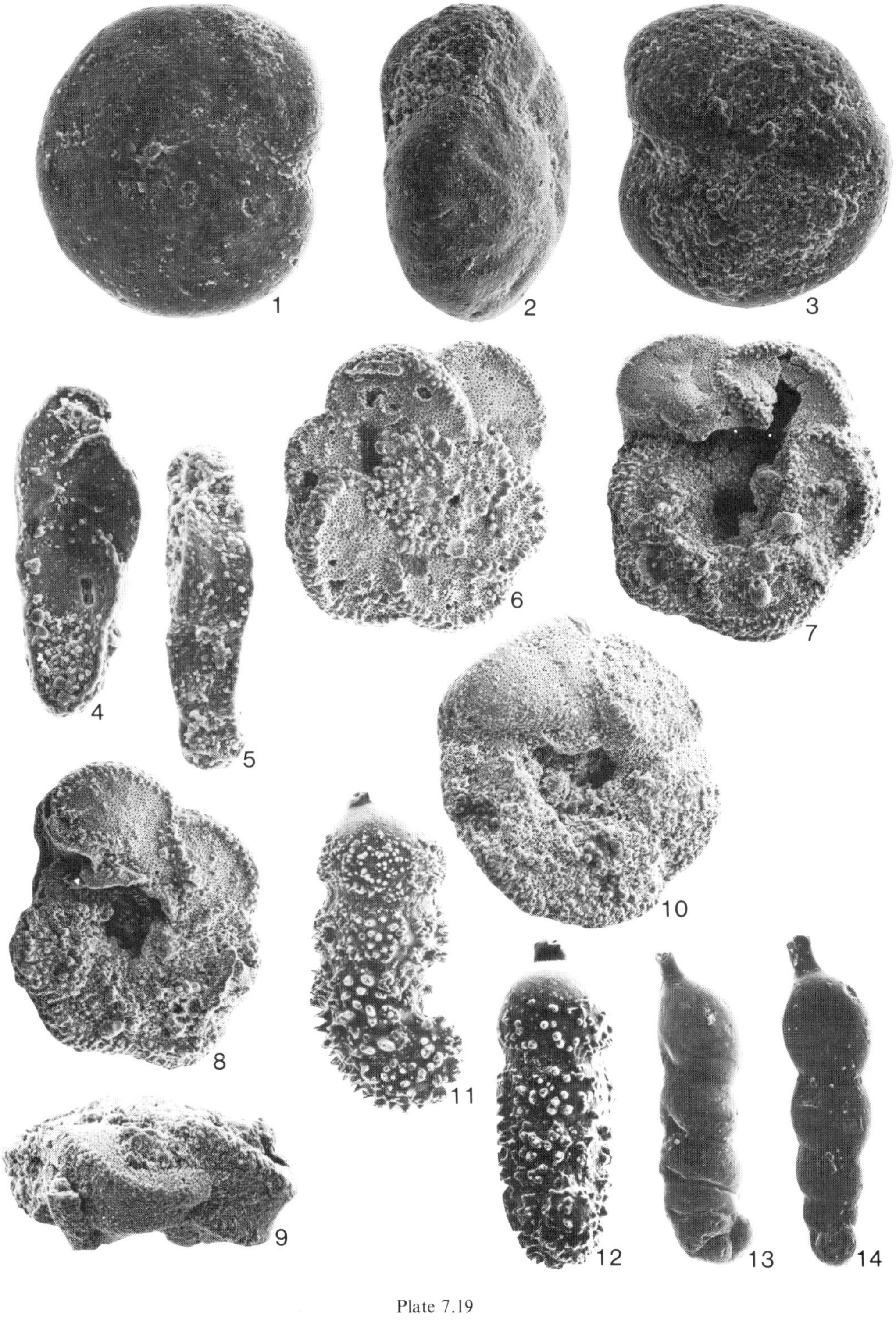

Plate 7.19

Neoflabellina praereticulata Hiltermann, 1952
Plate 7.20, Fig. 2 (X 60), Upper Chalk, Whitlingham, Norfolk, Upper Campanian. Description: test free, palmate, compressed, with angular to carinate edges; chambers initially coiled, uncoiling rapidly, chevron-shaped; sutures distinct, raised, slightly crenulated; aperture terminal, subcircular; test surface between sutures strongly ornamented by numerous short ridges, ornament on earliest chambers nodose.
Remarks: characterised by its markedly irregular, reticulate, ornament. Range: Upper Campanian to lower Lower Maastrichtian.

Neoflabellina reticulata (Reuss)
Plate 7.20, Fig. 3 (X 50), Upper Chalk, Trimingham, Norfolk, Lower Maastrichtian. = *Flabellina reticulata* Reuss, 1851. Description: test free, palmate to deltoid, sides flat, parallel, with angular to carinate edges; chambers initially coiled, rapidly uncoiling, chevron-shaped; sutures distinct, raised, crenulate; aperture terminal, sub-circular, on a short neck; test surface between sutures strongly ornamented by numerous ridges.
Remarks: differentiated from *N. praereticulata* Hiltermann by its more regular, reticulate, ornament, which extends even to the earliest chambers. Range: Maastrichtian.

Neoflabellina rugosa (d'Orbigny)
Plate 7.20, Fig. 1 (X 50), Upper Chalk, Catton Grove, Norfolk, Upper Campanian. = *Flabellina rugosa* d'Orbigny, 1840. Description: test free, palmate, sides flattened, parallel, with angular to slightly carinate margins. chambers initially coiled, rapidly becoming uncoiled, chevron-shaped, narrow, increasing uniformly in size; sutures distinct, raised, aperture terminal, sub-circular on short neck; test surface between sutures ornamented by one or two rows of distinct, raised papillae.
Remarks: the surface ornament makes this species distinctive, and many authors (e.g. Koch, 1977) have recognised several subspecies. Range: Campanian.

Osangularia cordieriana (d'Orbigny)
Plate 7.20, Figs 4–6 (X 100), Upper Chalk, Scratchells Bay, Isle of Wight, Upper Campanian. = *Rotalina cordieriana* d'Orbigny, 1840. Description: test free, biconvex, biumbonate, periphery acute, slightly carinate; 3–4 whorls, 8–10 chambers in final whorl; umbilical side possessing pronounced, raised, umbilical boss of clear calcite; sutures distinct, slightly obscure on spiral side at first; aperture a narrow, oblique, slit along the base of final chamber on umbilical side, then forming a 'V'-shaped bend at an oblique angle up apertural face.
Remarks: characterised by its almost equally biconvex, biumbonate test and its slightly carinate margin. Range: Campanian, but also rarely in Lower Maastrichtian.

Osangularia navarroana (Cushman)
Plate 7.20, Figs. 7, 8 (X 100), Upper Chalk, Trimingham, Norfolk, Lower Maastrichtian. = *Pulvinulinella navarrona* Cushman, 1938. Description: test free, equally biconvex; umbilical side possessing a moderately raised calcite boss; outline circular, periphery acute, bordered by a distinct keel; 2½ whorls, 10–12 chambers visible on umbilical side; sutures distinct, limbate, flush; aperture a narrow 'V'-shaped slit, occasionally separated into distinct interiomarginal and areal openings.
Remarks: the biconvex test and distinctly keeled margin serve to distinguish this species from the closely related *O. cordieriana* (d'Orbigny).

Osangularia whitei (Brotzen, 1936)
Plate 7.20, Figs. 9, 10 (X 140), Upper Chalk, Euston, Suffolk, Santonian. = *Eponides whitei* Brotzen, 1936. Description: test free, trochospiral, equally biconvex, biumbonate, margin acutely angled, becoming carinate; chambers distinct, 7–8 in final whorl; sutures straight to slightly oblique, flush; aperture a narrow, oblique slit towards the inner margin of the final chamber with a very slightly raised lip.
Remarks: very similar to *O. cordieriana* (d.Orbigny), but can be distinguished by its almost horizontal periphery, as compared to the almost sigmoid periphery of *O. cordieriana.* Range: Mid-Coniacian to lowermost Campanian.

Planularia crepidularis Roemer, 1842
Plate 7.20, Figs. 11, 12 (X 50), Speeton Clay, Speeton, Yorkshire, Hauterivian (C3). Description: test free, compressed, sub-parallel sides; initially coiled, becoming uncoiled, 5–7 chambers in a linear series; sutures distinct, raised; periphery carinate, normally with three keels; aperture radiate, on a small neck.
Remarks: there is a wide range of variation, mainly in the degree of coiling and the intensity of the ornamentation. In the Jurassic some workers refer it to *Lenticulina (Planularia) tricarinella* Reuss. Range: Jurassic to Lower Aptian.

Pleurostomella barroisi Berthelin, 1880.
Plate 7.20, Figs. 13, 14 (X 75), Bed XIII, Gault Clay, Folkestone, Kent, Upper Albian. Description: test free, elongate, 7–8 chambers, cuneate, later approaching uniserial; sutures depressed, markedly oblique; aperture terminal with a projecting hood drawn out into sharp, pointed, beak, bifid tooth on opposite side.
Remarks: both microspheric and megalospheric forms are known from the British succession. The distinct beak and bifid tooth separate this species from that which contains the microspheric *P. reussi* Berthelin and megalospheric *P. obtusa* Berthelin. Range: Upper Albian.

Plate 7.20

Praebulimina carseyae (Plummer)
Plate 7.20, Figs. 15, 16 (X 60), Upper Chalk, Caister St. Edmunds, Norfolk, Upper Campanian. = *Buliminella carseyi* Plummer, 1931. Description: test free, ovate, initial end rounded; generally composed of 4 whorls, 4 chambers per whorl; sutures distinct; apertural face of final chamber elongate, often flattened; aperture variable, subterminal, extending along interiomarginal suture.
Remarks: can be identified by its step-like outline, quadriserial chamber arrangement, and inflated chambers. Early forms in the Lower Campanian are smaller, slender, with less inflated chambers, while those in the Upper Campanian are more typical. Range: Campanian, and rarely in the lower Lower Maastrichtian.

Praebulimina laevis (Beissel)
Plate 7.21, Figs. 1, 2 (X 75), Upper Chalk, Overstrand (Fig. 1) and Trimingham (Fig. 2), Norfolk, Lower Maastrichtian. = *Bulimina laevis,* Beissel, 1891. Description: test free, subfusiform, large; initial end rounded, then rapidly flaring, until final whorl; 3–4 whorls, 4 chambers to a whorl; chambers only slightly inflated, with sutures flush to only slightly depressed; apertural face sub-terminal, aperture variable, comma-shaped, occasionally bordered by a thin, indistinct, lip.
Remarks: Beissel (1891), indicating a wide range of morphology, may have included more than one species in the initial definition. As a result, this species has been regularly confused with *P. carseyae* (Plummer) and *P. obtusa* (d'Orbigny). Range: uppermost Campanian and Maastrichtian.

Praebulimina obtusa (d'Orbigny)
Plate 7.21, Figs. 3 (X 60), 4 (X 50), Upper Chalk, Alum Bay, Isle of Wight, Upper Campanian. = *Bulimina obtusa* d'Orbigny, 1840. Description: test free, large, although two variants are known; variant one is bluntly pointed, rapidly flaring, 4–5 whorls, 4 slightly inflated chambers per whorl, sutures nearly flush; variant two is initially broadly rounded and then rapidly flaring, but almost parallel-sided in the last whorl; in both forms the loop-shaped aperture is set in a depression in the apertural face.
Remarks: the more inflated variant (Plate 7.21, Fig. 4) replaces the less inflated form (Plate 7.21, Fig. 3) in the lower levels of the middle Upper Campanian. Range: Upper Campanian to Lower Maastrichtian.

Praebulimina reussi (Morrow)
Plate 7.21, Figs. 5, 6 (X 120), Upper Chalk, Wells, Norfolk (Fig. 5) and Scratchells Bay, Isle of Wight (Fig. 6), Middle Campanian. = *Bulimina reussi* Morrow, 1934. Description: test free, ovate, cross-section sub-circular; 4–5 whorls, chambers triserial throughout, becoming rapidly inflated, elongate; sutures distinct, slightly depressed; aperture variable, normally a narrow terminal slit at inner margin of final chamber.
Remarks: a very variable species that has been much confused in the literature. Range: Santonian and Campanian.

Praeglobotruncana delrioensis (Plummer)
Plate 7.21, Figs. 7, 8 (X 125), Lower Chalk, Eastbourne, Sussex, Middle Cenomanian. = *Globorotalia delrioensis* Plummer, 1931. Description: test free, biconvex, low trochospiral; 2–2½ whorls, 5–6 chambers in final whorl; chambers on umbilical side triangular in shape, gently inflated, sutures radial, depressed; chambers on spiral side gently inflated, sutures curved, slightly beaded; periphery lobate, margins of chambers marked by an accumulation of pustules; primary aperture extraumbilical–umbilical, bordered by a narrow lip.
Remarks: differs from *P. stephani* (Gandolfi) in having a lower trochospire, and a less developed marginal band of pustules. Range: Lower to Middle Cenomanian.

Praeglobotruncana helvetica (Bolli)
Plate 7.21, Figs. 9, 10 (X 125), Middle Chalk, Beer, Devon, Turonian. = *Globotruncana helvetica* Bolli, 1945. Description: test free, plano-convex, 2–2½ whorls, 4–5½ chambers in the final whorl; chambers on umbilical side inflated, rugose, with radial, depressed, sutures; chambers on spiral side petaloid, with flat to concave surfaces, sutures raised; single keel usually forms edge to flat spiral surface; primary aperture extraumbilical–umbilical, bordered by a well-developed lip.
Remarks: this distinctive species normally has fewer chambers (4–5½ instead of 6–7) and is less ornamented than the typical, Tethyan, form. Range: Lower and Middle Turonian.

Praeglobotruncana stephani (Gandolfi)
Plate 7.21, Figs. 11–13 (X 125), Lower Chalk, Buckland Newton, Dorset, Upper Cenomanian. = *Globotruncana stephani* Gandolfi, 1942. Description: test free, moderate trochospiral; 2–2½ whorls, 5–7 chambers in the final whorl; chambers on umbilical side triangular in shape, slightly inflated, sutures radial, depressed; chambers on spiral side petaloid, flat to gently inflated, sutures curved, distinctly beaded; peripheral margin beaded; primary aperture extraumbilical–umbilical, bordered by a well-developed lip.
Remarks: the higher spired, and more ornamented, *P. gibba* Klaus is commonly found near the Cenomanian–Turonian boundary. Range: Middle Cenomanian to Lower Turonian.

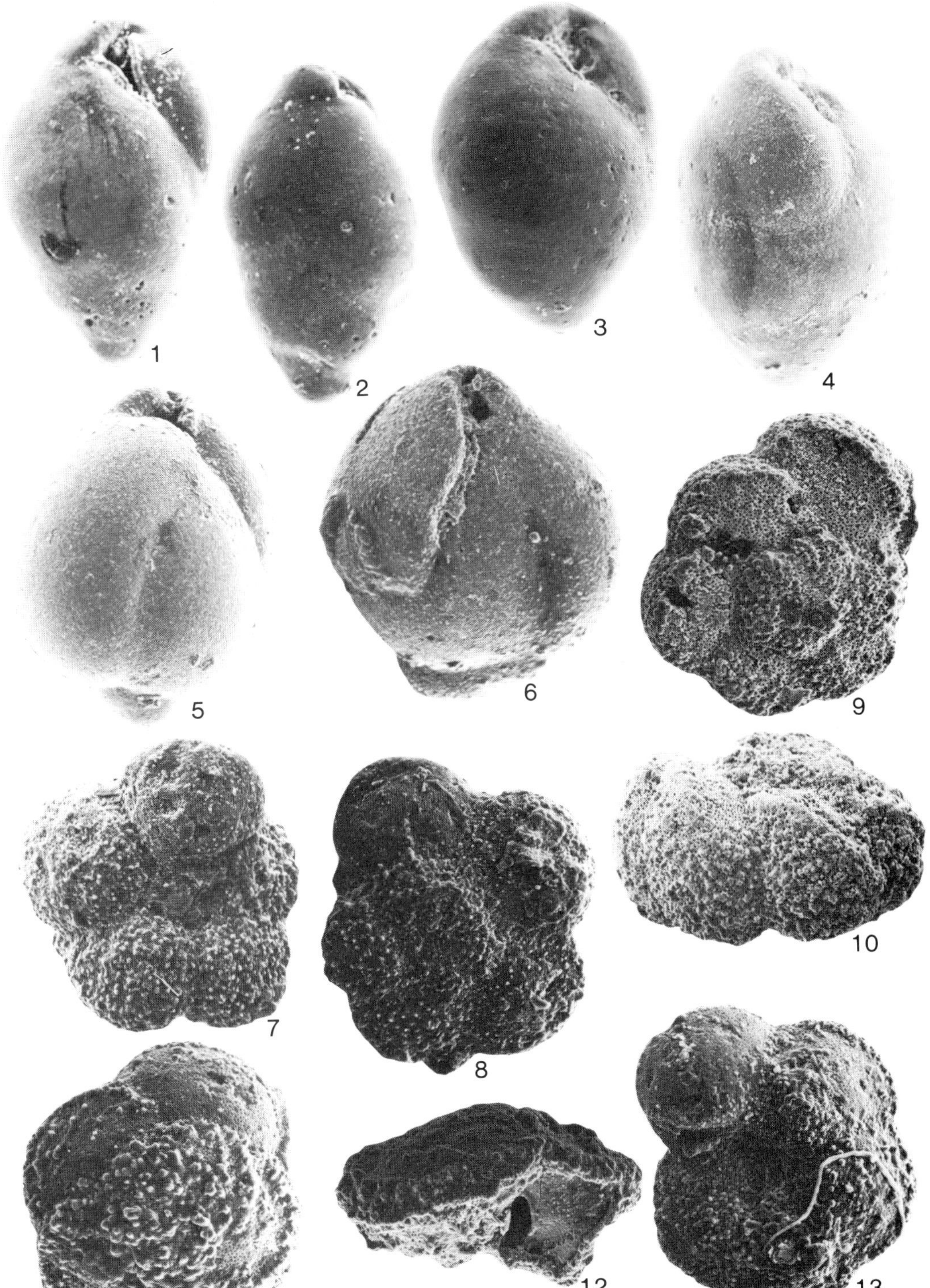

Plate 7.21

Pseudouvigerina cristata (Marsson)
Plate 7.22, Figs. 1, 2 (X 100), Upper Chalk, Trimingham, Norfolk (Fig. 1) and Overstrand, Norfolk (Fig. 2), Lower Maastrichtian. = *Uvigerina cristata* Marsson, 1878. Description: test free, elongate, cross-section triangular; 4–5 whorls, triserial; chambers slightly inflated, the margins possessing distinct, widely spaced, double vertical costae; sutures distinct, depressed; aperture terminal at end of short neck, occasionally with a lip; internal narrow columnellar tooth plate present.
Remarks: a very distinctively ornamented species. Range: Upper Campanian and Maastrichtian.

Pullenia quaternaria (Reuss)
Plate 7.22, Fig. 3 (X 100), Upper Chalk, Caister St. Edmunds, Norfolk, Upper Campanian. = *Nonionina quaternaria* Reuss, 1851. Description: test free, planispiral, involute coil; 4–5 chambers in the final whorl; outline sub-circular, weakly lobate; periphery broadly rounded; chambers triangular, curved, final chamber covering the umbilicus on either side; sutures distinct, flush to weakly depressed; aperture a low interiomarginal slit, bordered by a thick, imperforate, band.
Remarks: *P. quaternaria* is characterised by its moderately compressed test, curved sutures and low apertural face. Range: Middle Campanian to Maastrichtian.

Ramulina spandeli Paalzow, 1917
Plate 7.22, Fig. 4 (X 50), Speeton Clay, Speeton, Yorkshire, Hauterivian (C4). Description: test free, elongate, consisting of a single oviform chamber with long, stoloniferous tubes at either end; chamber and tube walls are coarsely hispid; aperture simple, rounded, at end of tubes.
Remarks: a distinctive species, even though it displays considerable variation in size and shape, and is always fragmentary. Range: Hauterivian and Lower Barremian.

Reussella kelleri Vasilenko, 1961
Plate 7.22, Figs. 5, 6 (X 80), Upper Chalk, Newton-by-Castleacre, Norfolk, Santonian. Description: test free, subtriangular in outline, triserial throughout, edges sub-acute to angular; chambers triangular, strongly overlapping; sutures distinct, curved, often raised; margins spinose, occasionally developing small flanges which project from the edges; apertural slit, perpendicular to the inner margin of the final chamber, surrounded by a slightly raised lip.
Remarks: the first *Reussella* to appear in the British succession. Range: Upper Turonian to Lower Campanian; extremely rare in the Middle Campanian.

Reussella szajnochae praecursor De Klasz and Knipscheer, 1954
Plate 7.22, Fig. 7 (X 100), Upper Chalk, Culver Cliff, Isle of Wight, Santonian. Description: test free, triangular, initial end pointed; test margins serrate, spinose; chambers triserial throughout, 5–6 whorls; aperture a slit-shaped opening extending up apertural face from mid-point of interiomarginal suture, with internal tooth plate, and external raised lip.
Remarks: this subspecies may be distinguished from *Reussella kelleri* Vasilenko by its smaller size, more spinose, generally non-carinate, test angles, and by its distinctly raised, limbate sutures. Range: Santonian to Middle Campanian.

Reussella szajnochae szajnochae (Grzybowski)
Plate 7.22, Fig. 8 (X 100), Upper Chalk, Overstrand, Norfolk, Lower Maastrichtian. = *Verneuilina szajnochae* Grzybowski, 1896. Description: test free, triangular, robust; test margins sharp, serrate, spinose, non-carinate; triserial throughout, 4–6 whorls; sutures distinct, raised, limbate, carinate, projecting from test angles to form spines; aperture an elongate, slit-shaped opening, extending up apertural face from the mid-point of the interiomarginal suture, and bordered by a distinct lip; there is an internal tooth-plate.
Remarks: *R. pseudospinosa* Troelsen from the Lower Maastrichtian is a common synonym of this subspecies. Range: Upper Campanian, absent in the Lower Maastrichtian but occurring rarely in the Upper Maastrichtian.

Rotalipora appenninica (Renz)
Plate 7.22, Figs. 9–11 (X 60), Lower Chalk, Eastbourne Sussex, Lower Cenomanian. = *Globotruncana appenninica* Renz, 1936. Description: test free, symmetrically biconvex, trochospiral; 2 whorls, 5–7 chambers in the final whorl; chambers on umbilical side triangular, sutures radial, depressed; chambers on spiral side petaloid, sutures distinct, raised; single keel; primary aperture extraumbilical–umbilical, secondary apertures umbilical initially, becoming sutural later.
Remarks: this species seems to appear later in the British succession than it does in the Tethyan localities, although Magniez-Jannin (personal communication) has recently found rare specimens in the upper part of Bed XIII, Gault Clay, at Folkestone. Range: uppermost Albian to Middle Cenomanian.

Rotalipora cushmani (Morrow)
Plate 7.22, Figs. 12, 13 (X 100), Lower Chalk, Buckland Newton, Dorset, Upper Cenomanian. = *Globorotalia cushmani* Morrow, 1934. Description: test free, biconvex, trochospiral; 2 whorls, 4½–8 chambers in the final whorl; chambers on umbilical side triangular, strongly inflated, sutures radial, depressed; chambers on spiral side semi-circular, sutures raised; single keel; primary aperture extraumbilical–umbilical, secondary apertures sutural, bordered by well-developed lips.
Remarks: a most distinctive species, of world-wide stratigraphic value. Range: Middle to Upper Cenomanian.

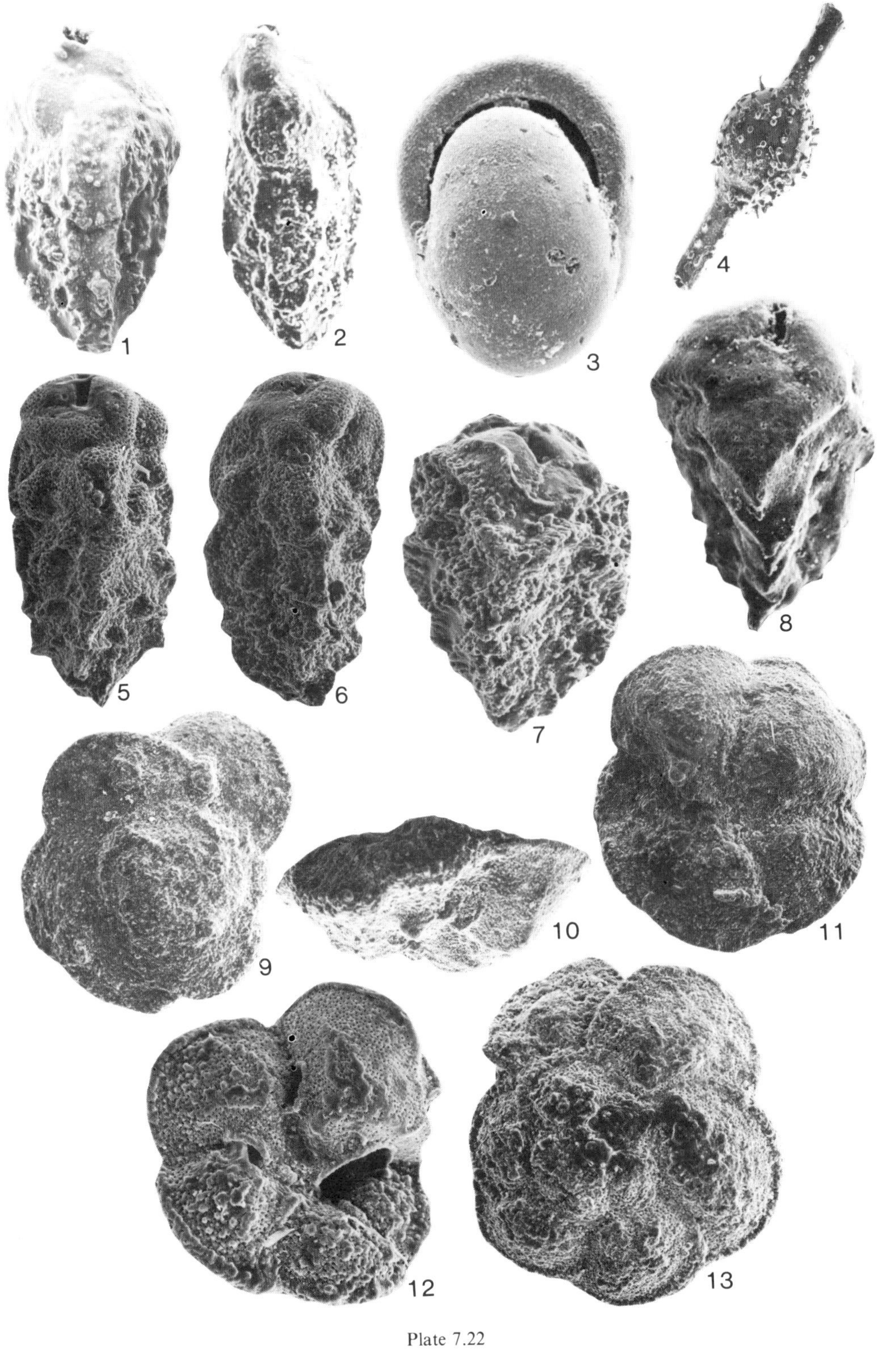

Plate 7.22

Rotalipora greenhornensis (Morrow)
Plate 7.23, Figs. 1–3 (× 100), Lower Chalk, Buckland Newton, Dorset, Upper Cenomanian. = *Globorotalia greenhornensis* Morrow, 1934. Description: test free, trochospiral, unequally biconvex; 2–2½ whorls, 7–10 chambers in final whorl, chambers on umbilical side trapezoidal, sutures curved, distinctly raised; chambers on spiral side crescentric, sutures raised; single keel; primary aperture extraumbilical–umbilical, secondary apertures umbilical, may be sutural in later chambers.
Remarks: differs from *R. brotzeni* (Sigal) in having a more pronounced asymmetrical test, and more chambers in the final whorl. Range: Middle and Upper Cenomanian.

Rotalipora reicheli Mornod, 1950
Plate 7.23, Figs. 4–6 (× 100), Lower Chalk, Southerham, Sussex, Lower/Middle Cenomanian boundary. Description: test free, trochospiral, strongly asymmetrical; 2–2½ whorls, 6–8 chambers in the final whorl; chambers on umbilical side triangular, inflated, sutures depressed; chambers on spiral side crescentric, sutures raised, beaded; single keel; primary aperture extraumbilical–umbilical, secondary apertures umbilical, peri-umbilical ridges on all chambers forming a rim around the umbilicus.
Remarks: this species is very close to *R. deeckei* (Franke) but can be distinguished by its more flattened spiral side and depressed sutures on the umbilical side. Range: occurs only near the Lower/Middle Cenomanian boundary; rarely in the lower Middle Cenomanian (Hart, 1979).

Rugoglobigerina rugosa (Plummer)
Plate 7.23, Figs. 7–9 (× 100), Upper Chalk, Overstrand, Norfolk, Lower Maastrichtian. = *Globigerina rugosa* Plummer, 1927. Description: test free, low trochospiral coil of 2½–3 whorls, 4½–5 chambers in the final whorl; outline subcircular, periphery rounded; chambers distinct, subglobular, sutures depressed, radial; surface of each chamber ornamented with meridional pattern of discontinuous ridges or costellae; primary aperture interiomarginal, umbilical.
Remakrs: *R. rugosa* is recognised by its low trochospiral coil, rapidly expanding chambers, and moderately large umbilicus. As a species it shows a great deal of variability, which has resulted in a complex taxonomic history. Range: Maastrichtian.

Saracenaria valanginiana Bartenstein and Brand, 1951
Plate 7.23, Fig. 10 (× 65), Speeton Clay, Speeton, Yorkshire, Ryazanian (D6AI). Description: test free, elongate, sub-triangular in cross-section; 2–3 close-coiled chambers followed by 3–5 slightly inflated chambers in a rectilinear series; sutures distinct, depressed; ornament comprises three keels at the angles and distinct, fine, ribs; aperture radiate, at peripheral margin, on a small neck.
Remarks: a distinctive species that shows very little variation. Range: Upper Ryazanian and Lower Valanginian.

Stensioina exsculpta exsculpta (Reuss)
Plate 7. 23, Figs. 11–13 (× 100), Upper Chalk, Euston, Suffolk, Santonian. = *Rotalia exsculpta* Reuss, 1860. Description: test free, plano-convex to biconvex, margins acutely angled; chambers distinct on dorsal side, sub-rectangular, sutures distinct, sharply raised, forming elevated septal ridges; sutures on ventral side flush to slightly depressed; aperture an elongate slit along the inner, ventral, margin of the final chamber.
Remarks: this subspecies is distinguished from the '*S granulata* (Olbertz) lineage' in having a more sharply angled periphery, and in having a more distinctive septal ornament on the spiral side. Range: Mid-Coniacian to Santonian; very rare in the lowermost Campanian offshore.

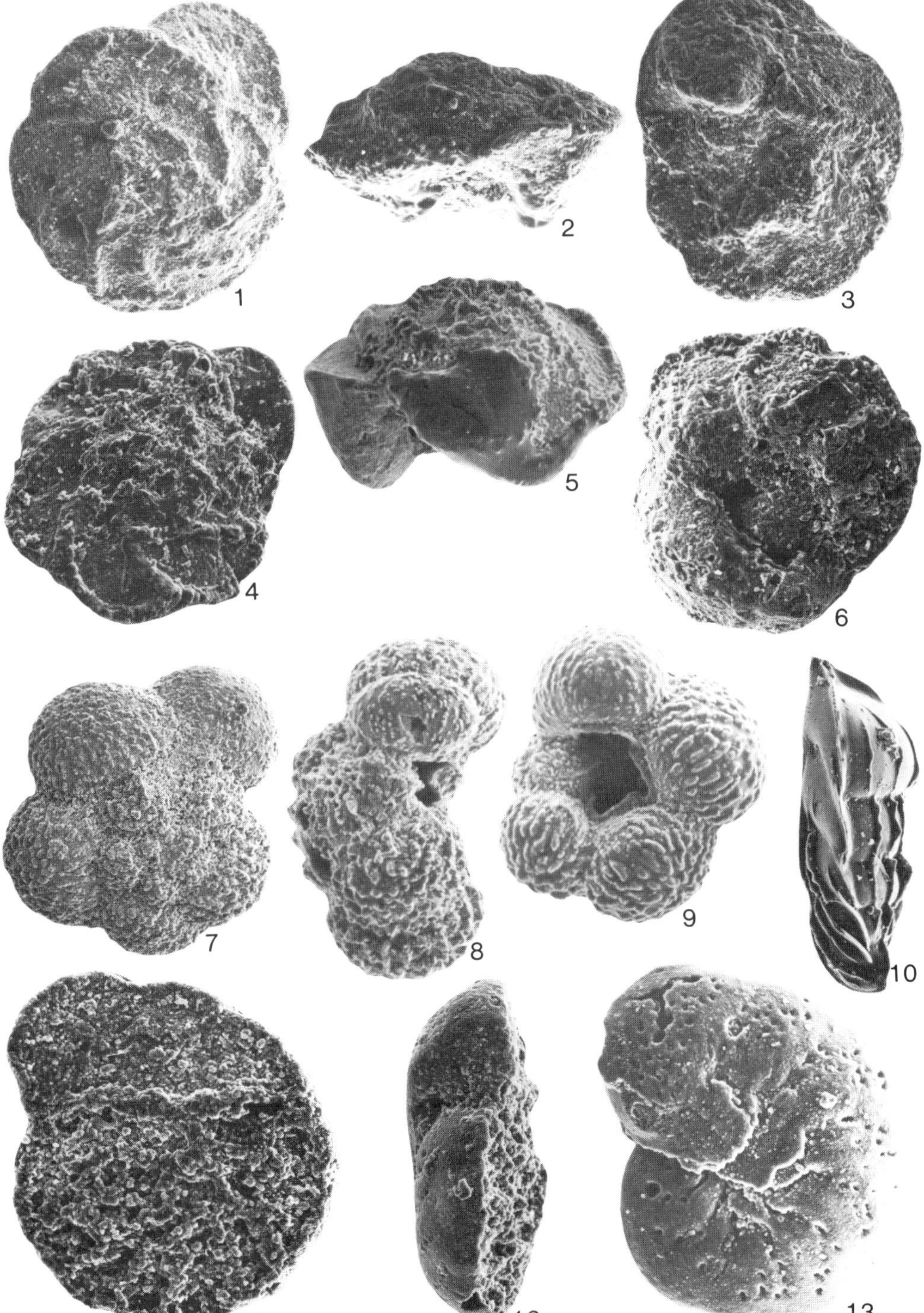

Plate 7.23

Stensioina exsculpta gracilis Brotzen, 1945
Plate 7.24, Figs. 1–3 (× 110), Upper Chalk, Stiffkey, Norfolk, Middle Campanian. Description: test free, biconvex, trochospiral, periphery acute; 2½–3 whorls, 10–12 chambers in final whorl; umbilical side weakly convex, involute; spiral side moderately to highly raised, evolute; sutures on umbilical side distinct, depressed; sutures on spiral side strongly elevated, forming a reticulate network with the apiral suture; aperture a low, interiomarginal opening, bordered by a distinct lip.
Remarks: the raised trochospiral coil, and sharply keeled margin, are characteristic of this subspecies. Range: Santonian to Middle Campanian.

Stensioina granulata granulata (Olbertz)
Plate 7.24, Figs. 4–6 (× 100), Upper Chalk, Helhoughton, Norfolk, Santonian. = *Rotalia exsculpta granulata* Olbertz, 1942. Description: test free, low trochospiral coil, plano-convex, periphery rounded; chambers distinct on ventral side only, 9–10 in final whorl; sutures indistinct on dorsal side, elevated as low, broad, septal ridges on the ventral side; aperture a narrow, interiomarginal slit, following the ventral edge of the final chamber; umbilicus shallow, covered by a calcareous plate.
Remarks: following the work of Trümper (1968) *S. prae-exsculpta* (Keller) has been included in the synonymy of the present subspecies. The first appearance of this subspecies in the British succession is extremely important (Bailey, 1978, unpublished thesis; Bailey and Hart, 1979). Range: Mid-Coniacian to basal Santonian.

Stensionina granulata polonica Witwicka, 1958
Plate 7.24, Figs. 7–9 (× 100), Upper Chalk, Oldstairs Bay, Kent, Santonian. Description: test free, low trochospiral coil, periphery narrow, sub-angular to rounded; chambers distinct on both sides, 9 in final whorl; sutures flush on dorsal side, slightly raised on ventral side, radial; aperture a narrow interiomarginal slit along the ventral side of the final cjamber; umbilicus extremely shallow.
Remarks: intermediate between the *S. granulata* (Olberz) and *S. exsculpta* (Reuss) lineages, as suggested by Trümper (1968). Range: Santonian.

Stensioina pommerana Brotzen, 1936
Plate 7.24, Figs. 10–12 (× 100), Upper Chalk, Alum Bay, Isle of Wight, Upper Campanian (Fig. 10), Caister St. Edmunds, Norfolk, Upper Campanian (Fig. 11), and Overstrand, Norfolk, Lower Maastrichtian (Fig. 12). Description: test free, plano-convex, trochospiral, periphery acute; 2½–3 whorls; umbilical side involute, domed, umbilical region covered by large, irregular chamber flap; chambers on umbilical side distinct, weakly inflated, sutures depressed; all sutures on spiral side elevated, producing a highly irregular, reticulate network; aperture a low interiomarginal arch bordered by a thick lip.
Remarks: easily distinguished by its domed shape, angled periphery, and large well-developed umbilical chamber flap. Range: rarely found in the Lower Campanian; Middle Campanian to Maastrichtian.

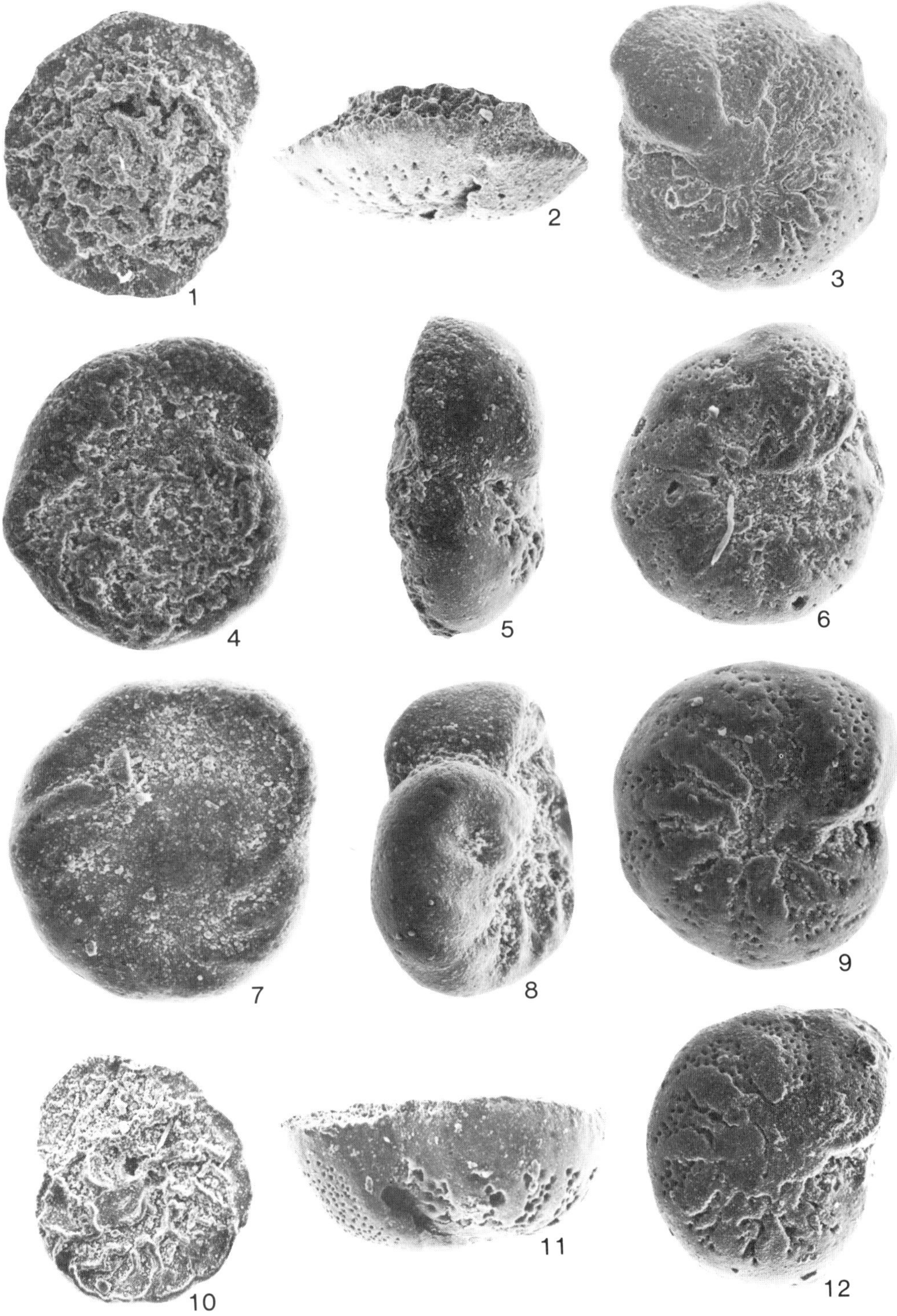

Plate 7.24

Vaginulina arguta Reuss, 1860
Plate 7.25, Figs. 1, 2 (× 15), Speeton Clay, Speeton, Yorkshire, Hauterivian C2A. Description: test free, robust, compressed, triangular in outline; dorsal margin nearly straight; ventral margin may be slightly lobulate; spherical proloculus followed by up to 14 chambers; sutures oblique, limbate, raised; aperture simple, terminal, at dorsal angle, on a small neck.
Remarks: this species has been the subject of much confusion in the past, particularly with relation to *V. kochi* Roemer. Albers (1952) has shown that only smooth forms belong in *V. kochi* while those with raised sutures belong in the present species. Range: Hauterivian to Lower Barremian.

Vaginulina mediocarinata Ten Dam, 1950
Plate 7.25, Fig. 3, (× 15), Bed IX, Gault Clay, Folkestone, Kent, Upper Albian. Description: test free, harp-shaped, rectangular in cross-section; fine but sharp discontinuous, vertical striations cover the chamber surfaces; sutures depressed; rounded proloculus with or without costae; edges of test strongly costate; aperture terminal, radiate.
Remarks: this species differs from other ornamented Albian vaginulinids in its discontinuous surface striations that run the entire length of the test. Range: Middle and Upper Albian.

Vaginulina riedeli Bartenstein and Brand, 1951
Plate 7.25, Figs. 4, 5 (× 20), Speeton Clay, Speeton, Yorkshire, Hauterivian (C8). Description: test free, elongate, robust, rectangular in cross-section; spherical proloculus followed by 4–7 slightly inflated chambers; peripheral angles of test acute, keeled; sutures oblique, slightly depressed; aperture radiate, at dorsal margin, on a small neck.
Remarks: closely related to *V. weigelti* Bettenstaedt, which may have evolved from *V. riedeli.* Range: Hauterivian.

Vaginulinopsis humilis praecursoria Bartenstein and Brand, 1951.
Plate 7.25, Fig. 6 (× 50), Speeton Clay, Speeton, Yorkshire, Valanginian (D3A). Description: test free, elongate, arcuate, oval in cross-section; initial coil followed by 4–6 chambers in a gently curving series; sutures in uncoiled part distinct, curved, raised; aperture radiate, at peripheral angle.
Remarks: this subspecies differs from *Lenticulina humilis* Reuss in having less prominent sutures, a less compressed test, and a rounder, not so tapering, proximal portion. Range: Valanginian and Hauterivian.

Vaginulinopsis scalariformis Porthault, 1970
Plate 7.25, Fig. 7 (× 35), Upper Chalk, Quidhampton, Wiltshire, Santonian. Description: test free, large, compressed ovoid in cross-section; initial portion a tight planispiral coil, rapidly becoming uniserial; chambers distinct, low; sutures distinct, sub-horizontal; test marked by very characteristic transverse septal ridges; aperture radiate, terminally positioned on the peripheral dorsal angle of the final chamber.
Remarks: recorded only from S.E. France and S. E. England; this highly distinctive species appears to have some stratigraphic value. Range: lowermost Santonian.

Valvulineria lenticula (Reuss)
Plate 7.25, Figs. 8, 9 (× 120), Upper Chalk, Arminghall, Norfolk (Fig. 8) and Catton Grove, Norfolk, (Fig. 9), Upper Campanian. = *Rotalia lenticula* Reuss, 1845. Description: test free, trochospiral, biconvex, periphery rounded; chambers indistinct except in last whorl; sutures poorly visible, except in later stages, radial, curving slightly at the margins; aperture a narrow, slit-like opening along the inner margin of the final chamber; umbilical area covered by a distinct apertural flap.
Remarks: Harris and McNulty (1957) have described this species (and its variants) in great detail. Range: Turonian to Lower Maastrichtian.

Whiteinella aprica (Loeblich and Tappan)
Plate 7.25, Figs. 10–12 (× 100), Middle Chalk, Beer, Devon, Turonian. = *Ticinella aprica* Loeblich and Tappan, 1961. Description: test free, trochospiral, biconvex to concavo-convex; 2–2½ whorls, 5–7 chambers in final whorl; chambers globular, pustulose, sutures radial, depressed; periphery lobate, with no keel; primary aperture tending to umbilical, bordered by a porticus.
Remarks: this species can be identified on the basis of its low spire. Range: Turonian.

Whiteinella baltica Douglas and Rankin, 1969
Plate 7.25, Figs. 13–15 (× 125), Upper Chalk, Quidhampton, Wiltshire, Santonian. Description: test free, trochospiral, biconvex to concavo-convex; 2–2½ whorls, 4–5 chambers in the final whorl; chambers globular, sutures depressed, radial to slightly curved; aperture umbilical, with evidence of a distinct flap in well-preserved specimens.
Remarks: this distinctive species, first described from Bornholm (Denmark), is very common in the British succession. Range: Coniacian and Santonian.

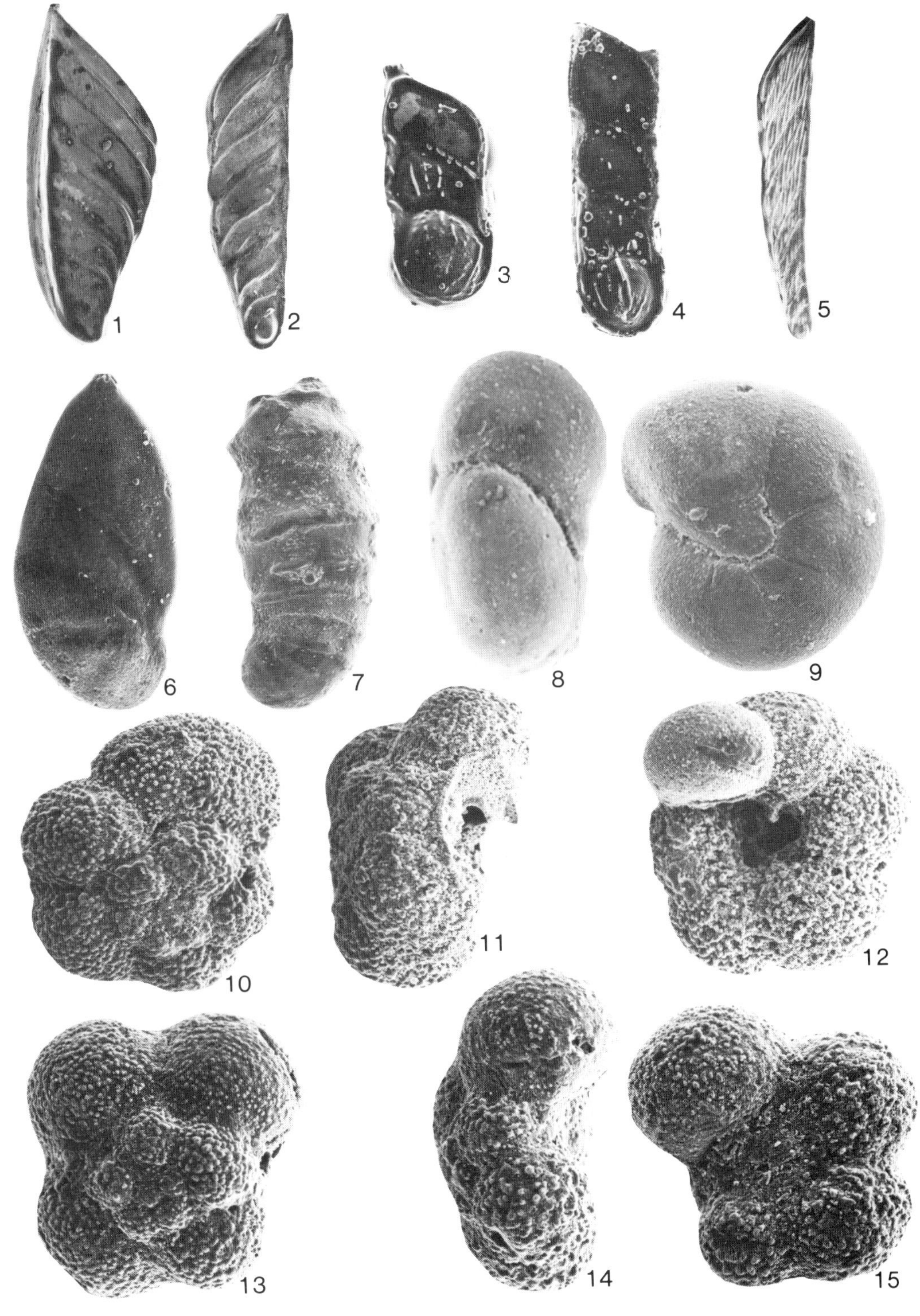

Plate 7.25

7.8 REFERENCES

Adams, C. G., Knight, R. H. and Hodgkinson, R. L. 1973. An unusual agglutinating forminifer from the Upper Cretaceous of England. *Palaeontology,* **16,** 637–544.

Albers, J. 1952. Taxonomie und Entwichlung einiger Arten von *Vaginulina* d'Orb, aus dem Barreme bei Hannover. *Mitt. geol. Staatsinst., Hamb.,* **21,** 75–112.

Ascoli, P. 1976. Foraminiferal and ostracod biostratigraphy of the Mesozoic–Cenozoic, Scotian Shelf, Atlantic Canada. *In:* 1st International Symposium on Benthonic Foraminifera of Continental Margins, Schafer, C. T. and Pelletier, B. R. (Eds.), *Maritime Sediments Spec. Publ. No.* **1,** 653–771.

Aubert, J. and Bartenstein, H. 1976. *Lenticulina (L.) nodosa;* additional observations in the worldwide Lower Cretaceous. *Bull. Centre Rech. Paul,* **10,** 1–33.

Bailey, H. W. 1975. A preliminary microfaunal investigation of the Lower Senonian at Beer, south-east Devon. *Proc. Ussher Soc.,* **3,** 280–285.

Bailey, H. W. and Hart, M. B. 1979. The correlation of the Early Senonian in Western Europe using Foraminiferida. *Aspekte der Kreide Europas,* IUGS, Series A, No. 6, 159–169.

Bandy, O. L. 1967, Cretaceous planktonic foraminiferal zonation. *Micropaleontology,* **13,** 1–31.

Barnard, T. 1958. Some Mesozoic adherent foraminifera from the Upper Cretaceous of England. *Palaeontology,* **1,** 116–124.

Barnard, T. 1963. Polymorphinidae from the Upper Cretaceous of England. *Palaeontology,* **5** (for 1962), 712–726.

Barnard, T. 1963. The morphology and development of species of *Marssonella,* and *Pseudotextulariella* from the chalk of England. *Palaeontology,* **6,** 41–45.

Barnard, T. and Banner, F. T. 1953. Arenaceous Foraminifera from the Upper Cretaceous of England. *Q. Jl. geol. Soc. Lond.,* **109,** 173–216.

Barr, F. T. 1962. Upper Cretaceous planktonic foraminifera from the Isle of Wight, England. *Palaeontology,* **4,** 552–580.

Barr, F. T. 1966a, The foraminiferal genus *Bolivinoides* from the Upper Cretaceous of the British Isles. *Palaeontology,* **9,** 220–243.

Barr, F. T. 1966b. Upper Cretaceous foraminifera from the Ballydeenlea Chalk, Co. Kerry, Ireland. *Palaeontology,* **9,** 492–510.

Bartenstein, H. 1952, Taxonomische Revision und Nomenklator ze Franz E. Hecht, Standard – Gliederung der Nordwestdeutschen Unterkreide nach Foraminiferen (1938). Teil 1. Hauterive – *Senckenbergiana leth.,* **33,** 173; Teil 2. Barreme – *ibid.,* 297.

Bartenstein, H. 1974. *Lenticulina (Lenticulina) nodosa* (Reuss, 1863) and its subspecies – world wide index foraminifera in the Lower Cretaceous. *Eclog. geol. Helv.,* **67,** 539–562.

Bartenstein, H. 1976a. Benthonic index foraminifera in the Lower Creatceous of the northern hemisphere between East Canada and North West Germany. *Erdol Kohle,* **29,** 254–256.

Bartenstein, H. 1976b, Practical applicability of a zonation with benthonic Foraminifera in the worldwide Lower Cretaceous. *Geol. Mijnb.,* **55,** 83–86.

Bartenstein, H. 1977. Stratigraphic parallelism of the Lower Cretaceous in the northern hemisphere. *Newsl. Stratigr.,* **6,** 30–41.

Bartenstein, H. 1978a. Parallelisation of the Lower Cretaceous stages in North West Germany with index ammonites and index microfossils. *Erdol Kohle,* **31,** 65–67.

Bartenstein, H. 1978b, Phylogenetic sequences of Lower Cretaceous benthic Foraminifera and their use in biostratigraphy. *Geol. Mijnb.,* **57,** 19–24.

Bartenstein, H. 1979. Worldwide zonation of the Lower Cretaceous using benthonic foraminifera. *Newsl. Stratigr.,* **7,** 142–154.

Bartenstein, H. and Bettenstaedt, F. 1962. Marine Unterkreide (Boreal und Tethys). *In* Simon, W. and Bartenstein, H. (Eds.) *Leitfossilien der Mikropaläontologie:* 225–297, pls. 33–41, Berlin.

Bartenstein, H. and Bettenstaedt, F. and Bolli, H. M. 1957, Die Foraminiferen der Unterkreide von Trinidad, B.W.I. Erster Teil: Cuche und Toco Formation. *Ecolg. geol. Helv.,* **50,** 5–67.

Bartenstein, H. and Brand, E. 1951. Mikropaläontologische Untersuchungen zur Stratigraphie des nordwest-deutschen Valendis. *Abh. senckenb. naturforsch. Ges.,* **485,** 239–336.

Bartenstein, H. and Kaever, M. 1973. Die Unterkreide von Helgoland und ihre mikropaläontologische Gliederung. *Senckenberg. leth.,* **54,** 207–264.

Beissel, L. 1891, Die Foraminifera der Aachener Kreide. *Abh. preuss. geol. Landesanst.,* part 3; 78 pp. 16 pls., Berlin.

Bellier, J.-P. 1971. Les Foraminifères Planctoniques du Turonien Type. *Rev. Micropalèont.,* **14,** 85–90.

Berthelin, G. 1880, Mémoire sur les Foraminifères de l'Etage Albien de Montcley (Doubs.). *Mém. Soc. géol. Fr.* (3), **1,** 1–84.

Bettenstaedt, F. 1952. Stratigraphische wichtige Foraminiferen Atren aus dem Barreme vorwiegend Nordwest–Deutschlands. *Senckenberg. leth.,* **33,** 263.

Breistroffer, M. 1947. Sur les Zones d'Ammonites dans ll'Albien de France et d'Angleterre. *Trav. Lab. géol. Univ. Grenoble,* **26,** 17–104.

Brotzen, F. 1934a, Vorläufiger Bericht über eine Foraminiferenfauna aud der schwedischen Schreibkreide. *Geol. För. Stockh. Förh.,* **56,** 77–80.

Brotzen, F. 1934b. Foraminiferen aus dem Senon Palästinas. *Z. dt. Ver. Palästinas,* (Leipzig), **57,** 28–72.

Brotzen, F. 1936. Foraminiferen aus dem Schwedischen untersten Senon von Eriksdal in Schonen. *Sver. geol. Unders. Afh.,* Ser. C, Årsbok 30, 1–206.

Brotzen, F. 1942. Die Foraminiferengattung *Gavelinella* nov. gen. unde die Systematik der Rotaliformes. *Sver. geol. Unders. Afh.,* Ser. C, **36,** (8), 1–60.

Brotzen, F. 1945. De Geologiska Resultaten fran Borrningarna vid Höllviken–Preliminär rapport Del 1: Kritan. *Sver. geol. Unders. Afh.,* Ser. C, **38** (7), 1–64.

Brotzen, F. 1948. The Swedish Paleocene and its foraminiferal fauna. *Sver. geol. Unders, Afh.,* Ser C,

42, (2), 1-140.
Burnaby, T. P. 1962, The Palaeoecology of the foraminifera of the Chalk Marl., *Palaeontology*, **4**, 599–608.
Burrows, H. W., Sherborn, C. D. and Bailey, G. 1890. The Foraminifera of the Red Chalk of Yorkshire, Norfolk and Lincolnshire. *Jl. R. microsc. Soc.*, 8, 549-566.
Busnardo, R. 1965, Le stratotype du Barrémien. 1 Lithologie et Mactofaune. *In:* Colloque sur le Crétacé inferieur. *Mém. Bur. Rech. géol. minier.*, **34**, 101-116.
Butt, A. A. 1966. Foraminifera of the Type Turonian. *Micropaleontology*, **12**, 162-182.
Caron, M. 1966. Globotruncanidae du Crétacé supérieur du synclinical de la Gruyére (Préalpes Medianes, Suisse). *Revue Micropaléont.*, **9**, 68-93.
Carter, D. J. and Hart, M. B. 1977, Aspects of mid-Cretaceous stratigraphical micropalaeontology. *Bull. Br. Mus. nat. Hist (Geol.)*, **29**, (1): 1-135.
Casey, R. 1961. The stratigraphical palaeontology of the Lower Greensand. *Palaeontology*, **3**, 487–621.
Casey, R. 1964. *In:* Dodson, M. H., Rex, D. C., Casey, R. and Allen P. Clauconite dates from the Upper Jurassic and Lower Cretaceous. *In:* The Phanerozoic Time-Scale. *Q. Jl. geol. Soc., Lond.*, **120S**, 145-158.
Casey, R. 1973. The ammonite succession at the Jurassic–Cretaceous boundary in eastern England. *In:* The Boreal Lower Cretaceous, Casey, R. and Rawson, P. F. (Eds.), *Geol. Jl. Spec. Iss. No.* 5, 193-266.
Chapman, F. 1891-1898. The Foraminifera of the Gault of Folkestone. *Jl. R. microsc. Soc.*, (Part 1) 1891, 565-575; (Part 2) 1892, 319-330; (Part 3) 1892, 749-758; (Part 4) 1893, 579-595; (Part 5) 1893, 153-163; (Part 6) 1894, 421–427; (Part 7) 1894, 645–654; (Part 8) 1895, 1-14; (Part 9) 1896, 581-591; (Part 10) 1898, 1–49. The whole work has been reprinted by Antiquariaat Jumk (1970).
Cooper, M. R. 1977. Eustacy during the Cretaceous; Its implications and importance. *Palaeogeogr. Palaeoclimat. Palaeoecol.*, **22**, 1–60.
Coquand, H. 1857. Position des *Ostrea columba* et *viauriculata* dans le groupe de la craie inférieur. *Bull. Soc. géol. Fr.*, (2) **14**, 745–66.
Dam, A. Ten, 1947. On foraminifera of the Netherlands No. 9. Sur quelques espèces nouvelles ou peu connues dans le Crétacé Inférieur (Albien) des Pays-Bas *Geol. Mijnb.*, 8, 25-29.
Dam, A. Ten, 1948a, Foraminifera from the Middle Neocomian of the Netherlands. *J. Paleont.*, **22**, 175-192.
Dam, A. Ten, 1948b. Les espèces du genre *Epistomina* Terquem, 1883. *Rev. Inst. fr. Pétrole*, **3**, 161-170.
Dam, A. Ten, 1950. Les Forminifères de l'Albien des Pays-Bas. *Mém. soc. géol. Fr.*, n.s. **29**, (63), 1–66.
Deroo, G. 1966. Cytheracea (Ostracodes) du Maastrichtien de Maastricht (Pays-Bas) et des règions voisines: résultats stratigraphiques et paléontologiques de leur étude. *Meded. geol. Sticht.*, C2(2), 197 pp.
Dilley, F. C. 1969. The foraminiferal fauna of the Melton Carstone. *Proc. Yorks. geol. Soc.*, **37**, 321-22.
Douglas, R. G. 1969a. Upper Cretaceous planktonic foraminifera in northern California. 1 – Systematics. *Micropaleontology*, **15**, 151-209.
Douglas, R. G. 1969b. Upper Cretaceous biostratigraphy of northern California. *Proc. First Int. Conf. planktonic Microfossils, Geneva, 1967*, **2**, 126-152.
Dumont, A. 1850. Rapport sur la carte gëologique du royaume. *Bull Acad. r. Belg. Cl. Sci.*, **16**, 351-73.
Eicher, D. L. 1965. Foraminifera and biostratigraphy of the Graneros Shale. *J. Paleont.*, **39**, 875-909.
Eicher, D. L. 1966, Foraminifera from the Cretaceous Carlile Shale of Colorado. *Contr. Cushman Fdn. foramin. Res.*, **17**, 16-31.
Eicher, D. L. 1967. Foraminifera from Belle Fourche Shale and equivalents Wyoming and Montana. *J. Paleont.*, **41**, 167-188.
Eicher, D. L. 1969. Cenomanian and Turonian planktonic foraminifera from the western interior of the United States. *Proc. First Int. Conf. planktonic Microfossils, Geneva, 1967*, **2**, 163-174.
Eicher, D. L. and Worstell, P. 1970. Cenomanian and Turonian foraminifera from the Great Plains, United States. *Micropaleontology*, **16**, 269-324.
Fletcher, B. N. 1973. The Distribution of Lower Cretaceous (Berriasian–Barremian) Foraminifera in the Speeton Clay. *In:* The Boreal Lower Cretaceous, Casey, R. and Rawson, P. F. (Eds.), *Geol. J. Spec. Iss. No. 5*, 161-168.
Frizzell, D. L. 1954. Handbook of Cretaceous Foraminifera of Texas. *Rep. Bur. econ. Geol. Univ. Tex.*, **22**, 1-232.
Fuchs, W. von, 1967. Die Foraminiferenfauna eines Kernes des höheren Mittel-Alb der Tiefbohrung DELFT 2 – Niederlands. *Jb. geol. Bundesanst. Wien*, **110**, 255--341.
Bawor-Biedowa, E. 1969. The genus *Arenobulimina* Cushman from the Upper Albian and Cenomanian of the Polish Lowlands. *Roczn. pol. Tow. geol.*, **39**, 73-104.
Gawor-Biedowa, E. 1972. The Albian, Cenomanian and Turonian Foraminifers of Poland and their Stratigraphic Importance. *Acta Palaeont. pol.*, **17**, 3-151.
Goel, R. K. 1965. Contributions à l'ètude de Foraminiféres du Crétacé supérieur de la Baisse-Seine. *Bull. Bur. Rech. géol. min.*, **5**, 49-157.
Hancock, J. M. 1976. The Petrology of the Chalk. *Proc. Geol. Ass.*, **86**, 499-536.
Hancock, J. M. and Kauffman, E. G. 1979. The great transgressions of the Late Cretaceous. *Jl. geol. Soc. Lond.*, **136**, 175-186.
Harris, R. W. and McNulty, C. L. Jnr., 1975. Notes concerning a Senonian Valvulinerian. *J. Paleont.*, **3**, 865-868.
Hart, M. B. 1973. A correlation of the macrofaunal and microfaunal zonations of the Gault Clay in South-east England. *In:* The Boreal Lower Cretaceous, Casey, R. and Rawson, P. F. (Eds.), *Geol. J. Spec. Iss. No. 5*, 267-288.
Hart, M. B. 1975. Microfaunal analysis of the Membury Chalk succession. *Proc. Ussher Soc.*, **3**, 271-279.
Hart, M. B. 1980a. A water depth model for the evolution of the planktonic Foraminiferida. *Nature*, **286**, 252-254.

Hart, M. B. 1980b. The recognition of mid-Cretaceous sea level changes by means of Foraminifera. *Cretaceous Res.*, **1**, 289–297.

Hart, M. B. and Bailey, H. W. 1979. The distribution of planktonic Foraminiferida in the mid-Cretaceous of N.W. Europe. *Aspekte der Kreide Europas*, IUGS., Ser. A, No. 6, 527–542.

Hart, M. B. and Bigg, P. J. (1981–in press), Anoxic events in the Late Cretaceous Chalk Seas of North West Europe. *In:* The Micropalaeontology of Shelf Seas, Neale, J. W. and Brasier, M. (Eds.).

Hart, M. B., Manley, E. C. and Weaver, P. P. E. 1979. A biometric analysis of an *Orbitolina* fauna from the Cretaceous succession at Wolborough, S. Devon. *Proc. Ussher. Soc.*, **4**, 317–326.

Hart, M. B. and Weaver, P. P. E. 1977. Turonian microbiostratigraphy of Beer, S.E. Devon. *Proc. Ussher Soc.*, **4**, 86–93.

Hinte, J. E. van, 1976. A Cretaceous time scale. *Bull. Am. Ass. Petrol. Geol.*, **60**, 498–516.

Jendryka-Fuglewicz, B. 1975. Evolution of the Jurassic and Cretaceous smooth-walled *Lenticulina* (Foraminiferida) of Poland. *Acta palaeont. pol.*, **20**, 99–197.

Jefferies, R. P. S. 1962, The palaeoecology of the *Actinocamax plenus* Subzone (Lowest Turonian) in the Anglo-Paris Basin. *Palaeontology*, **4**, 609–647.

Jefferies, R. P. S. 1963. The stratigraphy of the *Actinocamax plenus* Subzone (Turonian) in the Anglo–Paris Basin. *Proc. Geol. Ass.*, **74**, 1–33.

Kauffman, E. G. 1970. Population sytematics, radiometrics and zonation – a new biostratigraphy. *Proc. N. Am. Paleont. Conv.*, **1**, 612–666.

Kemper, E. 1973a. The Valanginian and Hauterivian stages in northwest Germany. *In:* Boreal Lower Cretaceous, Casey, R. and Rawson, P. F. (Eds.), *Geol. J. Spec. Iss. No. 5*, 327–344.

Kemper, E. 1973b. The Aptian and Albian stages in northwesr Germany. *In:* The Boreal Lower Cretaceous, Casey, R. and Rawson, P. F. (Eds.), *Geol. J. Spec. Iss. No. 5*, 345–360.

Kennedy, W. J. 1970. A correlation of the Uppermost Albian and the Cenomanian of south-west England. *Proc. Geol. Ass.*, **81**, 613–677.

Klaus, J. 1960a. Le 'Complexe schisteux intermédiare' dans le synclinal de la Gruyère (Préalpes médianes). Stratigraphie at micropalèontologie, avec l'étude spéciale des Globotruncanides de l'Albian, du Cenomanien, et du Turonien. *Eclog. geol. Helv.*, **52**, 753–851.

Klaus, J. 1960b. Etude biométrique at statisque de espèces du genre *Praeglobotruncana* dans le Cénomanien de la Breggia. *Eclog. geol. Hel.*, **53**, 285–308.

Klaus, J. 1960c. Rotalipores et Thalmanninelles d'un niceau des Couches Rouges de l'Anticlinal d'Ai. *Eclog. geol. Helv.*, **53**, 704–709.

Koch, W. 1977. Biostratigraphie in der Oberkreide und Taxonomie von Foraminiferen. *Geol. Jb.*, Reihe **A 38**, 128 pp.

Lecointre, G. 1959. Le Turonien dans sa région type, la Touraine. *C.r. Congr. Socs. Sav. Paris Sect. Sci.*, Colloque sur la Crétacé Supérieur Francais (Dijon, 1959), 415–423.

Loeblich, A. R. and Tappan, H. 1961. Cretaceous planktonic foraminifera. 1 – Cenomanian. *Micropalaeontology*, 7, 257–304.

Loeblich, A. R. and Tappan, H. 1964. Protista 2, Sarcodina, chiefly 'Thecamoebians' and Foraminiferida. *In:* Moore, R. C. (ed.), *Treatise on Invertebrate Palaeontology*, Part **C.1**, 1–510.

Loeblich, A. R. and Tappan, H. 1974. Recent advances in the classification of Foraminiferida. *In:* Hedley, R. H. and Adams, C. G. (Eds.), *Foraminifera*, **1**, 1–53.

McGugan, A. 1957. Upper Cretaceous Foraminifera from Northern Ireland. *J. Paleont.*, **31**, 329–348.

Magniez–Jannin, F. 1975, Les Foraminifères de l'Albien de l'Aube: paléontologie, stratigraphie, écologie. *Cah. Palèontol., CNRS*, 351 pp.

Malapris, M. 1965. Les Gavelinellidae et formes affines du gisement albien de Courcelles (Aube). *Revue Micropalèont.*, **8**, 131–150.

Malapris–Bizouard, M. 1974. Les premières Gavelinelles du Crétacé inférieur. *Bull. Inf. Gèol. Bass. Paris.* **40**, 9–23.

Marie, P. 1938. Sur quelques foraminifères nouveaux ou peu connus du Crétacé du Bassin de Paris. *Bull. Soc. gèol. Fr.*, (5), **8**, 91–104.

Marie, P. 1941, Les foraminifères de la Craie a *Belemnitella mucronata* du Bassin de Paris. *Mèm. Mus. natn. Hist. nat. Paris*, n. ser., **12**, 1–296.

Marks, P. 1967. *Rotalipora* et *Globotruncana* dans la Craie de Théligny (Cénomanien: Dépt. de la Sarthe). *Proc. K. ned. Akad. Wet.*, **B70**, 264–275.

Masters, B. A. 1977. Mesozoic Planktonic Foraminifera; a worldwide review and analysis. *In: Oceanic Micropalaeontology*, Ramsay, A. T. S. (Ed.), **1**, 301–731.

Michael, E. Die Evolution der Gavelinelliden (Foram) in der N.W.-deutschen Unterkreide. *Senckenberg. leth.* **47**, 411–459.

Neagu, T. 1965. Albian Foraminifera of the Rumanian Plain. *Micropaleonotology*, **11**, 1–38.

Neagu, T. 1969, Cenomanian Planktonic Foraminifera in the southern part of the Eastern Carpathians. *Roczn. pol. Tow. geol.*, **39**, 133–181.

Neale, J. W. 1974. Cretaceous. *In:* Rayner, D. H. and Hemingway, J. E. (eds.), *The Geology and Mineral Resources of Yorkshire. Yorks. geol. Soc.*, 225–243.

Obradovich, J. D. and Cobban, W. A. 1975. Late Cretaceous Time Scale. *In:* Caldwell, W. G. E. (Ed.) *The Cretaceous System in the Western Interior of North America, Geol. Assoc. Canada*, Spec. Pap., **13**, 40–52.

Olsson, R. K. 1964. Late Cretaceous Planktonic Foraminifera from New Jersey and Delaware. *Micropaleontology*, **10**, 157–188.

Orbigny, A. D. d'. 1842–1847. Pteropoda, Gasteropoda. *Palèontologie francaise, Terrains crètacès*, **2**, 456 pp., Atlas 236 pls., table, (1842). Brachipoda (includes Bivalvia). *loc. cit.* 4, 390 pp., Atlas pls. 490–599. (1847). Paris.

Pessagno, E. A. 1967. Upper Cretaceous planktonic foraminifera from the Western Gulf Coastal Plain. *Palaeontogr. am.*, **37**, 245–445.

Petters, S. W. 1977a. Upper Cretaceous planktonic Foraminifera from the subsurface of the Atlantic Coastal Plain of New Jersey. *J. foramin. Res.*, 7, 165–187.

Petters, S. W. 1977b. *Bolivinoides* evolution and Upper Cretaceous biostratigraphy of the Atlantic Coastal Plain of New Jersey. *J. Paleont.*, **51**, 1023–1036.

Plummer, H. J. 1931, Some Cretaceous Foraminifera in Texas. *Publs. Bur. econ. Geol. Univ. Tex.*, **3101**, 109–203.

Price, R. J. 1967. Palaeoenvironmental interpretations in the Albian of western and southern Europe, as shown by the distribution of selected foraminifera. *In: Proc. First Int. Symp. Benthonic Foraminifera of Continental Margins,* Schafer, C. T. and Pelletier, B. R. (Eds.) *Maritime Sediments, Spec. Publ. No. 1,* 625–648.

Price, R. J. 1977a. The stratigraphical zonation of the Albian sediments of north-west Europe, as based on foraminifera. *Proc. Geol. Ass.*, **88**, 65–91.

Price, R. J. 1977b. The evolutionary interpretation of the foraminiferida *Arenobulimina, Gavelinella,* and *Hedbergella* in the Albian of north-west Europe. *Palaeontology,* **20**, 503–527.

Rawson, P. F., Curry, D., Dilley, F. C., Hancock, J. M., Kennedy, W. J., Neale, J. W., Wood, C. J. and Worssam, B. C. 1978. A correlation of the Cretaceous rocks of the British Isles. Geol. Soc. Lond., Spec. Rpt., 9, 70 pp.

Renevier, E. 1974. Tableau des terrains sédimentaires. *Bull. Soc. vaud. Sci. mat.*, **13**, 218–252.

Rhys, G. H. 1975. A Proposed Standard Lithostratigraphic Nomenclature for the Southern North Sea. *In:* Woodland, A. W. (Ed.) *Petroleum and the Continental Shelf of N.W. Europe.* Vol. 1. Geology, 151–162, Academic Press, London.

Robaszynski, F., Amedro, F., Foucher, J. C., Gaspard, D., Magniez-Jannin, F., Manivit, H. and Sornay, J. 1980. Synthèse Biostratigraphique de l'Aptien au Santonian du Boulonnais a partir de sept groupes paléontologiques; Foraminifères, Nannoplancton, Dinoflagellés et Macrofaunas. *Rev. Micropaléont.*, **22**, 195–321.

Robaszynski, F. and Caron, M. 1979. Atlas of Mid-Cretaceous planktonic Foraminiferida (Boreal Sea and Tethys). *Cah. Micropaléont.*, Part 1, 1–185; Part 2, 1–181. (Published in both French and English).

Sazonov, N. T. 1951. On some little-known ammonites of the Lower Cretaceous. *Byull. mosk. Obshch. Ispyt. Prir.*, **56**, 57–63 (in Russian).

Schlanger, S. O. and Jenkyns, H. C. 1976. Cretaceous anoxic events: causes and consequences. *Geol. Mijnb.*, **55**, 179–184.

Schmid, F. 1959. La definition des limites Santonien–Campanien et Campanien inférieur–supérieur en France et dans le nord-ouest de l'Allemagne. *In: C.r. Congr. Socs. Sav. Paris, Sect. Sci.,* Colloque sur la Crétacé Supérieur Francais (Dijon, 1959), 535–546.

Schmid, F. 1967. Die Oberkreide–Stufen Campan und Maastricht in Limburg (Sudneiderlande, Nordostbelgien), bei Aachen und in Nordwest–deutschland. *Ber. dt. Ges. geol. Wiss.*, A. *Geol. Paläont.*, **12**, 471–478.

Séronie-Vivien, M. 1972. Contribution à l'etude de Sénonien en Aquitaine septentrionale, ses stratotypes: Coniacien, Santonien, Campanien. *Les Stratotypes Francais* **2**, CNRS, Paris, 195 pp.

Sliter, W. V. 1968, Upper Cretaceous Foraminifera from Southern California and North-West Baja California, Mexico. *Paleont. Contr. Univ. Kans.*, **49**, 1–141.

Stenestad, E. 1969. The genus *Heterohelix* Ehrenberg, 1843 (Foraminifera) from the Senonian of Denmark. *Proc. First Int. Conf. planktonic Microfossils, Geneva, 1967,* **2**, 614–661.

Terquem, O. 1883, Sur un nouveau genre de Foraminifères du Fuller's-earth de la Moselle. *Bull. Soc. géol. Fr.*, (3) **11**, 37–39.

Trümper, E. 1968. Variationsstatistische Untersuchungen an der Foraminiferen–Gattung *Stensioina* Brotzen. *Geologie, 17Bh.*, 59–103.

Vasilenko, V. P. 1961. (Upper Cretaceous foraminifers of the Mangyshlak Peninsula (descriptions, phylogenetic diagrams of some groups and stratigraphical analysis).). *Trudy vses. neft. nauchno–issled. geol.-razv. Inst.*, **171**, 1–487. (in Russian).

Voight, E. 1956. Zur Frage der Abgrenzung der Maastrict–Stufe. *Paläont. Z. Sonderh.*, **30**, 11–17.

Williams-Mitchell, E. 1948. The Zonal value of Foraminifera in the Chalk of England. *Proc. Geol. Ass.*, **59**, 91–112.

Ziegler, P. A. 1975. North Sea Basin History in the Tectonic Framework of North-Western Europe. *In:* Woodland, A. W. (Ed.), *Petroleum and the Continental Shelf of N.W. Europe,* Vol. 1. Geology; 131–148.

8
Palaeogene

J. W. Murray, D. Curry, J. R. Haynes, C. King

8.1 INTRODUCTION

Marine and marginal marine Palaeogene sediments on land are found only in southern England. They are preserved in two main areas, the Hampshire Basin centred on the Isle of Wight, and the London Basin comprising the lower valley and estuary of the River Thames. Most of the Palaeogene sediments are clastic sands and clays with occasional limestones. A major unconformity separates them from the underlying late Cretaceous chalk. The unconformity marks a period of uplift, slight deformation, and erosion of the chalk under subaerial and submarine conditions.

The onshore deposits represent a marginal development of a more complete succession in the North Sea and to a lesser extent in the English Channel (Smith and Curry, 1975). The oldest Palaeogene deposit, the Thanet Formation of late Palaeocene age, is found only in the area of the Thames estuary. It clearly represents a transgressive event from the adjacent North Sea Basin, and in contrast, the Reading Formation and the London Clay Formation were laid down throughout the area of the Hampshire and London Basins. Nevertheless, it is believed that inversion of the Mesozoic Weald Basin began to separate the two areas at some period in early Eocene times (Ziegler, 1975). The resultant shoal formed a partial to total barrier between the two basins for the remainder of the Palaeogene. This is reflected in the faunas of the Hampshire Basin which are mainly of northern affinity but from time to time show southern elements thought to have been introduced via the western English Channel. Depositional environments ranged from continental fluvial and lacustrine, through brackish marginal marine lagoons and deltas, to normal marine. The successions resulting from these varying environmental conditions, with faunas limited by environmental and geographic changes, are complex. Correlation within England is not always easy and that between England and mainland Europe is difficult.

The division of responsibilities between the contributors is as follows: planktonic and larger benthic species, D. Curry; Thanetian benthic species, J. R. Haynes; range chart for London Clay of London Basin, C. King; Eocene-Oligocene

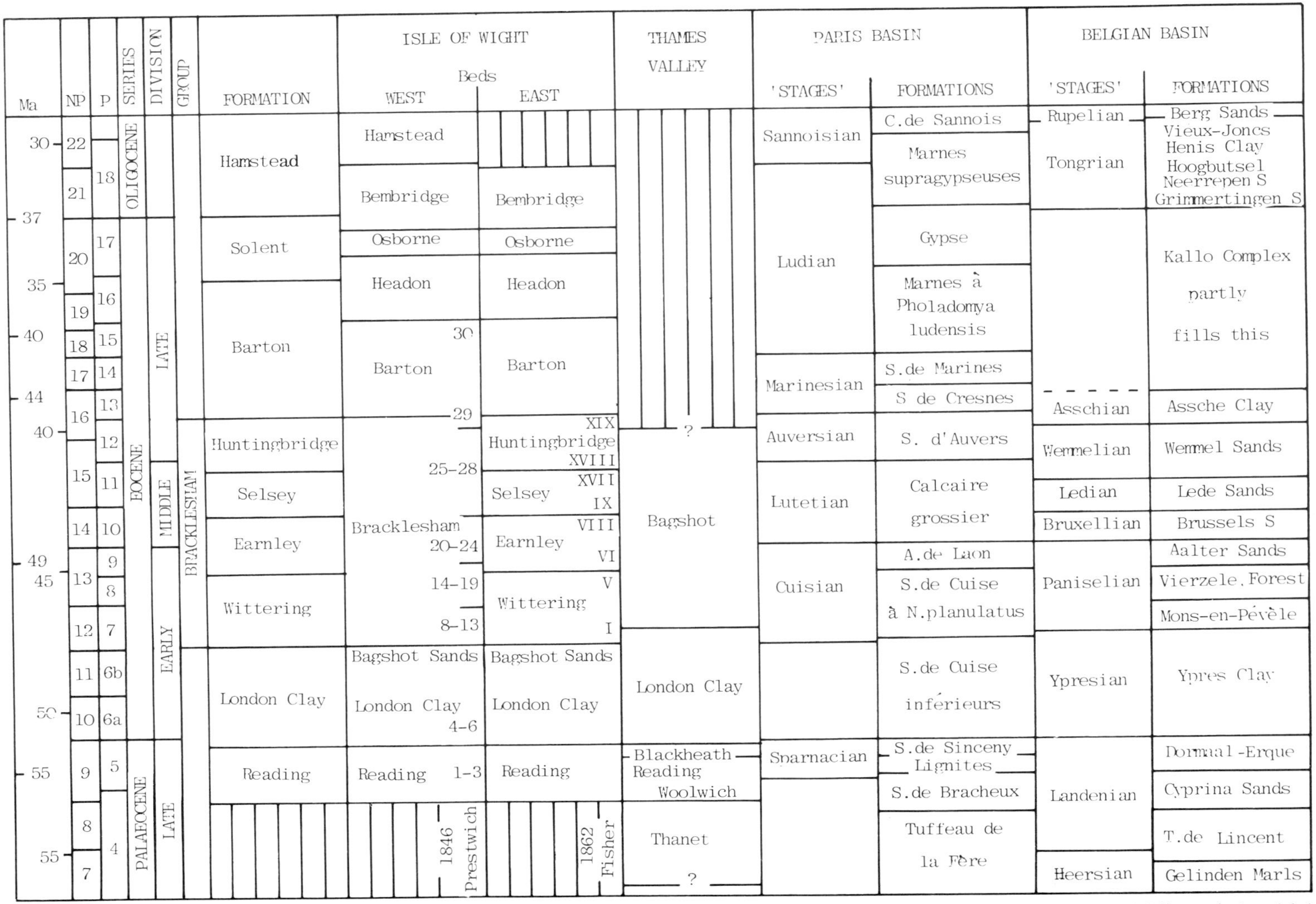

Fig. 8.1 – Stratigraphic summary (after Curry *et al.*, 1978). The radiometric age scales arethose of Hardenbol and Berggren (on left) and Odin *et al.*, (on right). NP = nannoplankton zone. P = planktonic foraminiferal zone. Vertical ruling indicates a gap in the sedimentary record.

benthic species and remainder of text, J. W. Murray.

8.2 STRATIGRAPHIC DIVISIONS

The history of stratigraphic studies on the Palaeogene of southern England extends back to the early part of the last century. The present state of knowledge has recently been reviewed (Curry *et al.*, 1978) and nomenclature in accordance with the international Stratigraphic Code has been introduced. Their recommendations are followed here (Fig. 8.1). In the opinion of the authors it is undesirable to use European stages because for the most part they are not soundly based; successions of varying facies and with innumerable stratigraphic gaps are ill-suited for the erection of satisfactory stages. However, the extent of the stages in the Paris Basin has been included on Fig. 8.1 in order that ranges of the foraminifera there may be compared with those in England.

Correlation between English successions is based on a great variety of biostratigraphic evidence (see Curry *et al.*, 1978). Correlation with the standard nannoplankton (NP) and planktonic foraminiferal (P) zones is based on relatively few observations so the presentation in Fig. 8.1 is only approximate. Because of the marginal marine environments of deposition neither of these groups is well represented in the successions and planktonic foraminifera, even when present, are rare except at some levels in the London Clay Formation.

It is beyond the scope of this chapter to discuss the stratigraphy in detail. Lithostratigraphic successions form the basis of the range charts (Figs. 8.4–8.9). Useful references to individual formations and localities are listed below:

London Basin:
- General, Field guides (Pitcher *et al.*, 1967; Blezard *et al.*, 1967).
- Thanet Formation (Burrows and Holland, 1897).
- London Clay Formation (Wrigley, 1924, 1940).

Hampshire Basin
- General, Field guides to the Isle of Wight (Curry *et al.*, 1966) and Southampton area (Curry and Wisden, 1958), White (1915, 1917, 1921).
- London Clay Formation, Whitecliff Bay, Alum Bay, (White, 1921).
- Lower Swanwick (Curry and King, 1965).
- Bracklesham Group, Whitecliff Bay, Alum Bay, (White, 1921), Selsey (Curry *et al.*, 1977).
- Barton Formation, Whitecliff Bay, Alum Bay (White, 1921), Barton (Burton, 1929, 1933).
- Solent Formation, Whitecliff Bay, (White, 1921; Stinton, 1971), Headon Hill, (White, 1921).
- Hamstead Formation, Whitecliff Bay, Alum Bay and Hamstead, (White, 1921).

8.3 COLLECTIONS OF PALAEOGENE FORAMINIFERA

British Museum (Natural History)

Adams:
- Bracklesham Beds, Whitecliff Bay and Selsey (Adams, 1962).

Barr and Berggren:
- Thanet Beds, Thanet (Barr and Berggren, 1964).

Bhatia:
- Headon, Bembridge and Hamstead Beds, Isle of Wight (Bhatia, 1955).

Bowen:
- London Clay, Alum and Whitecliff Bays; Barton Beds, Barton (Bowen, 1954).

Burrows and Holland:
- Thanet Beds (Burrows and Holland, 1897).
- Bracklesham Beds, Bracklesham (named but undescribed).

Burton:
- Barton Clay, Barton (named but undescribed).

Earland and Edwards:
- Bracklesham Beds, Bracklesham; Barton Beds, Barton (named but undescribed).

Haynes:
- type Thanetian (Haynes, 1956–1958).

Heron-Allen:
- shore sands from Selsey with reworked Bracklesham Beds material.

Murray and Wright:
- London Clay to Bembridge Marls, Whitecliff Bay; London Clay to Barton Beds, Alum Bay, London Clay, Swanwick; Bracklesham Beds, Bracklesham; Barton Beds, Barton; M. Headon Beds, Headon Hill and Colwell Bay; U. Hamstead Beds, Hamstead (Murray and Wright, 1974).

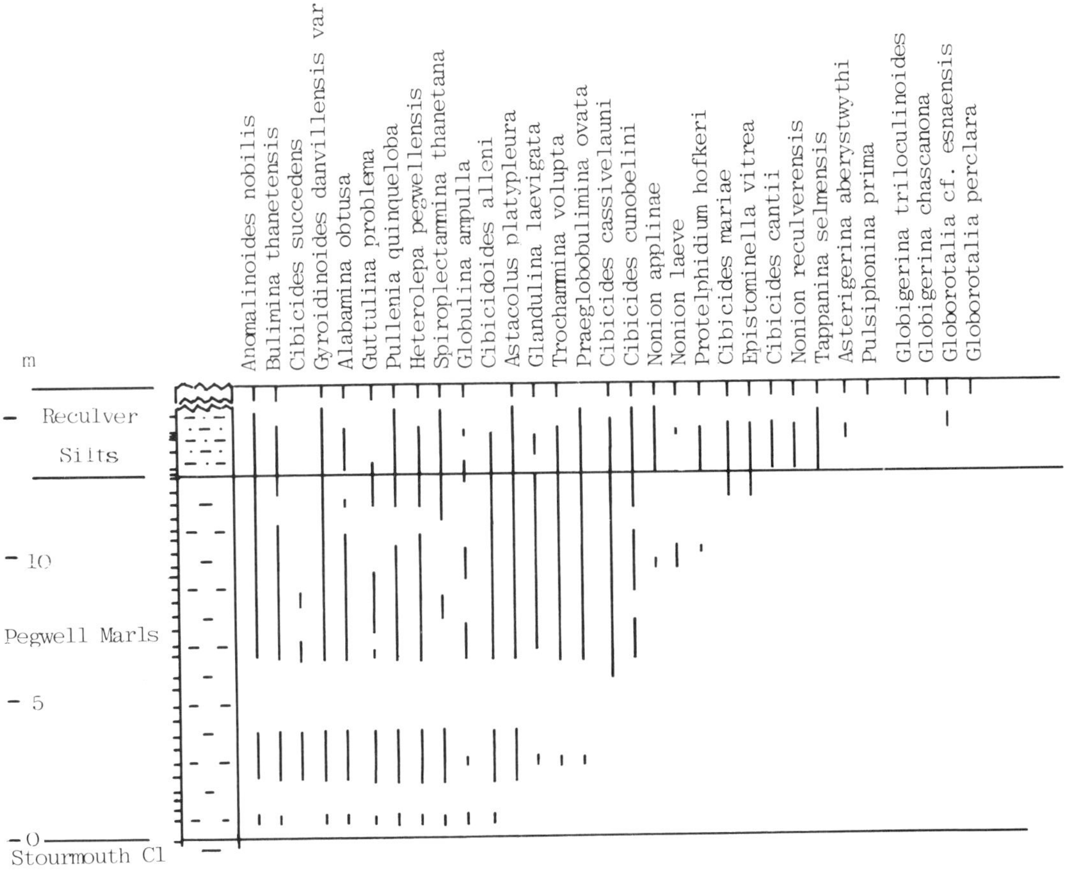

Fig. 8.4 – Range chart for the Thanet Formation (based on Haynes, 1956, 1958c). The main part of the succession is that exposed at Pegwell. At the top additional data on Reculver Silts occurrences at Reculver are given but this part of the succession is not to scale. The ticks to the left of the stratigraphic column show the position of the samples examined.

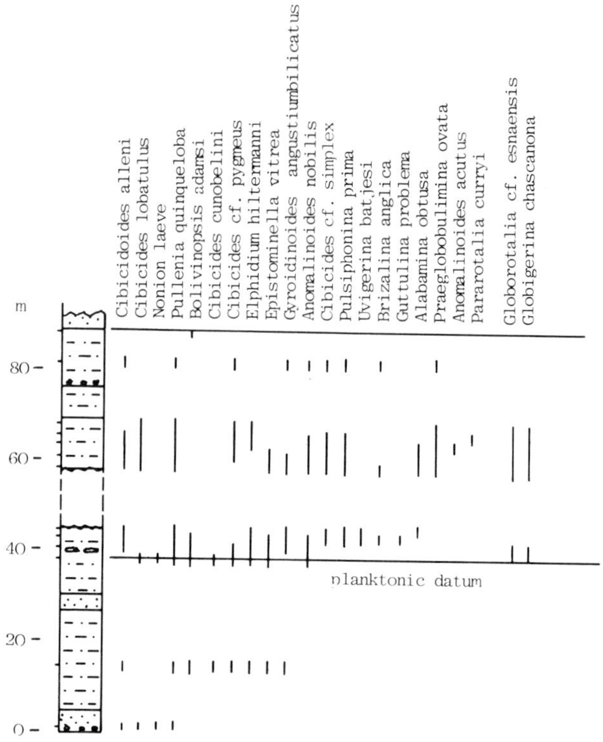

Fig. 8.5 – Range chart for the London Clay Formation, Whitecliff Bay, Isle of Wight.

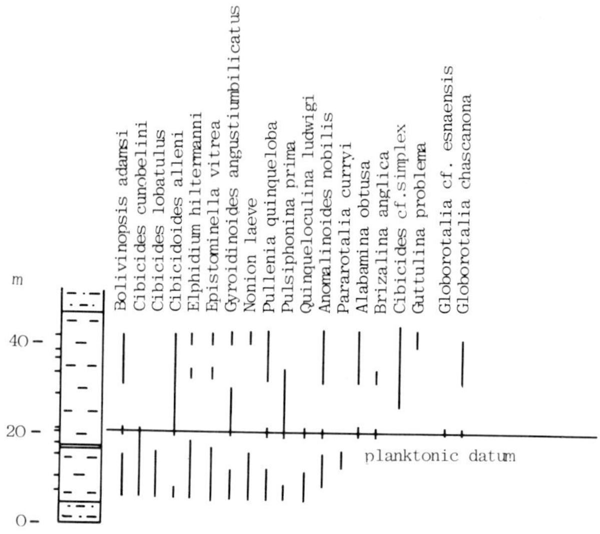

Fig. 8.6 – Range chart for the London Clay Formation, Isle of Wight.

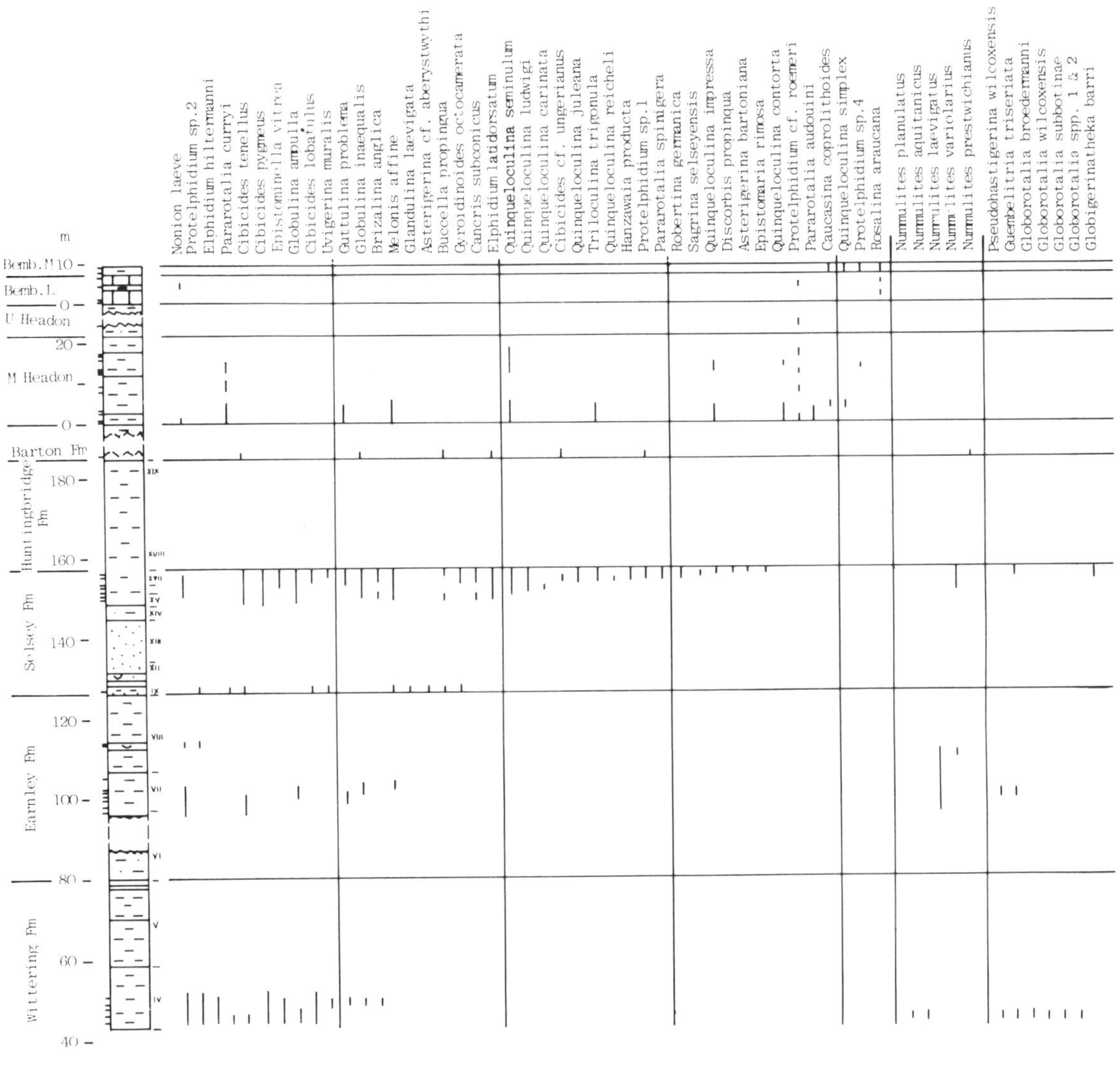

Fig. 8.7 – Range chart for the Bracklesham Group and Barton Formation, Whitecliff Bay, Isle of Wight (only the fossiliferous samples have been plotted).

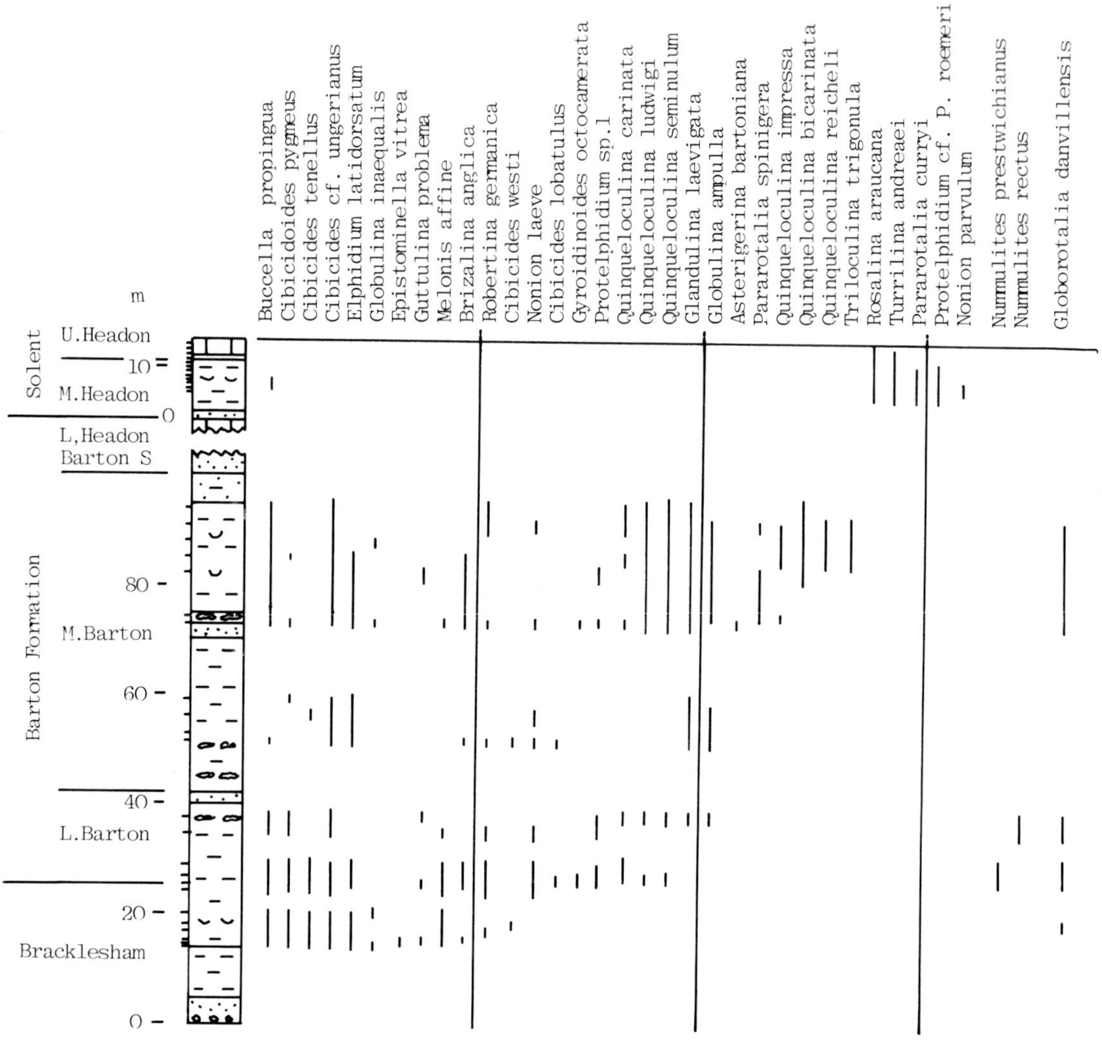

Fig. 8.8 – Range chart for the Bracklesham Group, Barton and Solent Formations, Alum Bay, Isle of Wight.

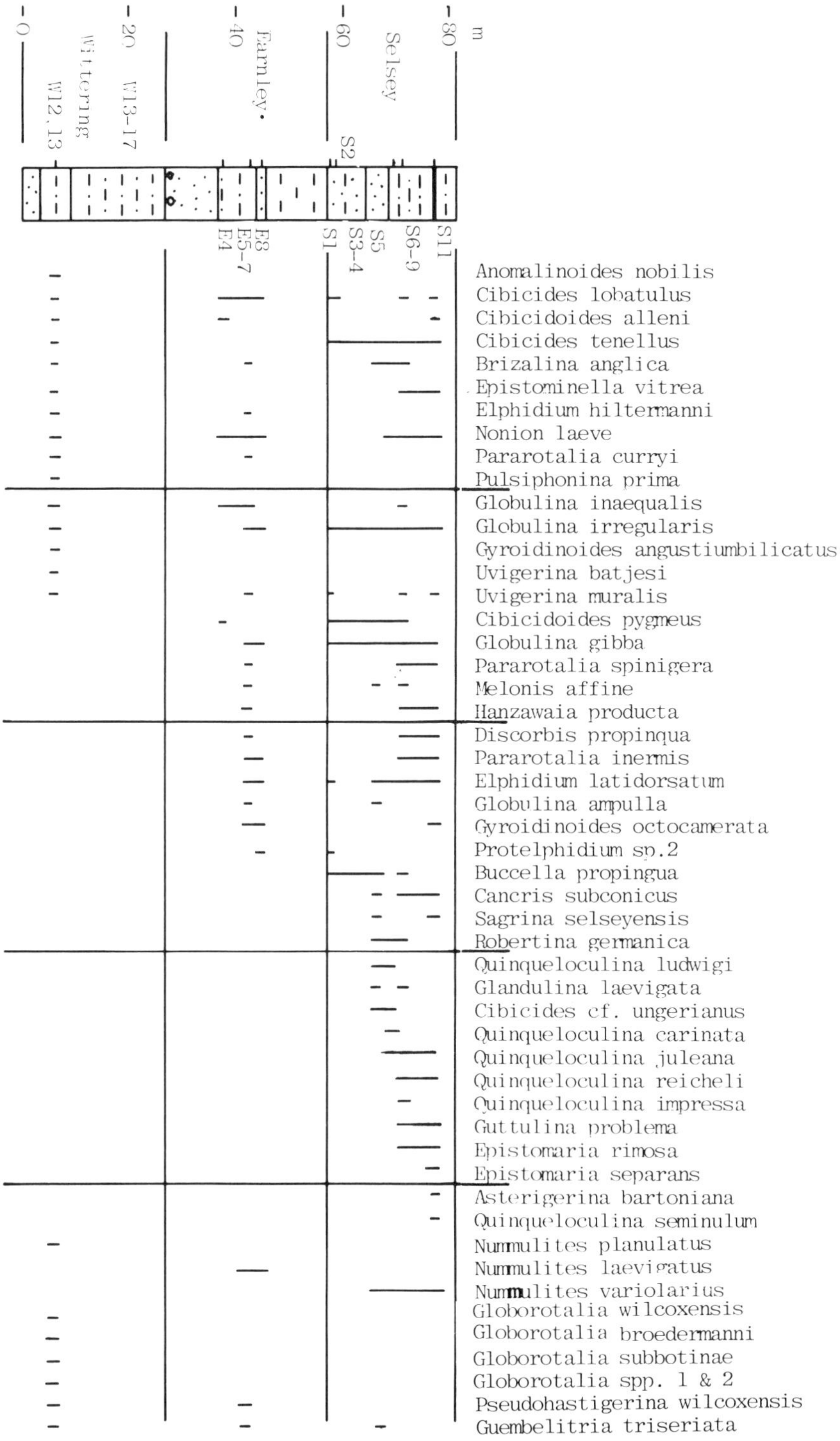

Fig. 8.9 – Range chart for the Bracklesham Group, Bracklesham–Selsey.

Sherborn and Chapman:
London Clay, Piccadilly, London (Sherborn and Chapman, 1886, 1889).
Venables:
London Clay, Bognor (Venables, 1962).

8.4 PUBLICATIONS ON FORAMINIFERA

The following list is not exhaustive; it includes only those papers of stratigraphical value.

Thanet Beds:
Burrows and Holland (1897), Haynes (1954, 1955, 1956, 1958a, b, c), Haynes and El-Naggar (1964), Wood and Haynes (1957), Barr and Berggren (1964, 1965), Berggren (1965), El-Naggar (1967), Brönnimann *et al.*, (1968).

London Clay:
Sherborn and Chapman (1886, 1889), Chapman and Sherborn (1889), Davis (1928), Kaasschieter (1961), Bignot (1962), Venables (1962), Brönnimann *et al.*, (1968), Wright (1972a, b), Murray and Wright (1974).
Bracklesham Beds:
Wrigley and Davis (1937), Curry (1937, 1962), Kaasschieter (1961), Blondeau and Curry (1963), Murray and Wright (1974).
Barton Beds:
Curry (1937), Bowen (1955), Kaasschieter (1961), Murray and Wright (1974).
M. Headon – U. Hamstead Beds:
Bhatia (1955, 1957), Brönnimann *et al.* (1968), Vella (1969), Murray and Wright (1974).
General papers dealing with correlation using Foraminifera:
Curry (1965, 1967), Curry *et al.* (1969, 1978),
Brönnimann *et al.* (1968).

8.5 PALAEOECOLOGY OF THE FORAMINIFERA

Palaeoecological interpretations are normally based on a comparison of the fossil assemblages with data for modern living assemblages. In doing this it is assumed that individual genera have not changed their ecological preferences through geological time. However, this assumption is not always valid and may on occasion be shown to be wrong because of inconsistency with other data. For instance, certain genera of the Nodosariacea, which in modern seas occupy normal marine shelf and slope habitats, in former times clearly also occupied shallow water including somewhat brackish environments (cf. Larsen and Jørgensen, 1977). Further problems arise through postmortem modification of assemblages (see Murray 1976b for a review).

There is now a vast quantity of information available on modern foraminifera and this has been summarised in Murray (1973) and Boltovskoy and Wright (1976). The following generalisations can be made concerning the distribution of shelf genera:

Genera occurring in water of normal salinity (∿ 35°/oo)
Asterigerina
Cancris
Discorbis
Globulina
Gyroidina
Melonis
Nonionella
Pullenia
Rosalina
Textularia
Uvigerina

Genera occurring in seawater with a salinity of 32 to 35°/oo (i.e. slight brackish tolerance). *indicates a somewhat greater brackish tolerance.
Bolivina
Brizalina
Buccella *
Bulimina
Cibicides
Globobulimina
Nonion *
Quinqueloculina

Genera common in brackish environments
Ammobaculites
Elphidium
Protelphidium

Palaeoecological interpretations of British Palaeogene foraminiferal assemblages have been made by Bhatia (1955, 1957), Haynes (1958c), Wright (1972a), and Murray and Wright (1974). From these sources it can be seen that almost all the Palaeogene deposits of southern England that have yielded foraminiferal assemblages have been deposited in shallow shelf seas (< 100 m deep) or marginal marine environments. The following environmental associations of dominant genera occur at different levels within the sequence:

Brackish marsh

Simple agglutinated genera; very low diversity. (Care must be taken not to confuse these with those shelf assemblages which have lost their calcareous component as a result of post-mortem solution).

Brackish lagoon

Protelphidium
Pararotalia (only *P. curryi*)
Ammobaculites
Buccella

Intertidal lagoon, salinity > 32‰

Rosalina araucana
Quinqueloculina reicheli
Turrilina acicula

Normal marine lagoon

Quinqueloculina
Cibicides

Shallow shelf of normal or near-normal salinity

Cibicides
Melonis
Globulina
Cancris
Elphidium
Buccella
Quinqueloculina
Asterigerina
Brizalina
Nonion
Nummulites
Alabamina
Anomalinoides

In Fig. 8.2 (pp. 238–9) a simple division into brackish and marine is shown for the foraminiferal-bearing parts of the succession.

As southeastern England was far removed from oceanic conditions during Palaeogene times environments were not generally favourable for planktonic foraminifera. For the most part these comprise small, immature tests brought into the depositional area from time to time through transport by water currents (Murray, 1976a). However, planktonic foraminifera are abundant at certain levels in the London Clay Formation of the London Basin where they are accompanied by other organisms suggesting water depths of >200 m.

One unique association of benthic foraminifera is that of Selsey Formation, Fisher Bed XVII at Whitecliff Bay and its equivalent (Fisher Bed 21, Selsey). The fauna includes species not found at any other level in the British Palaeogene: *Articulina* spp., *Arenagula kerfornei, Dendritina elegans, Epistomaria rimosa, Fasciolites fusiformis, Linderina brugesi, Orbitolites complanatus, Rotalia* spp., etc. This is interpreted as the product of a normal marine to hypersaline embayment, of warm temperature, and with a vegetational cover over part of the sedimentary substrate.

8.6 FORAMINIFERAL BIOSTRATIGRAPHY

No attempt has been made to erect a zonal scheme because the variable facies would make this exercise of doubtful value.

Reworked Cretaceous species occur at various levels and are easily recognised. Reworking of Cretaceous and probably Palaeocene planktonic species into the type Thanetian deposits has caused difficulties in dating the deposits (El-Naggar, 1967; Brönnimann, *et al.*, 1968). The presence of planktonic forms at particular stratigraphic levels (regardless of the species identified) has been used for correlation (Vella, 1969; Wright 1972b) on the grounds that such an incursion is likely to represent a simultaneous change in water mass movement throughout the depositional basin.

The influence of environment on the benthic species is very marked. They occur in deposits interpreted as normal marine inner to mid shelf, slightly brackish inner shelf, brackish lagoon, and hypersaline lagoon. The introduction of new species and the local extinction of species is invariably due to environmental (including biogeographical) change. Thus, all species are represented by partial ranges or a succession of partial ranges. Therefore, the stratigraphic distribution of species within a succession is mainly of local value for the purposes of correlation. The summary range charts (Figs. 8.2 and 8.3, pp. 238–241) list those species most useful for stratigraphic correlation in the Hampshire Basin and London Basin. The stratigraphic distribution of Palaeogene foraminifera in the southern part of the North Sea is shown in Fig. 11.1, Chapter 11, p. 296.

8.7 SPECIES ENTRIES

For each species the entry is arranged as follows: name used, primary synonym (common synonym where appropriate), description, remarks, palaeo-

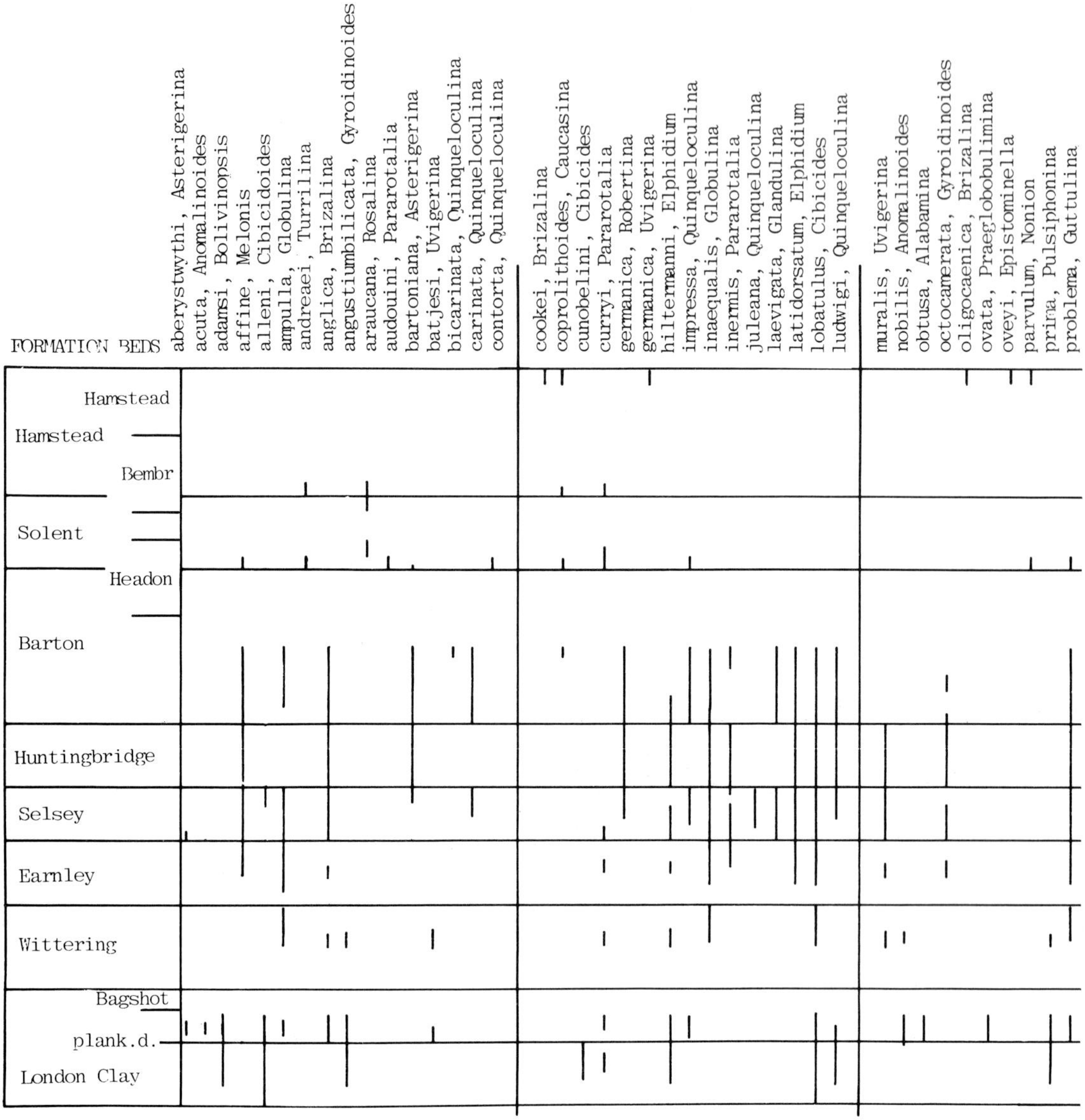

Fig. 8.2 – Summary range chart for the Hampshire Basin (based on figures 8.5 to 8.9 and data in Murray and Wright, 1974, and Wright, 1972).

ecology. PB = Paris Basin. Range based on Le Calvez (1970) = (L) or Murray and Wright (1974) = (MW). BB = Belgian Basin. Range based on Kaasschieter (1961) = (K), Batjes (1958) = (B), Blondeau (1972) = (Bl) or Le Calvez (1970) = (L). The ranges observed in the Paris and Belgian Basins are expressed in terms of local 'stages' mainly because this enables the entry to be short. The correlation of these 'stages' with the local lithostratigraphic successions and with that of southern England is shown in Fig. 8.1. TR = total range observed in Britain. The known range elsewhere may be longer. e = early, m = middle, l = late.

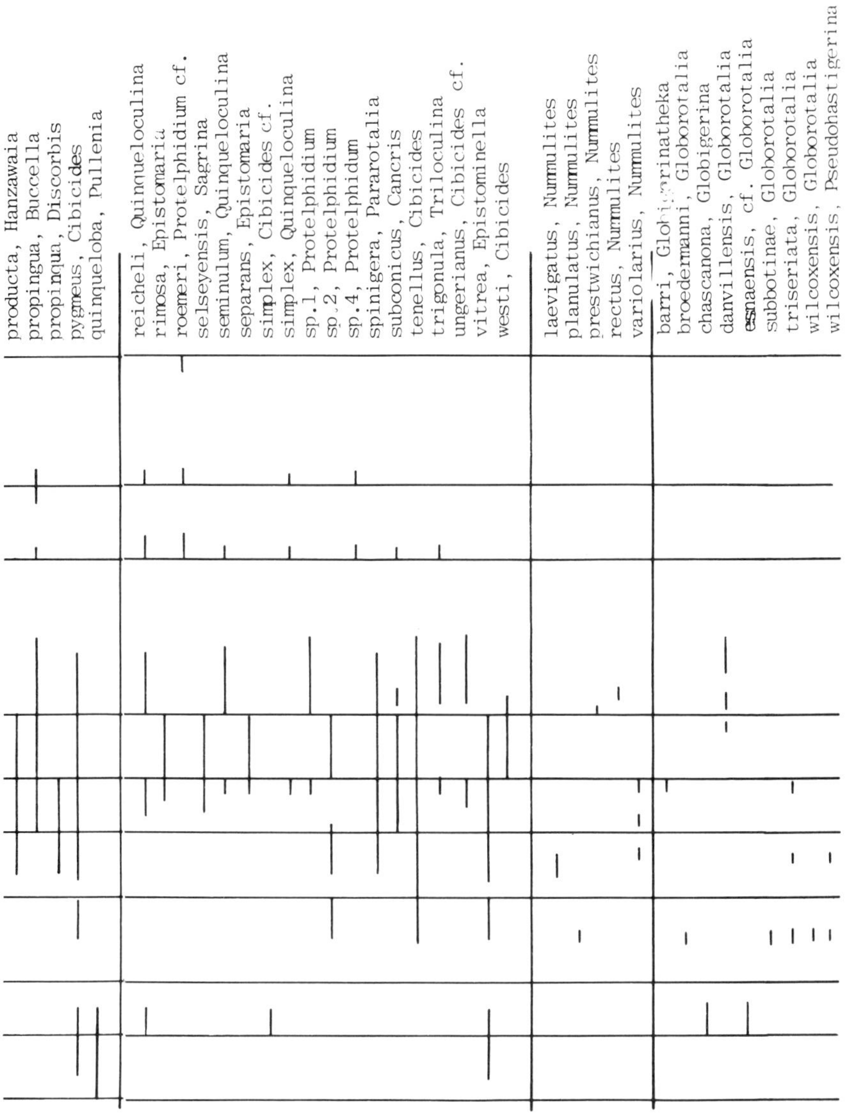

It should be noted that most of the species described are somewhat variable in size. The measurements given refer to the figured specimen.

The species are arranged alphabetically by genera within the suborders Textulariina and Miliolina. In the Rotaliina the order is alphabetical with the exception of the Glandulinidae and Polymorphinidae which are placed at the end because they are illustrated by line drawings (Figs. 8.10 and 8.11). The classification used is that of Loeblich and Tappan (1964).

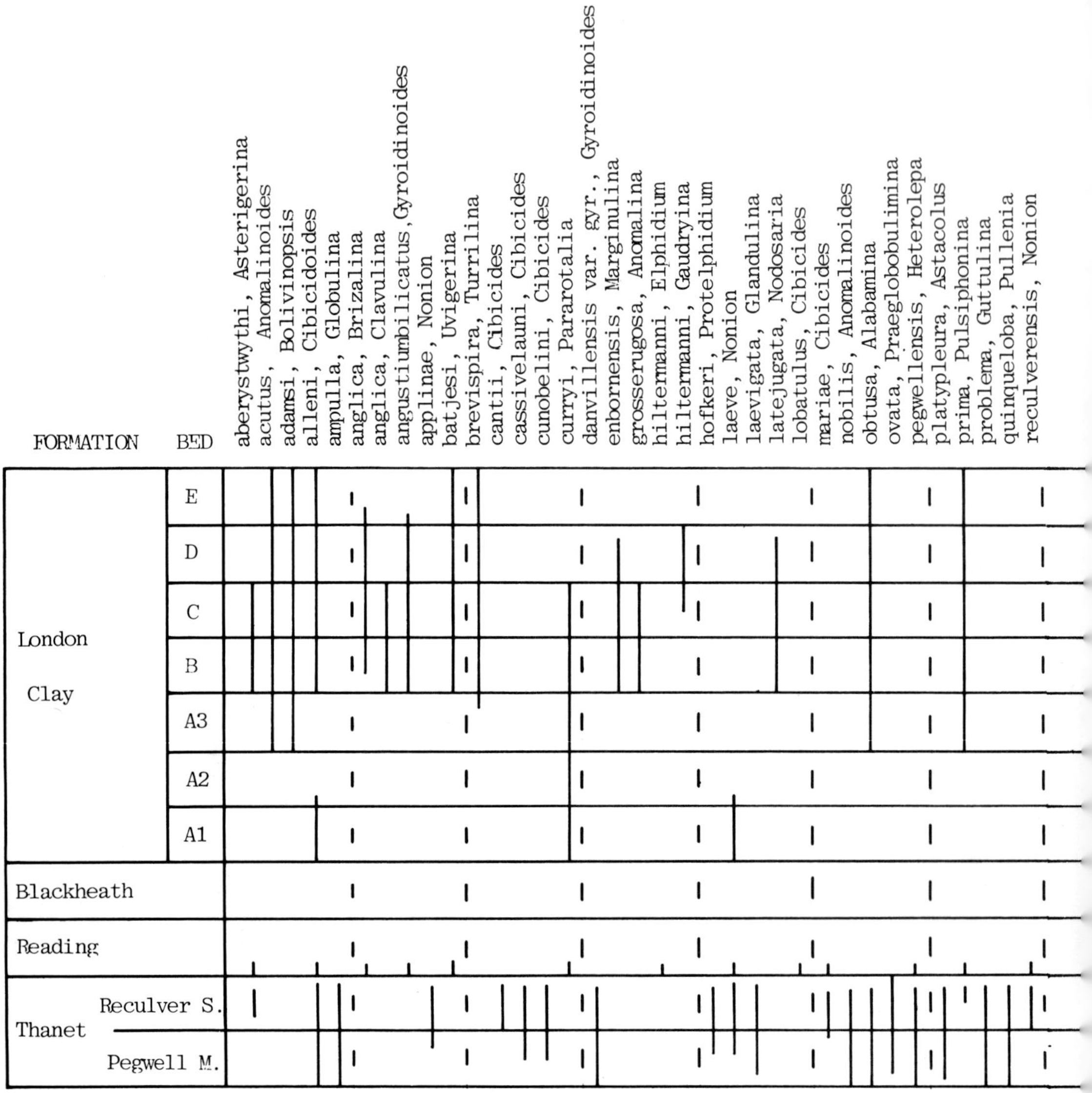

Fig. 8.3 – Summary range chart for the London Basin (based on Fig. 8.4 and data provided by Mr. C. King of Paleoservices Ltd.)

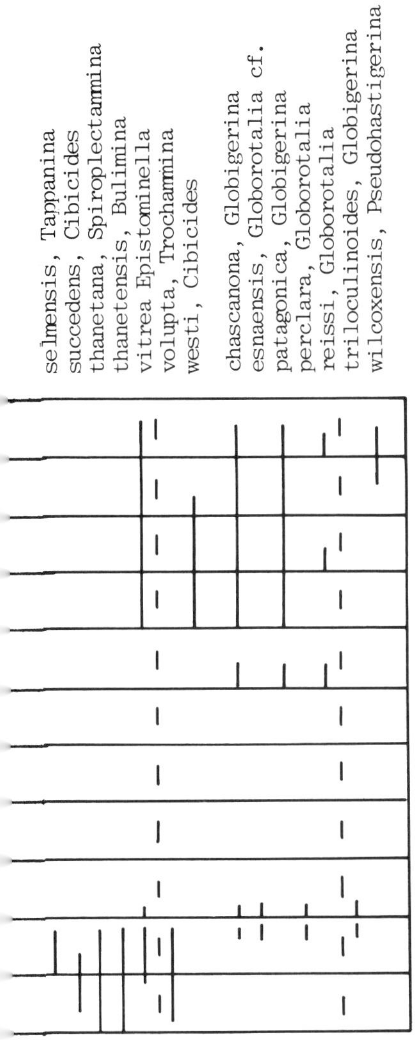

selmensis, Tappanina
succedens, Cibicides
thanetana, Spiroplectammina
thanetensis, Bulimina
vitrea Epistominella
volupta, Trochammina
westi, Cibicides
chascanona, Globigerina
esnaensis, Globorotalia cf.
patagonica, Globigerina
perclara, Globorotalia
reissi, Globorotalia
triloculinoides, Globigerina
wilcoxensis, Pseudohastigerina

8.7.1 Small Benthic Species

8.7.1.1 Suborder Textulariina

Bolivinopsis adamsi (Lalicker)

Plate 8.1, Fig. 1, × 66, length 600 μm. Fig. 2, × 74. London Clay Formation, Alum Bay. Figs. 8.2, 8.3, = *Spiroplectammina adamsi* Lalicker, 1935. Test initially planispiral then biserial, flaring, compressed; axial portion thick.
Remarks: PB no record. BB Ypresian–Paniselian (K). TR e Eoc *Palaeoecology:* normal marine to slightly brackish, inner to mid shelf, fine sediment substrate.

Clavulina anglica (Cushman)

Plate 8.1, Fig. 3, × 28, length 1.44 mm, Fig. 4, × 35. London Clay Formation, Hadley Wood, Middlesex. Fig. 8.3 = *Pseudoclavulina anglica* Cushman, 1936. Description: early portion of test triserial, triangular in section with flat or slightly concave sides; later portion uniserial; aperture terminal, round.
Remarks: confined to the London Basin. PB Lutetian (L). BB Bruxellian–Wemmelian (L). TR e Eoc. *Palaeoecology:* normal marine shelf.

Gaudryina hiltermanni (Meisl)

Plate 8.1, Figs. 5, 6, × 24, length 1.66 mm. London Clay Formation, Sheppey. Fig. 8.3 = *Gaudryina (Pseudogaudryina) hiltermanni* Meisl, 1959. Description: test triangular in cross section with slightly concave faces; aperture subterminal, round to oval.
Remarks: confined to the London Basin. PB, BB no record. TR e Eoc. *Palaeoecology:* normal marine, shelf.

Spiroplectammina thanetana (Lalicker)

Plate 8.1, Figs. 7–9, Fig. 7, × 40, Fig. 8 × 80, Fig. 9 × 40, length 1 mm. Thanet Formation, Reculver Silts, Reculver. Fig. 8.3 = *Spiroplectammina thanetana* Lalicker, 1955. Description: compressed rhomboid in section, periphery subacute; chambers low, up to 26 in microspheric (Fig. 7) and 16 in megalospheric (Fig. 9) generations; angle of taper ∿40° with 3–4 chambers in initial spire.
Remarks: differs from *S. palaeocenica* (Cushman) in having more chambers. PB Thanetian (Rouvillois, 1960). BB no record. TR 1 Palaeoc. *Palaeoecology:* normal marine, inner to mid shelf.

Trochammina volupta (Haynes)

Plate 8.1, Figs. 10–12, × 74, diam. 50 μm. Thanet Formation, Upper Pegwell Marls, Pegwell = *Gyroidinoides voluptus* Haynes, 1956. Description: globose with rounded periphery, 6–7 chambers in each whorl; aperture long and low, extending from umbilicus to periphery; wall of detrital grains, white, perforate.
Remarks: unusual wall structure and perforation makes generic assignment doubtful. PB, BB. Thanetian (Curry) TR 1 Palaeoc. *Palaeoecology:* normal marine inner shelf.

8.7.1.2 Suborder Miliolina

Quinquelculina bicarinata d'Orbigny, 1878

Plate 8.1, Figs. 13–15, Fig. 13 × 33, Fig. 14 × 40, Fig. 15 × 33, length 1.2 mm. Barton Formation, Middle Barton Beds, Alum Bay, Fig. 8.2. Description: test oval, subtriangular in cross section; wall smooth except at periphery where each chamber is ornamented with two keels; aperture round with a small tooth.
Remarks: the double peripheral keels together with the oval outline distinguish this species from all others. PB Lutetian–Marinesian (MW) BB Wemmelian (K) TR 1 Eoc. *Palaeoecology:* slightly brackish (32‰) to normal marine, inner shelf.

Quinqueloculina carinata d'Orbigny, 1826

Plate 8.1, Figs. 16–18, × 100, length 400 μm. Barton Formation, Middle Barton Beds (F), Barton. Fig. 8.2. Description: test oval, subtriangular in cross section, with subangular chamber margins; wall smooth; aperture rounded with a tooth.
Remarks: this species resembles *Q. seminulum* but differs in being smaller and in having a small tooth in the aperture. PB Cuisian–Ludian (L, MW) BB Ledian–Asschian (K) TR m–1 Eoc. *Palaeoecology:* slightly brackish 32‰ to normal marine, inner shelf.

Quinqueloculina contorta d'Orbigny, 1846

Plate 8.1, Figs. 19–21, Fig. 19 × 45, Fig. 20 × 60, Fig. 21 × 45. Solent Formation, Middle Headon Beds, Whitecliff Bay. Fig. 8.2. Description: test elongate, chambers subquadrangular in section giving a truncate periphery; aperture round with a tooth.
Remarks: this species is distinctive because of the truncate periphery but the forms with a rounded periphery grade into *Q. ludwigi.* PB Lutetian–Sannoisian (L) BB no record. TR 1 Eoc. *Palaeoecology:* normal marine, lagoonal.

Quinqueloculina impressa Reuss, 1851

Plate 8.1, Figs. 22–24, × 59, length 680 μm. Solent Formation, Middle Headon Beds, Whitecliff Bay. Fig. 8.2. Description: test oval, periphery rounded, surface smooth; aperture round with a small tooth.
Remarks: differs from other species of *Quinqueloculina* with a rounded periphery (*Q. ludwigi, Q. reicheli* and *Q. simplex*) in being oval rather than elongate. PB no record. BB Ledian–Asschian (K) Tongrian (B) TR e–1 Eoc. *Palaeoecology:* slightly brackish (32‰) to normal marine, inner shelf.

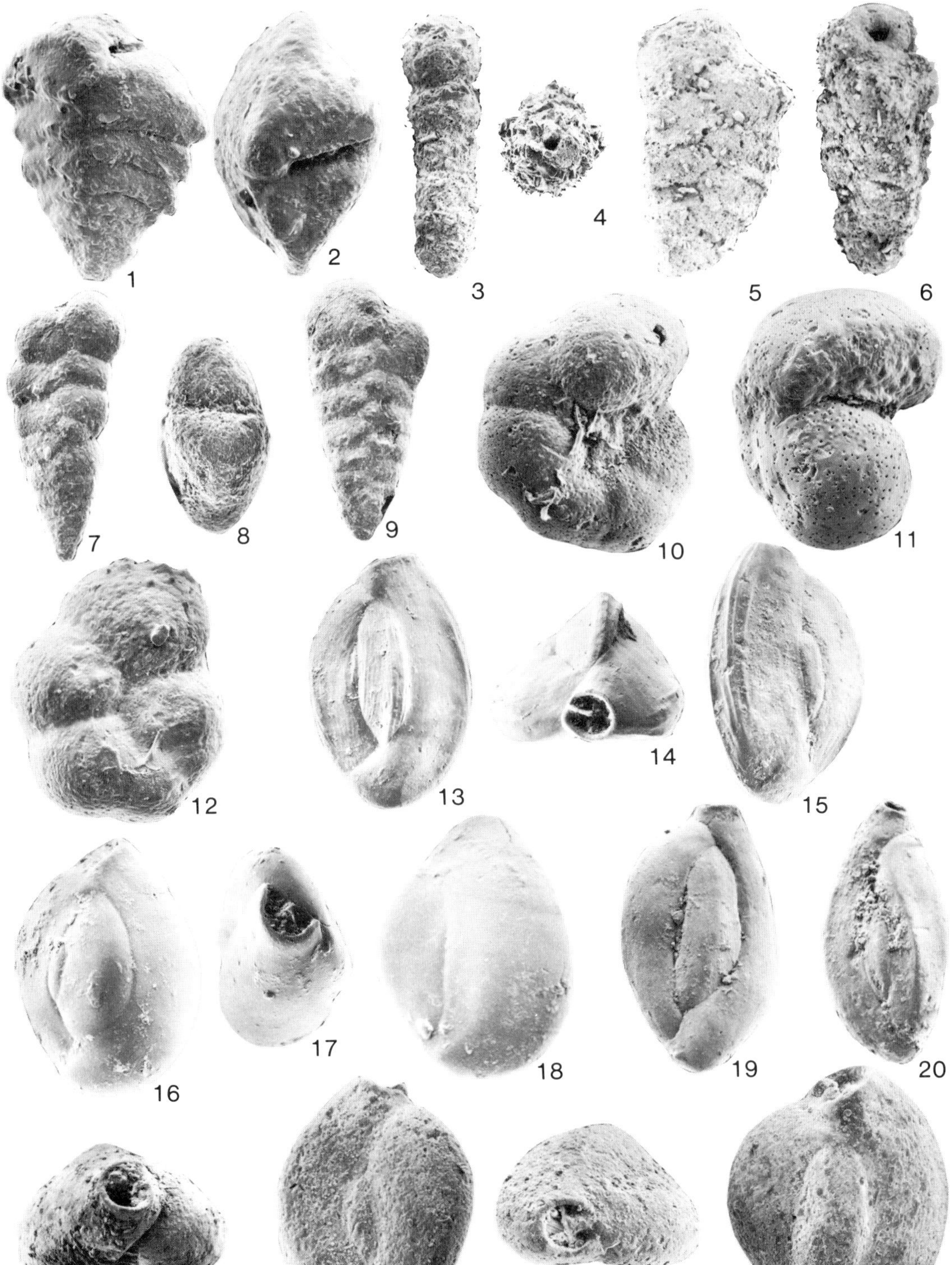

Plate 8.1

Quinqueloculina juleana (d'Orbigny, 1846)
Plate 8.2, Figs. 1-3, Fig. 1 × 50, Fig. 2 × 75, Fig. 3 × 50, length 800 μm. Bracklesham Group, Selsey Formation (Fisher Bed 21), Selsey. Fig. 8.2. Description: test elongate, chamber peripheries with keels; aperture round, with a tooth, situated on a short neck.
Remarks: the strongly developed keels, elongate test and aperture on a neck distinguish this species from all others. PB Auversian (L) BB Ledian–Asschian (K) Tongrian–Rupelian (L, B) TR m Eoc. *Palaeoecology:* slightly brackish (32‰) to normal marine, inner shelf, on muddy sand substrates.

Quinqueloculina ludwigi Reuss, 1866
Plate 8.2, Figs. 4–6, Fig. 4 × 118, Fig. 5 × 177, Fig. 6 × 118, length 340 μm. Bracklesham Group, Selsey Formation, Whitecliff Bay. Fig. 8.2. Description: test elongate oval, smooth, chambers rounded in section; aperture circular, with a bifid tooth, on a short neck.
Remarks: differs from *Q. simplex* in having chambers with a circular cross section and therefore depressed sutures. PB Lutetian–Sannoisian (L), BB Ledian–Asschian (K) Tongrian (L) Rupelian (B), TR e–l Eoc. *Palaeoecology:* slightly brackish (32‰) to normal marine, inner shelf, on muddy substrates.

Quinqueloculina reicheli Le Calvez, 1966
Plate 8.2, Figs. 7–9, Fig. 7 × 118, Fig. 8 × 135, Fig. 9 × 119, length 340 μm. Bracklesham Group, Selsey Formation, Whitecliff Bay. Fig. 8.2 = *Quinqueloculina (Scutuloris) reicheli* Le Calvez, 1966. Description: test oval, rounded in section, smooth; aperture the obliquely open end of the last chamber, sometimes with a flap.
Remarks: the oblique aperture is distinctive. PB Lutetian–Sannoisian (L, MW) BB Tongrian–Rupelian (L), TR e Eoc–e Olig. *Palaeoecology:* slightly brackish (32‰) to hypersaline, intertidal lagoonal, to inner shelf.

Quinqueloculina seminulum (Linné)
Plate 8.2, Figs. 10–12, × 25, length 1.6 mm. Solent Formation, Middle Headon Beds, Whitecliff Bay. Fig. 8.2 = *Serpula seminulum* Linné, 1758. Description: test oval, subtriangular in section with subangular chamber margins, smooth; aperture an oval opening with an elongate tooth.
Remarks: differs from *Q. carinata* in being larger and in having a big tooth in the elongate aperture. PB Sannoisian (L) BB Paniselian–Asschian (K) Tongrian–Rupelian (L, B) TR m–l Eoc.; Rec. *Palaeoecology:* modern forms are normal marine, inner shelf inhabitants.

Quinqueloculina simplex Terquem, 1882
Plate 8.2, Figs. 13–15, Fig. 13, × 125, Fig. 14, × 140, Fig. 15, × 115, Length 320 μm. Solent Formation, Middle Headon Beds, Whitecliff Bay. Fig. 8.2 = *Quinqueloculina simplex* Terquem, emend. Le Calvez, 1947. Description: test very elongate, rounded in cross section, smooth; aperture round with a small tooth.
Remarks: the very elongate form and small aperture are distinctive. PB Lutetian–Sannoisian (L, MW) BB Tongrian–Rupelian (B as *Scutuloris oblonga*) TR m Eoc.–e Olig. *Palaeoecology:* slightly brackish (32‰) to hypersaline, intertidal lagoonal to inner shelf.

Triloculina trigonula (Lamarck)
Plate 8.2, Figs. 16–18, × 87, length 460 μm. Solent Formation, Middle Headon Beds, Whitecliff Bay. Fig. 8.2 = *Miliolites trigonula* Lamarck, 1804. Description: test triloculine, of rounded triangular cross section; aperture a terminal arch with a small tooth.
Remarks: this is the only common species of *Triloculina* in the British Palaeogene. PB Cuisian–Sannoisian (L, MW) BB Paniselian–Asschian (K) TR m–l Eoc; Rec. *Palaeoecology:* normal marine, inner shelf.

8.7.1.3 Suborder Rotaliina
Alabamina obtusa (Burrows and Holland)
Plate 8.2, Figs. 19–21 × 80, diam. 500 μm. London Clay Formation, Alum Bay. Figs. 8.2 and 8.3 = *Pulvinulina exigua* (Brady) var. *obtusa* Burrows and Holland, 1897. Description: test nearly biconvex, umbilical side higher than spiral side; generally 5 chambers in final whorl; periphery subrounded; apertural face infolded, aperture an interiomarginal slit; diam. 0.2–0.6 mm.
Remarks: differs from *A. wilcoxensis* Toulmin in having a subrounded rather than subangular periphery, and in being more nearly biconvex rather than planoconvex. PB Thanetian (Curry) Cuisian (MW) BB Thanetian (Curry) Ypresian (K) TR 1 Palaeoc.–e Eoc. *Palaeoecology:* slightly brackish, inner shelf, on muddy substrates.

Anomalina grosserugosa (Gümbel)
Plate 8.2, Figs. 22–24, × 44, diam. 0.9 mm. London Clay Formation, Hadley Wood, Middlesex. Fig. 8.3. = *Truncatulina grosserugosa* Gümbel, 1868. Description: test trochospiral; spiral side flat, umbilical side convex, periphery rounded; aperture an interiomarginal slit; wall coarsely perforate.
Remarks: The large size of the test and the coarse wall pores are distinctive. PB Thanetian (L) BB Paniselian–Wemmelian (K) TR e Eoc. *Palaeoecology:* normal marine.

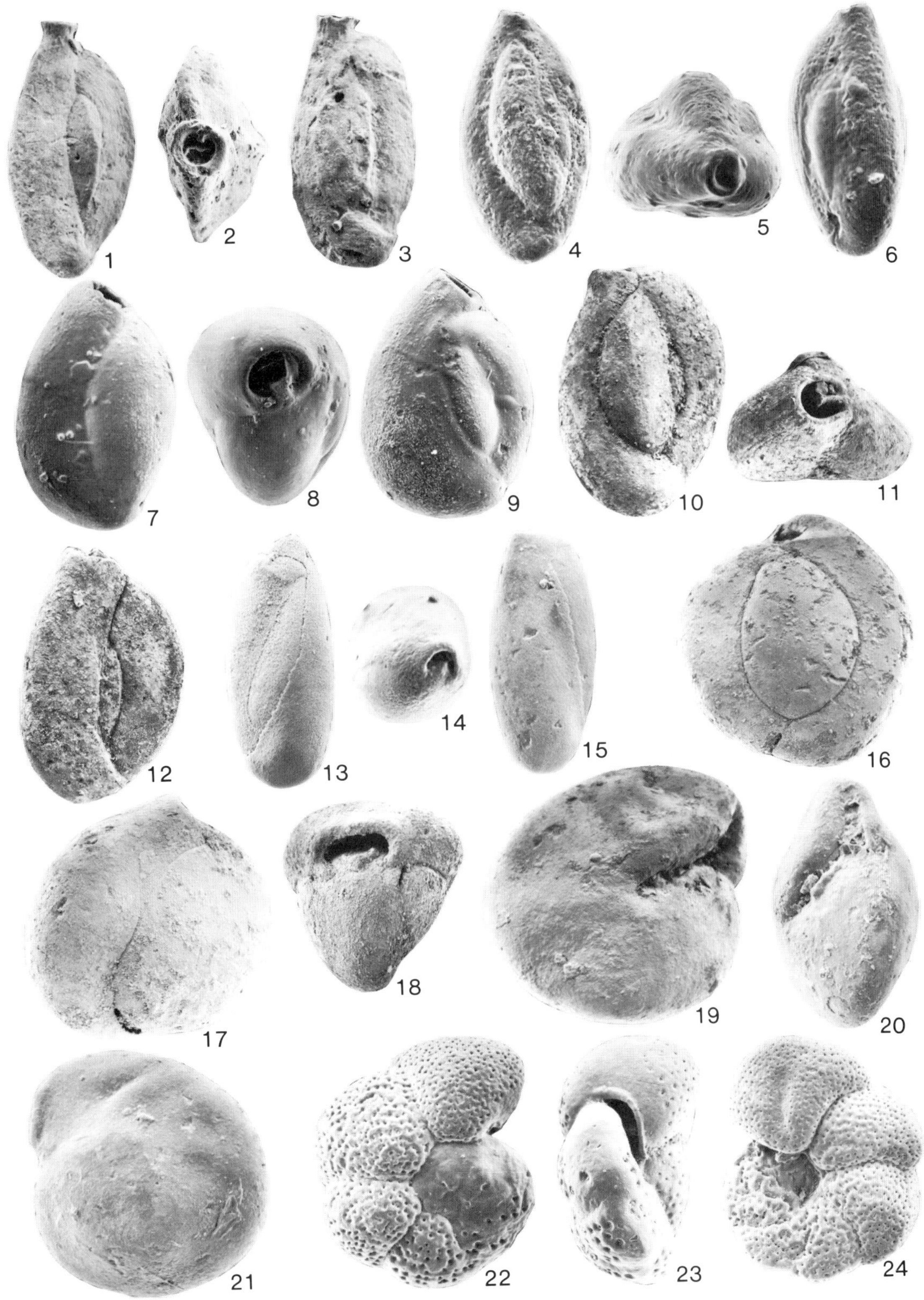

Plate 8.2

Anomalinoides acutus (Plummer)
Plate 8.3, Figs. 1–3, × 100, diam. 400 μm. London Clay Formation, Whitecliff Bay. Figs 8.2, 8.3 = *Anomalina ammonoides* (Reuss) var. *acuta* Plummer, 1926. Description: test trochospiral, very compressed; spiral side moderately involute with a thickened central boss; umbilical side involute with central boss surrounded by thickened ends of sutures; periphery acute; 13–15 narrow chambers in the last whorl; aperture an arch extending over the periphery and onto the umbilical side.
Remarks: differs from *A. nobilis* in its acute periphery and numerous chambers. PB. Thanetian (L) Cuisian (MW) BB Ypresian–Asschian (K) TR e Eoc. *Palaeoecology:* normal marine.

Anomalinoides nobilis Brotzen, 1948
Plate 8.3, Figs. 4–6, × 130, diam. 300 μm. London Clay Formation, Whitecliff Bay. Figs. 8.2, 8.3. Description: test low trochospiral, almost planispiral; spiral side partly involute with a shallow umbilicus; 7–9 chambers in final whorl; periphery rounded; aperture extending from periphery onto spiral side.
Remarks: this species superficially resembles *Melonis affine* but the latter is planispiral and involute on both sides. PB Cuisian (MW) BB no record, TR 1 Palaeoc.–e Eoc. *Palaeoecology:* normal marine to slightly brackish, inner–mid shelf, mainly on fine sediment substrates.

Astacolus platypleura (Jones)
Plate 8.3, Figs. 7, 8, × 22, length 1.8 mm. Thanet Formation, Lower Pegwell Marls, Pegwell. Fig. 8.3. = *Cristellaria platypleura,* Jones, 1852. Common synonym: *Lenticulina multiformis* Franke. Description: compressed, carinate; up to 13 chambers in close-coiled microspheric and up to 10 chambers in megalospheric generation, becoming high and astacoline; sutures limbate and raised with variable development of spiral cross-ornament.
Remarks: PB, BB no record. TR 1 Palaeoc. *Palaeoecology:* normal marine, mid and outer shelf, on silt and mud.

Asterigerina aberystwythi Haynes, 1956
Plate 8.3, Figs. 9–11, × 130, diam. 300 μm. Thanet Formation, Reculver Silts, Reculver. Fig. 8.3. Description: biconvex or subconoidal with spiral side raised and umbilical side flattened; periphery acute, 4 to 5 chambers visible on umbilical side, supplementary chamberlets small; aperture an arch at base of apertural face with tuberculate ornament below.
Remarks: differs from *A. bartoniana* in being smaller, having fewer chambers and in being flattened on the umbilical side. PB, BB no record. TR 1 Palaeoc. *Palaeoecology:* normal marine–?brackish, inner shelf.

Asterigerina bartoniana (Ten Dam)
Plate 8.3., Figs. 12–14, × 130, diam 300 μm. Barton Formation, Middle Barton Beds (D), Barton. Fig. 8.2. =*Rotalia granulosa* Ten Dam, 1944., renamed *Rotalia bartoniana,* Ten Dam, 1947. Description: test trochospiral, biconvex, umbilical side more convex than spiral side; 6–10 chambers in last whorl; umbilicus filled by calcite plug; granular ornament on umbilical side.
Remarks: see *A. aberystwythi.* PB Cuisian–Auversian (L, MW) BB Ypresian–Asschian (K) e Tongrian (B) TR m–1 Eoc. *Palaeoecology:* normal marine, inner shelf.

Brizalina anglica (Cushman)
Plate 8.3, Figs. 15, 16, Fig. 15, × 110, Fig. 16, × 220, length 360 μm. London Clay Formation, Alum Bay. Figs. 8.2, 8.3. = *Bolivina anglica* Cushman, 1936. Description: test biserial, elongate, about three times as long as broad; compressed, periphery rounded, aperture loop-shaped.
Remarks: this species is relatively broader than *B. oligocaenica.* PB Thanetian (L) BB Ypresian–Asschian (K). TR e–1 Eoc. *Palaeoecology:* normal marine, shelf.

Brizalina cookei (Cushman)
Plate 8.3, Figs. 17, 18, Fig. 17, × 155, Fig. 18, × 260, length 260 μm. Hamstead Formation, Upper Hamstead Beds, Cerithium Bed, Hamstead. Fig. 8.2. = *Bolivina cookei* Cushman, 1922. Description: test biserial, elongate, periphery subacute; numerous fine costae extending along most of the test length; last chambers smooth.
Remarks: the presence of costae distinguishes this from other species of *Brizalina.* PB Sannoisian (MW) BB Wemmelian–Asschian (K) Rupelian (L) TR m Olig. *Palaeoecology:* normal marine, shelf, mud substrate.

Brizalina oligocaenica (Spandel)
Plate 8.3, Figs. 19, 20, Fig. 19, × 100, Fig. 20, × 300, length 400 μm. Hamstead Formation, Upper Hamstead Beds, Cerithium Bed, Hamstead. Fig. 8.2. = *Bolivina oligocänica* Spandel, 1909. Description: test biserial, elongate, about 4 times as long as broad; periphery rounded; numerous chambers; sutures straight, oblique; aperture loop-shaped.
Remarks: this species is narrower and more elongate than *B. anglica.* PB Stampian (MW) BB Rupelian (L) TR m Olig. *Palaeoecology:* normal marine shelf, mud substrate.

Buccella propingua (Reuss)
Plate 8.3, Figs. 21–23, × 200, diam. 200 μm. Barton Formation, Middle Barton Beds (D), Barton. Fig. 8.2. = *Rotalia propingua* Reuss, 1856. Common synonym: *Ammonia propingua* (Reuss). Description: test trochospiral, biconvex, periphery subacute; sutures on spiral side oblique, those on umbilical side sigmoid, deeply incised in their inner part and ornamented with pustules; primary aperture an interiomarginal slit obscured by pustules.
Remarks: PB Lutetian–Auversian (Curry) Marinesian–Sannoisian (L, MW) BB Bruxellian–Asschian (K) 1 Tongrian–

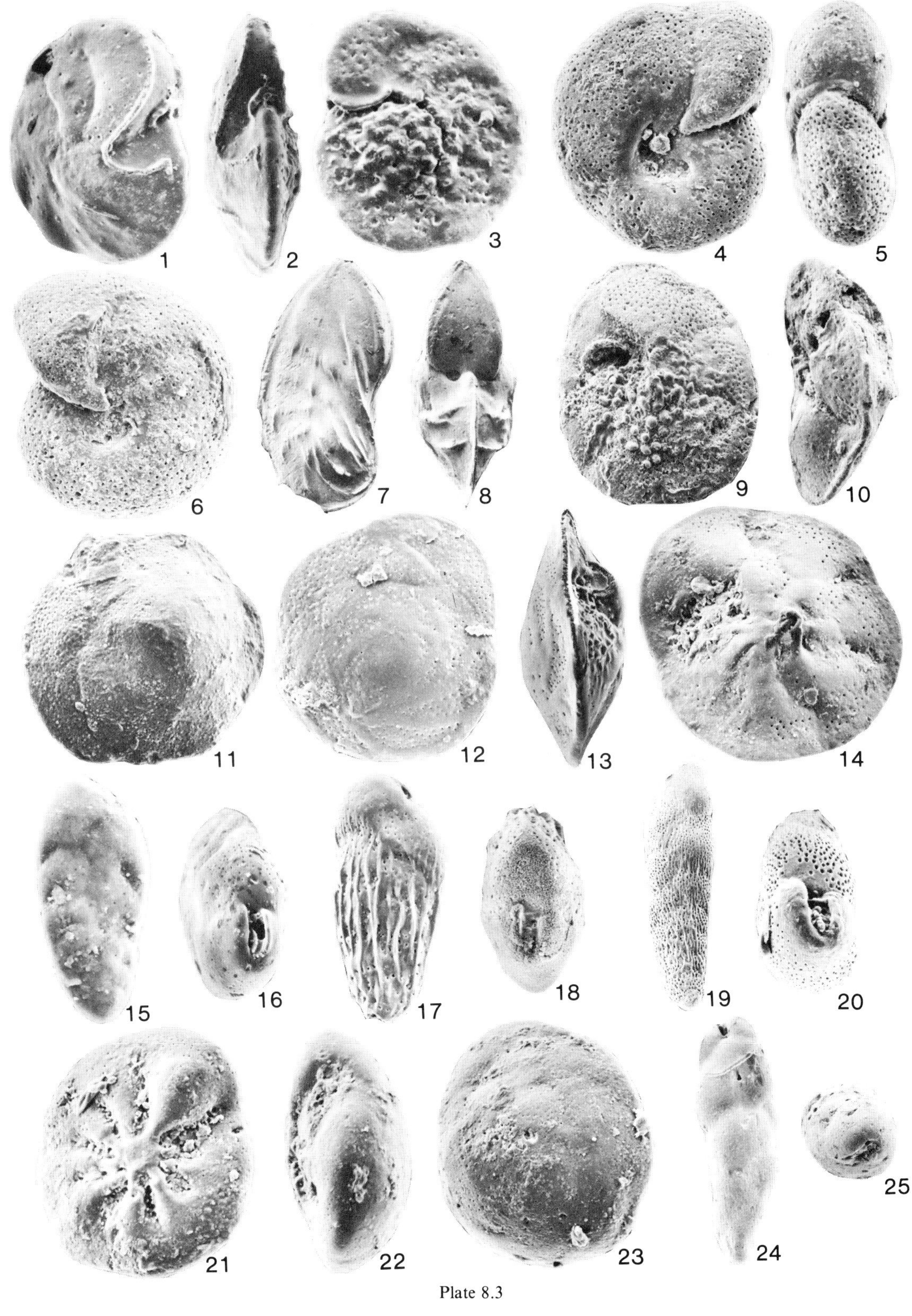

Plate 8.3

Rupelian (B) TR m Eoc.–1 Olig. *Palaeoecology:* slightly brackish inner to mid shelf, generally on fine sediment substrates.

Bulimina thanetensis Cushman and Parker, 1947
Plate 8.3, Figs. 24, 25, Fig. 24, × 100, Fig. 25, × 125, length 400 μm. Thanet Formation, Lower Pegwell Marls, Pegwell. Fig. 8.3. Description: elongate, initial part trigonal with marked spiral suture, adult rounded and parallel sided; chambers up to 30 in microspheric and up to 20 in megalospheric generations, last 4 making up ∿ half the test; aperture comma shaped with toothplate (simple trough with slight flange).
Remarks: *B. trigonalis* Ten Dam is smaller, *B. simplex* Ten Dam has a quadrangular aperture, *B. paleocenica* Brotzen is sharply triangular and *B. rosenkrantzi* Brotzen has an enlarged final whorl. PB, BB no record. TR 1 Palaeoc. *Palaeoecology:* marginal marine to mid shelf, probably tolerant of muddy bottoms with lowered oxygen levels.

Cancris subconicus (Terquem)
Plate 8.4, Figs. 1–3, × 130, diam. 300 μm. Bracklesham Group, Selsey Formation, Whitecliff Bay. Fig. 8.2. = *Rotalina subconica* Terquem, 1882. Common synonym: *Valvulineria subconica* (Terquem). Description: test trochospiral, ovate, periphery rounded; last chamber more elongate than the rest.
Remarks: PB Cuisian–Marinesian (L, MW) BB Ypresian–Asschian (K) TR m–1 Eoc. *Palaeoecology:* normal marine, inner to mid shelf, varied substrates.

Caucasina coprolithoides (Andreae)
Plate 8.4, Fig. 4, × 145, length 280 μm. Bembridge Formation, Bembridge Marls, Whitecliff Bay. Fig. 8.2. = *Bulimina coprolithoides* Andreae, 1884. (syn. *Buliminella carteri* Bhatia, 1955). Description: initial portion trochospiral, but main part of test triserial; aperture an elongate loop at the inner margin of the last chamber.
Remarks: juvenile forms of this species were named *Buliminella carteri* by Bhatia (1955). The initial trochospiral stage distinguishes *C. coprolithoides* from *Bulimina* species. PB Sannoisian (L. MW) BB Tongrian–Rupelian (L) TR 1 Eoc.–m Olig. *Palaeoecology:* slightly brackish, normal marine, slightly hypersaline, marginal lagoons, mainly muddy substrates.

Cibicides cantii Haynes, 1957
Plate 8.4, Figs. 5–7, × 135, diam. 300 μm, Thanet Formation, Reculver Silts, Reculver. Fig. 8.3. Description: planoconvex, periphery acute; chambers initially lobate becoming arcuate 6–8 in final whorl, wall coarsely perforate on spiral side; aperture extends from the periphery onto the spiral side.
Remarks: a small species. PB, BB no record. TR 1 Palaeoc. *Palaeoecology:* normal marine, ?brackish, inner shelf.

Cibicides cassivelauni Haynes, 1957
Plate 8.4, Figs. 8–10, × 73, diam. 550 μm. Thanet Formation, Upper Pegwell Marls, Pegwell. Fig. 8.3. Description: planoconvex with subangular periphery; 7–8 chambers in final whorl becoming arcuate on spiral side by end of second whorl; wall coarsely perforate on spiral side but on umbilical side only so on last few chambers; aperture extending from periphery onto spiral side.
Remarks: differs from *C. lobatulus* in lacking general development of large pores on umbilical side. PB, BB no record TR 1 Palaeoc. *Palaeoecology:* normal marine, inner–mid shelf.

Cibicides (Cibicidina) cunobelini Haynes, 1957
Plate 8.4, Figs. 11–13, × 80, diam. 500 μm. Thanet Formation, Reculver Silts, Reculver. Fig. 8.3. Description: equally biconvex or umbilical side more convex, periphery subrounded; 9–10 chambers in outer whorl of microspheric and 7–8 in megalospheric generation; sutures limbate on spiral side and swell into small bosses towards umbilicus but in later part of test are replaced by lappets along base of last few chambers; aperture extending from periphery onto spiral side.
Remarks: differs from *C.(C.) mariae* in being larger and in its convex spiral side. PB Thanetian–Sparnacian (L) Cuisian (MW) BB no record. TR 1 Palaeoc.–e Eoc. *Palaeoecology:* brackish–normal marine, inner shelf, on fine sediment.

Cibicides lobatulus (Walker and Jacob)
Plate 8.4, Figs. 14–16, × 100, diam. 360 μm. Bracklesham Group, Selsey Formation, Selsey. Figs. 8.2, 8.3. = *Nautilus lobatulus* Walker and Jacob, 1798. Description: test trochospiral, spiral side flat to concave, evolute, umbilical side convex, involute; 7–9 chambers in the outer whorl; periphery acute; aperture extending from the periphery along 2–3 chambers on spiral side.
Remarks: distinguished from *C. cantii* by its larger number of chambers and the absence of the coarse pores on the spiral side. *C. pseudoungerianus* generally has a more convex spiral side. PB Thanetian–Sannoisian (L,MW) BB Ypresian–Asschain (K) Tongrian–Rupelian (B) TR e–1 Eoc., Rec. *Palaeoecology:* marine, inner to mid shelf, often in areas disturbed by currents. *C. lobatulus* lives clinging or attached to firm substrates such as shells.

Cibicides (Cibicidina) mariae (Jones)
Plate 8.4, Figs. 17–19, × 160, diam. 250 μm. Thanet Formation, Reculver Silts, Reculver. Fig. 8.3. = *Rosalina mariae* Jones, 1852. Common synonym: *Cibicides newmanae* (Plummer). Description: concavo-convex, partially involute with well developed lappets; 8–9 chambers in outer whorl; suture limbate on spiral side, aperture extending from periphery onto spiral side.
Remarks: see *C. (C.) cunobelini*. PB, BB no record. TR 1 Palaeoc. *Palaeoecology:* normal marine–?brackish, inner shelf.

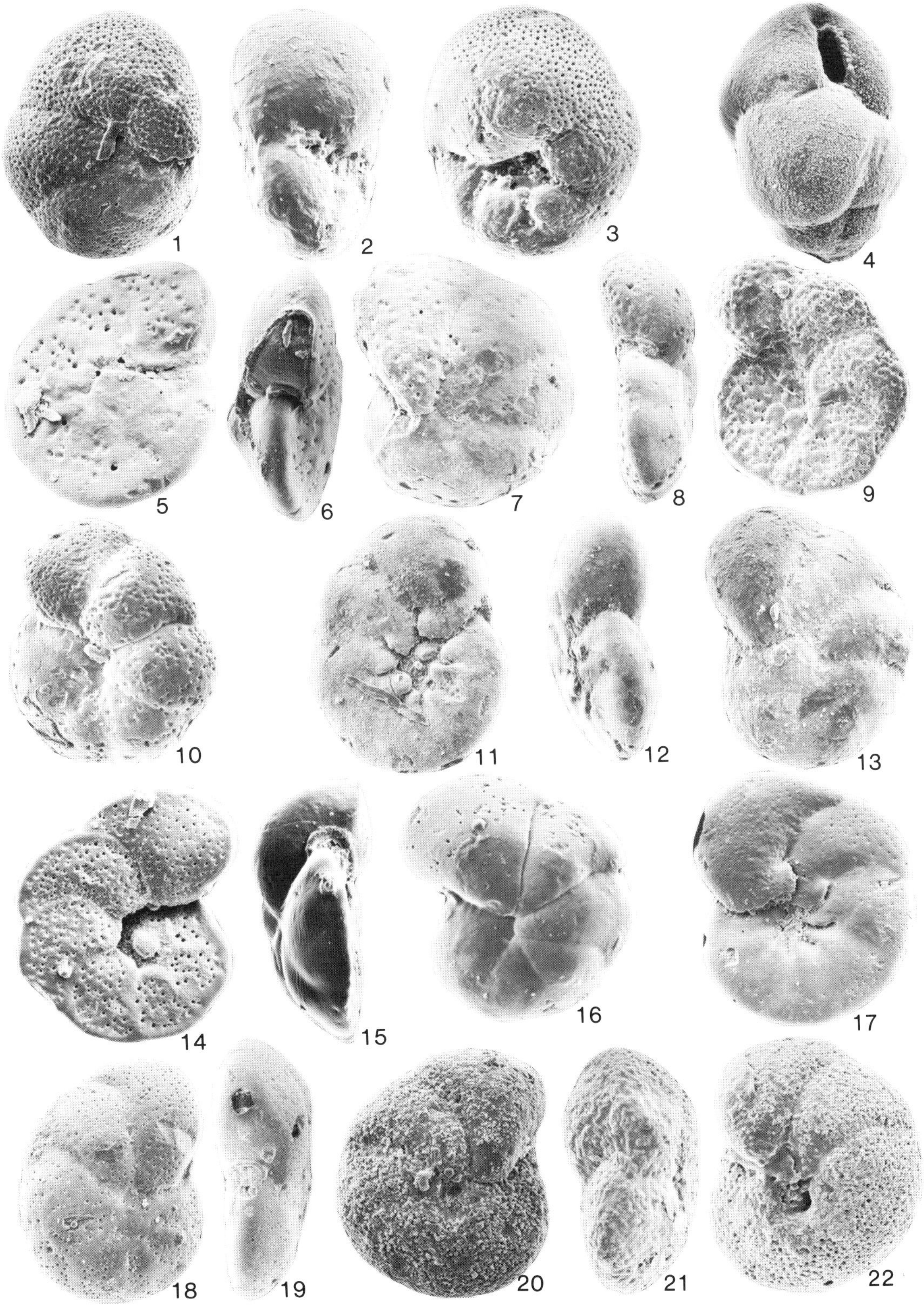

Plate 8.4

Cibicides cf. *simplex* Brotzen, 1948
Plate 8.4, Figs. 20–22, × 165, diam. 240 μm. London Clay Formation, Whitecliff Bay. Fig. 8.2. Description: test trochospiral, spiral side flat to convex, partially involute; umbilical side convex, involute; periphery rounded; 6–10 chambers in outer whorl; aperture extending from periphery along 2–3 chambers on the spiral side.
Remarks: PB, BB no record TR e Eoc. *Palaeoecology:* brackish, inner shelf, muddy sand substrates.

Cibicides (Cibicidina) succedens Brotzen, 1948
Plate 8.5, Figs. 1-3, × 90, diam. 440 μm. Thanet Formation, Lower Pegwell Marls, Pegwell. Fig. 8.3. Description: planoconvex or with spiral side slightly raised, bi-umbonate; periphery subangular; 8–10 chambers in final whorl; suture limbate on spiral side.
Remarks: PB Thanetian (L) BB no record. TR 1 Palaeoc. *Palaeoecology:* marginal marine to mid shelf, possibly tolerant of poorly aerated mud bottom.

Cibicides tenellus (Reuss)
Plate 8.5, Figs. 4–6, × 145, diam. 280 μm. Bracklesham Group, Selsey Formation, Whitecliff Bay. Fig. 8.2 = *Truncatulina tenella* Reuss, 1865. Description: test trochospiral, spiral side flat, evolute, umbilical side strongly convex, involute; 8–12 chambers in final whorl, periphery acute; aperture extending from the periphery onto the spiral side.
Remarks: *C. cantii, C. cassivelauni,* and *C. lobatulus* generally have fewer chambers and are less strongly convex on the umbilical side. PB Auversian–Marinesian (MW), BBD Ledian–Asschian (K) e Tongrian (B) TR e–1 Eoc. *Palaeoecology:* marine, inner to mid shelf.

Cibicides cf. *ungerianus* (d'Orbigny)
Plate 8.5, Fig.s 7–9, × 100, diam. 400 μm. Barton Formation, Middle Barton Beds (F), Fig. 8.2 = *Rotalina ungeriana* d'Orbigny, 1846. Description: test trochospiral, spiral side slightly convex, partly involute, umbilical side convex, involute, with a small open umbilicus; periphery rounded; 8–11 chambers in outer whorl; aperture extends from the periphery along 2–3 chambers on the spiral side; wall coarsely perforate; thickened bosses on the early portion of the spiral side.
Remarks: the distinctive feature is the thickening of the early portion of the spiral side with bosses. The Eocene forms have a more rounded periphery than the forms from the Neogene. PB Marinesian (MW). BB Paniselian–Asschian (K) Rupelian (B) TR m–1 Eoc. *Palaeoecology:* normal marine, inner or mid shelf.

Cibicides westi Howe, 1939
Plate 8.5, Figs. 10–12, × 135, diam. 300 μm. London Clay Formation, Ongar. Figs. 8.2, 8.3. Description: test trochospiral; spiral side flat, evolute; umbilical side strongly convex, involute; periphery acute; aperture a peripheral arch extending onto the spiral side.
Remarks: differs from *C. tenellus* in the absence of coarse pores on the umbilical side and in having relatively few coarse pores on the spiral side. PB Ypresian–Bartonian (L) Cuisian–Marinesian (MW) BB Ypresian–Wemmelian (K) TR m–1 Eoc. *Palaeoecology:* normal marine.

Cibicidoides alleni (Plummer)
Plate 8.5, Figs. 13-15, × 110, diam. 360 μm. London Clay Formation, Whitecliff Bay. Figs. 8.2, 8.3. = *Truncatulina alleni* Plummer, 1927. Common synonyms: *Cibicidoides proprius* Brotzen, 1948; *Cibicides proprius* (Brotzen). Description: test trochospiral, flattened, biconvex, spiral side evolute with thickened early part, umbilical side involute with shallow umbilicus; periphery subacute; 8–11 chambers in outer whorl; aperture a peripheral arch extending onto the spiral side; wall coarsely perforate.
Remarks: a less inflated biconvex form than *C. pygmeus.* PB Thanetian–Lutetian (L, MW) BB Ypresian–Paniselian (K). TR 1 Palaeoc.–e Eoc. *Palaeoecology:* brackish to normal marine, inner to mid shelf, varied substrates.

Cibicidoides pygmeus (Hantken)
Plate 8.5, Figs. 16–18, × 135, diam. 300 μm. Bracklesham Group, Selsey Formation, Whitecliff Bay. Fig. 8.2 = *Pulvinulina pygmea* Hantken, 1875. Common synonym: *Cibicides pygmeus.* Description: test trochospiral, biconvex, spiral side evolute, umbilical side involute; 8–10 chambers in outer whorl; periphery subacute, aperture extending from the periphery along the umbilical face of the last chamber; wall coarsely perforate.
Remarks: the biconvexity is more inflated than in *C. alleni.* PB no record, BB Wemmelian–Asschian (K) Rupelian (B), TR e–1 Eoc. *Palaeoecology:* marine, inner to mid shelf.

Discorbis propinqua (Terquem)
Plate 8.5, Figs. 19–20, × 100, diam. 400 μm. Bracklesham Group, Selsey Formation, Selsey. Fig. 8.2. = *Rosalina propinqua* Terquem, 1882 = *Discorbis propinqua* (Terquem) emend. Le Calvez, 1949. Description: test trochospiral, planoconvex, with flat umbilical side, periphery keeled; on the umbilical side the chambers have non-perforate extensions which fuse to form a star-shaped umbilical boss; sutures slightly depressed on spiral side, deeply depressed on umbilical side.
Remarks: PB Cuisian–Marinesian (L, MW) BB Bruxellian–Wemmelian (K), TR m Eoc. *Palaeoecology:* normal marine, inner shelf, varied substrates.

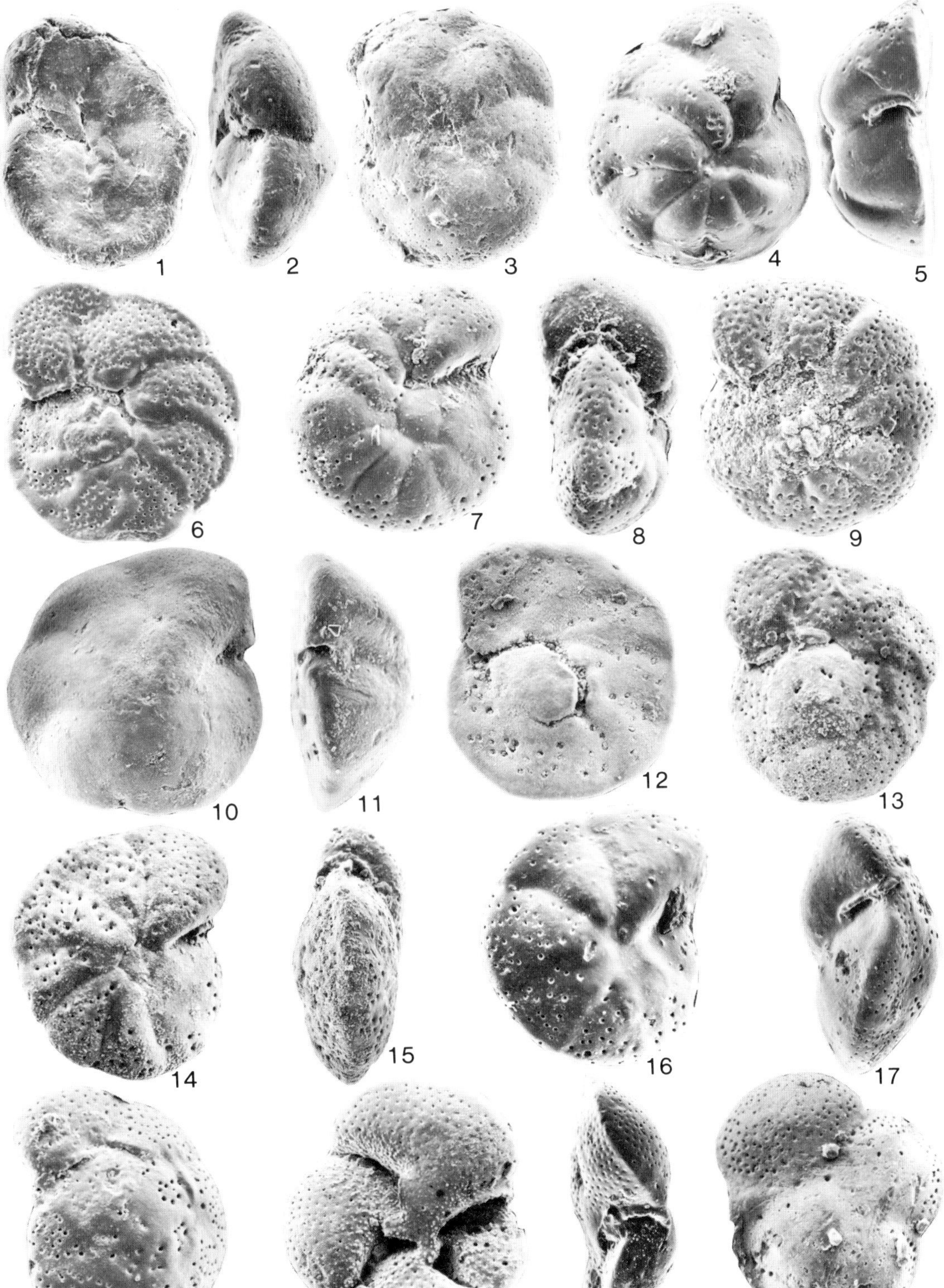

Plate 8.5

Elphidium hiltermanni Hagn, 1952
Plate 8.6, Figs. 1, 2, × 155, diam. 260 μm. Bracklesham Group, Wittering Formation, Whitecliff Bay. Figs. 8.2, 8.3. Description: test planispiral, somewhat compressed, periphery rounded; shallow umbilici; 6–7 chambers in last whorl; sutures depressed, retral processes short; aperture interiomarginal, of pores hidden by pustules which form rows along the apertural face and extend onto the adjacent part of the earlier test.
Remarks: resembles *E. latidorsatum* but has a much less inflated test. PB Cuisian–Sannoisian (L, MW), BB Bruxellian–Wemmelian (K) Rupelian (L), TR e–1 Eoc. *Palaeoecology:* brackish to normal marine, estuarine to inner shelf, varied substrates.

Elphidium latidorsatum (Reuss)
Plate 8.6, Figs. 3, 4, × 135, diam. 300 μm. Barton Formation, Barton Clay, Whitecliff Bay. Fig. 8.2 = *Polystomella latidorsata* Reuss, 1864. Description: test planispiral, involute, inflated, periphery rounded; sutures nearly flush with numerous small retral processes between which there are pore-like openings; aperture of pores hidden by pustular ornament which covers the apertural face and the adjacent areas of older test.
Remarks: differs from *E. hiltermanni* in having a very inflated test. PB Auversian–Sannoisian (L. MW) BB Ypresian–Wemmelian (K) TR m–1 Eoc. *Palaeoecology:* marine, inner to mid shelf.

Epistomaria rimosa (Parker and Jones)
Plate 8.6, Figs. 5–7, × 135, diam. 300 μm. Bracklesham Group, Selsey Formation, Selsey. Fig. 8.2. = *Discorbina rimosa* Parker and Jones, 1865. Description: test trochospiral, elongate in outline, periphery rounded, spiral side gently convex with deep sutures; umbilical side involute, internal walls give appearance of supplementary chambers around the umbilicus; primary aperture an interiomarginal slit, but secondary apertures are present along periphery, on the umbilical face, and in the sutures.
Remarks: PB Cuisian–Marinesian (L, MW) BB no record TR m Eoc. *Palaeoecology:* normal marine, shelf, varied substrates.

Epistomaria separans Le Calvez, 1949.
Plate 8.6, Figs. 8, 9, × 105, diam. 380 μm. Bracklesham Group, Selsey Formation, Selsey. Fig. 8.2. Description: test trochospiral, with convex spiral side, flat umbilical side and rounded periphery; sutures between chambers of final whorl deeply incised, giving the appearance of separation of the chambers; primary aperture an interiomarginal slit, secondary apertures in peripheral parts of incised sutures.
Remarks: PB Lutetian–Marinesian (L, MW) BB no record, TR m Eoc. *Palaeoecology:* normal marine shelf.

Epistominella oveyi (Bhatia)
Plate 8.6, Figs. 10–12, × 285, diam. 140 μm. Hamstead Formation, Upper Hamstead Beds, Cerithium Bed, Hamstead. Fig. 8.2. = *Pseudoparrella oveyi* Bhatia, 1955. Description: test trochospiral, 6–8 chambers in final whorl; planoconvex, with conical spiral side, periphery subrounded; sutures on spiral side slightly swept back, those on umbilical side radial; aperture comma-shaped.
Remarks: microspheric forms are more conical than the megalospheric forms. Differs from *E. vitrea* in being planoconvex, and in having less swept-back sutures on the spiral side. *E. oveyi* resembles and may represent juvenile *Caucasina coprolithoides.* PB Marinesian (L) BB Paniselian–Wemmelian (K) Rupelian (B), TR Olig. *Palaeoecology:* brackish to marine, shallow shelf, muddy substrate.

Epistominella vitrea Parker, 1953
Plate 8.6, Figs. 13–15, × 285, diam. 140 μm. London Clay Formation, Alum Bay. Figs. 8.2, 8.3. Description: test trochospiral, 5–7 chambers in final whorl; biconvex, periphery subrounded, sutures on spiral side backward curving, those on umbilical side radial; aperture comma-shaped.
Remarks: Haynes (1956) has recognised slight morphological differences between micro and megalospheric generations. See *E. oveyi* for differences. PB Cuisian–Lutetian (MW) BB no record. TR 1 Palaeoc.–1 Eoc.; Recent. *Palaeoecology:* slightly brackish to normal marine, inner to mid shelf, on muddy substrates.

Gyroidinoides angustiumbilicatus (Ten Dam)
Plate 8.6, Figs. 16–18, × 145, diam. 280 μm. London Clay Formation, Whitecliff Bay. Figs. 8.2, 8.3. = *Gyroidina augustiumbilicata* Ten Dam, 1944. Description: test trochospiral planoconvex, with strongly convex umbilical side and flat to slightly convex spiral side, periphery subacute; sutures radial on both sides; umbilical side with deep umbilicus; aperture an interiomarginal slit.
Remarks: differs from *G. octocamerata* in having a more highly convex umbilical side and in the orientation of the sutures. PB, BB no record. TR e Eoc. *Palaeoecology:* brackish to normal marine, shelf, muddy substrates.

Gyroidinoides danvillensis var. *gyroidinoides* (Bandy)
Plate 8.6, Figs. 18–20, × 135, diam. 300 μm. Thanet Formation, Upper Pegwell Marls, Pegwell. Fig. 8.3. = *Valvulineria danvillensis* var. *gyroidinoides* Bandy, 1949. Description: subglobose, umbilical side highest, periphery rounded; 8–9 chambers in final whorl; apertural face quadrangular, aperture interiomarginal.
Remarks: differs from *G. danvillensis* s.s. (Howe and Wallace) in having 8–9 chambers per whorl rather than 6. PB, BB no record. TR 1 Palaeoc. *Palaeoecology:* normal marine, inner to mid shelf, probably tolerant of poorly aerated muddy bottoms.

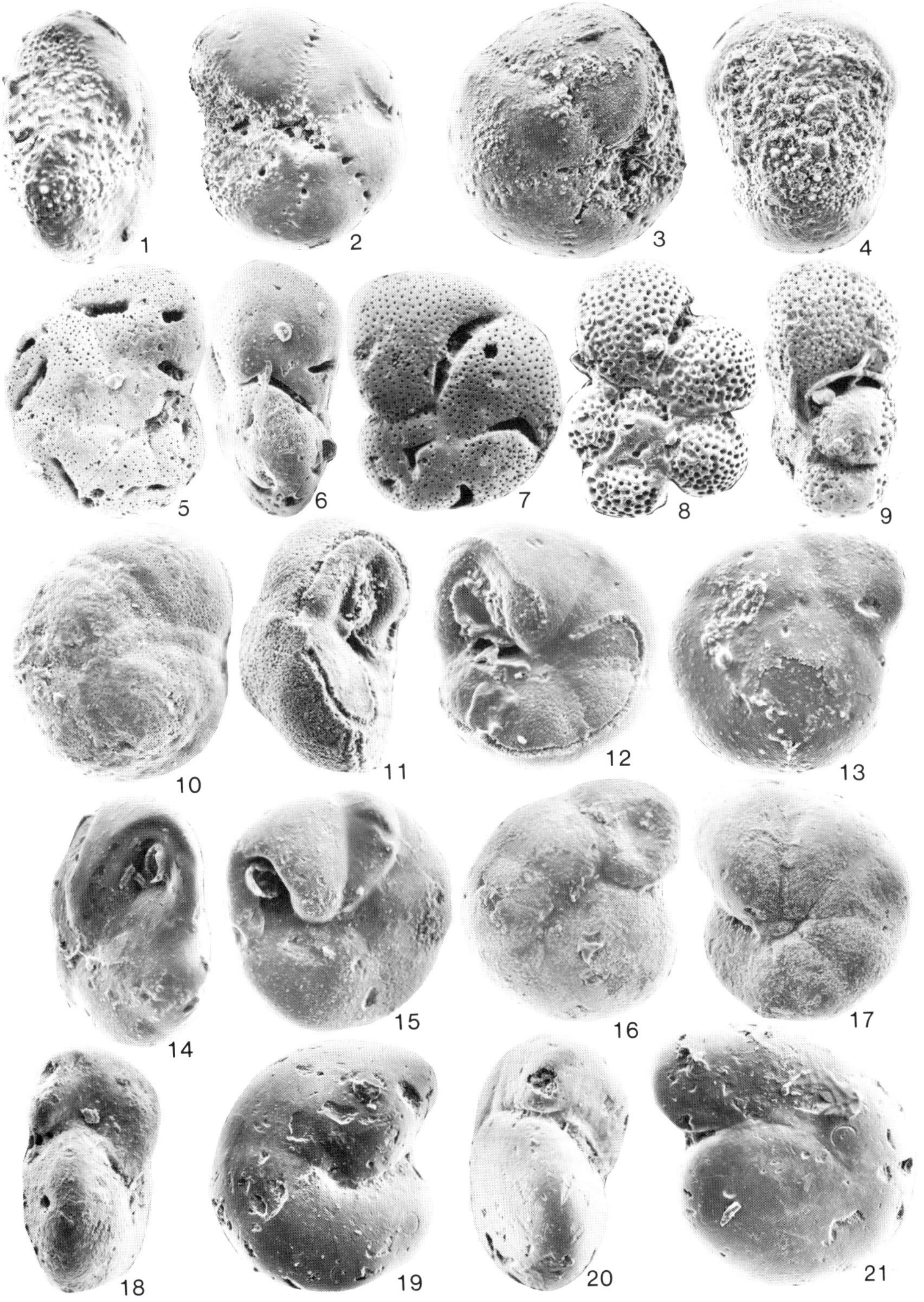

Plate 8.6

Gyroidinoides octocamerata (Cushman and Hanna)
Plate 8.7, Figs. 1–3, × 200, diam. 200 μm. Bracklesham Group, Selsey Formation, Selsey. Fig. 8.2. = *Gyroidina soldanii* d'Orbigny var. *octocamerata* Cushman and Hanna, 1927. Description: test trochospiral, planoconvex, with high convex spiral side, periphery rounded; sutures oblique on spiral side, radial on umbilical side; umbilical side with depressed umbilicus; aperture an interiomarginal slit.
Remarks: see *G. angustiumbilicatus*. PB Cuisian–Stampian (L), BB Ypresian–Asschian (K), TR m–l Eoc. *Palaeoecology:* normal marine, inner to mid shelf.

Hanzawaia producta (Terquem)
Plate 8.7, Figs. 4–6, × 105, diam. 380 μm. Bracklesham Group, Selsey Formation, Selsey. Fig. 8.2. = *Truncatulina producta* Terquem, 1882. = *Cibicides productus* (Terquem) emend. Le Calvez, 1949. Description: test trochospiral, planoconvex, with flat, fairly involute spiral side and convex involute umbilical side; periphery angular with keel; 6–7 chambers in outer whorl, the later ones having flaps on the spiral side; aperture extending from periphery on to spiral side.
Remarks: the sharp keel and the concave surface next to it on the umbilical side distinguishes this species from others of *'Cibicides'* type. PB Cuisian–Marinesian (L, MW) BB Ypresian–Asschian (K), TR e–l Eoc. *Palaeoecology:* normal marine, shelf.

Heterolepa pegwellensis (Haynes)
Plate 8.7, Figs. 7–9, × 160, diam. 250 μm. Thanet Formation, Upper Pegwell Marls, Pegwell. Fig. 8.3. = *Hollandina pegwellensis* Haynes, 1956. Description: biconvex or highest in spiral side, periphery acute and warped; up to 27 chambers in microspheric (7:9:9:–. or 7:9:8:– in successive whorls), becoming twice as long as high by 20th, up to 22 in megalospheric generation (7:8:7 or 7:7:7: in successive whorls), becoming twice as long as high by 13th; sutures flush, markedly swept back on spiral side; coarse pores on spiral side; aperture an interiomarginal slit.
Remarks: a very small species (<300 μm). PB, BB no record. TR l Palaeoc. *Palaeoecology:* normal marine, inner to mid shelf, probably tolerant of poorly aerated bottoms.

Marginulina enbornensis Bowen, 1954
Plate 8.7, Figs. 10, 11, × 21, length 1.9 mm. London Clay Formation, Hadley Wood. Fig. 8.3. Description: test initially planispiral then uncoiling, laterally compressed; aperture radiate on a neck, terminal; test commonly ornamented with longitudinal costae or tubercles.
Remarks: PB no record, BB Ypresian (K), TR e Eoc. *Palaeoecology:* normal marine.

Melonis affine (Reuss)
Plate 8.7, Figs. 12, 13, × 125, diam. 320 μm. Barton Formation, Middle Barton Beds (D), Barton. Fig. 8.2 = *Nonionina affinis* Reuss, 1851. Common synonym: *Nonion affine.* Description: test planispiral, involute, compressed, biumbilicate, periphery rounded; around 10 chambers in last whorl; aperture an interiomarginal arch.
Remarks: PB Lutetian–Sannoisian (L, MW), BB Ypresian–Asschian (K). Tongrian–Rupelian (B), TR m–l Eoc. *Palaeoecology:* marine, inner to mid shelf.

Nodosaria latejugata Gümbel, 1868
Plate 8,7, Fig. 14, × 15, length 2 6 mm. London Clay Formation, Hadley Wood. Description: test uniserial, chambers inflated, with stout longitudinal ribs that cross the depressed sutures; circular in cross section.
Remarks: PB no record, BB Paniselian (K), TR e Eoc. *Palaeoecology:* normal marine.

Nonion applinae Howe and Wallace, 1932
Plate 8.7, Figs. 15, 16, × 135, diam. 300 μm. Thanet Formation, Reculver Silts, Reculver. Fig. 8.3. Description: compressed, periphery rounded, apertural face high, oval; 7–9 chambers; sutures radial; umbilici small, granulate.
Remarks: PB, BB no record. TR l Palaeoc. *Palaeoecology:* normal marine–?brackish, inner shelf.

Nonion laeve (d'Orbigny)
Plate 8.7, Figs. 17, 18, × 100, diam, 400 μm. Bracklesham Group, Selsey Formation, Selsey. Figs. 8.2, 8.3. = *Nonionina laevis* d'Orbigny, 1826. Common synonym: *Elphidium laeve.* Description: test planispiral, compressed; periphery subangular to rounded; 10–14 chambers in last whorl; umbilici with umbilical boss; sutures depressed, ornamented with tubercules; aperture an interiomarginal slit obscured by tubercles.
Remarks: this form shows variation in the number of chambers, depth of depression of the sutures and the number and size of the umbilical bosses. PB Sparnacian–Ludian (L, MW). BB Ypresian–Asschian (K). TR l Palaeoc.–e Olig. *Palaeoecology:* brackish to normal marine, marginal marine to inner shelf.

Nonion parvulum (Grzybowski)
Plate 8.7, Figs. 19, 20, × 160, diam. 250 μm. Hamstead Formation, Upper Hamstead Beds, Cerithium Bed, Hamstead. Fig. 8.2. = *Anomalina parvula* Grzybowski, 1896. Description: test planispiral, involute, periphery rounded; sutures depressed; 5 chambers in last whorl; aperture an interiomarginal slit, normally obscured by pustulose material.
Remarks: superficially similar to *Elphidium hiltermanni* but with fewer chambers and an absence of retral processes. PB Sannoisian (L, MW) BB no record. TR l Eoc.–m Olig. *Palaeoecology:* brackish to normal marine, lagoonal, mainly muddy substrates.

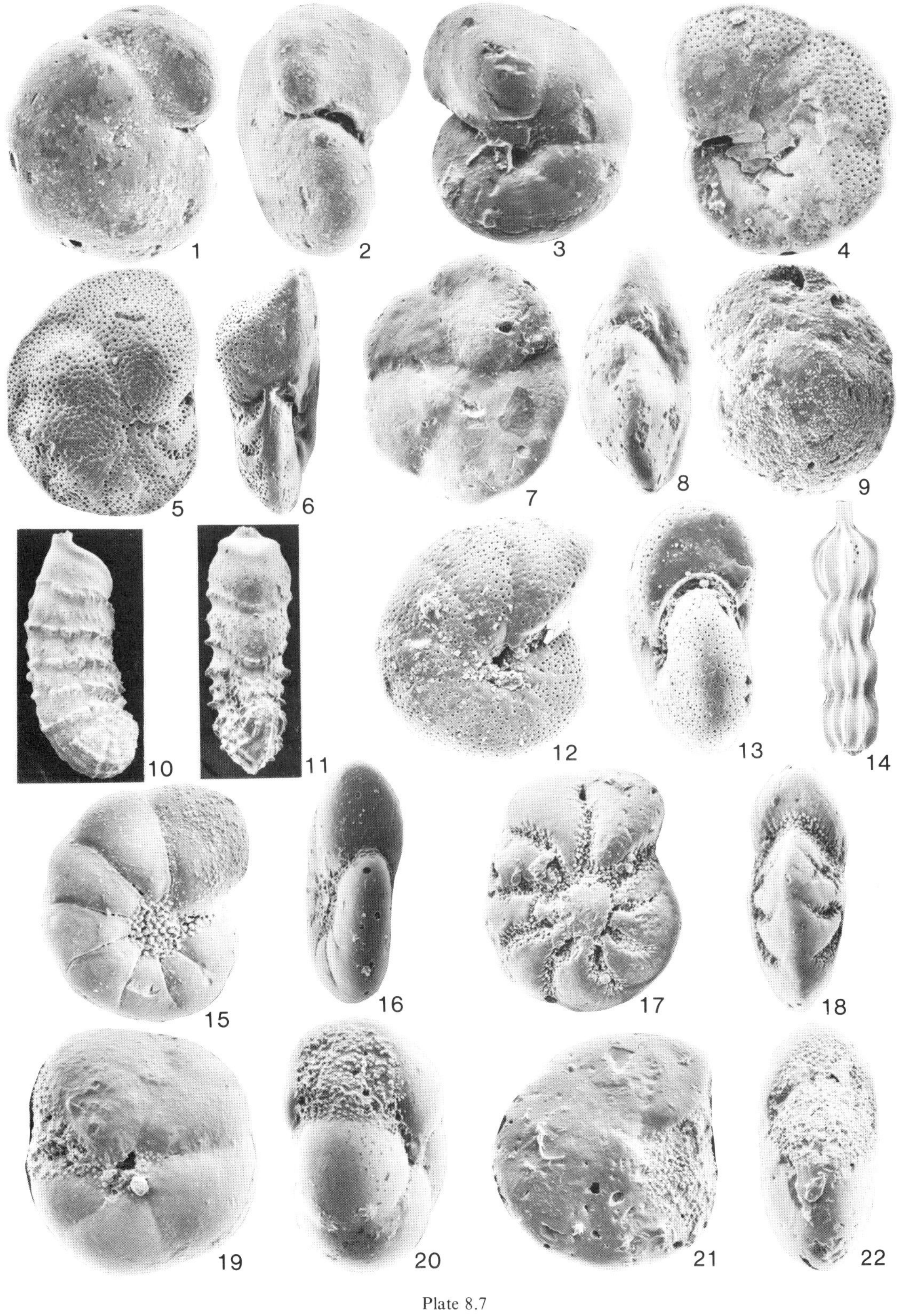

Plate 8.7

Nonion reculverensis Haynes, 1956
Plate 8.7, Figs 21, 22, × 135, diam. 300 μm. Thanet Formation, Reculver Silts, Reculver. Fig. 8.3. Description: planispiral, periphery entire, subangular; bi-umbonate; 8 chambers in outer whorl; sutures limbate, swept back, aperture obscured by granular pustules.
Remarks: a small species (<250 μm). Some individuals show incipient retral processes across the sutures,. PB, BB no record. TR 1 Palaeoc. *Palaeoecology:* normal marine–?brackish, inner shelf.

Pararotalia audouini (d'Orbigny)
Plate 8.8, Figs 1–3, × 57, diam. 700 μm. Solent Formation, Middle Headon Beds, Whitecliff Bay. Fig. 8.2. = *Rotalia audouini* d'Orbigny, 1826. Common synonym: *Pararotalia subinermis* Bhatia, 1955. Description: test trochospiral, unequally biconvex; 6–8 chambers in final whorl; periphery sharply keeled and with granular ornament; keel produced into spine-like outgrowths in latest chambers; umbilical side very convex with deep umbilicus largely filled with a calcite boss.
Remarks: differs from *P. inermis* (Terquem) in having a more angled outline in equatorial view, and from *P. spinigera* (Le Calvez) in having a sharper periphery and only short spines. PB Lutetian–Marinesian (L, MW) BB present but range not distinguished from other *Pararotalia* spp. (K) TR 1 Eoc. *Palaeoecology:* normal marine, inner shelf, varied substrates.

Pararotalia curryi Loeblich and Tappan, 1957
Plate 8.8, Figs. 4–6, × 135, diam. 300 μm. Solent Formation, Middle Headon Beds, Whitecliff Bay. Figs. 8.2, 8.3. Common synonym: *Rotalia canui* Cushman of some authors. Description: test trochospiral, biconvex, periphery subacute; spiral side convex; umbilical side convex with umbilical plug; 4–6 chambers in outer whorl, sometimes with a short peripheral spine.
Remarks: this species resembles *P. spinigera* but the latter is planoconvex and has more pronounced peripheral spines. PB Cuisian–Sannoisian (L, MW) BB Tongrian–Rupelian (L). TR e Eoc.–e. Olig. *Palaeoecology:* mainly brackish but extending into normal marine, inner shelf.

Pararotalia inermis (Terquem)
Plate 8.8, Figs. 7–9, × 57, diam 700 μm. Bracklesham Group, Selsey Formation, Selsey. Fig. 8.2. = *Rotalina inermis* Terquem, 1882 = *Pararotalia inermis* (Terquem) emend. Loeblich and Tappan, 1957. Description: test trochospiral, biconvex, periphery acute and keeled with occasional short spines; umbilical side with deep umbilicus around an umbilical plug, each chamber with a boss adjacent to the umbilicus; aperture on apertural face, with lip.
Remarks: differs from *P. audouini* in having a more rounded outline, being more equally biconvex, and in having a boss on the umbilical portion of each chamber. *P. spinigera* is planoconvex and much smaller. PB Lutetian–Marinesian (L, MW) BB no record, TR m–1 Eoc. *Palaeoecology:* normal marine, inner shelf, varied substrates.

Pararotalia spinigera (Le Calvez)
Plate 8.8, Figs. 10–12, × 115, diam. 350 μm. Bracklesham Group, Selsey Formation, Selsey. Fig. 8.2 = *Globorotalia spinigera* (Terquem) Le Calvez, 1949 (not *Rosalina spinigera* Terquem) = *Pararotalia spinigera* (Le Calvez) emend. Loeblich and Tappan, 1957. Description: test trochospiral, planoconvex, spiral side convex, umbilical side flat, periphery angled; umbilical side with umbilicus and plug; 5–6 chambers in outer whorl, the older ones bearing a short peripheral spine; aperture on apertural face, bearing a lip.
Remarks: this species resembes *P. curryi* but the latter is biconvex and has only feebly developed peripheral spines. PB Sparnacian–Ludian (L, MW), BB present but range not distinguished from that of other *Pararotalia* spp, TR e–1 Eoc. *Palaeoecology:* normal marine, inner shelf, varied substrates.

Praeglobobulimina ovata (d'Orbigny).
Plate 8.8, Fig. 13, × 120, length 340 μm. T. Fig. 8.10a, b, London Clay Formation, Whitecliff Bay. Figs. 8.2, 8.3. = *Bulimina ovata* d'Orbigny, 1846 = *Praeglobobulimina ovata* (d'Orbigny) emend. Haynes, 1954. Description: test triserial, elongate, with strongly overlapping chambers; aperture an elongate loop perpendicular to the basal margin of the apertural face.
Remarks: PB no record. BB Wemmelian–Asschian (K) TR 1 Palaeoc.–e Eoc. *Palaeoecology:* slightly brackish, inner shelf, mainly sandy clay substrates.

Protelphidium hofkeri Haynes, 1956
Plate 8.8, Figs. 14, 15, × 165, diam. 240 μm. Thanet Formation, Reculver Silts, Reculver. Fig. 8.3. Description: planispiral, periphery broadly rounded, lobate; 8–9 chambers in outer whorl; sutures backward curving, impressed and excavated; umbilici filled with granular calcite that extends along sutures; aperture a low slit.
Remarks: this is the most pustulate of all the *Protelphidium* spp. PB Thanetian (L) BB no record. TR 1 Palaeoc. *Palaeoecology:* normal marine–?brackish, inner shelf.

Protelphidium sp. 1
Plate 8.8, Figs. 16, 17, × 135, diam. 300 μm. Barton Formation, Middle Barton Beds (F), Barton. Fig. 8.2. = *Protelphidium* sp. 1 of Murray and Wright, 1974. Description: test planispiral, biconvex; periphery subacute; 8–10 chambers in outer whorl; umbilici with a small amount of tubercular ornament; aperture an interiomarginal row of pores.
Remarks: test more compressed and biconvex than in other *Protelphidium* species. PB Marinesian (MW), BB no record. TR m–1 Eoc. *Palaeoecology:* slightly brackish to normal marine, inner to mid shelf.

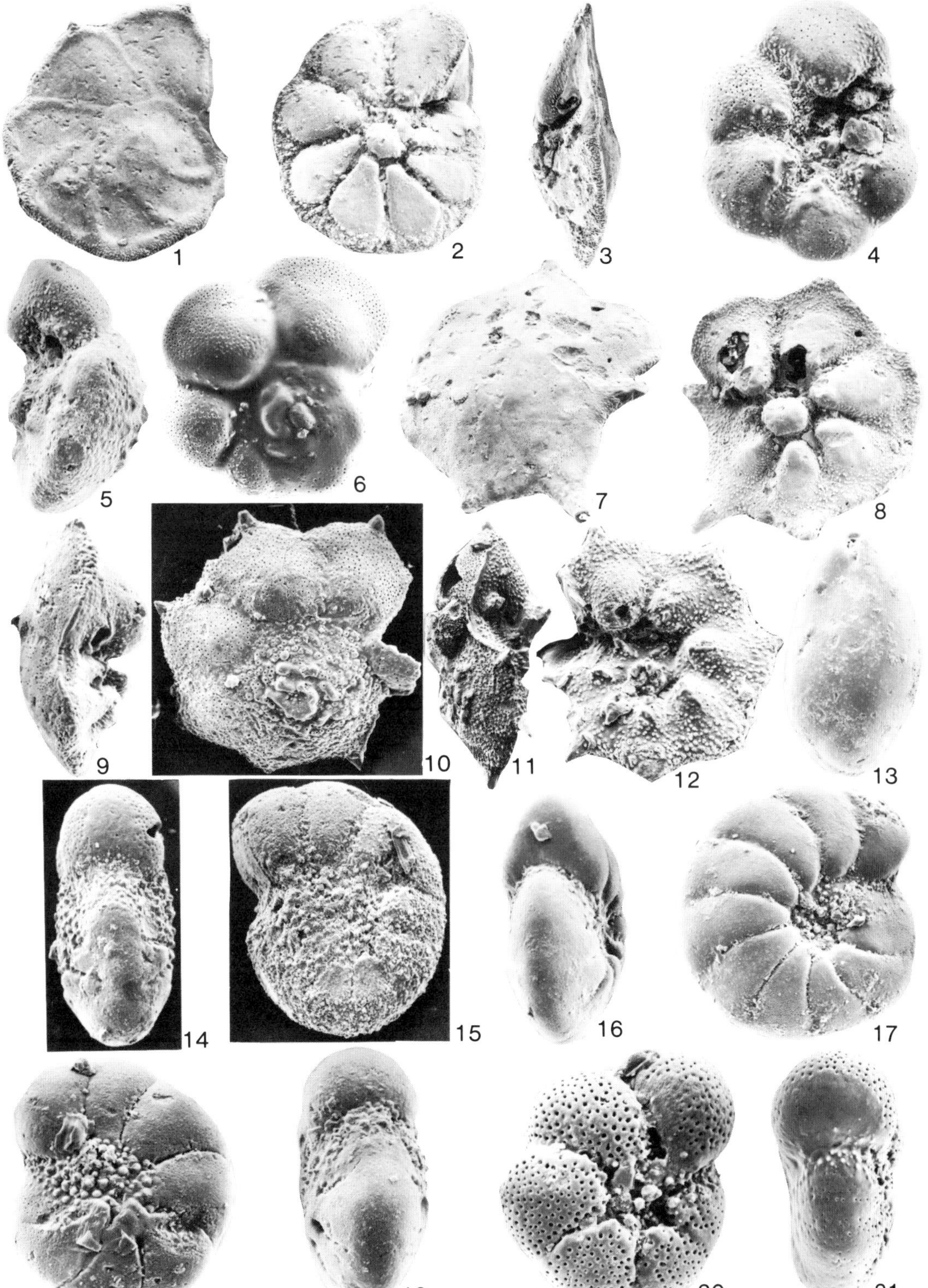

Plate 8.8

Protelphidium sp. 2
Plate 8.8, Figs. 18, 19, × 165, diam. 240μm. Bracklesham Group, Wittering Formation, Whitecliff Bay. Fig. 8.2 = *Protelphidium* sp. 2 of Murray and Wright, 1974. Description: test planispiral, biconvex; periphery rounded; 7–9 chambers in the outer whorl; sutures depressed and ornamented with tubercles except over the periphery; tubercules also ornament the umbilici; aperture an interiomarginal row of pores.
Remarks: this species has a more biconvex test than *P.* cf. *roemeri* or *Protelphidium* sp. 1 and is less coarsely perforate than *Protelphidium* sp. 4. PB, BB no record. TR e–m Eoc. *Palaeoecology:* brackish (20–30‰), estuarine.

Protelphidium sp. 4
Plate 8.8, Figs. 20, 21, × 250, diam. 160 μm. Solent Formation, Middle Headon Beds, Colwell Bay. Fig. 8.2. = *Protelphidium* sp. 4 of Murray and Wright (1974). Description: test planispiral, chambers inflated, periphery rounded and lobulate; 6–7 chambers in the outer whorl; sutures deeply depressed; umbilicus ornamented with tubercules; aperture an interiomarginal row of pores; wall pores coarse except on apertural face.
Remarks: this is a small species with inflated chambers and a relatively coarsely perforate wall. PB, BB no record. TR 1 Eoc.–e Olig. *Palaeoecology:* brackish to normal marine, estuarine and lagoonal.

Protelphidium cf. *P. roemeri* (Cushman)
Plate 8.9, Figs. 1, 2, × 160, diam. 250 μm. Bembridge Formation, Bembridge Marl, Whitecliff Bay. Fig. 8.2. = *Nonion roemeri* Cushman, 1936. Description: test planispiral, with fairly flat sides; periphery rounded and lobulate; 8–10 chambers in the outer whorl; tubercular ornament extends from the umbilici into the umbilical ends of the depressed sutures; aperture an interiomarginal row of pores.
Remarks: the flat-sided form and rounded periphery distinguishes this from other *Protelphidium* species. PB Sannoisian. (MW) BB 1 Olig. TR 1 Eoc.–m Olig. *Palaeoecology:* brackish, lagoonal.

Pullenia quinqueloba (Reuss)
Plate 8.9, Figs. 3, 4, × 135, diam. 300 μm. London Clay Formation, Alum Bay. Figs. 8.2, 8.3. = *Nonionina quinqueloba* Reuss, 1851. Description: test planispiral, involute, compressed with subrounded periphery; 5–6 chambers in last whorl; aperture an interiomarginal slit.
Remarks: superficially similar to *Nonion parvulum* but lacking the pustulose ornament around the aperture. PB Auversian–Marinesian (L). BB Ypresian–Asschian (K) Rupelian (B). TR 1 Palaeoc.–e Eoc.; Rec. *Palaeoecology:* normal marine, mid shelf, muddy substrate.

Pulsiphonina prima (Plummer)
Plate 8.9, Figs. 5–7, × 200, diam. 200 μm T. Fig. 8.10c, d. London Clay Formation, Alum Bay. Fig. 8.2 = *Siphonina prima* Plummer, 1927. Description: test trochospiral compressed, biconvex, periphery acute; 4–5 chambers in last whorl; aperture an arch close to the periphery.
Remarks: PB Thanetian–Cuisian (L, MW) BB Ypresian–Bruxellian (K). TR 1 Palaeoc.–e Eoc. *Palaeoecology:* slightly brackish to normal marine, inner shelf, mainly muddy substrates.

Robertina germanica Cushman and Parker, 1938
Plate 8.9, Fig. 8, × 110, length 360 μm. Barton Formation, Middle Barton Beds (C), Barton. Fig. 8.2. Description: test high trochospiral, elongate; chambers divided internally; aperture loop shaped; wall aragonitic.
Remarks: PB no record. BB Ledian–Asschian (K). TR m–1 Eoc. *Palaeoecology:* normal marine shelf.

Rosalina araucana d'Orbigny, 1839
Plate 8.9, Figs. 9–11, × 155, diam. 260 μm. Solent Formation, Upper Headon Beds. Headon Hill. Fig. 8.2. Common synonyms: *Discorbis araucanus, Valvulineria araucana.* Description: test trochospiral, planoconvex, spiral side gently convex; umbilical side has depressed, star-shaped umbilicus; 8 chambers in last whorl, final chamber large; aperture interiomarginal, umbilical.
Remarks: PB no record. BB Tongrian–Rupelian (B, as *Discorbis* sp.). TR 1 Eoc.–e Olig. *Palaeoecology:* brackish to normal marine, intertidal to shallow lagoonal on muddy substrates.

Sagrina selseyensis (Heron-Allen and Earland)
Plate 8.9, Figs. 12, 13, Fig. 12, × 100, final chamber missing, Fig. 13, × 160, length 400 μm. Bracklesham Group, Selsey Formation, Selsey. Fig. 8.2. = *Bigenerina selseyensis* Heron-Allen and Earland, 1909. Description: test initially biserial, (6–8 chambers), then uniserial with 2–5 chambers of oval cross section; aperture oval, terminal, with a raised lip.
Remarks: BB Ledian–Asschian (K). TR m–1 Eoc. *Palaeoecology:* normal marine, inner shelf.

Tappanina selmensis (Cushman)
Plate 8.9, Figs. 14. 15, Fig. 14, × 100, Fig. 15, × 120, length 400 μm. Thanet Formation, Reculver Silts, Reculver. Fig. 8.3 = *Bolivinita selmensis* Cushman, 1933. Description: test biserial with concave faces; aperture a narrow arch at base of the final chamber.
Remarks: PB Thanetian (L). TR 1. Palaeoc. *Palaeoecology:* normal marine.

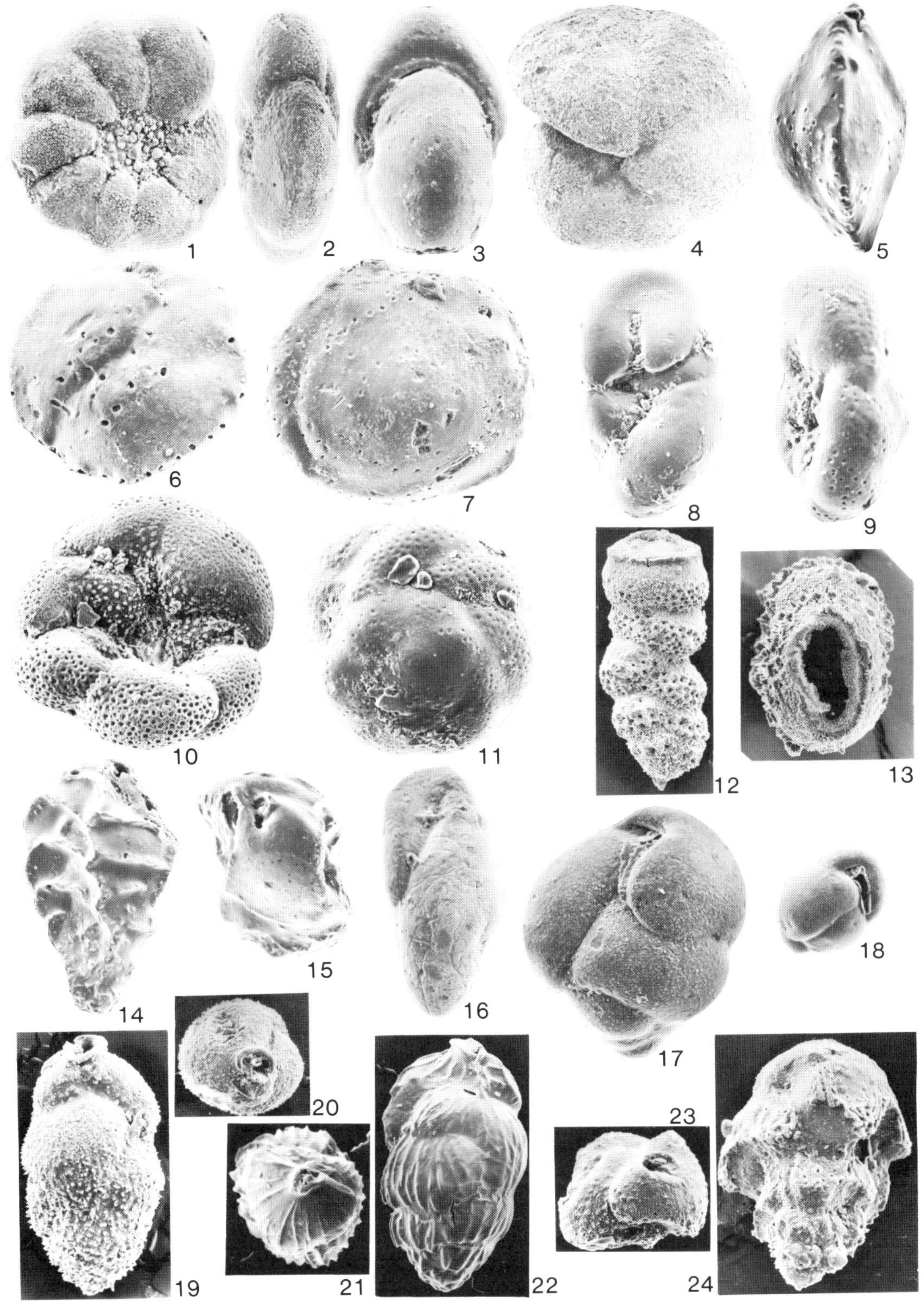

Plate 8.9

Turrilina andreaei Cushman, 1948
Plate 8.9, Fig. 16, × 155, length 260 μm. T. Fig. 8.10e. Solent Formation, Middle Headon Beds, Headon Hill. Fig. 8.2. = *Bulimina acicula* Andreae, 1884 (non *Bulimina acicula* Costa, 1856). Description: test a high trochospiral of around 6 whorls with many chambers; sutures depressed; aperture a small slit in a depressed area of the apertural face.
Remarks: PB Sannoisian (L, MW). BB Tongrian (L, as *T. acicula*) TR 1 Eoc.–e Olig. *Palaeoecology:* slightly brackish, normal marine, to hypersaline lagoons.

Turrilina brevispira Ten Dam, 1944
Plate 8.9, Figs. 17, 18, Fig. 17, × 200, Fig. 18, × 100, length 200 μm. London Clay Formation, Sheppey. Fig. 8.3. Description: a high trochospiral with 3 inflated chambers in the final whorl.
Remarks: PB Lutetian (MW). BB Ypresian (K). TR e Eoc. *Palaeoecology:* normal marine.

Uvigerina batjesi Kaasschieter, 1961
Plate 8.9, Figs. 19, 20, × 145, length 280 μm. London Clay Formation, Whitecliff Bay. Figs. 8.2, 8.3. Description: test initially triserial, becoming uniserial, almost circular in section; chambers inflated, sutures depressed in later part; aperture terminal, with a slight neck and lip.
Remarks: the circular cross section and absence of costae distinguish this species from *U. germanica* and *U. muralis*. PB no record. BB Ypresian–Paniselian (K). TR e Eoc. *Palaeoecology:* slightly brackish, inner to mid shelf, muddy substrates.

Uvigerina germanica (Cushman and Edwards)
Plate 8.9, Figs. 21, 22, × 135, Length 300 μm. Hamstead Formation, Upper Hamstead Beds, Cerithium Bed, Hamstead. Fig. 8.2. = *Angulogerina germanica* Cushman and Edwards, 1938. Description: test initially triserial becoming uniserial, triangular in section; chambers inflated, earlier ones ornamented with longitudinal costae, later ones smooth; aperture terminal, large, elliptical, with a lip.
Remarks: the presence of costae is distinctive. PB Sannoisian (L). BB Rupelian (B). TR m Olig. *Palaeoecology:* normal marine, shelf, muddy substrate.

Uvigerina muralis Terquem, 1882
Plate 8.9, Figs. 23, 24, × 200, length 200 μm. Bracklesham Group, Selsey Formation, Selsey. Fig. 8.2. Common synonym: *Angulogerina muralis* (Terquem). Description: test initially triserial, becoming uniserial; sides angular, formed of irregular chambers with truncated margins; sutures deep and wide; aperture small on a short neck.
Remarks: the triangular cross section is distinctive. PB Sparnacian–Marinesian (L, MW). BB Paniselian–Asschian (K). TR m–1 Eoc. *Palaeoecology:* normal marine, inner shelf, muddy substrate.

8.7.1.4 Planktonic species

Globigerina chascanona Loeblich and Tappan, 1957
Plate 8.10, Figs. 1–3, × 120, diam. 270 μm. London Clay Formation, Leca Works, Ongar, Essex (= *Globorotalia esnaensis* MW). Description: test tightly coiled, rather high-spired; 4½–5 chambers in last whorl, increasing slowly; surface pitted, with spines especially on the umbilical side; aperture an umbilical arch, with a narrow lip.
Remarks: a variable species, widespread in the London Clay. PB Ypresian (Varengeville, as *intermedia,* Bignot, 1963), BB Ypresian–Paniselian (K, as cf. *varianta*). TR 1 Palaeoc.–e Eoc.

Globigerina triloculinoides Plummer, 1926
Plate 8.10, Figs. 4, 5, × 120, diam. 210 μm. Thanet Formation, Reculver, Kent. Description: test a low trochospiral, spiral side almost flat; 3½ globular chambers in final whorl, rapidly increasing in size; surface with rather wide-spaced pore-pits; aperture a low arch, with a thick lip.
Remarks: rare in the Thanet Formation, and possibly derived. PB, BB not recorded, Danian of Denmark (Auctt) TR Palaeoc.

Globigerina patagonica Todd and Kniker, 1952
Plate 8.10, Figs. 6–8, 10–12, × 120, diam. 330 μm, London Clay Formation, Leca Works, Ongar Essex. Description: test low-spired, typically 3½ subglobular chambers in the final whorl; surface with well-defined pore-pits on all chambers; aperture a slightly twisted umbilical arch.
Remarks: Specimens with 3 chambers per whorl resemble *triangularis* White. PB no record, BB Ypresian (K, as *triloculinoides*), NW Germany (Berggren, 1969). TR e Eoc.

Globigerinatheka barri Brönnimann, 1952
Plate 8.10, Fig. 9, × 500, Fig. 13, × 120, diam. 280 μm. Amusium Bed, Bracklesham Group, Chilling, Hampshire (Curry *et al.*, 1968). Description: test almost spherical, initially trochospiral, but later chambers overlapping, the last covering nearly half of the test; multiple apertures are tiny, marginal in position, and have a thickened lip (Fig. 9).
Remarks: PB, BB no record. TR m–1 Eoc.

Globorotalia broedermanni Cushman and Bermudez, 1949
Plate 8.10, Figs. 14–16, × 120, diam. 240 μm. Bed W12, Wittering Formation, East Wittering, Sussex. Description: test a low trochospiral, approaching biconical, with 4½–5 radially compressed and somewhat angular chambers in the

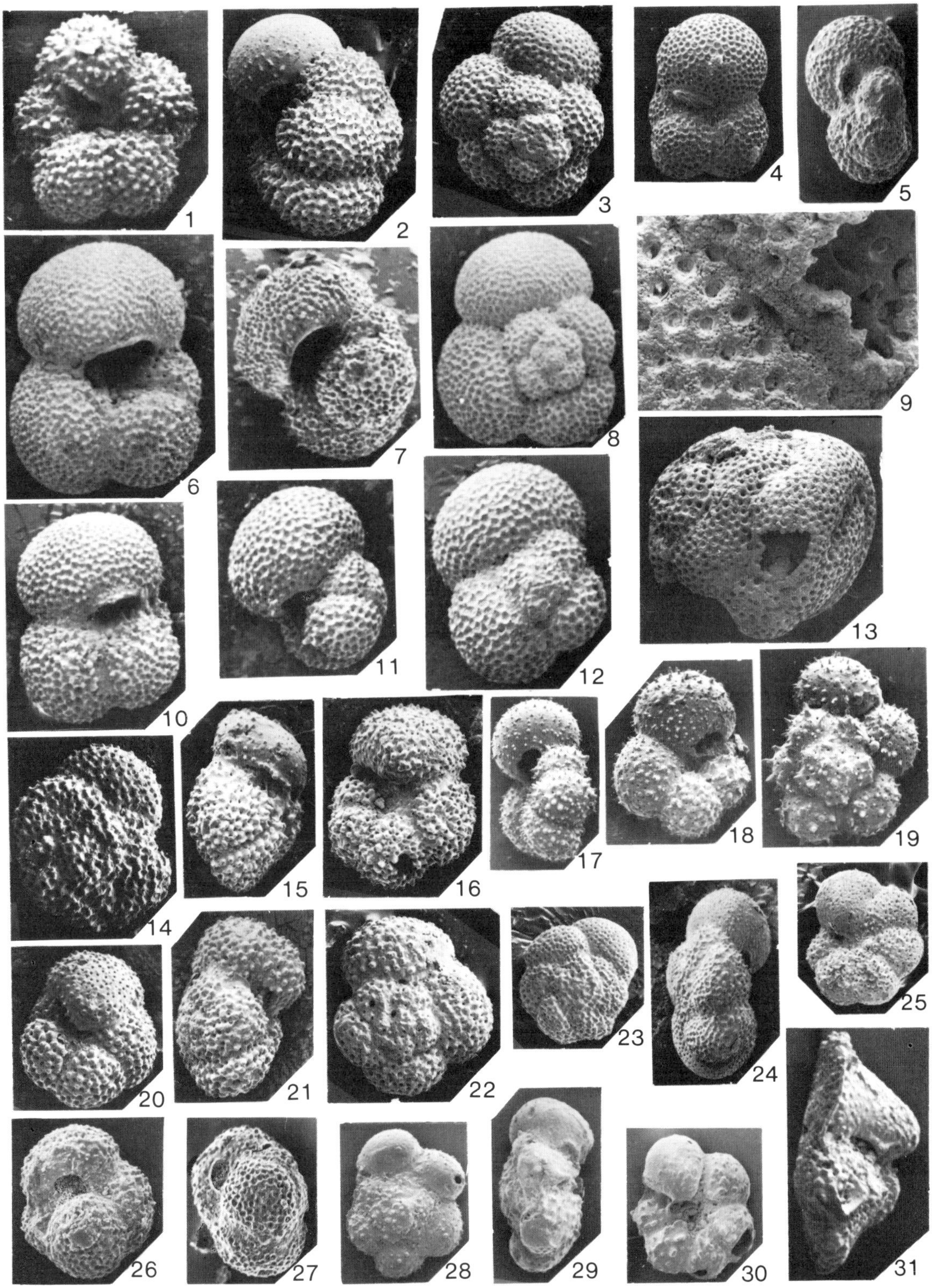
1
2
3
4
5
6
7
8
9
10
11
12
13
14
15
16
17
18
19
20
21
22
23
24
25
26
27
28
29
30
31

Plate 8.10

last whorl; periphery in side view subangular, surface hispid.
Remarks: PB Cuisian (L), BB no record. TR e–m Eoc.

Globorotalia danvillensis (Howe and Wallace)
Plate 8.10, Figs. 17–19, × 120, diam. 250 μm, Bed A_2, Barton Formation, Barton = *Globigerina danvillensis* Howe and Wallace, 1932 = *G.* cf. *angustiumbilicata,* Brönnimann *et al.* (L, MW). Description: test low trochospiral, with 4½–5½ almost spherical chambers in last whorl; surface progressively more spiny in earlier-formed chambers; aperture semicircular, without marked lip.
Remarks: The ornament is very characteristic. PB. Auversian, Marinesian (L), Stampian (? derived), BB Tongrian (Vieux-Joncs) (Curry). TR l Eoc.–m Olig.

Globorotalia cf. *esnaensis* (Leroy)
Plate 8.10, Figs. 26, 27, × 120, diam. 210 μm. Thanet Formation, Reculver, Kent = *Globigerina esnaensis* Leroy, 1953. Description: test subspherical, last whorl of 3½–4 rounded chambers which, viewed from the umbilical area, have a semi-oval to quadrate profile; surface pitted; aperture slightly extraumbilical, with lip.
Remarks: The characteristic quadrate profile seen in specimens from the early Eocene of Germany is only feebly shown in the British material, hence the use of the "cf." prefix. PB no record, BB Ypresian (Moorkens, 1968)? TR l Palaeoc.–e Eoc.

Globorotalia perclara Loeblich and Tappan, 1957
Plate 8.10, Figs. 28–30, × 120, diam. 210 μm. Thanet Formation, Reculver, Kent. Description: test trochospiral, almost planoconvex, last whorl of 5–6 subspherical chambers; sutures radial; periphery lobed; umbilicus open; aperture interiomarginal; ornament of pustules, becoming stronger on earlier chambers.
Remarks: PB Thanetian (Curry), BB no record. TR m Palaeoc.–e Eoc.

Globorotalia reissi Loeblich and Tappan, 1957
Plate 8.10, Figs. 23–25, × 120, diam. 180 μm. London Clay Formation, Leca Works, Ongar, Essex. Description: test low trochospiral, last whorl of 6 slightly compressed chambers; umbilicus shallow, small; aperture small, interiomarginal; ornament slight.
Remarks: specimens from Sheppey have a distinct keel, and were referred to *pseudoscitula* Glaessner (Brönnimann *et al.,* 1968). PB Ypresian (Varengeville), BB Paniselian (Brönnimann, *et al.,* 1968). TR e Eoc.

Globorotalia wilcoxensis Cushman and Ponton, 1932
Plate 8.10, Figs. 20–22, × 120, diam. 240 μm. Bed W12, Wittering Formation, East Wittering, Sussex. Description: test trochospiral, umbilical side very convex, final whorl of 4 subangular chambers, surface hispid, periphery lobate.
Remarks: PB (L, as *pseudotopilensis* Subbotina), BB no record. TR e Eoc.

Globorotalia subbotinae Morozova, 1939
(a) Plate 8.11, Fig. 1, (b) Plate 8.10, Fig. 31, × 120, diam. 290 μm. (a) Bed W12, Wittering Formation, East Wittering, Sussex, (b) Bracklesham Group, English Channel, 50° 11.5′N, 1° 31′W. Description: test subconical, last whorl of 4 angulate chambers, with a strong peripheral keel; surface very rugose, umbilicus small; aperture small, interiomarginal.
Remarks: Fig. 31 has an unusually strong keel and might be ascribed to *marginodentata* Subbotina. PB Cuisian (L), BB Paniselian? (K, fide Berggren, 1969). TR e Eoc.

Globorotalia sp. 1
Plate 8.11, Fig. 2, × 120, diam. 220 μm. Bed W12 Wittering Formation, East Wittering, Sussex. Description: low trochospiral, last whorl 4–5 somewhat flattened chambers, profile subangulate; surface finely papillose; aperture interiomarginal.
Remarks: probably related to ≥⁻ʃ∷→g√→∓′
Remarks: probably related to *G. chapmani* Parr. TR e Eoc.

Globorotalia sp. 2
Plate 8.11, Figs. 3–5, × 120, diam. 200 μm. Bed W12, Wittering Formation, East Wittering, Sussex. Description: low trochospiral, last whorl of 4–5 somewhat flattened chambers, profile rounded; surface coarsely pitted; umbilicus large; aperture subumbilical.
Remarks: this appears to be an undescribed species. TR e Eoc.

Guembelitria triseriata (Terquem)
Plate 8.11, Figs. 6, 7, × 120, length 220 μm. Bed W12, Wittering Formation, East Wittering, Sussex = *Textilaria triseriata* Terquem, 1882. Description: test elongate, triserial, chambers globular, smooth; aperture an interiomarginal arch with a twisted and thickened lip.
Remarks: PB Cuisian–Marinesian (L), BB no record. TR Eoc. (not earliest).

Pseudohastigerina wilcoxensis (Cushman and Ponton)
Plate 8.11, Figs. 8–10, × 120, diam. 290 μm. Bed W12, Wittering Formation, East Wittering, Sussex = *Nonion wilcoxensis* Cushman and Ponton, 1932. Description: planispiral, last whorl of 6 subglobular chambers, rapidly increasing; aperture a semicircular arch, with a marked lip; surface smooth except for the presence of pores.
Remarks: PB no record, BB Paniselian (K, as *micra*) TR e Eoc.

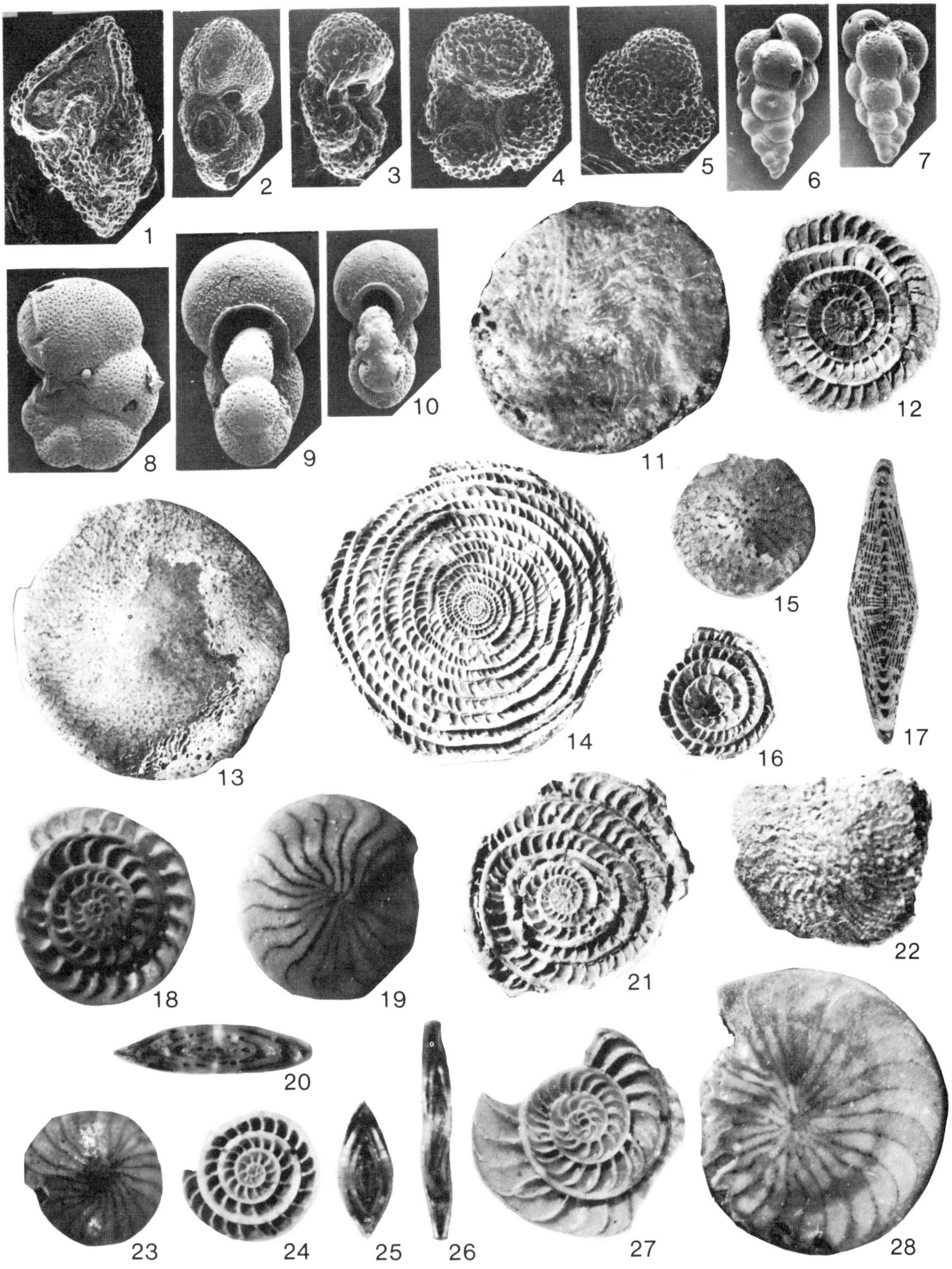

Plate 8.11

8.7.2 Larger benthic species

Nummulites aquitanicus Benoist, 1888
Plate 8.11, Figs. 21, 22 form B × 6, diam. 7 mm. Bed IV, Wittering Formation, Whitecliff Bay. Description: test biconical, septal filaments meandriform, pillars present both along and between filaments; spiral lamina of about 8 turns (B), chambers high.
Remarks: PB no record, BB one doubtful record. TR 1 Eoc.

Nummulites laevigatus (Bruguière)
Plate 8.11, Figs. 13–17, form B, Figs. 13, 14, 17, × 3; form A, Figs. 15, 16, × 6. Diam. (B) 15 mm, (A) 3.5 mm. Bed E6, Earnley Formation, Bracklesham. = *Camerina laevigata* Bruguière, 1792. Description: test biconical, septal filaments reticulate, pillars present, mostly along filaments; spiral lamina of 15 (B) or 5 (A) turns, chambers rather high, septa oblique.
Remarks: an important marker throughout the Anglo-Paris-Belgian area. PB Lutetian (Bl), BB Lutetian (Bl) TR m Eoc.

Nummulites planulatus (Lamarck)
Plate 8.11, Figs. 11, 12 form B, × 6, diam. 7 mm. Bed IV, Wittering Formation, Whitecliff Bay = *Lenticulites planulata* Lamarck, 1804. Description: test lenticular, septal filaments meandriform, pillars absent; spiral lamina of about 7 turns (B), chambers high, septa upright.
Remarks: an important marker throughout the Anglo-Paris-Belgian area. PB Cuisian (Bl), BB Paniselian (Bl), TR e Eoc.

Nummulites prestwichianus (Jones)
Plate 8.11, Figs 26–28 (form A, × 12, diam. 3.8 mm. = *Nummulina planulata* var. *prestwichiana* Jones, 1862. Base bed of Barton Formation, Alum Bay. Description: test discoidal, septal filaments radial to sinuate, spiral lamina of 4 turns (A), chamber height increasing rapidly; A and B generations indistinguishable externally.
Remarks: apart from one record in the English Channel, known only from the Hampshire Basin and southern USSR (Nemkov, 1968). TR m/1 Eoc. boundary.

Nummulites rectus Curry, 1937
Plate 8.11, Figs. 18–20 form A, × 12, diam. 2.7 mm. Barton Formation, Alum Bay, 10 m above base. Description: test lenticular, A and B forms indistinguishable externally, ratio diameter to thickness about 4; septal filaments sinuate, spiral lamina of 4–5 turns (A), chamber height increasing rather slowly; septa strongly curved.
Remarks: as for *N. prestwichianus.*

Nummulites variolarius (Lamarck)
Plate 8.11, Figs. 23–25 form A, × 12, diam. 1.9 mm. Bed XVII, Selsey Formation, Whitecliff Bay = *Lenticulites variolaria* Lamarck, 1804. Description: test lenticular, A and B forms almost indistinguishable externally, ratio diameter to thickness about 2.8, septal filaments radial, curving near margin, spiral lamina of 4 turns (A), septa oblique.
Remarks: an important marker in the Anglo-Paris-Belgian area. Range overlaps slightly those of *N. laevigatus* and *N. prestwichianus.* PB Lutetian–Auversian (Bl), BB Bruxellian–Ledian, Wemmelian? (Bl). TR m Eoc.

Glandulina laevigâta (d'Orbigny)
Fig. 8.11a, b, × 50, length 420 μm. Barton Formation, Middle Barton Beds (C), Barton. Fig. 8.2, 8.3. = *Nodosaria (Glandulina) laevigata* d'Orbigny, 1826. Description: test fusiform, pointed at both ends, circular in section; early chambers biserial in microspheric forms, but mainly or entirely uniserial; sutures flush; aperture terminal, radiate.
Remarks: PB Marinesian (L, MW). BB Paniselian–Asschian (K) Rupelian (B). TR 1 Palaeoc.–1 Eoc. *Palaeocology:* normal marine, inner to mid shelf.

Globulina ampulla (Jones)
Fig. 8.11c–e, × 50, length 460 μm. Barton Formation, Lower Barton Beds (A2), Barton. Figs. 8.2, 8.3. = *Polymorphina ampulla* Jones, 1852. Description: test oval, pointed at both ends, subcircular in cross section; few chambers, added in planes 144° apart; sutures very slightly depressed; aperture radiate.
Remarks: differs from *G. gibba* in being oval rather than globular; differs from *G. inaequalis* and *G. rotundata* in both profile and cross section. PB Thanetian–Auversian (L. MW). BB no record. TR 1 Palaeoc.–1 Eoc. *Palaeoecology:* normal marine, inner to mid shelf.

Globulina inaequalis (Reuss, 1850)
Fig. 8.11f, g, × 50, length, 300 μm. Bracklesham Group, Selsey Formation, Whitecliff Bay. Fig. 8.2. Description: test oval, broader at the base, compressed oval in cross section; sutures more or less flush; aperture radiate.
Remarks: differs from *G. gibba, G. ampulla* and *G. rotundata* in being compressed rather than round in cross section. PB Cuisian–Sannoisian (L, MW). BB Ypresian–Asschian (K). Tongrian–Rupelian (B as part of *G. gibba*) Rupelian (L). TR e–1 Eoc. *Palaeoecology:* normal marine, inner to mid shelf.

Guttulina problema d'Orbigny, 1826
Fig. 8.11h–j, × 50, length 450 μm). Barton Formation, Upper Barton Beds, (H), Barton. Figs. 8.2, 8.3. Description: test somewhat elongate, subtriangular in cross section; chambers arranged in a quinqueloculine series; sutures depressed; aperture radiate.

Remarks: PB Thanetian–Sannoisian (L, MW). BB Ypresian–Asschian (K). Tongrian–Rupelian B). TR 1 Palaeoc.–1 Eoc. *Palaeoecology:* normal marine, inner to mid shelf.

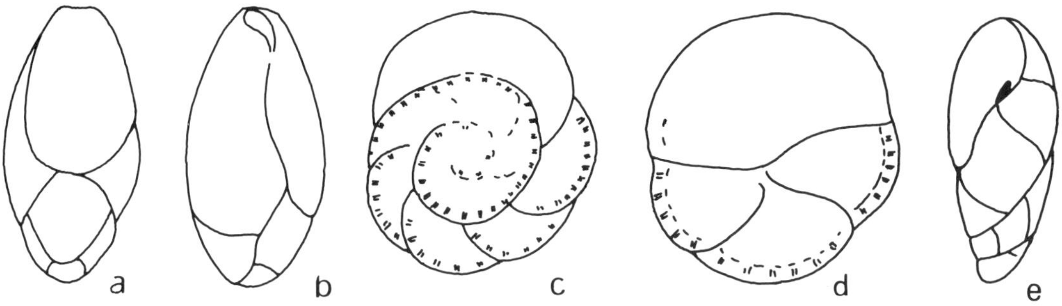

Fig. 8.10 a, b – *Praeglobobulimina ovata* (d'Orbigny), × 120; c, d: *Pulisphonina prima* (Plummer), × 130; e: *Turrilina andreaei* Cushman, × 130.

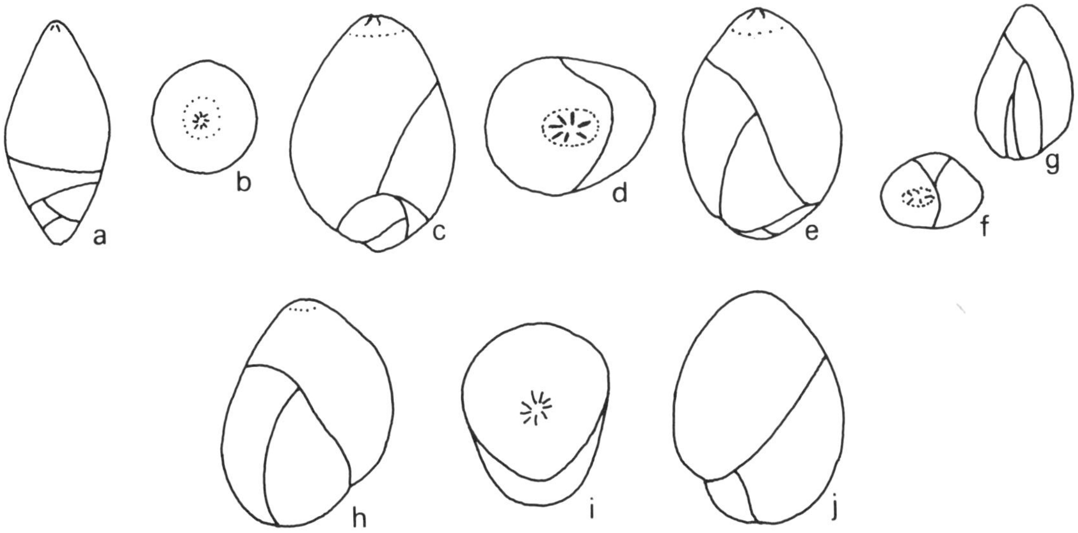

Fig. 8.11 a, b – *Glandulina laevigata* (d'Orbigny); c, d, e: *Globdulina ampulla* (Jones); f, g: *Globulina inaequalis* Reuss; h, i, j: *Guttulina problema* d'Orbigny.

Acknowledgements

Dr. C. G. Adams, British Museum (Natural History) kindly allowed access to the collections of Foraminifera. Mr. J. Jones and Mr. B. Evans printed the figures of plates 1-9 from negatives taken by J. W. Murray. For plates 10 and 11, electron microscope facilities were provided by Professor T. Barnard; the planktonic species illustrations were prepared by Mr. M. Gay and those of the nummulites by Mr. M. Gray (University College London). Additional negatives of Thanetian benthic species were provided by Dr. J. R. Haynes. Mrs. G. Wright typed the manuscript. We wish to thank them all.

8.8 REFERENCES

Adams, C. G. 1962. *Alveolina* from the Eocene of England, *Micropaleontology,* **8**, 45–54.

Barr, F. T. and Berggren, Q. A. 1964. Lower Tertiary planktonic Foraminifera from the Thanet Formation of England. *Int. geol. Congr.,* 22nd ser., 1964(3), 118–136.

Barr, F. T. and Berggren, W. A. 1965. Planktonic Foraminifera from the Thanet Formation (Paleocene) of Kent, England. *Stockh. Contr. Geol.,* **13**, 9–26.

Batjes, D. A. J. 1958. Foraminifera of the Oligocene of Belgium. *Mem. Inst. r. Sci. nat. Belg.,* **143**, 1–188.

Berggren, W. A. 1965. Further comments on planktonic Foraminifera of the type Thanetian. *Contr. Cushman Fdn.,* **16**, 125–127.

Berggren, W. A. 1969. Paleogene biostratigraphy and planktonic foraminifera of Northern Europe. *In* Brönnimann, O. and Renz, H. H. (Eds.) *Proc. first Intern. Conf. planktonic Microfossils,* Geneva 1967. E. J. Brill, Leiden, 121–59.

Bhatia, S. B. 1955. The foraminiferal fauna of the Late Palaeogene sediments of the Isle of Wight, England. *J. Paleont.,* **29**, 665–693.

Bhatia, S. B. 1957. The paleoecology of the late Palaeogene sediments of the Isle of Wight, England. *Contr. Cushman, Fdn.,* **8**, 11–28.

Bignot, G. 1962. Etude micropaléontologique de la formation de Varengeville du gisement Eocène du Cap d'Ailly (Seine–Maritime). *Rev. Micropaléont.,* **5**, 167–184.

Bignot, G. 1963. Foraminifères planctoniques et foraminifères remaniés dans la Formation de Varengeville. *Bull. Soc. géol. Normandie,* **53**, 1–12.

Blezard, R. G., Bromley, R. G., Hancock, J. M., Hester, S. W., Hey, R. W. and Kirkaldy, J. F. 1967. The London Region (North of the Thames). *Geol. Assoc. Guide.* **30A**, 1–34.

Blondeau, A. 1972. *Les Nummulites.* Vuibert, Paris, 255 pp.

Blondeau, A. and Curry, D. 1963. Sur la présence de *Nummulites variolarius* (Lmk) dans les diverses zones du Lutétien des bassins de Paris. de Bruxelles et du Hampshire. *Bull. Soc. géol. Fr.,* ser. 7, **5**, 275–277.

Boltovskoy, E. and Wright, R. 1976. *Recent Foraminifera.* Dr. W. Junk, The Hague, 515 pp.

Bowen, R. N. C. 1954. Foraminifera from the London Clay. *Proc. Geol. Assoc.,* **65**, 125–174.

Bowen, R. N. C. 1955. Smaller foraminifera from the Upper Eocene of Barton. *Micropaleontology,* **3**, 53–60.

Brönnimann, P., Curry, D., Pomerol, C. and Szöts, E. 1968. Contribution à la connaissance des Foraminifères planctoniques de l'Eocène (incluant le Paléocène) du Bassin anglo-franco-belge. *Mém. Bur. Rech. géol. min.,* **58**, 101–108.

Burrows, H. and Holland, R. 1897. The Foraminifera of the Thanet Beds of Pegwell Bay. *Proc. Geol. Assoc.* **15**, 19–52.

Burton, E. St. J. 1929. The horizons of Bryozoa (Polyzoa) in the Upper Eocene Beds of Hampshire. *Q. Jl. geol. Soc. Lond.,* **85**, 223–239.

Burton, E. St. J. 1933. Faunal horizons of the Barton Beds in Hampshire. *Proc. Geol. Assoc.,* **44**, 131–167.

Chapman, F. and Sherborn, C. D. 1889. The Foraminifera from the London Clay of Sheppey. *Geol. Mag.,* Dec. III, **6**, 497–239.

Curry, D. 1937. The English Bartonian nummulites. *Proc. Geol. Assoc.,* **48**, 229–246.

Curry, D. 1962. Sur la découverte de *Nummulites variolarius* (Lamarck) dans le Lutétien des bassins de Paris et du Hampshire. *C.r. Somm. Séanc. Soc. géol. Fr.,* **9**, 247.

Curry, D. 1965. The Palaeogene Beds of south-east England. *Proc. Geol. Assoc.,* **76**, 151–174.

Curry, D. 1967. Problems of correlation in the Anglo-Paris-Belgian Basin. *Proc. Geol. Assoc.,* **77**, 437–467 (for 1966).

Curry, D., Adams, C. G., Boulter, M. C., Dilley, F. C., Eames, F. E., Funnell, B. M. and Wells, M. K. 1978. A correlation of Tertiary rocks in the British Isles. *Geol. Soc. Lond., Special Report,* **12**, 72 pp.

Curry, D., Gulinck, M., Pomerol, C. 1969. Le Paléocène et l'Eocène, dans les Bassins de Paris, de Belgique et d' Angleterre. *Mém. Bur. Rech. géol. min.,* **69**, 361–369.

Curry, D., Hodson, F. and West, I. M. 1968. The Eocene succession in the Fawley transmission tunnel. *Proc. Geol. Assoc.,* **79**, 179–206.

Curry, D., King, A. D., King, C. and Stinton, F. C. 1977. The Bracklesham Beds (Eocene) of Bracklesham Bay and Selsey, Sussex. *Proc. Geol. Assoc.,* **88**, 243–254.

Curry, D. and King, C. 1965. The Eocene succession at Lower Swanwick Brickyard, Hampshire. *Proc. Geol. Assoc.,* **76**, 29–35.

Curry, D., Middlemiss, F. A. and Wright, C. W. 1966. The Isle of Wight, *Geol. Assoc. Guide,* **25**, 1–26.

Curry, D. and Wisden, D. E. 1958. Geology of the Southampton area including the coast sections at Barton, Hants., and Bracklesham, Sussex. *Geol. Assoc. Guide,* **14**, 1–16.

Davis, A. G. 1928. The geology of the City and South London Railway, Clapham–Morden extension. *Proc. Geol. Assoc.,* **39**, 339–352.

El-Naggar, Z. R. 1967. Planktonic Foraminifera in the Thanet Sands of England, and the position of the Thanetian in Paleocene stratigraphy. *J. Paleont.,* **41**, 575–586.

Haynes, J. 1954. Taxonomic position of some British Paleocene Buliminidae. *Contr. Cushman Fdn.*, **5**, 185–191.

Haynes, J. 1955. Pelagic Foraminifera in the Thanet Beds and the use of Thanetian as a stage name. *Micropaleontology*, **1**, 189.

Haynes, J. 1956. Certain smaller British Paleocene Foraminifera Part 1, Nonionidae, Chilostomellidae, Epistominidae, Discorbidae, Amphisteginidae, Globigerinidae, Globorotaliidae, and Gümbelinidae. *Contr. Cushman Fdn.*, **7**, 79–101.

Haynes, J. 1958a. Certain smaller British Paleocene Foraminifera Part III. *Contr. Cushman Fdn.*, **9**, 4–16.

Haynes, J. 1958b. Certain smaller British Paleocene Foraminifera Part IV. Arenacea, Lagenidea, Buliminidea and Chilostomellidae. *Contr. Cushman Fdn.*, **9**, 58–77.

Haynes, J. 1958c. Certain smaller British Paleocene Foraminifera Part V. Distribution. *Contr. Cushman Fdn.*, **9**, 83–92.

Haynes, J. and El-Naggar, Z. R. 1964. Reworked Upper Cretaceous and Danian planktonic foraminifera in the type Thanetian of England. *Micropaleontology*, **10**, 354–356.

Kaasschieter, J. P. H. 1961. Foraminifera of the Eocene of Belgium. *Mem. Inst. r. Sci. nat. Belg.*, **147**, 1–271.

Larsen, A. R. and Jørgensen, N. O. 1977. Palaeobathymetry of the Lower Selandian of Denmark on the basis of Foraminifera. *Bull. geol. Soc. Denmark*, **26**, 175–184.

Le Calvez, Y. 1970. Contribution à l'étude des foraminifères paléogènes du Bassin de Paris. *Cah. Paléont.*, 326 pp.

Loeblich, A. R. Jun., and Tappan, H. 1964. Sarcodina chiefly 'Thecamoebians' and Foraminiferida. *In* Moore, R. C. (ed.). *Treatise on invertebrate paleontology*. New York Geol. Soc. Amer., pt. C, Protista 2, 900 pp.

Moorkens, T. 1968. Quelques foraminifères planctoniques de l'Yprésien de la Belgique et du Nord de la France. *Bull. Bur, Rech, géol. min.*, **No. 58**, 109–29.

Murray, J. W. 1973. *Distribution and ecology of living benthic foraminiferids.* Heinemann Educational Books, London, 288 pp.

Murray, J. W. 1976a. A method of determining proximity of marginal seas to an ocean. *Mar. Geol.*, **22**, 103–119.

Murray, J. W. 1976b. Comparative studies of living and dead benthic foraminiferal distributions. In Hedley, R. H. and Adams, C. G. (Eds.) *Foraminifera*, **2**, 45–109.

Murray, J. W. and Wright, C. A. 1974. Palaeogene Foraminiferida and palaeoecology, Hampshire and Paris Basins and the English Channel. *Spec. Pap. Palaeontology*, **14**, 1–171.

Nemkov, G. 1968. Les Nummulites de l'U.R.S.S., leur évolution, systématique et distribution stratigraphique. *Mém. Bur. Rech. géol. min.*, **58**, 71–8.

Pitcher, W. S., Peake, N. B., Carreck, J. N., Kirkaldy, J. F. and Hancock, J. M. 1967. The London Region (South of the Thames). *Geol. Assoc. Guide*, **30B**, 1–32.

Rouvillois, A. 1960. Le Thanétien du Bassin de Paris. *Mém. Mus. nat. Hist.* n.s. ser C, **8**, 1–91.

Sherborn, C. D. and Chapman, F. 1886. On some microzoa from the London Clay exposed in the drainage works, Piccadilly, London, 1885. *Jl. R. microsc. Soc.*, **6**, 737–767.

Sherborn, C. D. and Chapman, F. 1889. Additional note on the foraminifera of the London Clay exposed in the drainage works, Piccadilly, London, 1885. *Jl. R. microsc. Soc.*, 483–488.

Smith, A. J. and Curry, D. 1975. The structure and geological evolution of the English Channel. *Phil. Trans, R. Soc.*, **A 279**, 3–20.

Stinton, F. C. 1971. Easter field meeting in the Isle of Wight. *Proc. Geol. Assoc.*, **82**, 403–410.

Vella, P. 1969. Correlation of base of Middle Headon Beds between Whitecliff Bay and Colwell Bay, Isle of Wight. *Geol. Mag.*, **106**, 606–608.

Venables, E. M. 1962. The London Clay of Bognor Regis, *Proc. Geol. Assoc.*, **73**, 245–271.

White, H. J. O. 1915. The geology of the country near Lymington and Portsmouth. *Mem. geol. Surv. G.B.*, Sheets 330–331, 78 pp.

White, H. J. O. 1921. A short account of the geology of the Isle of Wight. *Mem. geol. Surv. G.B.*, 219 pp.

Wood, A. and Haynes, S. J. 1957. Certain smaller British Paleocene Foraminifera Part II. *Cibicides* and its allies. *Contr. Cushman Fdn.*, **8**, 45–53.

Wright, C. A. 1972a. Foraminiferids from the London Clay at Lower Swanwick and their palaeoecological interpretation. *Proc. Geol. Assoc.*, **83**, 337–347.

Wright, C. A. 1972b. The recognition of a planktonic foraminiferid datum in the London Clay of the Hampshire Basin. *Proc. Geol. Assoc.*, **83**, 413–9.

Wrigley, A. 1924. Faunal divisions of the London Clay, illustrated by some exposures near London. *Proc. Geol. Assoc.*, **35**, 245–259.

Wrigley, A. 1940. The faunal succession in the London Clay, illustrated in some new exposures near London. *Proc. Geol. Assoc.*, **51**, 230–245.

Wrigley, A. and Davis, A. G. 1937. The occurrence of *Nummulites planulatus* in England, with a revised correlation of the strata containing it. *Proc. Geol. Assoc.*, **48**, 203–28.

Ziegler, P. A. 1975. Geologic evolution of North Sea and its tectonic framework. *Bull. Am. Ass. Petrol. Geol.*, **59**, 1073–97.

9

Neogene

M. Hughes, D. G. Jenkins

9.1.1 Benthic foraminifera

No Miocene strata containing foraminifera are present on land in the British Isles but off-shore oil prospecting and mapping have led to the recognition of sediments of this age on the British continental shelf area. Miocene benthic foraminifera are present in the western English Channel (Curry *et al.*, 1962, Murray *in* Curry *et al.*, 1965, Hughes *in* Warrington and Owens, 1977), the Celtic Sea (Hughes *ibid*), northwest Scottish waters (Hughes in preparation) and the North Sea (Dilley and Funnell *in* Curry *et al.*, 1978). No detailed descriptions of assemblages have been published.

A considerable amount of literature on Miocene foraminifera has been published by Continental workers since the early classic work of d'Orbigny (1846) on the Vienna basin, revised by Marks· (1951), which was followed by Reuss (1851, 1856, 1861, 1863a, b), Hosius (1892) and Clodius (1922) in Germany, and Hantken (1875) in Hungary. The first attempts at biostratigraphical zonation based on species ranges and assemblages was made by Staesche and Hiltermann (1940) in Germany and Ten Dam and Reinhold (1941a, b, c, 1942) in the Netherlands, to be followed by Indans (1958, 1962, 1965) and Langer (1963a, b, 1969) in West Germany and Batjes (1958) in Belgium. More recently Spiegler (1974) has published a comprehensive review of 532 species ranges through the Oligocene and Miocene of northwest Germany. In addition detailed analysis of the ranges of the species and sub-species of *Asterigerina* (Gramann, 1964), *Ehrenbergina* (Spiegler, 1973) and *Uvigerina* (Daniels and Spiegler, 1977) have been carried out. Miocene foraminiferal biostratigraphy is engaging workers also in Rumania (Gheorghian, 1971, 1974, 1975, Popescu, 1975) and Poland (Odrzywolska-Bienkowa, 1977). A review of the Aquitaine area has been completed recently (Poignant and Pujol, 1976, 1978).

The proceedings of the various meetings of the Congrés du Néogène Mediterranéan and the reports of the IGCP Project 124, The Northwest European Tertiary Basin, also contain biostratigraphical papers based on foraminifera. Many of the more recent works listed above also contain palaeoenvironmental interpretations and all are pertinent to the study of British off-shore Miocene assemblages.

The only Pliocene strata in Britain containing foraminifera are the Coralline Crag and the St Erth Beds. The foraminifera of the Coralline Crag in East Anglia were first described by Jones *et al.* (1866-1897). Amongst the other authors whose work was included in Jones' monograph were Burrows and Holland, who contributed the distribution lists (*ibid.* pp. 374-394) and the description of a number of new species. Carter (1951, 1957) initially examined the distribution of the foraminifera by size in order to determine the palaeoenvironment, and later discussed the genera *Alliatina* and *Alliatinella*, using the present day distribution of the former as further evidence for the palaeoenvironment of these beds. Wilkinson (1980) in a study of the ostracods, stratigraphy and palaeoenvironment briefly discusses the foraminifera.

The foraminifera of the Pliocene of the Netherlands and Belgium are comparable to those found in the Coralline Crag. A biostratigraphical framework for the Netherlands Plio-Pleistocene was first established by Ten Dam and Reinhold (1941d). Subsequently the results of further work were made available (Van Voorthuysen 1958, Toering and Van Voorthuysen 1973). Recently the separate studies in Belgium (De Meuter and Laga, 1977) and in the Netherlands (Doppert, 1975) have been unified, resulting in a biostratigraphy based on benthic foraminifera of the Neogene of these two areas (Doppert *et al.*, 1979).

The St Erth Beds in Cornwall contain foraminifera which were first recorded by Millett (1886a, b, 1895, 1897, 1898, 1902) [Literature note: an original letter from Millett to Burrows dated 12.4.1897, now in the Palaeontology Unit archives of the Institute of Geological Sciences, states that the list of species appended to the second paper (1886b) was complete and not, as the title suggests, only additions. This, according to Millett, was because very few reprints of the first paper were available, and explains the lack of subsequent references to the first paper (1886a)]. Macfadyen (1942) examined some material from St Erth in order to compare the assemblage with one obtained from a clay in Swansea Docks which he considered to be postglacial, without coming to any conclusions as to their relationship. The palaeoenvironment and stratigraphy were discussed by Funnell (*in* Mitchell 1965) as part of a study of the formation by Mitchell. Later the results of a more detailed examination involving new exposures (Mitchell 1973) included the study of a number of different fossil groups. The foraminifera were recorded by Margerel (*in* Mitchell 1973) as a detailed list of species, many figured, together with a discussion of the stratigraphy and palaeoecology.

The foraminifera of stratigraphically comparable formations of Redonian age in France have been described by Margerel (1970, 1972) as has the distribution of the genus *Faujasina*, unique to the St Erth Beds in the British Pliocene (Margerel, 1971).

9.1.2 Planktonic foraminifera

On the seabed of the Western Approaches of the English Channel Neogene deposits have been recorded and planktonic foraminifera were listed by Curry (*in* Curry *et al.*, 1962) and by Murray (*in* Curry *et al.*, 1965). Jenkins (1977) described and illustrated 35 planktonic foramineral species and subspecies from the Sealab Trial borehole in the English Channel, 110 km southwest of the Isles of Scilly at latitude 49°10.73′N, longitude 7°27.86′W in a water depth of 137-140 m (Fig. 9.1).

9.2 LOCATION OF COLLECTIONS OF IMPORTANCE

MIOCENE: The samples collected by the Bristol University group has been retained at Bristol, but the assemblages studied by Murray (*in* Curry *et al.*, 1965) are in the Murray collection, University of Exeter. The Institute of Geological Sciences Leeds collections contain the off-shore material which has been reported in the Institute publications.

PLIOCENE: The Coralline Crag slides of Jones and of Carter are at the British Museum (Natural History), London. The Burrows and Holland figured specimens are at the Institute of Geological Sciences, Leeds. The Millett Collection and others, of St Erth Beds foraminifera are in the British Museum (Natural History). The Mitchell collected material examined by Margerel has been deposited at the Institut des Sciences de la Nature, the University, Nantes.

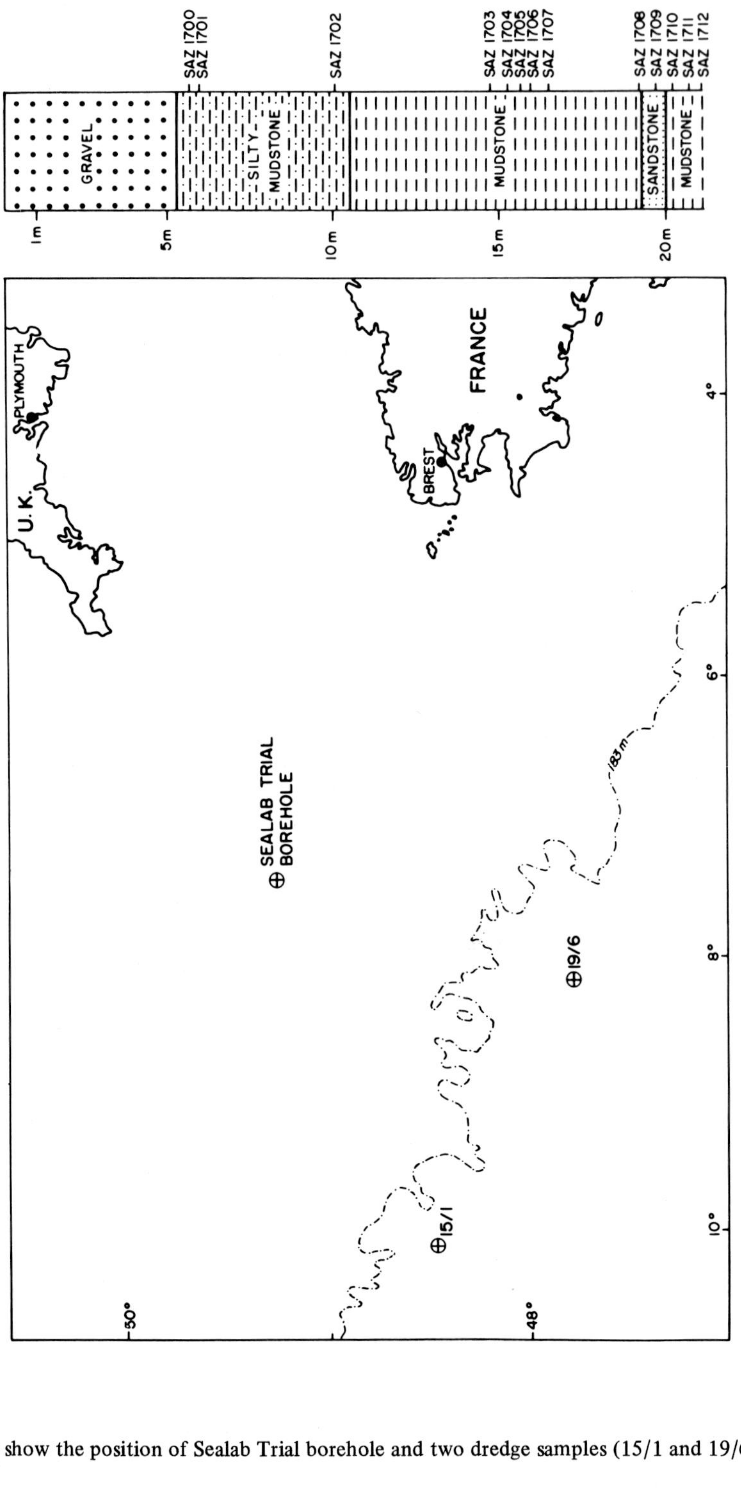

Fig. 9.1 – Map to show the position of Sealab Trial borehole and two dredge samples (15/1 and 19/6).

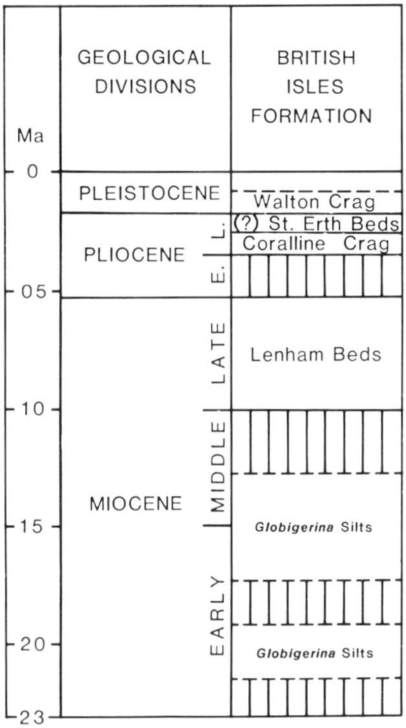

Fig. 9.2 – Correlation of British marine Neogene formations (based on Curry *et al.*, 1978, *Geol. Soc. Lond., Spec. Rept.*, 12).

9.3 STRATIGRAPHIC DIVISIONS

The boundaries of the Miocene and Pliocene Series have been discussed recently (Curry *et al.*, 1978; see Fig. 9.2). The Miocene is divided into Early- nannoplankton zones NN1-lower 5, planktonic foraminifera zones N4-8; Middle N5-10, N7-15 and Late-N11-part 12, N16-17. The Pliocene is divided into Early-NN part 12-14, N18-19 and Late NN15-17, N20-21.

The Miocene strata of the western English Channel are known collectively as the Globigerina Silts, and are shown (*ibid*) as being of two distinct ages. The lower belongs to the Early Miocene and is dated as NN2, N5 and the upper spans the Early-Middle Miocene boundary being dated as NN4-5, N7-10. These have been correlated (*ibid*) with the Belgian stages Houthalenian and Anversian respectively. In the literature on the western English Channel the stage names which have been used are the international stages Aquitainian (NN1-2, N4-5), Burdigalian (NN3-4, N6-7) and Langhian (NN5, N8-9) (= Vindobonian of Curry *et al.*, 1965).

In the northern Scottish sea area the German stage names Hemmoorian (Early Miocene, NN2-part 5,) and Reinbeckian (Middle Miocene, NN middle 5) have been used (nannoplankton correlation of Hinsch *et al.*, 1978).

The Lenham Beds, dated as Late Miocene (Curry *et al.*, 1978), contain no recorded foraminifera and need not be considered further.

Cambridge (1977) regards both the St Erth Beds and the Coralline Crag as Scaldisian (Pliocene) on molluscan evidence. The Coralline Crag is dated as Gedgravian and regarded as the older formation. Margerel (*in* Mitchell, 1973) has dated the St Erth Beds as Redonian. Curry *et al.* (1978) have modified Cambridge (1977) to the extent of placing both formations in the Late Pliocene.

9.4 FAUNAL ASSOCIATIONS AND FACIES

In the western English Channel the practical problem in both single sample stations and in boreholes is to differentiate the Miocene Globigerina Silts from Recent silts. In boreholes one is frequently directly superimposed upon the other without any discernable interface. Benthic species which are not restricted to the Miocene elsewhere, but do not extend into the Recent, are important in differentiating these two lithologically similar silts.

Murray (*in* Curry *et al.*, 1965) listed the benthic species occurring in the Miocene Globigerina Silts of the western English Channel. The most frequent species present in the Aquitainian were *Brizalina scalprata miocenica* (Macfadyen), *Florilus grateloupi* (d'Orbigny) and *Trifarina bradyi* Cushman which were present at every station placed in that stage. The first two named were present also at all stations recognised as Vindobonian (= Langhian) together with *Cibicides dutemplei* (d'Orbigny) and *Elphidium crispum* (Linné). Different subordinate species were recorded for each Stage but it was noted that the two faunas were very similar.

Six assemblages have been recognised from the Sealab Trial borehole (Hughes 1977, unpublished thesis, in prep. see Fig. 9.1) and dated as Burdigalian (Jenkins 1977). The benthic assemblages, for convenience named after the dominant species, can be related to the planktonic subzones in the Trial borehole.

Subzone	Assemblages
	Gyroidina aff. *parvus*
	Bolivina hebes – Bulimina alazanensis
Globorotalia praescitula	*Bulimina elongata*
	Bolivini hebes – Bulimina alazanensis
Globorotalia semivera	*Cibicides ungerianus* *Astrononion perfossum* *Pararotalia* sp.

The benthic assemblages are facies controlled and a return to a pre-existing palaeoenvironmental condition may result in the return of a previous assemblage as shown by the *B. hebes–B alazanensis* assemblage. The three assemblages in the *semivera* Subzone represent a gradually increasing water depth from inner to outer shelf. The lithology is a glauconitic sandstone with extensive recrystalisation of the fossils. The rocks of the *praescitula* Subzone are a silty, greenish grey mudstone. *B. alazanensis,* a living species, with a frequency of 10–15% over a range of levels in the borehole succession, is considered to be significant in determining the palaeobathymetry. The species is found at the present time only in water depths greater than 400 m (Pflum and Frerichs, 1976) and a middle bathyal regime is postulated for this assemblage. *B. alazanensis* has not been recorded from elsewhere in the *Globigerina* Silts, but neither have rocks of Burdigalian age except on the continental slope (Jenkins, 1977). The two remaining assemblages are regarded as outer shelf. It would appear that a marked change in water depth and lithology coincide with the Subzone boundary in this borehole succession. A number of benthic species are restricted to the Miocene; three of these, *Astrononion perfossum, Unicosiphonia zsigmondyi* and *Virgulinella pertusa,* are present in the Trial borehole, whilst the last named has been recorded from a number of other western English Channel stations (Hughes *in* Warrington and Owens 1977).

Curry (*in* Curry *et al.*, 1962) postulated that the Miocene sea in the western English Channel was a gulf with no connection with the Miocene basin situated in the region of the present North Sea. Murray (*in* Curry *et al.*, 1965) agreed with this and suggested a water depth slightly greater than the present day 128 m. A greater water depth is indicated for the early part of the *praescitula* Subzone, at least in the area of the Trial borehole (see above), but Hughes (1977 unpublished thesis, in preparation) agrees with both Curry and Murray in considering that the English Channel and the North Sea were not connected during the Miocene.

Two Miocene benthic assemblages from northwest Scottish waters are present in IGS borehole 77/7 (see IGS 1978, p. 23 for lithostratigraphical log). The younger assemblage dominated by *Cibicides peelensis* Ten Dam and Reinhold and *Ehrenbergina healyi* is placed in the Reinbeckian (Middle Miocene) Stage and the older assemblage with high frequencies of *Asterigerina staeschei* and *Elphidium inflatum* are considered as Hemmoorian Stage (Early Miocene). *Uvigerina semiornata* and *Florilus boueanus* (d'Orbigny) are common to both assemblages; the latter replacing *F. dingdeni* recorded in the western English Channel. These are typical middle shelf assemblages (Hughes, in preparation).

A characteristic benthic foraminiferal assemblage of the Coralline Crag has been recorded recently by Wilkinson (1980). The most frequent species are *Cibicides lobatulus* (Walker and Jacob), *Ammonia batavus* (Hofker), *Elphidium macellum* (Fichtel and Moll) and *Quinqueloculina seminulum* (Linné) very common, *Textularia sagittula* Defrance and *Pararotalia serrata* common. Amongst the species present which are restricted to the Pliocene, *Monspeliensina pseudotepida, Alliatina excentrica* and *Alliatinella gedgravensis* Carter are rare. The last two named species appear to be relatively more common in the Lower rather than the Upper Gedgravian of Carter (1957).

The evidence of a high degree of sorting convinced Carter (1951) that the assemblages do not totally represent the original biocoenoses. A change in composition of the assemblages at the e/f 'zonal' boundary of Prestwich (1871) indicates an alteration in conditions of deposition, and that this boundary is not of any great stratigraphical importance. Later Carter (1957) revised his opinion and considered that this change had some stratigraphical significance. The ostracod evidence (Wilkinson, 1980) indicates a considerable variation in current velocity and Wilkinson follows Harmer's interpretation (1898, 1900, 1902) that none of the 'zones' of Prestwich has any stratigraphical significance.

The total geographical isolation of the St Erth Beds has resulted in considerable importance being attached to the stratigraphical and palaeoenvironmental interpretations based on the foraminifera, as well as other groups of fossils. Margerel (*in* Mitchell, 1973) recorded over 100 species, but noted that there were only five which he would consider as characteristic of the formation: *Quinqueloculina cliarensis* (Heron-Allen and Earland), *Q. seminulum, Monspeliensina pseudotepida, Ammonia* sp. and *Faujasina subrotunda.* However it should be noted that the form of *Buccella frigida* (referred to as *B.* aff. *frigida* herein) present in this formation is recognisably different from the Pleistocene species. Funnell (*in* Mitchell, 1965) determined the palaeoenvironment by analysing the number of species in terms of faunal dominance and faunal diversity. The result from the former study suggested a water depth of not more than 20 m, whilst the diversity factor indicated 60-100 m. It was noted that the conditions around Cornwall were very different from the Gulf of Mexico where the criteria for the interpretation of the water depth were originally established. Funnell came to the conclusion that the fauna originated in shallow water of less than twenty metres, but was deposited in depths of up to 100 m. Margerel had nothing to add to this conclusion.

9.5 PALAEOECOLOGY OF THE PLANKTONIC FORAMINIFERA

The 35 species and subspecies in the Early Miocene Sealab sequence of the English Channel suggest that there were oceanic conditions with normal salinity prevailing at this time. It had been suggested by Curry *et al.* (1965) that the Miocene foraminifera of the Western Approaches indicated a sea temperature similar to the present day sea-surface temperature of 15-18°C. The presence of *Globigerinoides sacculifer* in the Sealab samples indicated a higher range of temperatures, because it lives in the present day oceans in waters of 15-30°C. Confirmation of this higher estimate has come from an oxygen isotope analysis of the tests of *G. sacculifer;* there was a rise of temperature from 15.5°-16°C in the two lowermost samples to 22°C in the highest sample in the *G. trilobus* Zone (Jenkins and Shackleton, 1979).

There is a species diversity gradient of planktonic foraminifera from about 22 species in the tropics to one in the freezing polar seas. Species diversity changes in the Sealab borehole tend to parallel changes in palaeotemperature (Jenkins and Shackleton, 1979).

9.6 INDEX SPECIES

The benthic species have been studied by M. J. Hughes and the planktonic species by D. G. Jenkins.

Benthic foraminifera are recorded from the Miocene of three offshore boreholes and on-land sections in the Crags and St Erth Beds (Fig. 9.3).

Thirty five species and subspecies of planktonic foraminifera were recorded from 13 samples taken from the Sealab Trial borehole (Jenkins, 1977) and 18 have been chosen as diagnostic in the 20.06 m of cored Lower Miocene silty mudstone, mudstone and thin sandstone (Figs. 9.1 and 9.4). A record of some of the species is also shown in two Middle Miocene dredge samples in Fig. 9.4; the *P. glomerosa curva* Zone sample from Lat. 47° 48′N, Long. 8° 09.50′W and the *G. mayeri mayeri* Zone sample from Lat. 48° 27′N, Long. 10° 07′W.

The classification used is that of Loeblich and Tappan (1968). The illustrated specimens have been deposited in the collections of the Institute of Geological Sciences, Leeds.

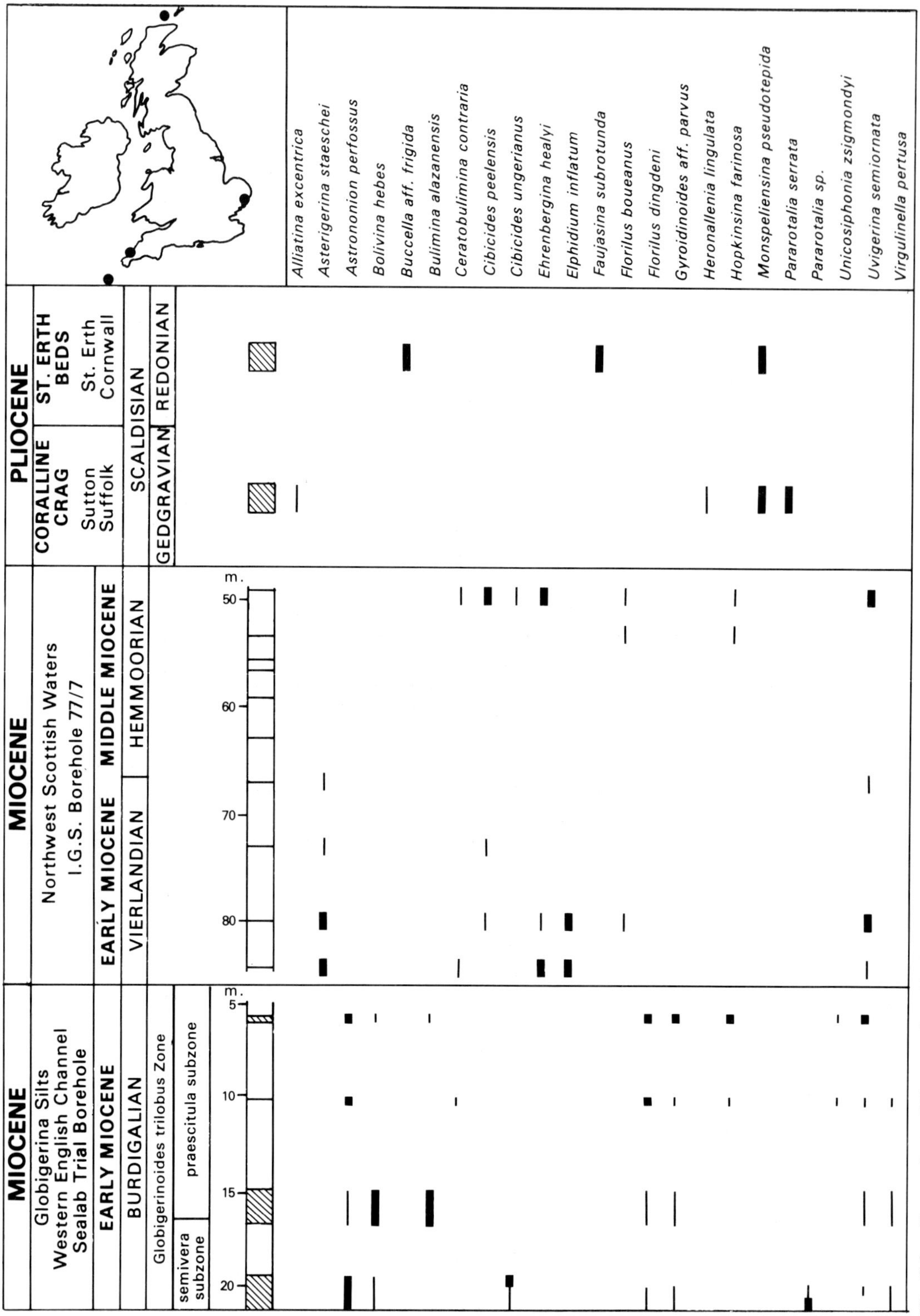

Fig. 9.3 – Range chart of Neogene benthic foraminifera.

SERIES	Lower Miocene													Middle Miocene	
ZONE	G. trilobus trilobus zone													*P. glomerosa curva*	*G. mayeri mayeri*
SUBZONE	*G. semivera* subzone					*G. praescitula* subzone								ZONE	
SEALAB TRIAL BOREHOLE SAMPLES LAT. 49° 10·73′ N. LONG. 7° 27·86′ W.	SAZ 1712	SAZ 1711	SAZ 1710	SAZ 1709	SAZ 1708	SAZ 1707	SAZ 1706	SAZ 1705	SAZ 1704	SAZ 1703	SAZ 1702	SAZ 1701	SAZ 1700	19/6	15/1
DEPTH BELOW SEAFLOOR (M)	21·06	20·70	20·20	19·70	19·20	16·50	15·95	15·65	15·25	14·73	10·06	5·98	5·68	SAMPLE	
PLANKTONIC FORAMINIFERA															
Cassigerinella chipolensis	X	X	X	X	X	X	X	X	X	X	X	X	X	X	
Globigerina bradyi							X cf.		X	X cf.		X cf.		X	X
G. ciperoensis angulisuturalis	X cf.	X	X	X	X	X	X	X	X	X	X	X			
G. ciperoensis ciperoensis	X cf.	X	X	X	X	X	X	X	X	X	X	X			
G. woodi	X	X	X	X	X	X	X	X	X	X	X	X	X	X cf	
Globigerinella praesiphonifera											X		X		
Globigerinoides ruber		X	X	X		X	X	X		X	X		X	X	
G. saculifer	X	X	X			X	X	X	X	X	X	X	X		
G. trilobus trilobus	X	X	X	X	X	X	X	X	X	X	X	X	X	X	X
Globoquadrina dehiscens						X cf.	X			X	X			X	X
Globorotalia obesa		X cf.	X	X	X	X	X	X	X		X	X	X	X	
G. peripheroronda		X			X cf.	X cf.	X cf.	X	X	X	X	X	X	X cf.	
G. praescitula						X	X	X	X	X	X	X	X	X	X
G. pseudocontinuosa	X	X	X	X	X	X	X	X		X	X cf.	X			
G. saginata									X		X cf.				
G. semivera	X	X	X	X	X	X	X	X cf.		X	X cf.	X		X	
G. zealandica			X cf.				X		X				X cf.		
Protentella clavaticamerata						X	X				X		X		

Fig. 9.4 – Range table of Neogene planktonic foraminifera.

9.6.1 Benthic species

Alliatina excentrica (di Napoli Alliata)
Plate 9.1, Fig. 1, × 75, Fig. 2, × 90 (both MPK 2655); Pliocene, Coralline Crag; Sudbury Park Pit, Suffolk, colln. I. P. Wilkinson. Fig. 9.3. = *Cushmanella excentrica* di Napoli Alliata, 1952. Description: dextrally trochospiral; up to 10 chambers in final whorl, increasing rapidly in height; small accessory chambers at umbilical margins on both sides of test; apertures, slit at base, and oval perforation sometimes closed by a thin calcareous plate, in the centre, of the apertural face; a groove extends across the apertural face from the ventral umbilicus to the areal aperture.
Remarks: differs from *Alliatinella gedgravensis* Carter in being less strongly trochospiral, the areal aperture is central rather than laterally offset, less strongly arched apertural groove.

Asterigerina staeschei Ten Dam and Reinhold, 1941
Plate 9.1, Fig. 3 (MPK 2640) × 105, Fig. 4 (MPK 2641) × 80, Fig. 5 (MPK 2642) × 70; Lower Miocene; Figs. 3, 4 IGS borehole 77/7 at 80 m (Sample CSC 2533), Fig. 5 IGS borehole 77/9 at 74.4 m (Sample CSC 2525); Fig. 9.3. Description: biserially trochospiral; biconvex; umbilical side higher than spiral side; keeled; umbilical chambers large, nearly triangular, together forming a dome higher than peripheral chambers which are narrow; strong granular ornament around aperture.
Remarks: differs from *A. guerichi* (Franke) in the greater convexity of the umbilical side, in the well developed keel and the greater development of ornament around the aperture.

Astrononion perfossus (Clodius)
Plate 9.1, Fig. 6 (MPK 2643) × 125, Fig. 7 (MPK 2644) × 135; Lower Miocene; Sealab Trial borehole at 5.48 m (Sample CSB 2704); Fig. 9.3. = *Nonionina perfossa* Clodius, 1922 (syn. *Nonion nanum* Van Voorthuysen, 1950). Description: involute planispiral; compressed; 9 chambers in final whorl; periphery rounded; small deep umbilici; distinctive opening at peripheral extremity of infilling at inner ends of each suture; apertural face imperforate.

Bolivina hebes Macfadyen, 1930
Plate 9.1, Figs. 8 (MPK 2645) × 230, Fig. 9 (MPK 2646) × 85; Lower Miocene; Sealab Trial borehole, Fig. 8 at 6.00 m (Sample CSB 2705), Fig. 9 at 10.06 m (Sample SAZ 1702); Fig. 9.3. Description: spatulate; biserial; broad, gently tapering with rounded initial end in side view; broad in end view; 6–8 pairs of chambers; basal margin of chambers strongly dentate; chamber surface strongly reticulate; aperture oval with collar; internal toothplate visible.

Buccella aff. *frigida* (Cushman)
Plate 9.1, Fig. 10 (MPK 2649) × 135, Fig. 11 (MPK 2647) × 160, Fig. 12 (MPK 2648) × 150; Pliocene, St Erth Beds; St Erth, Cornwall (Sample SAA 2107); Fig. 9.3. = *Pulvinulina frigida* Cushman, 1922. Description: trochospiral; 6–7 chambers in final whorl; periphery sharply angled with slight keel; umbilical side covered with granular ornament.
Remarks: differs from Pleistocene *B. frigida* in the extension of the granular ornament to the entire ventral surface and in its sharper periphery.

Bulimina alazenensis Cushman, 1927
Plate 9.1, Fig. 13 (MPK 2650) × 300, Fig. 14 (MPK 2651) × 90; Lower Miocene; Sealab Trial borehole, Fig. 13 at 14.70 m (Sample CSB 2707), Fig. 14 at 10.06 m (Sample SAZ 1702); Fig. 9.3. Description: triserial; test small, tapering, with 12 strong longitudinal costae; aperture deeply recessed with prominent elevated rim of fixed border and toothplate.
Remarks: differs from *B. rostrata* Brady in presence of notches in the costae where they cross the sutures and the absence of a strong apical spine.

Ceratobulimina aff. *contraria* (Reuss)
Plate 9.1, Fig. 15 (MPK 2653) × 130, Fig. 16 (MPK 2652) × 120; Lower Miocene; Sealab Trial borehole at 5.98 m (Sample SAZ 1710); Fig. 9.3. = *Rotalina contraria* Reuss, 1851. Description: dextrally trochospiral; 7–8 rapidly enlarging chambers in the final whorl; deeply umbilicate; aperture situated in deep broad groove extending from umbilicus up the apertural face.
Remarks: differs from the Oligocene type in the lower height and relatively greater width of the apertural face (see Langer 1969, p. 62).

Cibicides ungerianus (d'Orbigny)
Plate 9.1, Fig. 17 (MPK 2730) × 60, Fig. 18 (MPK 2731) × 65, Lower Miocene; Sealab Trial borehole, Fig. 17 at 6.00 m (Sample CSB 2705), Fig. 18 at 5.98 m (Sample SAZ 1701); Fig. 9.3. = *Rotalina ungeriana* d'Orbigny, 1846. Description: trochospiral; compressed; 12–14 chambers in final whorl; periphery acute, sometimes slightly keeled; sutures curved; early whorls on evolute spiral side covered with coarse granulations; coarsely perforate; aperture, at base of final chamber on the umbilical side and extending along the spiral suture.
Remarks: differs from *Heterolepa dutemplei* (d'Orbigny) in the greater compression and the dorsal granulation.

Ehrenbergina healyi Finlay, 1947
Plate 9.1, Fig. 19 × 100, Fig. 20 × 125 (both MPK 2654); Lower Miocene; IGS borehole 77/7 at 80 m (Sample CSC 2533); Fig. 9.3. Description: biserial; initial end enrolled with short spines; ventral side of later chambers strongly inflated; margin serrated; aperture extending from base of final chamber up the apertural face parallel to dorsal margin, with distinct lip on ventral edge.
Remarks: differs from *E. serrata* Reuss in the strong inflation of the ventral side.

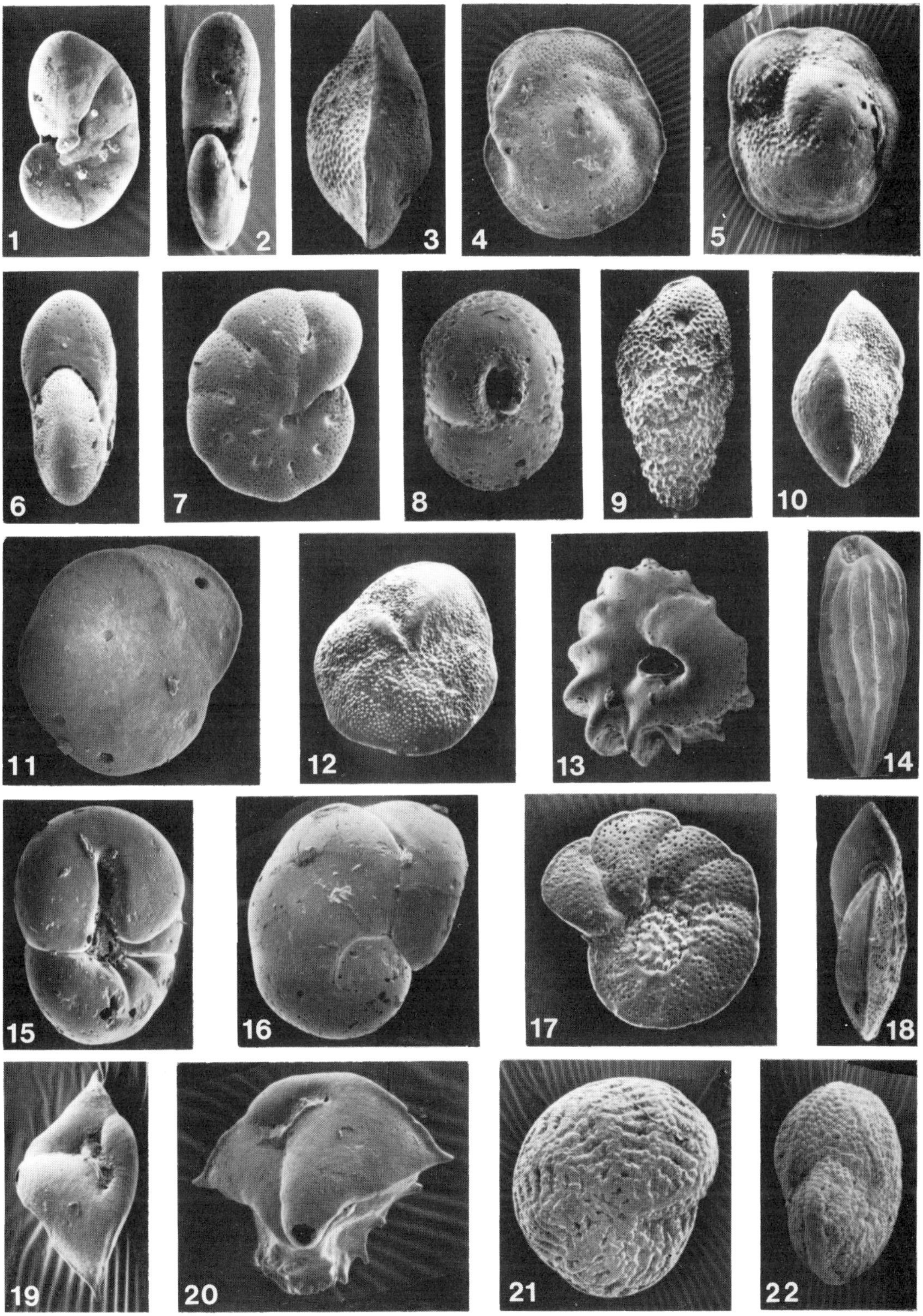

Plate 9.1

Elphidium inflatum (Reuss)
Plate 9.1, Fig. 21 (MPK 2657), Fig. 22 (MPK 2658) both × 70; Lower Miocene; IGS borehole 77/9 at 73.4 m (Sample CSC 2525); Fig. 9.3. = *Polystomella inflata* Reuss, 1861. Description: involute, planispiral; slightly compressed; 10 chambers in final whorl; 15–20 thick sutural bridges; sutural pores appear as slits; apertural face low; aperture, indistinct pores at base of final chamber;
Remarks: the only specimens available are somewhat corroded, the type description refers to a smooth wall. The preservational features, however, may be common in high latitude specimens.

Faujasina subrotunda Ten Dam and Reinhold, 1941
Plate 9.2, Fig. 1 (MPK 2659) × 75, Fig. 2 (MPK 2660) × 70, Fig. 3 (MPK 2661) × 80; Pliocene, St Erth Beds, St Erth, Cornwall (Sample SAA 2107); Fig. 9.3. Description: trochospiral; plano-convex; periphery bluntly keeled; all chambers visible on flat spiral side, only those of the final whorl on convex side; 10–12 chambers in final whorl; sutural bridges well developed and numerous; surface ornament finely granular; apertural face broad with granular ornament; aperture, pores at base of final chamber.
Remarks: differs from *F. carinata* d'Orbigny and *F. compressa* Margerel in its greater thickness.

Florilus dingdeni (Cushman)
Plate 9.2, Fig. 4 (MPK 2662) × 45, Fig. 5 (MPK 2663) × 60; Lower Miocene; Sealab Trial borehole at 10.06 m (Sample SAZ 1702); Fig. 9.3. = *Nonion dingdeni* Cushman, 1936. Description: evolute; planispiral; compressed; 10–11 chambers in final whorl, increasing rapidly in height; periphery narrowly rounded; sutures depressed; wide umbilici infilled with granular calcite which extends into the sutural depressions.
Remarks: differs from *F. boueanus* (d'Orbigny) in that the granular infilling extends into the sutural depressions.

Gyroidinoides aff. *parvus* (Cushman and Renz)
Plate 9.2, Fig. 6 (MPK 2664) × 150, Fig. 7 MPK 2665) × 170, Fig. 8 (MPK 2666) × 140; Lower Miocene; Sealab Trial borehole; Figs. 6 and 8 at 15.25 m (Sample SAZ 1704), Fig. 7 at 5.98 m (Sample SAZ 1701); Fig. 9.3. = *Gyroidina parva* Cushman and Renz, 1941. Description: small, trochospiral; spiral side flat or slightly domed; on spiral side final whorl lower than previous one at the margin; 5½–6 chambers in final whorl; umbilical area depressed; umbilicus closed; periphery rounded; outline lobate; outline of apertural face rounded; aperture extends from umbilicus to periphery with lip on final chamber extending for about three-quarters of distance from the umbilicus.
Remarks: differs from type in its lesser height of test; the less marked lowering of the final whorl and the wider apertural face; from other northern European Miocene species in its smallness and rounded periphery.

Heronallenia lingulata (Burrows and Holland)
Plate 9.2, Figs. 9, 10 (MPK 2667) both × 170; Pliocene, Coralline Crag; Sudbury Park Pit, Suffolk; colln. I. P. Wilkinson; Fig. 9.3. = *Discorbina lingulata* Burrows and Holland, 1896. Description: trochospiral; concavo-convex; compressed; 5 chambers in final whorl rapidly increasing in height; periphery bluntly carinate; each chamber surface on spiral side bears a raised boss which together with the limbate sutures present a distinctive ornament with high relief.

Hopkinsina farinosa (Hantken)
Plate 9.2, Fig. 11 (MPK 2668) × 70; Lower Miocene, Sealab Trial borehole at 5.48 m (Sample CSB 2704); Fig. 9.3. = *Uvigerina farinosa* Hantken, 1875. Description: elongate; triserial becoming loosely biserial; biserial portion half or more of test length in adult specimens; sutures depressed becoming wide and deep in biserial part; ornament finely pustulate, with fine spines along the sutures frequently bridging the wide sutures of the biserial part; aperture terminal on neck with lip; internal toothplate clearly visible.
Remarks: differs from *Uvigerina tenuipustulata* Van Voorthuysen and *Angulogerina gracilis* (Reuss) in the greater development and looseness of the biserial part and also from the latter in the coarser ornament.

Monspeliensina pseudotepida (Van Voorthuysen)
Plate 9.2, Fig. 12 (MPK 2670) × 70, Fig. 13 (MPK 2671) × 70, Fig. 14 (MPK 2672) × 60; Pliocene, St Erth Beds; St Erth, Cornwall (Sample SAA 2107); Fig. 9.3. = *Streblus beccarii* (Linné) *pseudotepidus* Van Voorthuysen, 1950. Description: trochospiral; biconvex; 6–7 chambers in final whorl; periphery broadly rounded; umbilicus deep and narrow; sutures on umbilical side deeply excavated, in well preserved specimens umbilicus and inner end of sutures covered by plates extending back from subsequent chambers; aperture at base of final chamber midway between umbilicus and periphery; accessory apertures on spiral side of intersection of septal and spiral sutures.
Remarks: differs from *Ammonia* spp. in the presence of the accessory apertures on the spiral side.

Pararotalia serrata (Ten Dam and Reinhold)
Plate 9.2, Fig. 15 (MPK 2673) × 85, Fig. 16 (MPK 2674) × 70, Fig. 17 (MPK 2675) × 80; Pliocene, Coralline Crag; Sudbury Park Pit, Suffolk; colln. I. P. Wilkinson; Fig. 9.3. = *Rotalia serrata* Ten Dam and Reinhold, 1941. Description: trochospiral; 8–9 chambers in final whorl; periphery sharply angled with short stout spines at the centre of the periphery of each chamber; outline stellate; chambers ventrally very convex with deep sutures and umbilicus, the latter filled with a calcite boss.

Unicosiphonia zsigmondyi (Hantken)
Plate 9.2, Fig. 18 (MPK 2677) × 165, Fig. 19 (MPK 2676) × 22; Lower Miocene; Sealab Trial borehole, Fig. 18 at 5.48

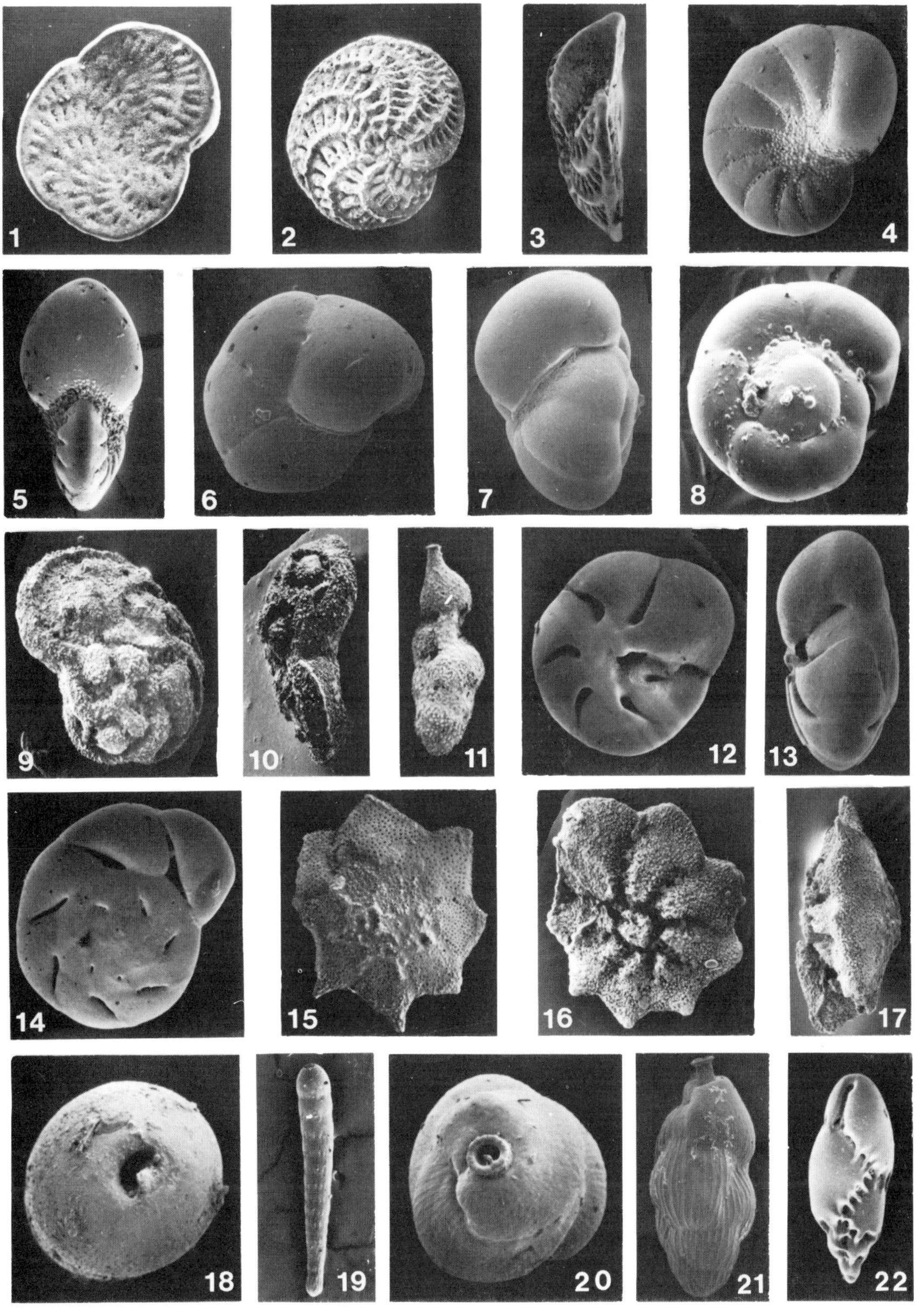

Plate 9.2

m (Sample CSB 2704), Fig. 19 at 10.06 m (Sample SAZ 1702); Fig. 9.3. = *Nodosaria (Dentalina) zsigmondyi* Hantken, 1868 (syn. *Rectobolivina marentinensis* Ruscelli, 1952). Description: large; straight to slightly arcuate; biserial becoming uniserial; biserial part restricted to first two pairs of chambers; circular in section; chambers with basal crenulations extending back over suture; wall smooth; aperture terminal with internal toothplate.
Remarks: frequently broken but individual chambers may be recognised by crenulations and very prominent internal columella process; under reflected light wall appears translucent with perforations clearly visible.

Uvigerina semiornata d'Orbigny, 1846
Plate 9.2, Fig. 10 (MPK 2669) × 80, Fig. 21 (MPK 2656) × 60; Lower Miocene; Sealab Trial borehole, Fig. 20 at 10.05 m (Sample CSB 2706), Fig. 21 at 10.06 (Sample SAZ 1702); Fig. 9.3. Description: elongate; triserial; rounded in section; costae prominent on early chambers and lower part of later chambers, upper part of final chamber smooth; aperture terminal on a stout neck with phialine lip and prominent internal toothplate.
Remarks: the various subspecies described from the north German Miocene do not appear to have differentiated from the central type in the two areas investigated by this writer.

Virgulinella pertusa (Reuss)
Plate 9.2, Fig. 22 (MPK 2678) × 50; Lower Miocene; Sealab Trial borehole at 16.50 m (Sample SAZ 1707); Fig. 9.3. = *Virgulina pertusa* Reuss, 1861. Description: elongate: triserial later biserial; rounded in section; two morphological types, either uniformly tapering test or an abrupt change from narrow early whorls to wide parallel sided chambers; very long digitate retral processes developed across sutures and onto previous chambers, early chambers may be completely obscured; surface smooth; aperture long and narrow extending from the base of final chamber in the direction of growth.
Remarks: very fragile but fragments of the distinctive retral processes serve to record its presence.

9.6.2 Planktonic Species

Cassigerinella chipolensis (Cushman and Ponton)
Plate 9.3, Fig. 1 × 125. Fig. 9.4. = *Cassidulina chipolensis* Cushman and Ponton, 1932. Description: very small (c. 0.19 mm diameter), biserially arranged chambers continuing to spiral in same plane, biumbilicate, periphery broadly rounded.
Remarks: designated a planktonic taxon mainly because of its widespread geographic distribution; the genus became extinct in the Middle Miocene. Range: Britain; Lower Miocene *G. trilobus* Zone to Middle Miocene *P. glomerosa curva* Zone; France, Upper Aquitainian (Jenkins, 1966) and in the Upper Oligocene of the Aquitaine Basin (Pujol, 1970). Total range: Late Eocene to Middle Miocene.

Globigerina bradyi Wiesner, 1931
Plate 9.3, Fig. 2, × 125. Fig. 9.4 (syn *Globigerinoides parva* Hornibrook, 1961). Description: very small test (c. 0.16 mm diameter), high spired and lipped umbilical aperture and rare supplementary apertures on the spiral side as in the figured specimen.
Remarks: related to *Globigerinita glutinata* (Egger) which has a lower spired test and an umbilical bulla; living at present in cooler waters of 0–10°C. Range: Britain; Lower Miocene *G. trilobus* Zone to Middle Miocene *G. mayeri mayeri* Zone; France, Upper Oligocene to Lower Miocene in the Aquitaine Basin (Pujol, 1970). Total range: Upper Oligocene to Recent.

Globigerina ciperoensis angulisuturalis Bolli, 1957
Plate 9.3, Fig. 3, × 125. Fig. 9.4. Description: five chambers in the final whorl with relatively deeply cut sutures on the umbilical side.
Remarks: differs from *G. ciperoensis ciperoensis* and *G. ciperoensis angustiumbilicata* in having incised sutures. Range: Britain; Lower Miocene *G. trilobus* Zone; France, type Lower Aquitanian (Jenkins, 1966), Upper Oligocene to Lower Miocene in the Aquitaine Basin (Pujol, 1970, Bizon *et al.*, 1972). Total range: Upper Oligocene–Lower Miocene.

Globigerina ciperoensis ciperoensis Bolli, 1954
Plate 9.3, Fig. 4, × 75. Fig. 9.4. Description: five chambers in the final whorl increasing slowly in size and with an umbilical aperture.
Remarks: it lacks the lipped aperture of *G. ciperoensis angustiumbilicata.* Range: Britain; Lower Miocene *G. trilobus* Zone; France, Lower Miocene in the Aquitaine Basin (Pujol, 1970). Total range: Upper Oligocene to Lower Miocene.

Globigerina woodi Jenkins, 1960
Plate 9.3, Figs, 5, 6, × 75. Fig. 9.4. Description: 3–4 chambers in the final whorl increasing slowly in size; a large rimmed aperture and coarse wall.
Remarks: it differs from *G. bulloides* in having a higher arched aperture and coarser wall structure and from its descendant *G. decoraperta* in having a much lower spired test. Range: Britain; Lower Miocene *G. trilobus* Zone to the Middle Miocene *P. glomerosa curva* Zone; France, type Aquitanian–Burdigalian (Jenkins, 1966) and the Lower to Upper Miocene in the Aquitaine and Mediterranean Basins (Bizon *et al.*, 1972).

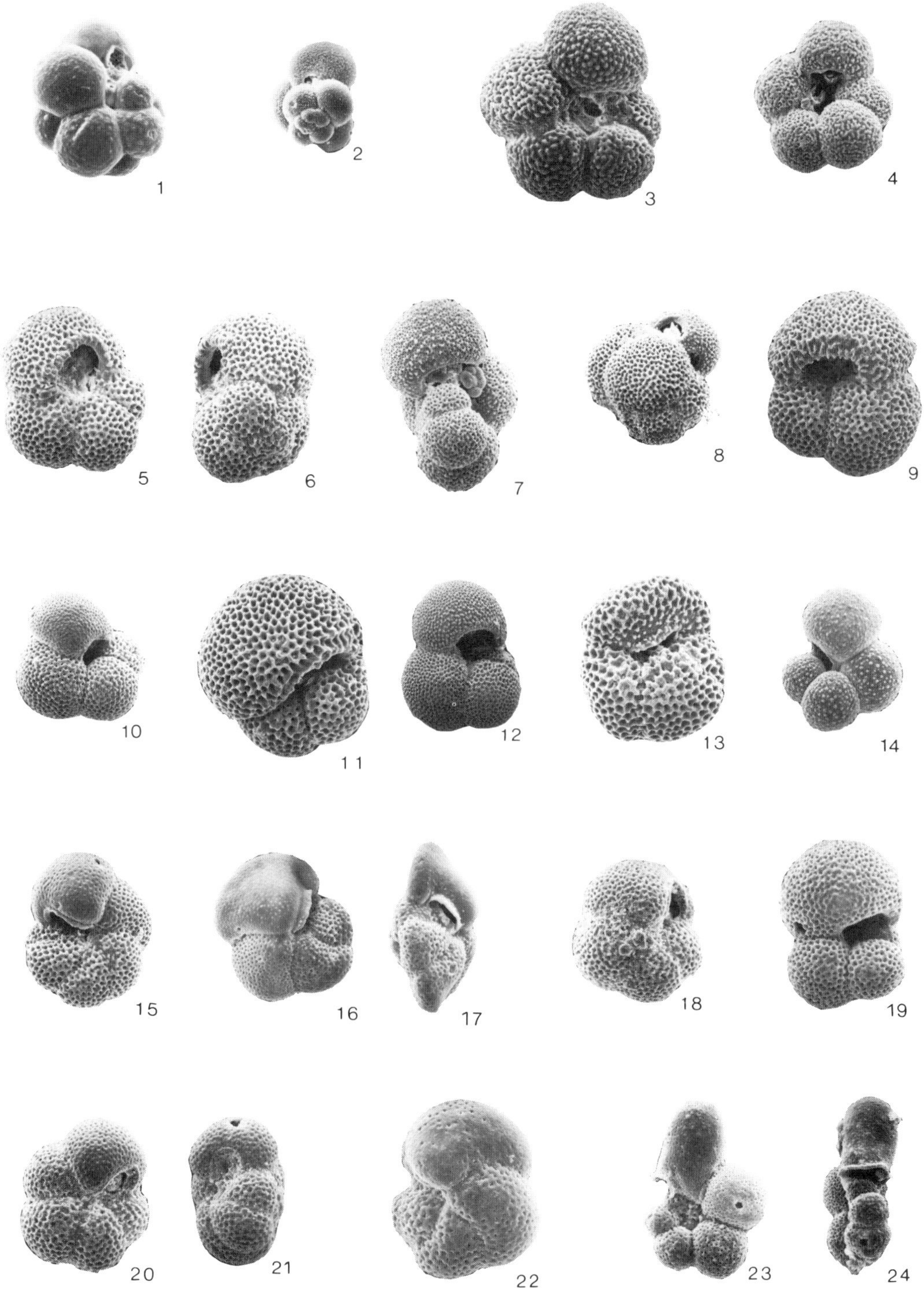

Plate 9.3

Globigerinella praesiphonifera (Blow)
Plate 9.3, Fig. 7, × 75. Fig. 9.4. = *Hastigerina (H.) siphonifera praesiphonifera* Blow, 1969. Description: 5 chambers in final whorl, nearly planispirally coiled; a finely pitted wall with small spine bases.
Remarks: differs from its trochospirally coiled ancestor *Globorotalia obesa* in having 5 chambers in the final whorl instead of 4, and from its descendant *G. aequilateralis* which has a planispirally coiled adult test. Range: Britain; Upper *G. trilobus* Zone of the Lower Miocene. Total range: Lower to Middle Miocene (Blow, 1969).

Globigerinoides ruber (d'Orbigny)
Plate 9.3, Figs. 8, 9, Fig. 8, × 50, Fig. 9, × 75. Fig. 9.4. = *Globigerina rubra* d'Orbigny, 1839. Description: small rounded apertures positioned opposite the sutures, normally on a high spired test as in Fig. 8.
Remarks: differs from other Miocene–Recent *Globigerinoides* in the position of the apertures and in having a high spiral test. Living at present in the upper 50 m of oceanic water within a temperature range of 10–30°C. Range: Britain; Lower Miocene *G. trilobus* Zone to Middle Miocene *P. glomerosa curva* Zone; France, Lower Miocene to Middle Pliocene in the Aquitaine Basin (Pujol, 1970). Total range: Lower Miocene–Recent.

Globigerinoides sacculifer (Brady)
Plate 9.3, Fig. 10, × 50. Fig. 9.4. = *Globigerina sacculifer* Brady, 1877. Description: three to four chambers in the final whorl with a sac-like final chamber.
Remarks: the distinctive final chamber distinguishes it from other species of the genus and its relatively thin wall from *Sphaeroidinella dehiscens;* it lives today in the upper 50 m of oceanic water within a temperature range of 14.5–30°C with maximum numbers found in the 24–30°C range. Range: Britain; Lower Miocene *G. trilobus* Zone; France, Mediterranean and Aquitaine Basins from the Lower Miocene to Pleistocene (Pujol, 1970). Total range: Lower Miocene to Recent.

Globigerinoides trilobus (Reuss)
Plate 9.3, Figs. 11, 12, Fig. 11, × 75, Fig. 12 × 37.5. Fig. 9.4. = *Globgerina triloba* Reuss, 1850. Description: three chambers in the final whorl with a low arched umbilical aperture and one supplementary aperture on the spiral side; final chamber can be enlarged (as illustrated in Fig. 11) during the evolutionary development towards its descendant *G. bisphericus* Todd.
Remarks: distinguished from its immediate ancestor *Globigerina woodi connecta* in the possession of a supplementary aperture on the spiral side; *G. trilobus* is related to *G. sacculifer* but is distinguished by not having the sac-like chamber. Range: Britain; Lower Miocene *G. trilobus* Zone to Middle Miocene *G. mayer mayeri* Zone; France, type Upper Burdigalian (Jenkins, 1966) and in the Lower to Middle Miocene in the Aquitaine Basin (Pujol, 1970). Total range: Lower Miocene to Recent.

Globoquadrina dehiscens (Chapman, Parr and Collins)
Plate 9.3, Fig. 13, × 75. Fig. 9.4. = *Globorotalia dehiscens* Chapman, Parr and Collins, 1934. Description: normally 4 chambers in the final whorl, quadrate in umbilical view, with a low arched aperture at base of final chamber.
Remarks: distinguished from *G. altispira* which is high spired with 5 chambers in the final whorl and from *G. globosa* which has more inflated chambers and is less quadrate in outline. Range: Britain; Lower Miocene *G. trilobus* Zone to Middle Miocene *G. mayeri mayeri* Zone; France, type Aquitanian–Burdigalian (Jenkins, 1966) and in the Aquitaine Basin from Lower to Upper Miocene. Total range: Lower Miocene to Lower Pliocene.

Globorotalia obesa Bolli, 1957
Plate 9.3, Fig. 14, × 75. Fig. 9.4. Description: 4 chambers in the final whorl increasing slowly in size, low-arched aperture, and a hispid wall.
Remarks: it is distinguished from its immediate descendant *G. praesiphonifera* in only having 4 chambers in the final whorl and has a low trochospiral test. Range: Britain; Lower Miocene *G. trilobus* Zone to Middle Miocene *P. glomerosa curva* Zone;France, type Burdigalian (Jenkins, 1966) and from the Upper Oligocene to Upper Miocene in the Aquitaine Basin (Pujol, 1970). Total range: Lower to Upper Miocene.

Globorotalia peripheroronda Blow and Banner 1966
Plate 9.3, Fig. 15, × 75. Fig. 9.4. Description: 5–6 chambers in the final whorl, rounded periphery with strongly recurved sutures on the spiral side and a low-arched lipped aperture.
Remarks: distinguished from its immediate descendant *G. mayeri* which has a large open aperture and nearly radial sutures. Range: Britain, Lower Miocene *G. trilobus* Zone to Middle Miocene *P. glomerosa curva* Zone; France, from the Lower to Middle Miocene in the Aquitaine Basin (Pujol, 1970). Total range: Lower to Middle Miocene.

Globorotalia praescitula Blow, 1959
Plate 9.3, Figs. 16, 17, × 100. Fig. 9.4. Description: 4–4½ chambers in the final whorl with strong, recurved sutures on spiral side, acutely angled rounded periphery with a low arched lipped aperture.
Remarks: distinguished from most descendants such as *G. archaeomenardii, G. miozea* and *G. praemenardii* in not possessing a keeled or incipient keeled periphery, and from *G. scitula* in having a more coarsely pitted wall, a more rapid increase in chamber size and in having a more bi-convex shape in side view. Range: Britain; Lower Miocene *G. trilobus* Zone to Middle Miocene *G. mayeri mayeri* Zone; France, Lower to Middle Miocene in the Aquitaine Basin (Pujol, 1970). Total range: Lower to Middle Miocene.

Globorotalia pseudocontinuosa Jenkins, 1967
Plate 9.3, Fig. 18, × 75. Fig. 9.4. Description: 4 chambers in final whorl, coarsely ornamented wall and a large comma-shaped aperture.
Remarks: distinguished from *G. semivera* which has 5 chambers in the final whorl and from *G. obesa* which has a low arched aperture and a pustular wall. Range: Britain; Lower Miocene *G. trilobus* Zone; France, Upper Burdigalian and in one Helvetian (?) sample. Total range: Lower Miocene.

Globorotalia saginata Jenkins, 1966
Plate 9.3, Fig. 19, × 100. Fig. 9.4. Description: 3½ chambers in the final whorl, chambers increase rapidly in size, low arched aperture.
Remarks: distinguished from *G. obesa* by having a more rapid increase in chamber size and in having a more coarsely ornamented wall structure. Range: Britain; Lower Miocene *G. trilobus* Zone; France, type Aquitanian-Burdigalian (Jenkins, 1966). Total range: Middle Oligocene to Middle Miocene.

Globorotalia semivera (Hornibrook)
Plate 9.3, Figs. 20, 21, × 75. Fig. 9.4 = *Globigerina semivera* Hornibrook, 1961. Description: 5 chambers in the final whorl, large open aperture, radial sutures and coarse wall.
Remarks: differs from *G. pseudocontinuosa* in having 5 chambers in the final whorl, and from *G. peripheroronda* in having radial sutures and a larger more open aperture. Range: Britain; Lower Miocene *G. trilobus* Zone to Middle Miocene *P. glomerosa curva* Zone; France, type Aquitanian–Burdigalian (Jenkins, 1966), and as *G. acrostoma* by Pujol (1970) from the Lower–Middle Miocene in the Aquitaine Basin. Total range: Upper Oligocene–Middle Miocene.

Globorotalia zealandica Hornibrook, 1958
Plate 9.3, Fig. 22, × 100. Fig. 9.4. Description: 4 chambers in the final whorl, with a flattened spiral and convex umbilical side, high arched aperture and a coarse wall.
Remarks: differs from *G. pseudocontinuosa* in having a more rapid increase in chamber size in the final whorl and in having recurved sutures on the spiral side. Range: Britain; Lower Miocene *G. trilobus* Zone. Total range: Lower Miocene.

Protentella clavaticamerata Jenkins, 1977
Plate 9.3, Figs. 23, 24, × 100. Fig. 9.4. Description: small test (holotype is 0.26 mm diameter), planispirally coiled, final chambers increasing fairly rapidly in size and becoming radially elongate, aperture centrally placed, lipped at base of final chamber.
Remarks: the smaller planispirally coiled, involute test distinguishes it from *G. praesiphonifera*. Range: Britain; Upper *G. trilobus* Zone, Lower Miocene. Total range: Upper Lower Miocene.

9.7 REFERENCES

Batjes, D. A. J. 1958. Foraminifera of the Oligocene of Belgium. *Mém. Inst. r. Sci. Nat. Belg.*, **143.**

Bizon, G., Bizon, J., Aubert, J. and Oertli, H. 1972. Atlas des principaux foraminifères plantconiques du bassin méditerranéen. Oligocène á Quaternaire. Paris: *Editions Technip*, 1–316.

Blow, W. H. 1969. Late Middle Eocene to Recent planktonic foraminiferal biostratigraphy. In Brönniman, P., and Renz, H. H. (eds.) *Proceedings of the first international conference in planktonic microfossils.* Leiden: E. J. Brill, **1**, 199–421.

Cambridge, P. G. 1977. Whatever happened to the Boytonian? A review of the marine Plio-Pleistocene of the southern North Sea basin. *Bull. geol. Soc. Norfolk,* **29**, 23–45.

Carter, D. J. 1951. Indigenous and exotic foraminifera in the Coralline Crag of Sutton, Suffolk. *Geol. Mag.*, **88**, 236–248.

Carter, D. J. 1957. The distribution of the foraminifera *Alliatina excentrica* (di Napoli Alliata) and the new genus *Alliatinella. Palaeontology,* **1**, 76–86.

Clodius, G. 1922. Die Foraminiferen des obermiozänen Glimmertons in Norddeutschland mit besonderer Berücksichtigung der Aufschlüsse in Mecklenburg. *Arch. Ver. Freunde Naturg. Mecklenb.*, **75**, 76–145.

Curry, D., Adams, C. G., Boulter, M. C., Dilley, F. C., Eames, F. E., Funnell, B. M. and Wells, M. K. 1978. A correlation of Tertiary rocks in the British Isles. *Geol. Soc. Lond., Special Report,* **12.**

Curry, D., Martini, E., Smith, A. J. and Whittard, W. F. 1962. The geology of the western approaches of the English Channel. I. Chalky rocks from the upper reaches of the continental slope. *Phil. Trans. R. Soc.*, **B245**, 267–290.

Curry, D., Murray, J. W. and Whittard, W. F. 1965. The geology of the western approaches of the English Channel. III. The *Globigerina* Silts and associated rocks. *Colston Pap.*, **17**, 239–264.

Daniels, C. H. von and Spiegler, D. 1977. *Uvigerinen* (Foram.) im Neogen Nordwestdeutschlands (Das Nordwestdeutsche Tertiärbecken, Beitrag Nr. 23). *Geol. Jb.*, **A40**, 3–59.

De Meuter, F. J. and Laga, P. G. 1977. Lithostratigraphy and biostratigraphy based on benthonic foraminifera of the Neogene deposits of northern Belgium. *Bull. Soc. belge Géol. Paléont. Hydrol.*, **85**, 133–152.

Doppert, J. W. C. 1975. Foraminiferenzonering van het Nederlandse Onder-Kwartair en Tertiair. 114–118. in *Toelichting bij geologische overzichtskaarten van Nederland.* Zagwijn, W. H. (Editor). Haarlem, Netherlands.

Doppert, J. W. C., Laga, P. G. and De Meuter, F. J. 1979. Correlation of the biostratigraphy of marine Neogene deposits, based on benthonic foraminifera, established in Belgium and the Netherlands. *Meded. Rijks geol. Dienst,* **31-1.**

Gheorghian, M. 1971. Sur quelques affleurements de dépots ottnangiens de Roumanie et sur leur contenu microfaunique. *Mem. Inst. Geol. Bucarest,* **14,** 103-121.

Gheorghian, M. 1974. Considerations of the genus *Hidina* (order Foraminiferida, Eichwald 1830). *Dari Seama Sedint. Inst. Geol.,* **60/3,** 23-31.

Gheorghian, M. 1975. Asupra biostratigrafiei depozitelor Miocene din Romania (Stadiul 1974) [On biostratigraphy of Miocene deposits in Roumania (Stage 1974)]. *Ibid.* **61/4,** 85-104 (in Roumanian).

Gramann, F. 1964. Die Arten der Foraminiferen-Gattung *Asterigerina* d'Orb. im Tertiär NW-Deutschlands. *Paläont. Z.,* **38,** 207-222.

Hantken, M. von, 1875. Die Fauna der *Clavulina szaboi* Schichten. I. Foraminiferen. *Mitt. Jb. K. ung. geol. Anst.,* **4,** 1-39.

Harmer, F. W. 1898. The Pliocene deposits of the east of England: the Lenham Beds and the Coralline Crag. *Q. Jl. geol. Soc. Lond.,* **54,** 308-356.

Harmer, F. W. 1900. On a proposed new classification of the Pliocene deposits of the east of England. *Rep. Br. Ass. Advmt. Sci.,* Dover 1899, 751-753.

Harmer, F. W. 1902. A sketch of the Later Tertiary history of East Anglia. *Proc. Geol. Asso.,* **17,** 416-479.

Hinsch, W., Kaever, M. and Martini, E. 1978. Die Fossilführung des Erdfalls von Nieheim (SE-Westfalen) und seine Bedeutung für Paläeogeographie im Campan und Moizän. *Paläont. Z.,* **52,** 219-245.

Hosius, A. 1892. Beiträge zur Kenntnis der Foraminiferen-Fauna des Miocens. *Verh. naturh. Ver. preuss. Rheinl.,* **49,** 148-197.

Indans, J. 1958. Mikrofaunistische Korrelationen im marinen Tertiär der Niederrheinischen Bucht. *Fortschr. Geol. Rheinld Westf.,* **1,** 223-238.

Indans, J. 1962. Foraminiferen-Fauna aus dem Moizän des Niederrheingebietes. *Ibid.,* **6,** 19-82.

Indans, J. 1965. Mikrofaunistisches Normalprofil durch das marine Tertiär der Neiderrheinsichen Bucht. *Forschr.-Ber. Nordrhein-West,* **1484.**

Institute of Geological Sciences. 1978. IGS Boreholes 1977. *Rep. Inst. Geol. Sci.,* **78/21.**

Jenkins, D. G. 1966. Planktonic foraminifera from type Aquitanian-Burdigalian of France. *Contr. Cushman Found.,* **17,** 1-15.

Jenkins, D. G. 1977. Lower Miocene planktonic foramini fera from a borehole in the English Channel. *Micropaleontology,* **23,** 297-318.

Jenkins, D. G. and Shackleton, N. 1979. Parallel changes in species diversity and palaeotemperature in the Lower Miocene. *Nature,* **278,** 50-51.

Jones, T. R., Parker, W. K. and Brady, H. B. 1866-1897. *A monograph of the foraminifera of the Crag.* Palaeontographical Society, London.

Langer, W. 1963a. Einige wenig bekannte Foraminiferen aus dem mittleren und oberen Moizän des Nordseebeckens. *Neues Jb. Geol. Paläont. Abh.,* **117,** 169-184.

Langer, W. 1963b. Bemerkungen zur Stratigraphie nach Foraminiferen im mittleren und oberen Miozän von N-und NW-Deutschland. *Neues Jb. Geol. Paläont. Mh.,* 543-558.

Langer, W. 1969. Beitrag zur Kenntnis einiger Foraminiferen aus dem mittleren und oberen Miozän des Nordsee-Beckens. *Neues. Jb. Geol. Paläont. Abh.,* **133,** 23-78.

Macfadyen, W. A. 1942. A post-glacial microfauna from Swansea Docks. *Geol Mag.,* **79,** 133-146.

Margerel, J. P. 1970. Les foraminifères des marnes a 'Nassa Prismatica' due Bosq d'Aubigny. *Bull. Soc. belge Géol. Paléont. Hydrol.,* **79,** 133-156.

Margerel, J. P. 1971. Le genre *Faujasina* d'Orbigny dans le Plio-Pléistocène du Bassin Nordique Européen. *Revue Micropaléont.,* **14,** 113-120.

Margerel, J. P. 1972. Les foraminifères du Néogène de l'Ouest de la France intéret paléoécologique, paléogéographique et stratigraphique. *Bull. Soc. géol. Fr.,* (7), **14,** 121-126.

Marks, P. 1951. A revision of the smaller foraminifera from the Miocene of the Vienna Basin. *Contr. Cushman Fdn., foramin. Res.,* **2,** 33-73.

Millett, F. W. 1886a. Notes on the fossil foraminifera of the St Erth Clay pits. *Trans, R. geol. Soc. Corn.,* **10,** 213-216.

Millett, F. W. 1886b. Additional notes on the foraminifera of St Erth Clay. *Ibid.,* **10,** 221-226.

Millett, F. W. 1895. The foraminifera of the Pliocene beds of St Erth. *Ibid.,* **11,** 655-661.

Millett, F. W. 1897. The foraminifera of the Pliocene beds of St Erth in relation to those of other deposits. *Ibid.,* **12,** 43-46.

Millett, F. W. 1898. Additions to the list of foraminifera from the St Erth Clay. *Ibid.,* **12,** 174-176.

Millett, F. W. 1902. Notes on the *Faujasinae* of the Tertiary beds of St Erth. *Ibid.,* **12,** 719-720.

Mitchell, G. F. 1965. The St Erth Beds – an alternative explanation. *Proc. Geol. Ass.,* **76,** 345-366.

Mitchell, G. F. 1973. The Late Pliocene marine formation of St Erth, Cornwall. *Phil Trans. R. Soc.,* **B226,** 1-37.

Odrzywolska-Bienkowa, E. 1977. Wybrane profile Miocenu Opolszczyzny w swietle baden mikropaleontologicznych [Selected Miocene profiles in the Opole region in the light of micropalaeontological investigations]. *Przegl. Geol.,* **1,** 12-16 (in Polish).

Orbigny, A. d', 1846. *Foraminifères fossiles du bassin tertiaire de Vienne (Autriche).* Paris. Gide et Comp.

Pflum, C. E. and Frerichs, W. E. 1976. Gulf of Mexico deep-water foraminifera. *Spec. Publs. Cushman Fdn.,* **14.**

Poignant, A. and Pujol, C. 1976. Nouvelles données micropaléontologiques (foraminifères planctoniques et petits foraminifères benthiques) sur le stratotype de l'Aquitanien. *Géobios,* **9,** 607-663.

Poignant, A. and Pujol, C. 1978. Nouvelles données micropaléontologiques (foraminfères planctoniques et petits foraminifères benthiques) sur le stratotype bordelais du Burdigalien. *Ibid.,* **11,** 655-712.

Popescu, G. 1975. Etudes des foraminfères due Miocène inférieur et moyen du nord-ouest de la Transylvanie. *Mem. Inst. Geol. Bucarest,* **23.**

Prestwich, J. 1871. On the structure of the Crag-beds of Suffolk and Norfolk with some observations on their organic remains. I. The Coralline Crag of Suffolk. *Q. Jl. geol. Soc. Lond.*, **27**, 115–146.

Pujol, C. 1970. Contribution à l'étude des Foraminifères planktoniques néogènes dans le Bassin Aquitaine. *Bull. Inst. Géol. Bassin Aquitaine*, **9**, 201–219.

Reuss, A. E. 1951. Uber die fossilen Foraminiferen und Entomostraceen der Septarienthone der Umgegend von Berlin. *Z. dt. geol. Ges.*, **3**, 49–92.

Reuss, A. E. 1856. Beiträge zur Charakteristik der Tertiärschichten des nördlichen und mittleren Deutschlands. *Sber. Akad. Wiss. Wien.*, **18**, 197–273.

Reuss, A. E. 1861. Beiträge zur Kenntnis der tertiären Foraminiferenfauna. *Ibid.*, **42**, 355–370.

Reuss, A. E. 1863a. Les foraminiferes du Crag d'Anvers. *Bull. Acad. r. Sci. Belg., Cl. Sci.* **(2), 15**, 137–162.

Reuss, A. E. 1863b. Beiträge zur Kenntnis der tertiären Foraminiferen-Fauna. *Sber. Akad. Wiss. Wien.*, **48**, 36–71.

Spiegler, D. 1973. Die Entwicklung von *Ehrenbergina* (Foram.) im höheren Tertiär NW-Deutschlands. *Geol. Jb.*, **A6**, 3–23.

Spiegler, D. 1974. Biostratigraphie des Tertiärs zwischen Elbe und Weser/Aller (Bentische Foraminiferen, Oligo-Miozän). (Das Nordwestdeutsche Tertiärbecken) Beitrag Nr. 2. *Geol. Jb.*, **A16**, 27–69.

Staesche, K. and Hiltermann, H. 1940. Mikrofaunen aus dem Tertiär Nordwestdeutschlands. *Abh. Reichsstelle Bodenforsch.*, NF 201, 1–26.

Ten Dam, A. and Reinhold, T. 1941a. The genus *Darbyella* and its species. *Geologie Mijnb.*, **3**, 108–111.

Ten Dam, A. and Reinhold, T. 1941b. Nonionidae as Tertiary index foraminifera. *Ibid.*, **3**, 209–212.

Ten Dam, A. and Reinhold, T. 1941c. Asterigerinen als index-Foraminiferen fuer das Nordwest-Europaeische Tertiaer. *Ibid.*, **3**, 220–223.

Ten Dam, A. and Reinhold, T. 1941d. Die Stratigraphische Gliederung des Niederländischen Plio-Plistozäns nach Foraminiferen. *Meded. Geol. Sticht.* **C-V-1.**

Ten Dam, A. and Reinhold, T. 1942. Die Stratigraphische Gliederung des Neiderländischen Oligo-Miozäns nach Foraminiferen (mit ausnahme von S. Limburg). *Ibid.*, **C-V-2.**

Toering, K. and Van Voorthuysen, J. H. 1973. Some notes about a comparison between the Lower Pliocene foraminiferal faunae of the south-western and north-eastern parts of the North Sea basin. *Revue Micropaléont.*, **16**, 50–58.

Van Voorthuysen, J. H. 1958. Les foraminifères Mio-Pliocenes et Quaternaires due Kruisschans. *Mém. Inst. r. Sci. nat. Belg.*, **142.**

Warrington, G. and Owens, B. (Compilers), 1977. Micropalaeontological biostratigraphy of offshore samples from south-west Britain. *Rep. Inst. Geol. Sci.*, **77/7.**

Wilkinson, I. P. 1980. Coralline Crag ostracods and their environmental and stratigraphic significance. *Proc. Geol. Ass.*, **91**, 291–306.

10
Quaternary

B. M. Funnell

10.1 INTRODUCTION

The earliest systematic descriptions of British Quaternary foraminifera were made by Jones (1865), and Jones, Parker, Brady, H. B., Burrows and Holland *et al.* (1866–1897). The first of these references listed foraminifera from the Hoxnian Nar Valley Clay, whilst the multi-authored Palaeontographical Society Monograph on the 'Crag Foraminifera' eventually illustrated many species from the Pliocene and early Pleistocene deposits of East Anglia. It was commenced before, and completed after, the voyage of H.M.S. Challenger and H. B. Brady's monograph of the Challenger collections of Recent foraminifera.

There followed a lapse of many years before any further studies of British Quaternary foraminifera were made. Macfadyen (1932, 1933, 1942, 1952) then returned to the subject, publishing a number of papers, firstly on miscellaneous Crag foraminifera, and subsequently on Post-Glacial (i.e. Flandrian) assemblages. His papers post-date the early work of Cushman on world-wide collections of Recent foraminifera, but still rely heavily on the taxonomy of late 19th and early 20th century U.K. investigators of Recent species. In 1941 Ten Dam and Reinhold published the first account of early Quaternary foraminifera from the Netherlands using Cushman's classification and nomenclature. Van Voorthuysen (1949, 1955) continued and developed Ten Dam and Reinhold's work in the Netherlands, introducing extensive use of quantitative methods, until his retirement in 1972. The present author started stratigraphical work using early Pleistocene Crag foraminifera in 1957 as a Ph.D. student. Funnell and others (listed in the References) have subsequently also investigated foraminifera from a number of interglacial and post-glacial deposits in and around the British Isles.

10.2 LOCATION OF IMPORTANT COLLECTIONS

Some slides of materials examined by Jones, Parker *et al.* (1866–97) are still preserved at the British Museum (Natural History) and the Geological Museum, South Kensington, but are of historical interest only. Hypotypes of species investigated by Funnell (unpublished thesis) are lodged at the Sedgwick Museum, Cambridge.

10.3 STRATIGRAPHICAL DIVISIONS

The Quaternary period in the British Isles has been divided into stages based on pollen assemblage biozones. Fifteen stages have so far been recognised in the Pleistocene epoch, whilst the Holocene epoch comprises one stage only. In general the stages represent alternating cool and temperate conditions in the early Quaternary, and alternating glacial and interglacial conditions in the late Quaternary. The earliest evidence for periglacial ground-ice conditions (ice wedges) is found in the Baventian Stage. The earliest evidence of glacial deposits is found in the Anglian Stage.

Since 1952 the base of the Quaternary in the British Isles has been taken at the base of the Red Crag Formation, and this practice is continued in this account. Thus defined it probably corresponds to the base of the Reuverian 'c' (pollen) stage of the Netherlands' succession. The immediately following cool Praetiglian (pollen) Stage of the Netherlands sequence, in which the Pacific foraminiferal species *Elphidium oregonensis* appears, is not recognisable in the British Isles, where *E. oregonensis* has not so far been found. The Praetiglian, commonly now taken as the base of the Quaternary in the Netherlands, seems most likely to be equivalent in age to, or to slightly post-date, the Butley-Crag towards the top of the Red Crag member of the Red Crag Formation. The upper limit of the Quaternary extends to the present day. Table 10.1 shows the full sequence of Quaternary stages so far recognised in the British Isles, together with a selected list of some geological formations to which they correspond, and an indication of References to Quaternary foraminifera of the relevant age.

10.4 BRIEF SUMMARY OF FAUNAL ASSOCIATIONS AND FACIES

Most Quaternary foraminiferal associations in the British Isles are of intertidal or inner sublittoral facies. Cold and warmer water associations tend to occur repetitively in successive cool and warmer stages. Deeper water associations occur in continental shelf, especially North Sea, boreholes (Greogry and Bridge 1979, Gregory and Harland 1978, Harland *et al.,* 1978 and Hughes *et al.,* 1977), and in the Netherlands. Cold, late Quaternary deeper water associations are found above sea-level in areas subject to post-glacial isostatic uplift.

Climatic and water depth influences exert a dominating control on British Quaternary foraminiferal assemblages. Appearances of new species by evolution or immigration, and disappearances by extinction or emigration are subordinate. Therefore true stratigraphic index species are few in number and most fine stratigraphic discrimination necessarily depends on other criteria such as pollen analysis or ^{14}C dating. Palaeoenvironmental interpretation however, based exclusively on living species, is well-developed.

Characteristic faunal associations include:

(a) Polymorphinid-textulariid containing assemblages, reminiscent of the late Neogene (Pliocene) faunas of the Coralline Crag, characterise the Red Crag Formation (Red and Ludham Crag members). These contain strong Lusitanian faunal elements, appear to indicate water depths of about 50 m, and are typically deposited in sand-wave sedimented deposits. As well as polymorphinid and textulariid species, a variety of cibicidid and rosalinid species are usually present, and the larger and more heavily calcified species of *Elphidium;* the species *Pararotalia serrata* is also typical.

(b) *Elphidiella* – dominated associations are typical of all the post-Ludhamian/pre-Anglian stages.

Temperate (interglacial) stages contain higher percentages of *Ammonia beccarii,* which in particular dominates the inter-tidal assemblages of the Chillesford Crag.

Cool (glacial) stages contain the highest percentages of *Elphidiella,* together with significant percentages of such cold-water species of *Elphidium* as *E. frigidum* and *Protelphidium orbiculare.* Typically *A. beccarii* is almost or totally unrepresented in the cold stages.

(c) Post-Anglian associations are all very similar to modern assemblages from comparable environments. Textulariid genera such as *Trochammina* and *Jadammina* dominate high intertidal salt-marsh assemblages, *Protelphidium* and *Elphidium* species, together in temperate (interglacial) periods with *A. beccarii,* are typical of lower intertidal tidal flats. The pattern is clear in assemblages from

Table 10.1
Stratigraphical subdivisions of the Quaternary: (f = freshwater, m = marine)

Epochs	Stages	Formations	Members	References
Holocene	Flandrian			Adams and Haynes 1965; Coles and Funnell 1981; Culver and Banner 1978; Haynes *et al.* 1977; Jansen *et al.* 1979; Lees 1975; Macfadyen 1933, 1942, 1952, 1955; Murray and Hawkins 1976; Peacock *et al.* 1978.
Pleistocene	Devensian		(glacial)	Haynes *et al.* 1977; Jansen *et al.* 1979; Lord 1980; Peacock *et al.*, 1977, 1978.
	Ipswichian			
	Wolstonian		(glacial)	
	Hoxnian			Fisher *et al.* 1969; Jones 1865; Lord and Robinson 1978; van Voorthuysen 1955.
	Anglian		(glacial)	Macfadyen 1932.
		Cromer Forest-bed	Bacton (f)	
	Cromerian		Mundesley (m)	
			West Runton (f)	
	Beestonian		Runton (f/m)	
	Pastonian		Paston (m)	
	Pre-Pastonian		Sheringham (f)	
	Bramertonian	Norwich Crag	Sidestrand Chillesford	Funnell 1961, 1980; Funnell and West 1962, 1977; Funnell *et al.* 1979; West *et al.* 1980.
	Baventian		Norwich Crag	
	Antian			
	Thurnian			
	Ludhamian	Red Crag	Ludham Crag	Beck *et al.* 1972; Funnell 1961; Funnell and West 1977.
	Pre-Ludhamian		Red Crag	

the Hoxnian, Ipswichian and Flandrian temperate (interglacial) stages. Deeper water associations, usually, but not always of Devensian (Late-Glacial) age, are dominated by or contain high proportions of *Elphidium clavatum, Bulimina* spp., *Cassidulina* spp. and even *Uvigerina* spp.

10.5 INDEX SPECIES

10.5.1 Suborder Rotaliina

Ammonia beccarii (Linné)
Plate 10.1, Figs. 1–3, × 50, = *Nautilus beccarii* Linné, 1758. Description: 9–11 chambers in final whorl; rounded periphery; margins of umbilical radial sutures thickened and crenulate; often an umbilical boss.
Remarks: confined to temperate (interglacial) stages in the North Sea basin.

Cibicides lobatulus (Walker and Jacob)
Plate 10.1, Figs. 4–6, × 60, = *Nautilus lobatulus* Walker and Jacob, 1798. Description: 6–9 chambers in final whorl; periphery usually acute, margin usually lobate; spiral side flat, slightly convex or irregular, umbilical side convex.
Remarks: less common in cold (glacial) stages in the North Sea basin, but the only species of *Cibicides* which is tolerant of arctic conditions.

Cibicides subhaidingerii Parr, 1950
Plate 10.1, Figs. 7–9, × 70. Description: circa 9 chambers in final whorl; periphery rounded, margin smooth; spiral side slightly convex, umbilical side strongly convex; coarsely perforate.
Remarks: although this species is now apparently restricted to the Pacific, it occurs in temperate (interglacial) stages in the early Pleistocene of the North Sea basin.

Elphidiella hannai (Cushman and Grant)
Plate 10.1, Figs. 13–14, × 65, = *Elphidium hannai* Cushman and Grant, 1927. Description: 9–14 chambers in final whorl; periphery rounded, margin smooth sometimes becoming slightly lobate; planispiral biconvex; bifurcating canal system opens as pairs of pores along radial sutures.
Remarks: the most common and ubiquitous species in littoral and inner sublittoral environments of the early Pleistocene of the North Sea basin. Not known from the British Isles area after the Anglian stage, it is confined to the northern Pacific at the present-day.

Elphidium frigidum Cushman, 1933
Plate 10.1, Figs. 17–19, × 70. Description: circa 8–9 chambers in final whorl; rounded periphery, margin usually becoming increasingly lobate; planispiral, flattened umbilical areas often finely granular with granulations extending along radial sutures; indistinct retral processes and single pore openings from canal system along sutures.
Remarks: consistently present in early Pleistocene as small percentages, from Butley Crag onwards; more abundant in cold (glacial) stages. Present-day distribution Arctic, and, if *E. subarcticum* is included in synonomy, also east coast of N. America.

Elphidium haagensis van Voorthuysen, 1949
Plate 10.1, Figs. 20–21, × 75. Description: circa 12 chambers in final whorl; periphery rounded, but tendency to keel, margin smooth; planispiral, biconvex; radial sutures curved, central line of chambers thickened forming radial ribs; umbilical boss or bosses.
Remarks: not known later than the Antian stage of the early Pleistocene of the North Sea basin.

Elphidium pseudolessonii Ten Dam and Reinhold, 1941
Plate 10.1, Figs. 15–16, × 70. Description: circa 18 chambers in the final whorl; periphery acute, margin smooth occasionally becoming slightly lobate; planispiral, biconvex; retral processes strong and numerous, usually extending across full width of chambers, umbilical tubercles.
Remarks: Similar to the modern *E. crispum/macellum,* it differs in having less well-developed umbilical bosses, less strong and consistent development of a peripheral keel, and more variable development of retral processes, with anterior edges of chambers sometimes slightly inflated. Occurs commonly in the early Pleistocene from the Butley Crag onwards. At first accompanied by *E. crispum*, but from Thurnian stage onwards usually alone; last known from the Pre-Pastonian stage, but may possibly occur later.

Pararotalia serrata (Ten Dam and Reinhold)
Plate 10.1, Figs. 10–12, × 70, = *Rotalia serrata* Ten Dam and Reinhold, 1941. Description: circa 9 chambers in the final whorl; periphery acute with margin showing tendency to but not actual development of radial spines; biconvex with single strong umbilical boss on umbilical side.
Remarks: representative of a much larger number of Lusitanian, Pliocene-relict species found in the Polymorphinid–textulariid-containing assemblages of the Red Crag Formation. Resembles the modern Mediterranean *Rotalia calcar* but in *P. serrata* the spines are never more than incipiently developed.

Protelphidium orbiculare (H. B. Brady)
Plate 10.1, Figs. 22–24, × 75, = *Nonionina orbicularis* H. B. Brady, 1881. Description: circa 8 chambers in final whorl; periphery broadly rounded; planispiral, strongly biconvex; radial sutures incised.
Remarks: particularly characteristic of transitional temperate (interglacial)/cold (glacial) conditions in the North Sea basin. Present-day distribution Arctic and colder temperate waters.

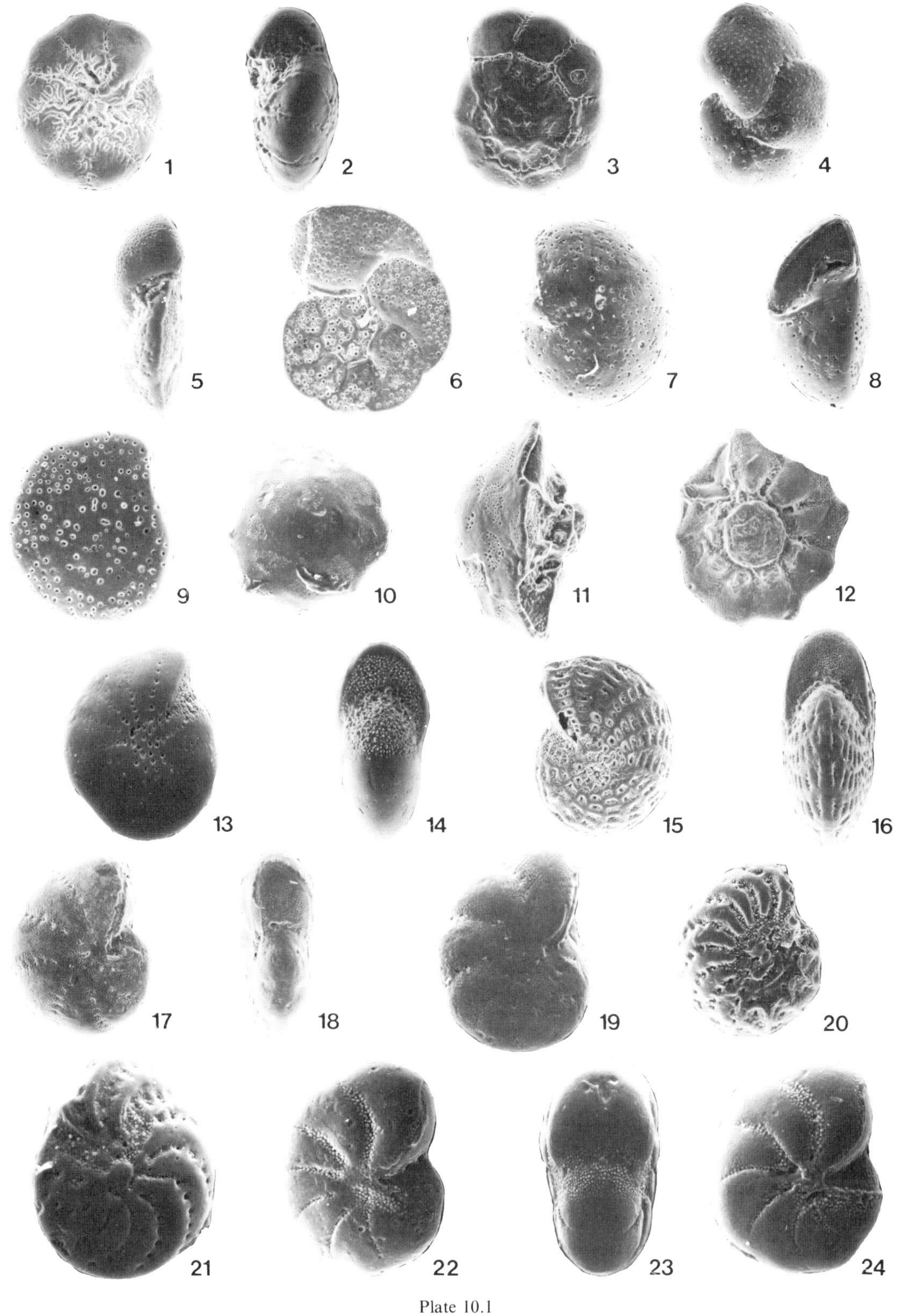

Plate 10.1

10.6 REFERENCES

Adams, T. D. and Haynes, J. 1965. Foraminifera in Holocene marsh cycles, at Borth, Cardiganshire (Wales). *Palaeontology,* **8,** 27–38.

Beck, R. B., Funnell, B. M. and Lord, A. R. 1972. Correlation of Lower Pleistocene Crag at depth in Suffolk. *Geol. Mag.,* **109,** 137–139.

Coles, B. P. L. and Funnell, B. M. 1981. Holocene palaeoenvironments of Broadland, England. *In Holocene marine sedimentation in the North Sea Basin.* Internat. Assoc. Sediment. Spec. Publ. No. 4.

Culver, S. J. and Banner, F. T. 1978. Foraminiferal assemblages as Flandrian palaeoenvironmental indicators. *Palaeogeog., Palaeoclimatol., Palaeoecol.,* **24,** 53–72.

Fisher, M. J., Funnell, B. M. and West, R. G. 1969. Foraminifera and pollen from a marine interglacial deposit in the Western North Sea, *Proc. Yorks. geol. Soc.,* **37,** 311–320.

Funnell, B. M. 1961. The Palaeogene and Early Pleistocene of Norfolk. *Trans. Norfolk Norwich Nat. Soc.,* **19,** 340–364.

Funnell, B. M. 1980. Palaeoenvironmental Analysis of the Dobb's Plantation section, Crostwick (and comparison with type localities of the Norwich and Weybourne Crags). *Bull Geol. Soc. Norfolk,* **31,** 1–10.

Funnell, B. M. and West, R. G. 1962. The Early Pleistocene of Easton Bavents, Suffolk. *Q. Jl. Geol. Soc. Lond.,* **117,** 125–141.

Funnell, B. M. and West, R. G. 1977. Preglacial Pleistocene deposits of East Anglia. *In* Shotton, F. W. (ed.) *British Quaternary Studies – Recent Advances.* Oxford University Press, 247–265.

Funnell, B. M., Norton, P. E. P. and West, R. G. 1979. The crag at Bramerton, near Norwich, Norfolk. *Phil. Trans. Roy. Soc.,* **B287,** 489–534.

Gregory, D. and Bridge, V. A. 1979. On the Quaternary foraminiferal species *Elphidium? ustulatum,* Todd, 1957: its stratigraphic and palaeoecological implication. *J. Foramin. Res.,* **9,** 70–75.

Gregory, D. and Harland, R. 1978. The late Quaternary climatostratigraphy of IGS Borehole SLN 75/33 and its application to the palaeoceanography of the north-central North Sea. *Scott. J. Geol.,* **14,** 147–155.

Harland, R., Gregory, D. M., Hughes, M. J. and Wilkinson, I. P. 1978. A late Quaternary bio-climatostratigraphy for marine sediments in the north-central part of the North Sea. *Boreas,* **7,** 91–6.

Haynes, J. R., Kiteley, R. J., Whatley, R. C. and Wilks, P. J. 1977. Microfaunas, microfloras and the environmental stratigraphy of the Late Glacial and Holocene in Cardigan Bay. *Geol. J.,* **12,** 129–158.

Hughes, M. J., Gregory, D. M., Harland, R. and Wilkinson, I. P. 1977. Late Quaternary foraminifera and dinoflagellate cysts from boreholes in the UK sector of the North Sea between 56° and 58° N. *In* Holmes, R. Quaternary deposits of the central North Sea, 5. *Rep. Inst. Geol. Sci.,* 77/14, 36–46.

Jansen, J. H. F., Doppert, J. W. C., Hoogendoorn-Toering, K., Jong, D. E., J. and Spaink, G. 1979. Late Pleistocene and Holocene deposits in the Witch and Fladen Ground area, northern North Sea. *Neth. J. Sea. Res.,* **13,** 1–39.

Jones, T. R. 1865. Microzoa of the deposits of the Nar Valley. *Geol. Mag.,* **2,** 306–307.

Jones, T. R., Parker, W. K., Brady, H. B., Burrows and Holland, 1866–1897. A Monograph of the Foraminifera of the Crag. *Mon. Palaeontogr. Soc.*

Lees, B. J. 1975. Foraminiferida from Holocene sediments in Start Bay. *Jl. Geol. Soc. Lond.,* **131,** 37–49.

Lord, A. R. 1980. Interpretation of the Lateglacial marine environment of North-West Europe by means of Foraminifera. *In* Lowe, J. J., Gray, J. M. and Robinson, E. (eds.) *The Lateglacial environment of the British Isles and possible correlations with North-West Europe.* Pergamon, Oxford. 103–114.

Lord, A. R. and Robinson, J. E. 1978. Marine Ostracoda from the Quaternary Nar Valley Clay, west Norfolk. *Bull geol. Soc. Norfolk,* **30,** 113–118.

Macfadyen, W. A. 1932. Foraminifera from some Late Pliocene and Glacial Deposits of East Anglia. *Geol. Mag.,* **69,** 481–497.

Macfadyen, W. A. 1933. The Foraminifera of the Fenland Clays at St. Germans, near King's Lynn. *Geol. Mag.,* **70,** 182–191.

Macfadyen, W. A. 1942. A post-glacial microfauna from Swansea Docks. *Geol. Mag.,* **79,** 133–146.

Macfadyen, W. A. 1952. Foraminifera and other microfauna from Post-glacial clays from the middle Bure valley, Norfolk. *Roy. Geogr. Soc. Res. Mem.,* **2,** 59–62.

Macfadyen, W. A. 1955. Appendices 1 and 2, Foraminifera. *In* Godwin, H. Studies of the post-glacial history of British vegetation XIII. The Meare Pool region of the Somerset Levels. *Phil. Trans. R. Soc.,* **B239,** 185–190.

Mitchell, G. F., Penny, L. F., Shotton, F. W. and West, R. G. 1973. A correlation of Quaternary deposits in the British Isles. *Geol. Soc., Lond., Special Report* **4,** 1–99.

Murray, J. W. and Hawkins, A. B. 1976. Sediment transport in the Severn Estuary during the past 8000–9000 years. *Jl. Geol. Soc. Lond.,* **132,** 385–398.

Peacock, J. D., Graham, D. K., Robinson, J. E. and Wilkinson, I. 1977. Evolution and Chronology of Lateglacial Marine Environments at Lochgilphead, Scotland. *In* Gray, J. M. and Lowe, J. J. (eds.) *Studies in the Scottish Lateglacial Environment.* Pergamon, Oxford. pp. 89–100.

Peacock, J. D., Graham, D. K. and Wilkinson, I. P. 1978. Late-Glacial and post-Glacial marine environments at Ardyne, Scotland, and their significance in the interpretation of the history of the Clyde Sea area. *Rep. Inst. Geol. Sci.,* 78/77, 1–25.

Ten Dam, and Reinhold, T. 1941. Die stratigraphische Gliederung des Neiderlandischen Plio- Plistozans nach Foraminiferen. *Meded. geol. Sticht. C-V,* **1,** 1–66.

Van Voorthuysen, J. H. 1949. Foraminifera of the Icenian of the Netherlands. *Verh. geol.-mijnb. Genoot. Ned. Kolon geol. ser.,* **15,** 63–68.

Van Voorthuysen, J. H. 1955. *In* Baden-Powell, D. F. W.

Report on the marine fauna of the Clacton Channels. *Q. Jl. geol. Soc., Lond.,* **111**, 301–305.

West, R. G., Funnell, B. M. and Norton, P. E. P. 1980. An Early Pleistocene cold marine episode in the North Sea: pollen and faunal assemblages at Covehithe, Suffolk, England. *Boreas,* **9**, 1–10.

11
North Sea Cainozoic

C. King, H. W. Bailey, A. D. King, R. W. Meyrick, V. L. Roveda

11.1 INTRODUCTION

This chapter has been prepared by Mr. C. King and colleagues of Paleoservices Ltd.

During the past fifteen years the Cenozoic succession in the North Sea has been penetrated by several hundred boreholes in the search for oil and gas. The Cenozoic sequences in neighbouring onshore areas (southern England, Belgium, Holland, north Germany and Denmark) are now seen to be marginal portions of a major sedimentary basin (the North Sea Basin). The sediments thicken fairly regularly towards the centre of the basin, where over 3000 m of Tertiary and Quaternary sediments are present (Ziegler 1975, Fig. 16). In the deepest, central, part of the basin, seafloor depths were probably below 200 m during most of the Cenozoic. Basin filling, leading to progressive shallowing, began during the Late Palaeocene and has continued to the present day.

The Cenozoic sediments in the North Sea are mainly marine, with nearshore and deltaic facies developed locally near the margins of the basin. Early Palaeocene (Danian) carbonates are very widespread, but rifting and rapid subsidence in the mid-Palaeocene led to a drastic change in the depositional environment, and the Late Palaeocene to Pleistocene sediments are essentially clastic, with minor limestone beds. Nearshore and deltaic sands occur at several levels in the succession near the margins of the basin, and thick submarine fan and turbidite sands are present in the Palaeocene and Eocene of the northern North Sea; otherwise clays and claystones predominate. In the centre of the basin sedimentation was almost uninterrupted, but towards the margins unconformities and discontinuities develop, and condensed sequences with local reworking occur at some levels.

11.2 FORAMINIFERA

Very little has been published on the offshore Cenozoic foraminifera in the U.K. sector, except for brief notes by Dilley (in Curry *et al.* 1978) and Berggren (1969). An extensive study of the agglutinating faunas by Gradstein and Berggren is in press. Rasmussen (1974, 1978) gives some details of Cenozoic microfaunas in wells drilled in the Danish sector of the North Sea.

In the central North Sea Basin, Late Palaeocene to mid-Miocene faunas are dominated by agglutinating foraminifera of 'flysch basin' type, including the genera *Cyclammina, Trochammina, Ammodiscus, Recurvoides, Glomospira, Rhabdammina* and *Bathysiphon.* This fauna is believed to reflect rapid deposition of fine-grained organic-rich clastic sediments in a basin with restricted water circulation. Calcareous foraminifera are often absent or poorly preserved (due either to diagenetic solution or to deposition below the carbonate compensation depth (CCD)). The sequences in this area are difficult to relate directly to more marginal sequences, although some key calcareous benthic and planktonic species do occur. Most of the agglutinating species are long-ranging, but a few do appear to have relatively short ranges and can be used for correlation.

Away from the centre of the basin, the agglutinating foraminiferal assemblage grades into assemblages dominated by calcareous benthic and planktonic forms. These spread throughout the basin in Late Miocene times, and are comparable with the onshore faunas from Denmark, north Germany, Holland and Belgium. Correlation with these areas is generally straightforward, due to the presence of most of the key benthic species recorded from onshore sequences.

Zonal schemes proposed recently for parts of the Cenozoic sequences in Belgium, Holland and north Germany (Doppert, Laga and de Meuter 1979, Daniels and Spiegler 1974) can, in general, be recognised in the offshore sequences.

Planktonic foraminifera are generally more diverse and more abundant than in the onshore sequences, especially in the Miocene and Pliocene, and detailed study should eventually permit closer links between the onshore sequences and the standard planktonic zonation. Larger foraminifera (e.g. *Nummulites*), which occur in the onshore Eocene successions, are not recorded with certainty in the North Sea. Links with the Cenozoic faunas of southern England are in general rather remote (except for parts of the Early Eocene) due to the very marginal situation of the English outcrops.

11.3 RANGE CHART

The range chart (Fig. 11.1) gives the vertical distribution of some of the species used in subdivision and correlation of the North Sea sections. Several points must be emphasised:

(1) Material available for study is mainly from ditch cuttings, and the earliest occurrence of most species is usually indefinite due to caving of sediments downhole. This is shown by the dashed lines extending the ranges downwards.

(2) The species on the chart are chosen for their short stratigraphic range and wide distribution. No single section yields all these species, and faunas from different areas have been put together to give a composite chart.

(3) A good correlation with onshore sequences can be made at most levels, but stratigraphic boundaries can generally be placed only approximately.

(4) Sample collection in offshore wells does not generally begin until several hundred metres below the seafloor and the youngest Quaternary sediments are often unsampled. The youngest faunas included in the chart are therefore Early Pleistocene (Icenian).

11.3.1 Palaeocene

The top of the Cretaceous is sharply defined almost everywhere by the highest occurrence of *Pseudotextularia elegans* and *Racemiguembelina fructicosa*, associated with *Heterohelix* spp. and *Globotruncana* spp. This assemblage is typical of the Late Maastrichtian.

Early Palaeocene (Danian) is represented largely by chalks, which contain a planktonic assemblage dominated by *Globigerina triloculinoides, G. daubjergensis, Globorotalia compressa* and *G. pseudobulloides,* as in the onshore sequence in Denmark (Rasmussen 1978). Benthic foraminifera are usually less abundant. At the Early/Late Palaeocene boundary, benthic foraminifera are often abundant, including *Gavelinella beccariiformis, G. danica, Osangularia expansa, Matanzia varians* and *Lenticulina multiformis,* and these also occur at higher levels in the Palaeocene. The assemblage also includes many other species recorded from the Danish Kerteminde Marl (Hofker 1966) and the Swedish, Dutch and German Palaeocene (Wick 1943, Brotzen 1948). In the Late Palaeocene the assemblage is usually dominated by agglutinating foraminifera; among these, *Spiroplectammina spectabilis* is especially characteristic.

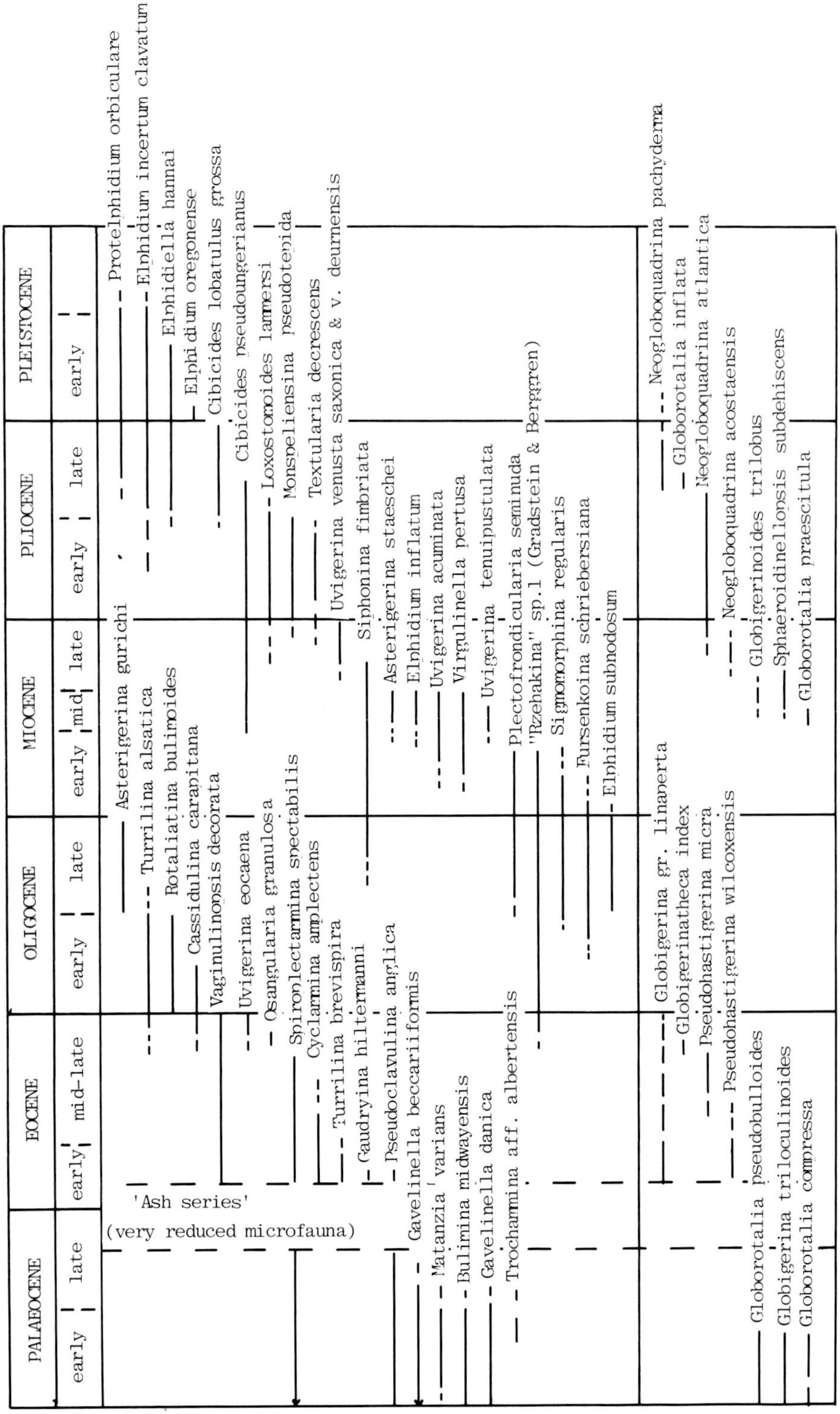

Fig. 11.1 – Range chart of North Sea Cenozoic foraminifera.

The Palaeocene/Eocene boundary, as defined in the dinoflagellate zonation, is placed within the 'Ash Series' (Knox and Harland 1979), a sequence of laminated shales, tuffaceous clays and tuffs which extends almost throughout the North Sea Basin. It was deposited in a very restricted marine environment, and although diatoms are often abundant, foraminifera are very scarce. Poorly preserved agglutinants and rare *Protelphidium* sp. have been found.

11.3.2 Eocene

In the claystones immediately above the 'Ash Series' calcareous foraminifera are present almost throughout the basin. The fossils often occur in a red claystone and are usually distinctively red-stained. The assemblage is characterised by abundant planktonic species (*Globigerina* gr. *linaperta* and *Acarinina* spp., with occasional *Pseudohastigerina wilcoxensis*) associated with benthic species including *Marginulina decorata, Turrilina brevisspira,* and *Gaudryina hiltermanni.* It was formerly assigned to the Palaeocene, due to identification of *G.* gr. *linaperta* as *G.* cf. *triloculinoides* (Walmsley 1975), but can be correlated with the Early Eocene Rosnaes Clay of Denmark (Berggren 1960, Dinesen 1972), the Unter Eozän 3 of Germany (Staesche and Hiltermann 1940), and parts of the Ypresian of Belgium and Holland and the London Clay Formation of England.

The overlying Eocene beds in the centre of the basin contain mainly long-ranging agglutinating faunas, but in the southern part of the U.K. sector, calcareous benthic and planktonic foraminifera occur. These permit an approximate correlation with the Dutch, Danish and German onshore sequences, although neither the Early/Middle or Middle/Late Eocene boundaries can be precisely defined. In the upper part of the Eocene the species *Osangularia granulosa* and *Globigerinatheka index* are widespread, and their co-occurrence forms a useful marker horizon. *G. index* is known onshore only in Denmark, where it is abundant in the Late Eocene Sovind Marl (Berggren 1969).

11.3.3 Oligocene

Rotaliatina bulimoides and *Turrilina alsatica* are very widespread in the Early Oligocene, enabling correlation with the Rupelian (Batjes 1958). In the Late Oligocene *Asterigerina gurichi, Elphidium subnodosum* and *Sigmomorphina regularis* occur in association in marginal areas. This fauna can be correlated with the onshore Chattian; in the central part of the basin calcareous benthic species are rare, and agglutinants are dominant.

11.3.4 Miocene

The highest occurrences of *Plectofrondicularia seminuda* and *'Rzehakina'* sp. 1 (Gradstein and Berggren in press) (*Sigmoilina tenuis* of Rasmussen 1974) are approximately coincident, and form basin-wide marker horizons within the Early Miocene. The occurrence of *'Rzehakina'* sp. 1 becomes progressively restricted towards the basin margins, and it is not recorded at all in onshore sequences. *P. seminuda* characterises the Early Miocene of Denmark (Kristoffersen 1972) and the Vierlandium-Lower Hemmoorium of Germany (Spiegler 1974).

The uppermost part of the Early Miocene, and the Middle Miocene, generally have a diverse and abundant planktonic fauna, with *Globorotalia praescitula, Globigerinoides trilobus, Sphaeroidinellopsis subdehiscens* and *Globigerina druryi,* similar to assemblages recorded from the Antwerp Sands of Belgium (Hooyberghs and De Meuter 1972). These are associated with *Asterigerina staeschei, Elphidium inflatum, Uvigerina acuminata, U. tenuipustulata* and *Virgulinella pertusa,* which are key benthic species of the Antwerp Sands, the Danish Hodde Formation and the German Hemmoorium and Reinbekium stages (Batjes 1958, Spiegler 1974, Kristoffersen 1972).

The Late Miocene is characterised by *Uvigerina pigmea* and *U. venusta;* a sequence of *Uvigerina* species and subspecies similar to those recorded by Daniels and Spiegler (1977) can be recognised. The Late Miocene faunas are similar to the faunas from the Gramium/Langenfeldium of Germany and the Gram Formation of Denmark. *Neogloboquadrina acostaensis* is common at some levels.

11.3.5 Pliocene

Early Pliocene benthic species include *Monspeliensina pseudotepida, Loxostomoides lammersi* and *Textularia decrescens,* and compare closely with Early Pliocene faunas of Holland (van Voorthuysen 1950). The planktonic species *Neogloboquadrina atlantica* is often common. Late Pliocene faunas are very homogenous in composition across the basin, and are characterised by *Cibicides lobatulus grossa, Cassidulina laevigata, Elphidium incertum clavatum, Bulimina marginata, Sigmoilopsis schlumbergeri* and *Buccella frigida.* They compare closely with Late Pliocene faunas recorded in the Netherlands (ten Dam and Reinhold 1941, van Voorthuysen 1950). The planktonic species *Neogloboquadrina pachyderma* is present at some levels. *Globorotalia inflata* occurs within the Late Pliocene.

11.3.6 Pleistocene

The Pliocene/Pleistocene boundary in the Netherlands is placed by van Voorthuysen, Toering and Zagwijn (1972) at the base of a zone with *Elphidium oregonense.* This species occurs offshore in the Dutch and Danish sectors of the North Sea (Rasmussen 1974, 1978) but has not been recorded in the U.K. sector. Here a tentative boundary is taken at the highest occurrence of *Cibicides lobatulus grossa.* Overlying this level, *Elphidiella hannai* is often abundant, associated with a benthic assemblage which includes *Cassidulina laevigata* and *Elphidium* spp. Units in which this assemblage occurs can be correlated with the onshore Icenian (Early Pleistocene).

11.4 REFERENCES

Batjes, D. A. J. 1958. Foraminifera of the Oligocene of Belgium. *Mém. Inst. roy. Sci. Nat. Belg.,* **143,** 188 pp.

Berggren, W. A. 1960. Some planktonic Foraminifera from the Lower Eocene (Ypresian) of Denmark and Northwestern Germany. *Stockholm Contr. Geol.,* **5,** 41–108.

Berggren, W. A. 1969. Palaeogene Biostratigraphy and Planktonic foraminifera of Northern Europe. *Proc. First Int. Conf. on Planktonic Microfossils,* **1,** 121–160. Geneva.

Brotzen, F. 1948. The Swedish Palaeocene and its Foraminiferal fauna. *Sverig. geol. Unders.* Ser C. **493,** 140 pp.

Curry, D., Adams, C. G., Boulter, M. C., Dilley, F. C., Eames, F. E., Funnell, B. M. and Wells, M. K. 1978. A correlation of Tertiary rocks in the British Isles. *Geol. Soc. Lond. Special Report* No. 12, 72 pp.

Dam, A. ten 1944. Die stratigraphische gliederung des niederlandischen Palaozäns und Eozäns nach Foraminiferen. *Meded. geol. Sticht.* Ser. C. No. 3, 142 pp.

Dam, A. ten and Reinhold, T. 1941. Die stratigraphische gliederung des Niederlandischen Plio-Pleistozäns nach foraminiferen. *Meded. geol. Sticht.* Ser. C. No. 1, 66 pp.

Dam, A. ten and Reinhold, T. 1942. Die stratigraphische gliederung des niederlandischen Oligo-Miozäns nach Foraminiferen. *Meded. geol. Sticht.* Ser. C. No. 2, 106 pp.

Daniels, C. H. von and Spiegler, D. 1977. Uvigerinen (Foram.) im Neogen Nordwestdeutschlands. *Geol. Jb.* **A40,** 3–59.

Dinesen, A. 1972. Foraminiferselskaber fra de jyske Eocaene formationer. *Dansk. geol. Foren., Arsskrift for 1971,* 70–78.

Doppert, J. W., Laga, P. G. and De Meuter, F. T. 1979. Correlation of the biostratigraphy of marine Neogene deposits, based on benthonic foraminifera, established in Belgium and the Netherlands. *Meded. Rijks. Geol. Dienst.,* **31,** 1–8.

Gradstein, F. M. and Berggren, W. A. (in press). 'Flysch-Type' Agglutinated Benthic Foraminiferal Assemblages in Upper Cretaceous and Palaeogene Sediments of the East Newfoundland Basin, Labrador Sea and North Sea – Palaeogeographic and Palaeoecologic Implications. *Mar. Micropalaeont.*

Hofker, J. 1966. Maastrichtian, Danian and Palaeocene Foraminifera. *Palaeontographica. Suppl.,* **10,** 375 pp.

Hooyberghs, H. J. F. and De Meuter, F. J. C. 1972. Biostratigraphical and interregional correlation of the 'Miocene' deposits of Northern Belgium based on planktonic foraminifera; the Oligocene-Miocene boundary on the southern edge of the North Sea Bsin. *Meded. Kon. Acad. Wetensch.,* **34,** 47 pp.

Knox, R. W. and Harland, R. 1979. Stratigraphical relationships of the early Palaeogene ash-series of NW Europe. *J. geol. Soc. Lond.,* **136,** 463–470.

Kristoffersen, F. N. 1972. Foraminiferzonering i det Jyske Miocene. *Dansk. geol. Foren., Arsskrift for 1971,* 79–85.

Rasmussen, L. B. 1974. Some geological results from the first five Danish exploration wells in the North Sea. *Danm. geol. Unders.,* (3), **42,** 1–46.

Rasmussen, L. B. 1978. Geological aspects of the Danish North Sea sector. *Danm. geol. Unders.,* **(3), 44,** 1–85.

Spiegler, D. 1974. Biostratigraphie des Tertiärs zwischen Elbe unde Weser/Aller (Benthische Foraminiferen, Oligo-miozän). *Geol. Jb.,* **A16,** 27–69.

Staesche, K. and Hiltermann, H. 1940. Mikrofaunen aus dem Tertiär Nordwestdeutschlands. *Reichsstelle f. Bodenforsch. No. 201.* 26 pp.

Voorthuysen, J. H. van 1950. The quantitative distribution of the Pleistocene, Pliocene and Miocene Foraminifera of boring Zaandam (Netherlands). *Med. Geol. Sticht.,* N.S., **4,** 51–72.

Voorthuysen, J. H. van, Toering, K. and Zagwijn, W. H. 1972. The Plio-Pleistocene boundary in the North Sea Basin. Revision of its position in the marine beds. *Geol. en Mijn.,* **51,** 627–639.

Walmsley, P. J. 1975. The Forties Field: in *Petroleum and the Continental Shelf of North West Europe,* **1,** 477–485. London.

Wick, W. 1943. Mikrofaunistische Untersuchen des tieferen Tertiärs über einem Salzstock in der Nahe von Hamburg. *Abh. senck. naturforsch. Ges.,* **468,** 40 pp.

Ziegler, W. H. 1975. Outline of the Geological History of the North Sea. In *Petroleum and the Continental Shelf of North West Europe.,* **1,** 165–190. London.

12
An outline of faunal changes through the Phanerozoic

J. W. Murray

Within the small area of the British Isles there is a fairly complete Phanerozoic succession, much of which is represented by fossiliferous marine strata. Rocks older than Carboniferous have so far yielded only sparse foraminiferal faunas but diverse assemblages are known from the Carboniferous, Jurassic, Cretaceous, Palaeogene, Neogene and Quaternary. The succession of faunas is described below. However, although the changes observed are partly due to evolution, they are also the consequence of changing environments. The assemblages described all come from shelf seas, or marginal marine lagoons and estuaries.

Our knowledge of pre-Carboniferous faunas is incomplete. *Saccamminopsis*, from the Ordovician of Girvan, is the oldest recorded genus (Chapter 2). Apart from this, simple agglutinated forms are sometimes found in the insoluble residues of limestones which have been processed for conodonts.

The shallow shelf seas of the Carboniferous provided habitats favourable for benthic foraminifera and the faunas are rich and diverse. The majority of genera are of the suborder Fusulinina, especially superfamily Endothyracea, but there are also representatives of the superfamilies Parathuramminacea (e.g. *Saccamminopsis*) and Fusulinacea (e.g. *Millerella*). The suborder Miliolina makes its first appearance with *Eosigmoilina* and *Nodosigmoilina*. All these forms are confined to limestones or calcareous mudstones (Chapter 3) believed to have been deposited in water of near normal marine salinity. They are absent from rocks believed to have been deposited in hypersaline environments. In the late Carboniferous simple agglutinated forms are found in the marine bands.

In much of Britain deposits of Permian age are of continental origin but in northern England normal marine and very slightly hypersaline deposits have yielded foraminiferal faunas (Chapter 4). There are simple tubular Miliolina: *Agathammina, Orthovertella* and *Calcitornella;* simple Textulariina: *Ammobaculites, Ammodiscus, Hyperammina;* and the Rotaliina are represented by the nodosariaceans *Dentalina, Geinitzina, Nodosaria* and *Frondicularia*. The Miliolina alone occur in the slightly hypersaline and patch reef deposits.

The marine assemblages include representatives of all the groups listed above. These Permain assemblages are of very low diversity as compared with the Carboniferous and none of the Carboniferous genera are present in them. However, the Permian assemblages do include early representatives of the Nodosariacea which were to become so important during the early part of the Mesozoic.

With the exception of a few organic linings of foraminifera seen in palynological preparations from Triassic mudstones, the record is confined to rocks of the Rhaetian. A few Palaeozoic genera linger on: *Nodosinella, Stacheia* and *Agathammina,* but the most noticeable aspect of the fauna is the incoming of essentially Jurassic genera and species (Chapter 5). Among these are *Lingulina*, especially representatives of the *Lingulina tenera* plexus, *Nodosaria, Dentalina, Marginulina,* and *Eoguttulina.*

Jurassic sediments range from moderately deep basinal deposits to shallow shelf sands, silts and carbonates to marginal marine and continental deposits. At the present time sampling for foraminifera has been concentrated on the clays. Some of the sands have been looked at but few of the carbonates have received attention. So our knowledge of Jurassic faunas is not truly representative.

The early Jurassic transgression which drowned much of the British Isles brought with it a fauna which diversified into the newly available niches (Chapter 6). The common genera are, Lituolacea: *Textularia, Haplophragmoides, Ammobaculites, Trochammina* and *Triplasia.* Miliolacae: *Ophthalmidium* and *Nubecularia,* Nodosariacea: *Nodosaria, Dentalina, Frondicularia, Lingulina, Lenticulina, Marginulina, Vaginulina, Vaginulinopsis, Marginulinopsis, Saracenaria, Citharina, Eoguttulina,* and *Tristix,* Spirillinacea: *Spirillina,* Cassidulinacea: *Involutina* and *Paalzowella,* Robertinacea: *Epistomina, Conorboides* and *Reinholdella.*

Representatives of the Nodosariacea occur in assemblages from most environments. An abundance of Miliolacea (*Ophthalmidium, Nubecularia, Nubeculariella*) and Spirillinacea (*Spirillina*) is thought to indicate well oxygenated shallow, normal marine conditions. Assemblages of small agglutinated genera (*Trochammina, Haplophragmoides, Ammobaculites*) may be indicative of brackish water and perhaps less oxygenated bottom conditions. Larger Lituolacea (*Ammobaculites coprolithiformis, Triplasia* spp.) are found in deeper marine waters. Among the Robertinacea, *Epistomina* is from normal marine conditions but *Reinholdella,* when occurring alone, may indicate waters of low oxygen content (Chapter 6).

All the major faunal changes in the Jurassic can be related to changes in the environments of deposition. In each stage there is an introduction of new species. Even though many of the forms are relatively long-ranging, the faunas of each stage are fairly distinct.

The lower Cretaceous faunas of the Ryazanian to Aptian are essentially Jurassic in character. *Lenticulina muensteri* continues through to the basal Albian. The following genera, which were present in the Jurassic, are found: *Ammobaculites, Textularia, Trochammina, Nodobacularia, Citharina, Saracenaria, Lenticulina, Marginulopsis, Frondicularia, Planularia, Vaginulina, Lingulina, Epistomina,* and *Conorboides.* New arrivals in the Hauterivian include Lituolacea: *Tritaxia,* Miliolacea: *Wellmanella* and Nonionacea: *Gavelinella.* During the Barremian and Aptian, many of the Nodosariacea die out and after this time they never again dominate the benthic assemblages.

There are major additions to the fauna in the Albian, Lituolacea: *Arenobulimina, Dorothia, Flourensina,* Nonionacea: *Lingulogavelinella.* These herald the incoming of the Upper Cretaceous fauna.

The other major event recorded in the Lower Cretaceous succession is the appearance, in the Barremian, of the first planktonic foraminifera (*Hedbergella infracretacea*). This is joined in the Albian by *Hedbergella delrioensis, H. planispira, Guembelitria cenomana, H. moremani* and *Globigerinelloides bentonensis* (Chapter 7).

The Upper Cretaceous, typified by chalk deposition, is characterised by diverse benthic assemblages dominated by Textulariina and Rotaliina.

New genera include the following:

Cenomanian:	Lituolacea, *Pseudotextulariella*
Turonian:	Discorbacea, *Valvulineria,* Nonionacea, *Globorotalites*
Coniacian:	Buliminacea, *Eouvigerina,* Nonionacea, *Stensioina*
Santonian:	Buliminacea, *Cibicides*
Campanian:	Cassidulinacea, *Loxostomum,* No-

dosariacea, *Neoflabellina.*

Evolutionary relationships have been proposed for a number of benthic groups including *Bolivinoides* (Barr, 1966), *Arenobulimina* (Carter and Hart, 1977; Price, 1977), *Flourensina* (Carter and Hart, 1977) and *Gavelinella* (Price, 1977).

The late Cretaceous planktonic fauna underwent both diversification and increase in number of individuals. The *Hedbergella* and *Guembelitria* groups continued on from the Albian. In the Cenomanian *Rotalipora* appeared and became extinct, and *Praeglobotruncana* and *Whiteinella* made their appearance. The first *Globotruncana* are found in the Coniacian. *Rugoglobigerina* at the top of the Campanian and *Abathomphalus* in the Maastrichtian. The late Cretaceous chalks are richer in planktonic foraminifer than any other part of the land based stratigraphic succession in Britain.

At the end of the Cretaceous the whole of Britain became emergent and underwent erosion. The oldest Palaeogene deposits (Thanet Formation) are of late Palaeocene age. All the Palaeogene successions are of variable facies and most of them are shallow water shelf, marginal marine or continental. The only deeper water (mid/outer shelf) deposits are found in the London Clay of the London basin.

The foraminiferal faunas are very different from those of the Cretaceous. Among the new genera are: Buliminacea: *Bulimina, Uvigerina, Turrilina, Praeglobobulimina,* Discorbacea: *Asterigerina, Cancris, Discorbis,* Rotaliacea: *Elphidium, Protelphidium, Pararotalia, Nummulites,* Cassidulinacea: *Caucasina,* Nonionacea: *Alabamina, Anomalinoides,* Miliolacea: *Orbitolites, Fasciolites.*

Because of the rapid facies variations many of the extinctions and new appearances are due to environmental causes.

The Palaeogene planktonic faunas are generally sparse. The genera represented include *Globigerina, Globorotalia, Pseudohastigerina, Guembelitria* and *Globigerinatheka.* Most of the Cretaceous genera were extinct by the Palaeogene but reworked tests are common at certain Palaeogene levels even in non-marine deposits (Chapter 8).

The Neogene is represented by the Globigerina Silts of the English Channel, the Crags of East Anglia and the St. Erth Beds. The benthic assemblages are quite similar to modern day faunas not only at generic level but also at specific level. Extant forms include *Florilus grateloupi, Cibicides dutempli, C. lobatulus, Elphidium crispum, Bulimina alazanensis* and *Textularia sagittula.* The planktonic faunas of the Globigerina Silts are moderately diverse but they are rare in the rest of the deposits (Chapter 9).

The Quaternary assemblages of the warmer intervals are closely similar to the modern ones and are dominated by shallow water forms (Chapter 10). During the cooler intervals the genus *Elphidiella* was common together with cold-water species of *Elphidium* and *Protelphidium.*

The principal changes are listed below (– – – = change in fauna).

Quaternary Neogene	benthic fauna of modern aspect
	– – – – – – – – – –
Palaeogene	common Miliolacea, Discorbacea, Buliminacea, Rotaliacea, Cassidulinacea, Nonionacea. Planktonic forms generally rare.
	– – – – – – – – – –
Cretaceous (Albian–Maastrichtian)	common Lituolacea, Nodosariacea, Buliminacea, Discorbacea, Cassidulinacea, Nonionacea. Planktonic forms are common.
	– – – – – – – – – –
(Ryazanian–Aptian)	common Lituolacea, Nodosariacea, Nonionacea. Incoming of planktonic fauna.
Jurassic	common Lituolacea, Nodosariacea, Spirillinacea, Cassidulinacea, Robertinacea. Subordinate Miliolacea.
Triassic	start of 'Jurassic' faunas.
	– – – – – – – – – –
Permian	common Ammodiscacea, Lituolacea, Miliolacea, Nodosariacea
	– – – – – – – – – –
Carboniferous	common Parathuramminacea, Endothyracea, subordinate Fusulinacea.
	– – – – – – – – – –
Pre- Carboniferous	mainly Ammodicacea.

From this it can be seen that the really major events in the foraminiferal faunas of Britain took place at the end of the Carboniferous, Permian, early Cretaceous, end Cretaceous and end Palaeogene.

12.1 REFERENCES

Barr, F. T. 1966. The foraminiferal genus *Bolivinoides* from the Upper Cretaceous of the British Isles, *Palaeontology*, **9**, 220–243.

Carter, D. J. and Hart, M. B. 1977. Aspects of mid-Cretaceous stratigraphical palaeontology. *Bull. Brit. Mus. (Nat. Hist.)*, Geol. Ser., **29**, 1–135.

Price, R. J. 1977. The evolutionary interpretation of the Foraminiferida *Arenobulimina*, *Gavelinella* and *Hedbergella* in the Albian of North-West Europe. *Palaeontology*, **20**, 503–527.

General Index

Index of genera and species